Handbook of Metallurgy

Handbook of Metallurgy

Editor: Darren Wang

NY RESEARCH PRESS

New York

Published by NY Research Press
118-35 Queens Blvd., Suite 400,
Forest Hills, NY 11375, USA
www.nyresearchpress.com

Handbook of Metallurgy
Edited by Darren Wang

International Standard Book Number: 978-1-63238-870-4 (Hardback)

Cataloging-in-Publication Data

Handbook of metallurgy / edited by Darren Wang.
    p. cm.
Includes bibliographical references and index.
ISBN 978-1-63238-870-4
1. Metallurgy. 2. Physical metallurgy. 3. Metals. 4. Chemical engineering. 5. Materials science. I. Wang, Darren.
TN665 .H36 2022
669--dc23

Contents

Preface

Every book is a source of knowledge and this one is no exception. The idea that led to the conceptualization of this book was the fact that the world is advancing rapidly; which makes it crucial to document the progress in every field. I am aware that a lot of data is already available, yet, there is a lot more to learn. Hence, I accepted the responsibility of editing this book and contributing my knowledge to the community.

Metallurgy is a field of material science and engineering that focuses on the study of physical and chemical properties of metallic elements. It studies their inter-metallic compounds and mixtures, known as alloys. It is primarily used to separate metals from their ores. The discipline of metallurgy is sub-divided into physical and chemical metallurgy. The Physical field systematically considers physical properties of metals and alloys, while chemical metallurgy is concerned with obtaining metals from their ores and analyzing their reactions. Metallurgy is further subdivided into ferrous and non-ferrous metallurgy which are respectively known as black and colored metallurgy. This book unravels the recent studies in the field of Metallurgy. While understanding the long-term perspectives of the topics, the book makes an effort in highlighting their impact as a modern tool for the growth of the discipline. The extensive content of this book provides the readers with a thorough understanding of the subject.

While editing this book, I had multiple visions for it. Then I finally narrowed down to make every chapter a sole standing text explaining a particular topic, so that they can be used independently. However, the umbrella subject sinews them into a common theme. This makes the book a unique platform of knowledge.

I would like to give the major credit of this book to the experts from every corner of the world, who took the time to share their expertise with us. Also, I owe the completion of this book to the never-ending support of my family, who supported me throughout the project.

Editor

Effect of (0001) Strain on the Electronic and Magnetic Properties of the Half-Metallic Ferromagnet Fe$_2$Si

Wang Chen,[1] Ruijie Li,[2] and Yanhui Liu[1]

[1]State Key Laboratory for Seismic Reduction/Control & Structural Safety (Cultivation), Guangzhou University, Guangzhou 510006, China
[2]School of Materials Science and Engineering, Guizhou Minzu University, Guiyang 550025, China

Correspondence should be addressed to Yanhui Liu; liuyanhui2012@163.com

Academic Editor: Marco Cannas

The electronic and magnetic properties of the half-metallic ferromagnet Fe$_2$Si under (0001) strain have been evaluated by the first-principles density functional theory method. The spin-up band structure shows that bulk Fe$_2$Si has metallic character, whereas the spin-down band structure shows that bulk Fe$_2$Si is an S-L indirect band gap of 0.518 eV in the vicinity of Fermi surface. Indirect-to-direct band gaps and an unstable-to-stable transition are observed in bulk Fe$_2$Si as strain is applied. In the range -11% to 11% (excluding zero strain), bulk Fe$_2$Si has stable half-metallic ferromagnetism, the spin polarization at the Fermi surface is 100%, and the magnetic moment of the Fe$_2$Si unit cell is 4.0 μB. The density distribution shows that the spin states of bulk Fe$_2$Si mainly come from the Fe^1-3d and Fe^3-3d states, indicating that bulk Fe$_2$Si has spin-polarized ferromagnetism. The half-metallic ferromagnetism of bulk Fe$_2$Si is mainly caused by d–d exchange and p–d hybridization, which are not sensitive to strain. It is very important to investigate the effect of changes in the lattice constant on the half-metallic ferromagnetic properties of bulk Fe$_2$Si.

1. Introduction

Half-metallic ferromagnets (HMFs) have attracted much attention because of their potential applications in spintronics [1]. Like a normal ferromagnet, HMFs have two different spin channels [2–4]. They have an energy gap in one spin direction at the Fermi level, whereas the other spin is strongly metallic, which results in complete spin polarization of the conduction electrons. Such materials generally have a high Curie temperature and close to 100% spin polarization [5, 6]. Therefore, half-metallic ferromagnetic materials will undoubtedly be an ideal semiconductor spin electron injection source. This shows that half-metallic ferromagnetic research is important and has application prospects. In addition, it will promote the rapid development of semiconductor spin electronics.

Bulk Fe$_2$Si has not been well investigated. Chen and Tan [7] used X-ray diffraction (XRD) to study the structure of Fe$_2$Si thin films. They found that the film thickness is affected by the base material, and the magnetic, electric, and optical characteristics are affected by the film structure. The structure, lattice parameter, phonon spectrum, and reflectance spectrum of hexagonal Fe$_2$Si have been investigated by first-principles calculations [8]. The results show that Fe$_2$Si is a ferromagnetic material and it has a spin-polarized half-metal-like band structure. However, there have been no studies associating the magnetism and mechanical properties of Fe$_2$Si. Therefore, in the present work, the properties of bulk Fe$_2$Si under (0001) strain were calculated using plane-wave pseudopotential methods based on density functional theory (DFT), and the results were analyzed in detail.

2. Calculation Method

There are four Fe atoms and two Si atoms in the Fe$_2$Si unit cell. The fractional coordinates of the three nonequivalent Fe atoms are (0, 0, 0), (0, 0, 0.5), and (0.333, 0.667, 0.78). The fractional coordinate of the Si atoms is (0.333, 0.667, 0.28). Fe$_2$Si belongs to space group $P - 3m1$ (number 164). The lattice parameters are $a = b = 4.052$ Å and $c = 5.086$ Å. The crystal plane angles are $\alpha = \beta = 90°$ and $\gamma = 120°$. The (0001) strain line changes with the lattice constant c of the hexagonal Fe$_2$Si structure with strain ranging from -12% to 12%.

The calculations were performed with the plane-wave pseudopotential method implemented in the Cambridge sequential total energy package [9] to calculate the effect of (0001) strain on the electromagnetic mechanism of half-metallic ferromagnet Fe_2Si. The generalized gradient approximation [10] with the revised approximation of the Perdew–Burke–Ernzerhof scheme was used for the exchange–correlation potential. All of the possible structures were optimized by the Broyden–Fletcher–Goldfarb–Shanno algorithm [11, 12]. Geometry optimization was performed to fully relax the structures until self-consistent field (SCF) convergence per atom. The tolerance in the SCF calculation was 5.0×10^{-6} eV/atom. The ultrasoft pseudopotentials were expanded within a plane-wave basis set with 330 eV, and the iteration convergence accuracy was 5.0×10^{-7} eV. The energy of bulk Fe_2Si was calculated based on optimization of the structural system, with the minimum energy structure which was chosen as the stable structure. Sampling of the Brillouin zone (BZ) was performed with an $8 \times 8 \times 6$ k-point mesh according to the Monkhorst–Pack method [13].

3. Results and Discussion

3.1. Electronic Structure

3.1.1. Band Structure.

Figure 1 shows the effect of (0001) strain on the band structure of bulk Fe_2Si near the Fermi energy. Figure 1(c) shows the band structure without strain. It shows that the band structure of spin-up electrons has metallic character and the band structure of spin-down electrons has semiconductor character. The valence band maximum is 0.164 eV at point S and the conduction band minimum is 0.682 eV at the BZ point L. Thus, Fe_2Si forms a band gap of 0.518 eV in the vicinity of Fermi surface (spin-down). The half-metallic gap [14] is determined by the minimum difference between the lowest energy of the spin conductive bands with respect to the Fermi level and the absolute value of the highest energy of the spin valence bands. Therefore, the half-metallic gap of bulk Fe_2Si is 0.164 eV.

Under (0001) plane strain, the lattice constant strain of bulk Fe_2Si changes its band structure, which leads to an energy change of the conduction band bottom and the valence band top. The valley near the bottom of the conduction band is divided into two groups of the degenerate valley, forming the conduction band edge and secondary conduction band edge. The energy shift of the valence band changes the light and heavy hole band into two groups of peaks, forming the valence band edge and the secondary valence band edge. The band gap (E_g) of the strain is determined by the heavy cavities and degenerate valleys. Figures 1(a) and 1(b) show the band structure of bulk Fe_2Si under compressive strain, which changes to a direct band gap of 0.338 eV at point L (spin-down). The strain reaches −12%; then bulk Fe_2Si becomes metallic. Figures 1(d)–1(f) show the band structure under tensile strain, which is the same as the band structure when applying compressive strain. This indicates that, under strain, bulk Fe_2Si is first a stable half-metallic ferromagnet and then changes to metallic with increasing strain. In conclusion, the band structure of bulk Fe_2Si can be changed using compressive and tensile strain, verifying that strain is an effective way to control the band structure.

3.1.2. Density of States.

Figure 2 shows the total density of states (TDOS) and partial density of states (PDOS) of bulk Fe_2Si under various strains. The plotted energy range is −6 to 5 eV, and lower lying semicore states are omitted for clarity. Only the density of states (DOS) distribution near the Fermi level determines the magnetic properties. Hence, we concentrate on the DOS in the vicinity of the Fermi level, which is set to zero. To investigate the effect of strain on the subelectron structure of the system, the Si-s, Si-p, Fe-s, and Fe-d PDOS were investigated under various strains. The configuration of the extranuclear electrons of Si is $3s^2 3p^2$, and the configuration of the extranuclear electrons of Fe is $3d^6 4s^2$. Figure 2 shows that the DOS near the Fermi level mainly comes from Fe-3d spin states, and the contributions of Si-3p spin states are small. There is large exchange splitting between the spin-up and spin-down bands of the Fe-3d states, which leads to a large localized spin magnetic moment at the Fe atoms and results in polarization of the Fe-3d bands away from the Fermi level. Hybridization of 3p states provided by Si atoms with 3d electrons determines the degree of occupation of the p–d orbitals. Therefore, the half-metallic ferromagnetism of bulk Fe_2Si under various strains is mainly caused by d–d exchange and p–d hybridization.

The spin-up DOS is mainly distributed below the Fermi level. The spin-down DOS has two main peaks, which are located on both sides of the Fermi level. The spin-up and spin-down TDOS distributions near the Fermi surface are asymmetric; that is, the number of electrons in the spin-up and spin-down is quite different, which is the main contribution to the magnetic properties. Figures 2(b), 2(d), and 2(e) show that the Fermi level is completely located in the band gap of the spin-down energy band and the spin polarization of bulk Fe_2Si at the Fermi level is 100%, so the spin polarization of bulk Fe_2Si can be improved by strain.

3.2. Magnetic Properties

3.2.1. Magnetic Mechanism Analysis.

The magnetic moments of all of the atoms in the Fe_2Si unit cell under various strains are given in Table 1. The total magnetic moment of the Fe_2Si unit cell under strains of −12%, −7%, 0%, 5%, 9%, and 12% are 9.29 μB, 4.01 μB, 3.54 μB, 4.00 μB, 4.00 μB, and 4.05 μB, respectively. This shows that bulk Fe_2Si has stable half-metallic ferromagnetism under strain of −11% to 11% (excluding zero strain), which agrees well with Figures 1 and 2. If half-metals are in the same applied magnetic field, larger magnetic moments possibly result in stronger spin-correlation scattering of conductive electrons, and the variation of the resistance is then larger. Therefore, bulk Fe_2Si under strain of −12% and 12% will have more magnetoresistance than under no strain. Table 1 shows that the magnetic moment of bulk Fe_2Si mainly comes from the Fe^1-3d and Fe^3-3d states.

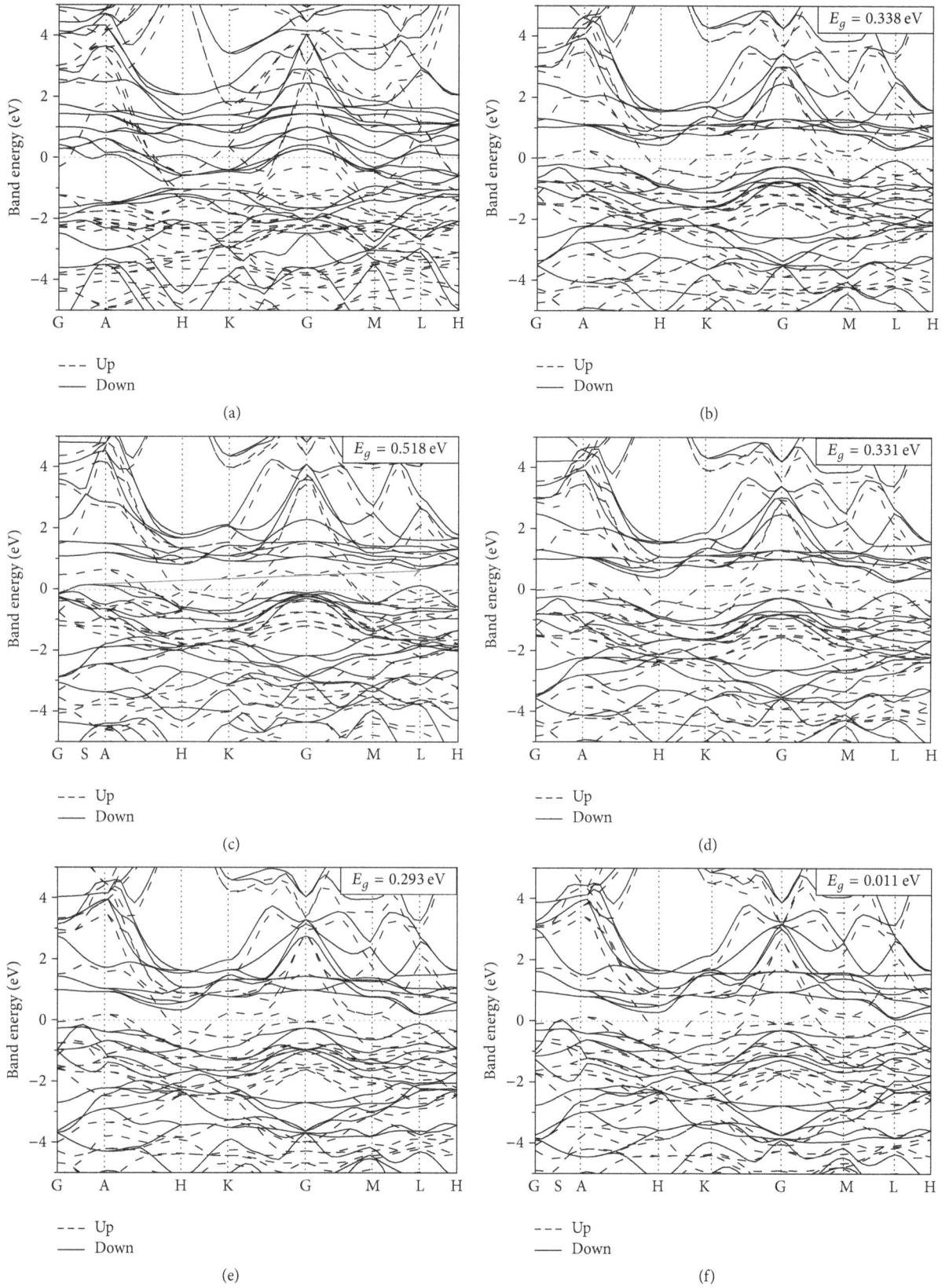

FIGURE 1: Band structures of Fe$_2$Si under various strains: (a) −12%, (b) −7%, (c) 0%, (d) 5%, (e) 9%, and (f) 12%. The band structure of spin-down electrons has indirect bandgap semiconductor character. The valence band maximum is 0.164 eV at point S and the conduction band minimum is 0.682 eV at the BZ point L. The blue arrow is from the top of valence band point to the bottom of the conduction band.

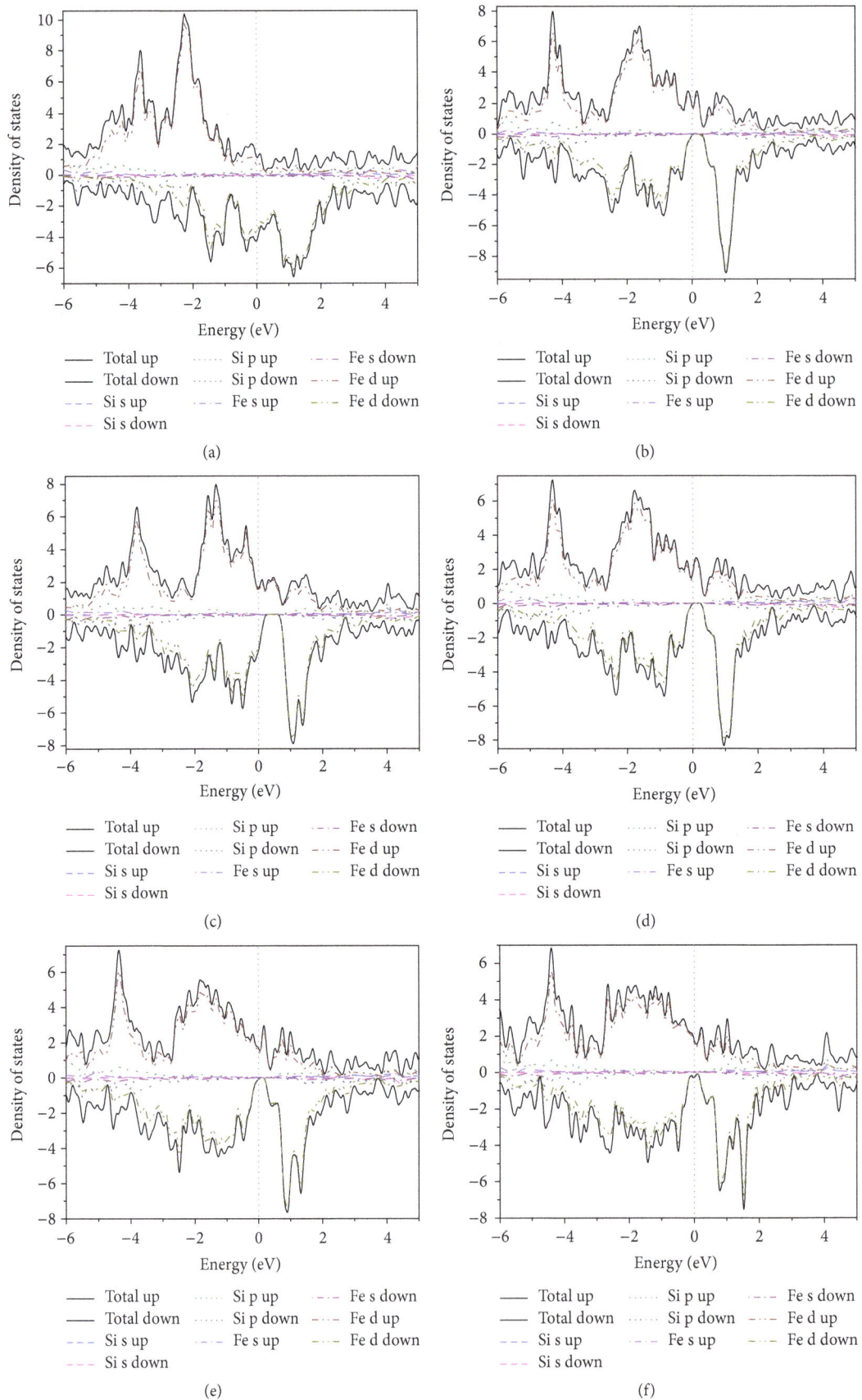

FIGURE 2: DOS of Fe_2Si under various strains: (a) −12%, (b) −7%, (c) 0%, (d) 5%, (e) 9%, and (f) 12%.

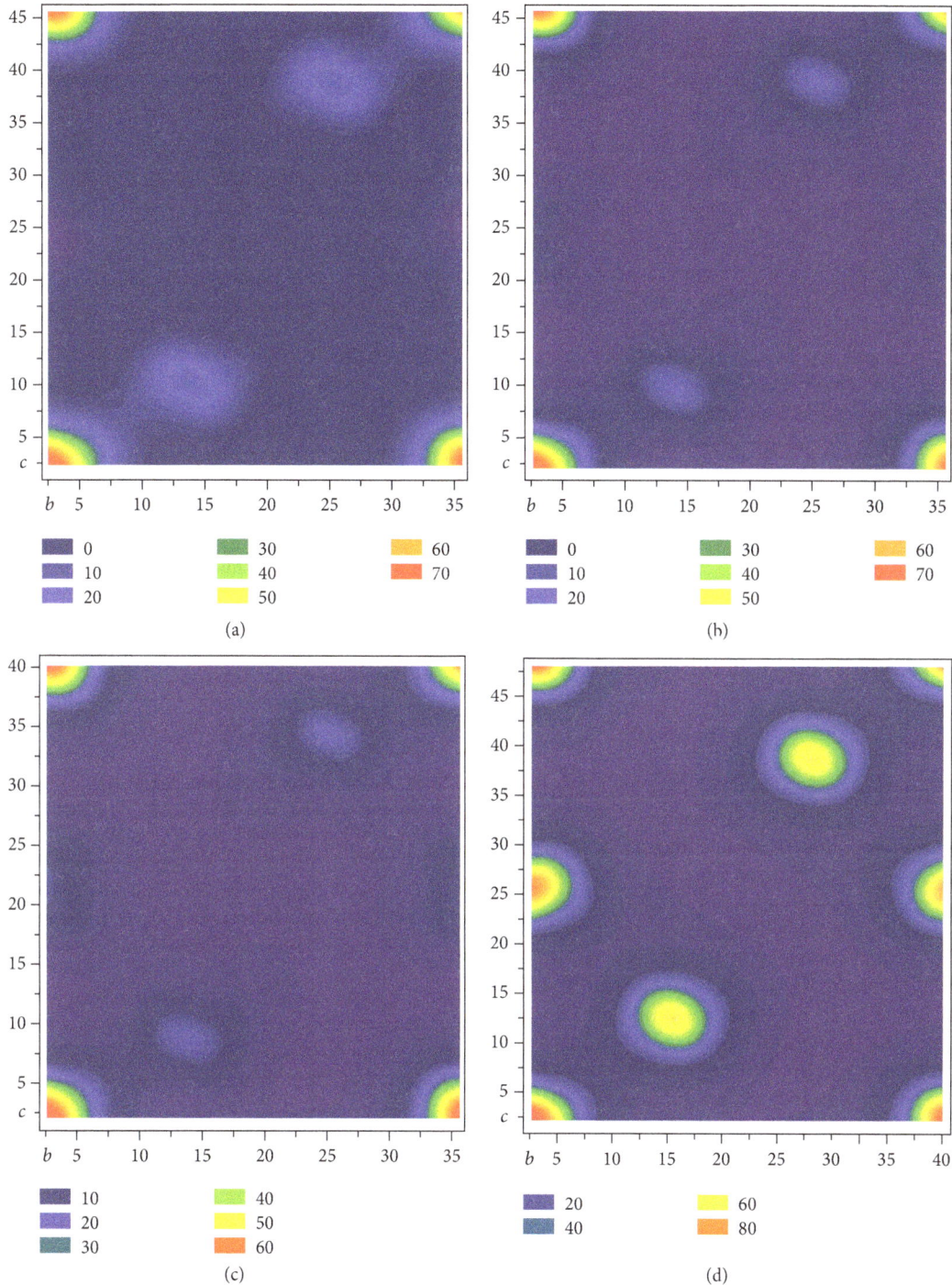

FIGURE 3: Spin density distributions of bulk Fe_2Si under various strains: (a) 0%, (b) −11% to 11% (excluding zero strain), (c) 12%, and (d) −12%.

In the range −11% to 11% (excluding zero strain), the total magnetic moment of bulk Fe_2Si is $4.0\,\mu B$ and the spin polarization is 100% from the DOS map, indicating that the half-metallic ferromagnetic properties of bulk Fe_2Si are not sensitive to strain. In practical application in an optional electronic device, a small change in the lattice constant may have a significant effect on the electron transport properties at the Fermi level [15]. Therefore, it is very important to investigate the effect of changes in the lattice constant on the half-metallic ferromagnetic properties of bulk Fe_2Si.

3.2.2. Distribution of the Magnetic and Electric Charge. Figure 3 shows the spin density distributions of bulk Fe_2Si under various strains. The results show that the Fe atoms are surrounded by a high potential, whereas the Si atoms are surrounded by a low potential. The charge of Fe_2Si under

TABLE 1: Magnetic moments of all of the atoms in the Fe_2Si unit cell under various strains.

Species	Ion	-12% Spin (μB)	-7% Spin (μB)	0% Spin (μB)	5% Spin (μB)	9% Spin (μB)	12% Spin (μB)
Si	1	−0.34	−0.09	−0.06	−0.09	−0.09	−0.09
Si	2	−0.34	−0.09	−0.06	−0.09	−0.09	−0.09
Fe	1	2.81	2.83	2.80	2.81	2.75	2.69
Fe	2	2.58	0.06	−0.62	0.07	0.23	0.40
Fe	3	2.29	0.65	0.74	0.65	0.60	2.57
Fe	4	2.29	0.65	0.74	0.65	0.60	2.57

various strains is mainly localized on the Fe atoms and there is little charge on the Si atoms, indicating that bulk Fe_2Si has spin polarization ferromagnetism, which is consistent with the results in Table 1. The charge cloud distribution of the Si atom is affected by the Fe atom, making the distribution along the $(11\,\overline{2}\,0)$ crystal face. Charge transfer of the atoms in the system under stress is because of Fe d–d exchange and Fe-d–Si-p hybridization. In the range −11% to 11% (excluding zero strain), the charge density between atoms is relatively low. Figure 3 shows that the spin states of bulk Fe_2Si mainly come from the Fe^1-3d and Fe^3-3d states. Bulk Fe_2Si maintains its spin polarization ferromagnetism in the strain range.

4. Conclusion

The effect of (0001) strain on the electronic structure and magnetic properties of bulk Fe_2Si has been investigated by the first-principle pseudopotential method based on DFT. The spin-up band structure shows that bulk Fe_2Si has metallic character, whereas the spin-down band structure shows that bulk Fe_2Si is an S-L indirect band gap in the vicinity of Fermi surface. The half-metallic gap of bulk Fe_2Si is 0.164 eV. Applying strain, bulk Fe_2Si is first a stable half-metallic ferromagnet, and it then becomes metallic with further increasing strain. The DOS near the Fermi level mainly comes from Fe-3d spin states, and the DOS of Fe mainly comes from the Fe^1-3d and Fe^3-3d states, which are the source of the ferromagnetic properties of bulk Fe_2Si. The magnetic moment and charge transfer of bulk Fe_2Si are mainly caused by Fe d–d exchange and Fe-d–Si-p hybridization. In the strain range −11% to 11% (excluding zero strain), bulk Fe_2Si has stable half-metallic ferromagnetism and the spin polarization at the Fermi level is 100%, indicating that the spin polarization and half-metallic ferromagnetism of bulk Fe_2Si can be improved by strain.

Conflicts of Interest

The authors declare that they have no conflicts of interest.

Acknowledgments

This work was mainly supported by the Cultivated Project of Outstanding Young Teachers of Guangdong Province (no. YQ2015124). This support is greatly appreciated.

References

[1] H. Ohno, "Making nonmagnetic semiconductors ferromagnetic," *Science*, vol. 281, no. 5379, pp. 951–956, 1998.

[2] W.-H. Xie and B.-G. Liu, "Half-metallic ferromagnetlsm in ternary transltion-metal compounds based on ZnTe and CdTe semiconductors," *Journal of Applied Physics*, vol. 96, no. 6, pp. 3559–3561, 2004.

[3] R. Y. Oeiras, F. M. Araújo-Moreira, and E. Z. Da Silva, "Defect-mediated half-metal behavior in zigzag graphene nanoribbons," *Physical Review B - Condensed Matter and Materials Physics*, vol. 80, no. 7, Article ID 073405, 2009.

[4] K. Hasegawa, M. Isobe, T. Yamauchi et al., "Discovery of ferromagnetic-half-metal-to-insulator transition in K2Cr8O16," *Physical Review Letters*, vol. 103, no. 14, Article ID 146403, 2009.

[5] R. A. De Groot, F. M. Mueller, P. G. V. Engen, and K. H. J. Buschow, "New class of materials: Half-metallic ferromagnets," *Physical Review Letters*, vol. 50, no. 25, pp. 2024–2027, 1983.

[6] S. M. Watts, S. Wirth, A. Von Molnár, A. Barry, and J. M. D. Coey, "Evidence for two-band magnetotransport in half-metallic chromium dioxide," *Physical Review B - Condensed Matter and Materials Physics*, vol. 61, no. 14, pp. 9621–9628, 2000.

[7] Y. T. Chen and Y. C. Tan, "The optical, magnetic, and electrical characteristics of Fe_2Si thin films," *Journal of Alloys and Compounds*, vol. 615, pp. 946–949, 2014.

[8] C. P. Tang, K. V. Tam, S. J. Xiong, J. Cao, and X. Zhang, "The structure and electronic properties of hexagonal Fe2Si," *AIP Advances*, vol. 6, no. 6, Article ID 065317, 2016.

[9] M. D. Segall, P. J. D. Lindan, M. J. Probert et al., "First-principles simulation: ideas, illustrations and the CASTEP code," *Journal of Physics Condensed Matter*, vol. 14, no. 11, pp. 2717–2744, 2002.

[10] J. P. Perdew, K. Burke, and M. Ernzerhof, "Generalized gradient approximation made simple," *Physical Review Letters*, vol. 77, no. 18, pp. 3865–3868, 1996.

[11] C. G. Broyden, "The convergence of a class of double-rank minimization algorithms. II. The new algorithm," *Journal of the Institute of Mathematics and Its Applications*, vol. 6, pp. 222–231, 1970.

[12] D. Goldfarb, "A family of variable-metric methods derived by variational means," *Mathematics of Computation*, vol. 24, pp. 23–26, 1970.

[13] H. J. Monkhorst and J. D. Pack, "Special points for Brillouin-zone integrations," *Physical Review. B. Solid State*, vol. 13, no. 12, pp. 5188–5192, 1976.

2

Core/Shell Structure of TiO$_2$-Coated MWCNTs for Thermal Protection for High-Temperature Processing of Metal Matrix Composites

Laura Angélica Ardila Rodriguez ⓘ **and Dilermando Nagle Travessa**

Federal University of São Paulo (UNIFESP), Institute of Science and Technology, Laboratory of Advanced Metals and Processing, São José dos Campos, SP, Brazil

Correspondence should be addressed to Laura Angélica Ardila Rodriguez; laardilar88@gmail.com

Academic Editor: Ilia Ivanov

The production of metal matrix composites with elevated mechanical properties depends largely on the reinforcing phase properties. Due to the poor oxidation resistance of multiwalled carbon nanotubes (MWCNTs) as well as their high reactivity with molten metal, the processing conditions for the production of MWCNT-reinforced metal matrix composites may be an obstacle to their successful use as reinforcement. Coating MWCNTs with a ceramic material that acts as a thermal protection would be an alternative to improve oxidation stability. In this work, MWCNTs previously functionalized were coated with titanium dioxide (TiO$_2$) layers of different thicknesses, producing a core-shell structure. Heat treatments at three different temperatures (500°C, 750°C, and 1000°C) were performed on coated nanotubes in order to form a stable metal oxide structure. The MWCNT/TiO$_2$ hybrids produced were evaluated in terms of thermal stability. Thermogravimetric analysis (TGA), X-ray diffraction (XRD), scanning electron microscopy (SEM), Fourier transform infrared spectroscopy (FTIR), Raman spectroscopy (RS), and X-ray photoelectron spectroscopy (XPS) were performed in order to investigate TiO$_2$-coated MWCNT structure and thermal stability under oxidative atmosphere. It was found that the thermal stability of the TiO$_2$-coated MWCNTs was dependent of the TiO$_2$ layer morphology that in turn depends on the heat treatment temperature.

1. Introduction

Since its discovery by Iijima in 1991 [1], carbon nanotubes (CNTs) have been very attractive for different applications due to their unique properties and structure. CNTs have been used as a reinforcement phase in several types of composite materials, including polymeric and ceramic matrix, and now are being used in high-performance metal matrix composites [2]. To take full advantage of CNTs as a reinforcement material, it is very important to preserve its structure as perfect as possible, during the composite processing. Furthermore, a good dispersion in the matrix and the formation of an interface that allows the load transfer between matrix and reinforcement are also important issues. It is well known that CNTs can support temperatures up to 2500°C in inert atmosphere, but in presence of oxygen, they can be completely decomposed at temperatures between 400 and 600°C, depending on their structure and purity level. In presence of metals like Al, it can form compounds like Al$_4$C$_3$ [3–5].

When processing metal matrix composites (MMCs), CNTs are commonly exposed to high temperatures and consequently preserving their integrity is a challenging task. Those processing routes include powder metallurgy [6–14], high-energy ball milling [15, 16], friction stir processing [17], spark plasma sintering [18], plasma spraying [19–21], and more recently laser sintering [22]. However, the obtained results are still not completely satisfactory, because the extreme processing conditions, either thermal or impact, can damage the CNT structure and cause a detrimental effect on the final mechanical properties of the composite. Bakshi et al. [23], for instance, reported that during plasma spraying processing of Al/MWCNT composites, due to the interaction of the CNTs with the plasma plume, many defects were generated, which led to the formation of Al_4C_3 needles in contact with the aluminum matrix. Ci et al. [5] reported that this reaction for carbide formation usually occurs at amorphous regions of the nanotubes, at structural defects, or at the open sides of their extremities. Consequently, structural defects of CNTs are the precursors for the formation of nanoscale Al_4C_3 layers that are brittle and sensitive to moisture contact. Park and Lucas [24] found in a $SiC_p/6061$ Al composite that the interface formed by the reaction of SiC reinforcement and molten aluminum was Al_4C_3. The disintegration of Al_4C_3 occurred in less than 120 h when exposed to a wet environment, and failure during fracture toughness testing occurs predominantly at the Al_4C_3 particle/matrix interface. For the above reasons, protecting CNTs during the MMC fabrication, without negatively affecting their integrity, is a prerequisite in order to keep their properties.

For severe processing conditions, such as high-power laser, used for synthesizing Inconel 718 super alloy reinforced by carbon nanotubes, Chen et al. [22] coated the CNT wall with NiP by electroless plating process, aiming to improve the CNTs to matrix adherence, preventing direct laser radiation and increasing the wettability. The authors reported a homogeneous dispersion of CNTs and a good interfacial bonding with the matrix. Jo et al. [25] coated carbon nanofibers with TiO_2 by the sol-gel method and dispersed it in aluminum matrix composites by the liquid pressing process, reporting that the coating not only protects the carbon nanofiber but also improves the mechanical properties of the composite.

In this context, the idea of creating a protection system for CNTs that will be subjected to high temperatures during the production of MMCs is promising. This procedure can guarantee a better structural stability and wettability with the matrix. It can even create an additional hardening effect in the composite, with the introduction of hard TiO_2 particles along with the CNTs.

The sol-gel technique is an ideal method to deposit layers in nanostructures and has been showed to be a feasible technique to coat CNTs without adverse effects in their intrinsic properties. In some works [26, 27], it has been reported successful TiO_2 coating on CNTs by the sol-gel processing route for photocatalysis and electrocatalysis applications. In these works, the obtained coating is amorphous and crystallizes at temperatures around 500°C. Eder and Windle [28] employed different metallic oxides to coat CNTs, obtaining crystalline coatings after calcinations at temperatures between 250°C and 400°C [29]. They found that Al_2O_3 was the best coating to avoid CNT oxidation. Similar result was found by Inam et al. [30] who obtained Al_2O_3-coated CNTs by atomic laser deposition.

In the present work, the thermal stability of TiO_2-coated MWCNTs, obtained by the sol-gel technique, was evaluated. This thermal stability was evaluated as a function of the coating layer thickness and for different coating structures, obtained from different calcination temperatures. Fourier transform infrared spectroscopy (FTIR), field

emission gun-scanning electron microscopy (FEG-SEM), X-ray photoelectron spectroscopy (XPS), and X-ray diffraction were employed in order to evaluate the morphology, structure, and crystalline phase of the TiO_2 layer obtained. Raman spectroscopy was employed to evaluate the integrity of the CNTs before and after the coating process. Finally, thermogravimetric analysis was employed to assess the TiO_2-coated MWCNT stability in the O_2 atmosphere.

2. Materials and Methods

2.1. Carbon Nanotube Functionalization. Baytubes® C 150 P multiwalled carbon nanotubes (MWCNTs), produced by Bayer MaterialScience-Germany by the CVD process, with purity greater than or equal to 95%, were used. The MWCNTs have internal and external diameters of the order of 4 and 13 nm, respectively, with a length greater than 1 μm. The MWCNTs were acid treated in 60 mL of 1 : 3 ($v : v$) of analytical grade nitric acid 65% PA and sulfuric acid PA-ACS for 6 hours under magnetic stirring. The resulting mixture was washed with deionized water until a neutral pH was reached and dried at 80°C for 15 hours.

2.2. Synthesis of the TiO_2-Coated MWCNTs. The route to produce TiO_2 shell supported over the MWCNT core was carried out according to the modified surfactant wrapping sol-gel method reported by Gao et al. [26] and is shown in a flowchart in Figure 1. This procedure, when repeated (Cycles I, II, and III), is able to produce thicker TiO_2 coatings.

In the first cycle (Cycle I), MWCNTs were mixed with 2% wt. of sodium dodecyl sulphate (Synth) in aqueous solution and kept in an ultrasonic bath for 3 hours. The final concentrations of MWCNTs in Milli-Q water were 4, 6, and 8 mg/mL. Then, 20 mL of absolute ethyl alcohol 99.8% P.A. was added to the aqueous solution and stirred for additional 0.5 hours, in order to form a MWCNT solution. In parallel, titanium isopropoxide (Aldrich) was mixed to 15 mL of absolute ethyl alcohol 99.8% P.A. and glacial acetic acid 99.8% P.A. under stirring for 0.5 hours, to form a TiO_2 precursor solution. Then the TiO_2 precursor solution was added dropwise in the CNT solution under vigorous stirring that was kept for 2 hours in order to complete the coating reaction. The final percentage of MWCNTs, related to the total weight of the MWCNT-coated TiO_2 hybrid nanocomposite, was estimated to be 16, 23, and 28 by weight (wt.%), when the concentrations of MWCNTs in Milli-Q water were 4, 6, and 8 mg/mL, respectively. Ammonium hydroxide P.A. (Synth) was added dropwise until pH 9 was reached. Finally, 10 mL of absolute ethyl alcohol was added, and the mixture was kept under stirring for another 0.5 hours. Subsequently, the solution was vacuum filtered using a cellulose acetate membrane and washed with 200 mL of absolute ethyl alcohol. The drying process was carried out for 15 hours in an oven at 60°C to obtain the powdered core-shell structure of TiO_2-coated MWCNTs, named CI (Cycle I).

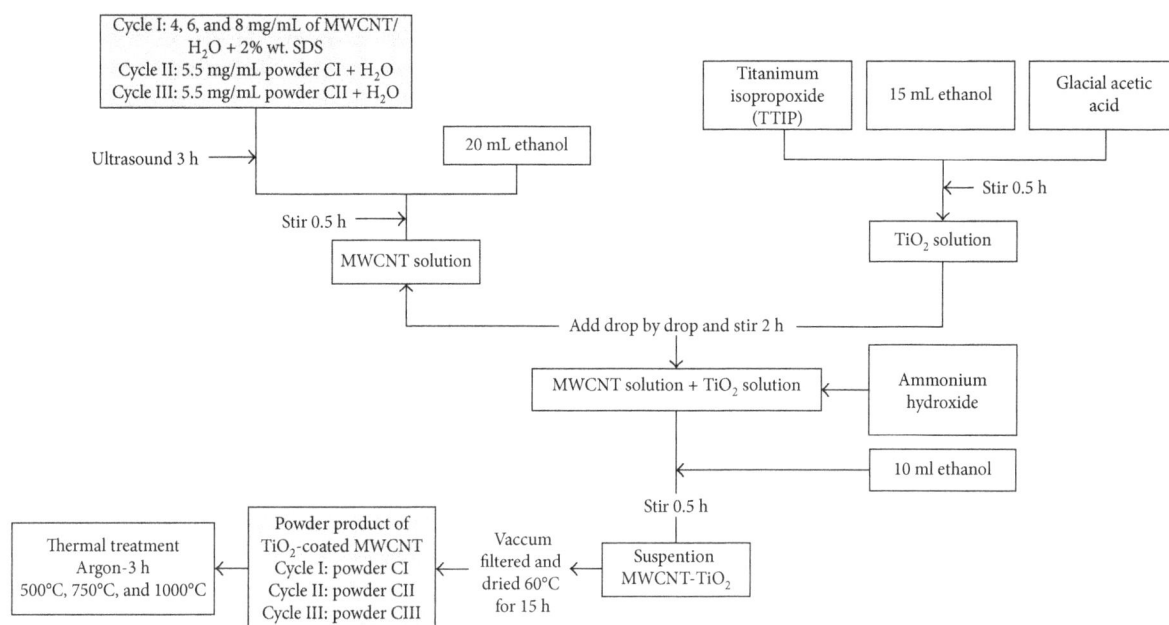

FIGURE 1: Flowchart describing the carbon nanotube-coating process with TiO_2.

For the second cycle (Cycle II), powder CI (obtained in Cycle I with 28 wt.% of MWCNTs) was mixed with Milli-Q water (5.5 mg/mL) and dispersed in an ultrasonic bath for 3 hours. The obtained suspension was subjected to the same process described above for the CI powder. For the product of the Cycle II (named CII), the final percent of MWCNTs, related to the total weight of the MWCNT-coated TiO_2 nanocomposite, was estimated from the mass balance of the precursors to be 8.23 wt.%. For the third cycle (Cycle III), the overall process is repeated, starting from the powder CII and the final percent of MWCNTs, related to the total weight of the MWCNT-coated TiO_2 nanocomposite, was estimated to be 2.32 wt.%.

As the obtained TiO_2 coating is amorphous and unstable, powders obtained from Cycles I, II, and III were heat treated at 500°C, 750°C, and 1000°C for 3 hours in an argon atmosphere, in order to crystallize the coating layer and to reduce its pore size and surface area.

2.3. Samples Analysis. The starting (pristine) MWCNTs, the functionalized MWCNTs, and the hybrid MWCNT/TiO_2 nanocomposites after calcination were characterized by X-ray diffraction, the later to evaluate the crystallinity degree of the coating. X-ray analysis was performed in a Rigaku X-ray diffractometer, model Ultima IV, using Cu Kα radiation ($\lambda = 1.54178$ Å), voltage of 40 kV, and current of 30 mA. Multiple detectors (fast detection mode) were used at steps of 0.01° and speed of 5°/min, resulting in a high signal level. The nature of the surface molecular groups of MWCNTs before and after functionalization and calcined TiO_2 coating was characterized by a Shimadzu IR-affinity-1 FTIR spectrophotometer, in the region of 4000 to 500 cm^{-1} with a resolution of 4 cm^{-1} after 32 accumulations. The data were acquired in transmittance mode, using the KBr pellet technique. Complementary information about the molecular structure of the functionalized and TiO_2-coated MWCNTs, after calcination, was obtained by X-ray photoelectron spectroscopy (XPS), using a Kratos Axis Ultra XPS, operating in an ultra-high vacuum (approximately 10^{-7} Pa). A monochromatic Al X-ray source was used, with energy of 1486.5 eV and power of 150 W, given by the voltage of 15 kV. The emitted photoelectrons were collected in a hemispherical analyzer with 15 μm of spatial resolution. The energy resolution of the equipment is 0.58 eV. Analysis from the samples was performed in the survey mode, followed by scans located in the regions of C 1s, O 1s, and Ti 2p, corresponding to the locations of the carbon, oxygen, and titanium peaks, respectively. Raman spectroscopy was performed in samples using a Horiba LabRAM microscope with 514 nm lasers, in order to identify typical scatters from carbon structures, as well as from the coating structure after calcination. Spectra were collected from 3 accumulations of 30 seconds each, in the range of 50 to 2000 cm^{-1}.

The microstructure of the MWCNT/TiO_2 core/shell nanocomposites was analyzed using a Tescan model Mira 3 field emission gun-scanning electron microscope (FEG-SEM).

The thermal stability of the samples was evaluated from thermal gravimetric analyzes, using a NETZSCH STA 449 F1 Jupiter thermal analysis equipment (TG-DSC/DTA) with a heating rate of 10°C/min in the range of 200 to 1000°C, in synthetic air (20% O$_2$ and 80% N$_2$) atmosphere with a flow of 20 mL/min.

3. Results and Discussion

3.1. Scanning Electron Microscopy. The FEG-SEM images of nanotubes after sol-gel TiO_2-coating processing are presented in Figure 2. It can be observed that high-quality deposits have been obtained. For coating Cycle I, the

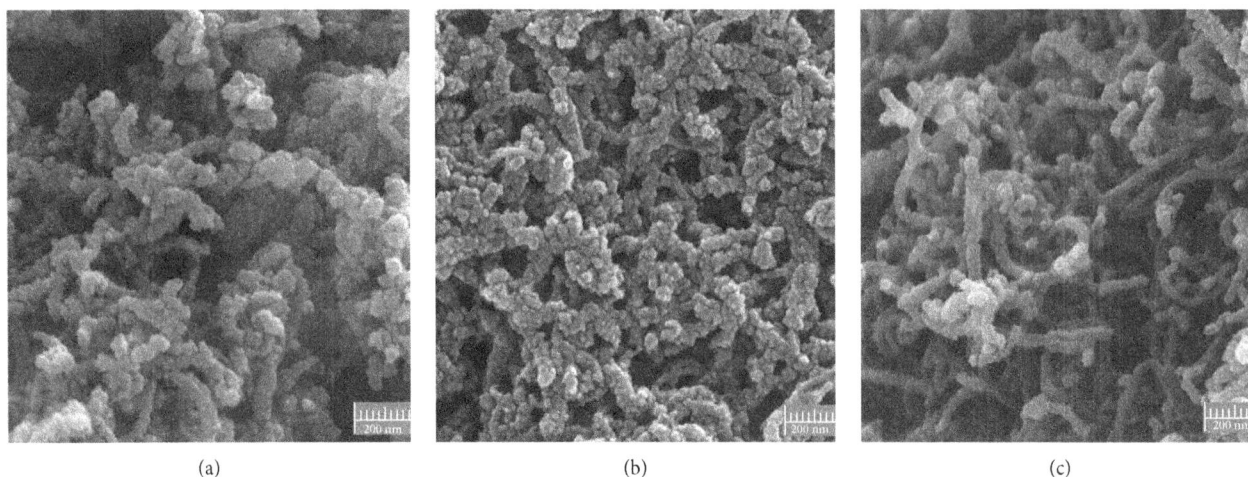

(a)   (b)   (c)

FIGURE 2: FEG-SEM images of core/shell structures of TiO$_2$-coated MWCNT nanocomposites, obtained from different MWCNT concentrations in water during the sol-gel process: (a) 4 mg/mL, (b) 6 mg/mL, and (c) 8 mg/mL.

effect of varying the final concentration of MWCNTs in aqueous solution (4, 6, and 8 mg/mL) is also observed. From Figure 2(a), it is observed that the lower the MWCNT concentration (4 mg/mL), the thicker is the TiO$_2$ coating. Besides the coating formation, a large amount of isolated TiO$_2$ nanoparticles are formed at this lower MWCNT concentration, as less superficial area of substrate (core) is available for the coating formation. Figure 2(b) shows that when the MWCNT concentration in the aqueous solution increases to 6 mg/mL, the MWCNTs are completely coated, similarly to the previous sample, but with a smaller amount of isolated TiO$_2$ nanoparticles. Finally, when the concentration of MWCNTs in water increases to 8 mg/mL, the nanotubes are observed to be completely coated with a thin layer of TiO$_2$ and no isolated particles are present. In such a condition, it seems that the mass relation between MWCNT core and TiO$_2$ shell is optimized.

Once a mass relation of MWCNTs to TiO$_2$ that results in a homogeneous coating and less isolated particles was established, attempts to obtain thicker deposits of TiO$_2$ over the nanotubes surface have been made by increasing the number of coating cycles. Figure 3 shows the aspect of MWCNTs submitted to two coating cycles and calcined at 500 and 1000°C. It can be easily observed that the second cycle produces much thicker layers not uniformly covering the nanotubes. The same is observed in Figure 3(b). However, when comparing the coatings after calcination at 500 and 1000°C (Figures 3(a) and 3(b), resp.), the changes in the layer morphology treated at higher temperature are easily seen. The layer becomes denser and particles size increase from few nanometers to the order of hundred nanometers, forming a necklace arrangement. These changes are related to the anatase to rutile phase transformation and are accompanied by a significant reduction in the specific surface area [28].

3.2. FTIR Analysis. After the acid functionalization, functional groups containing oxygen such as hydroxyl (-OH),

carboxyl (-COOH), and carbonyl (C=O) are expected to form over the nanotubes surface. In order to verify the successful formation of such groups, FTIR spectroscopy was employed. The results are presented in Figure 4 and summarized in Table 1.

From Figure 4 and Table 1, it can be observed that acid functionalization was successful, as several functional groups were formed on the MWCNT surface. The acid treatment breaks few C=C bonds, forming other C, H, and O groups tightly attached to the nanotubes [32]. Furthermore, structural defects eventually present are sites for functional groups to form. The functional groups formed are a basis for the TiO$_2$ layer anchoring in the subsequent coating treatment, as observed by the presence of the band at 500 cm^{-1} in the coated sample, associated with the titanium Ti-O-Ti stretch bonding from the titanium oxide formed after calcination [34].

3.3. X-Ray Photoelectron Spectroscopy. Although the TiO$_2$ formation was identified by FTIR analysis, its tight bonding to the MWCNT surface has to be verified by other techniques. XPS measurements were made to study the chemical state of the MWCNT/TiO$_2$ core/shell nanocomposite. The formation of TiO$_2$ attached to the surface of the functionalized nanotubes is clearly shown by the data from Ti2p binding energy (Figure 5(a)). After deconvolution, the measured energies for Ti^{+4} bonding were 459.56 eV for Ti2p 3/2 and 465.28 eV for Ti2p 1/2. The difference between these two peaks was 5.72 eV. Theoretically, the TiO$_2$ bonding energies (Ti2p 3/2 and Ti2p 1/2) are centered at 458.7 and 464.4 eV, and their difference is 5.7 eV. The small positive displacement observed in this work suggests that an electronic interaction between TiO$_2$ and the MWCNT substrate was established [37], resulting from a close interaction between the nanotubes and the oxide coating.

Figure 5 also shows the analysis for (Figure 5(b)) O 1s and (Figure 5(c)) C 1s energy binding of the TiO$_2$-coated

(a)   (b)

FIGURE 3: FEG-SEM images of core/shell structures of TiO_2-coated MWCNT composites with two coating cycles thermal treated at (a) 500°C and (b) 1000°C.

MWCNT sample. The O 1s spectrum can be adjusted to three peaks, where the main peak at 530.9 eV corresponds to the Ti-O network oxygen. The other two peaks with binding energies of 531.7 and 532.9 eV are assigned to the O-H and C=O groups, respectively. Similarly to the Ti 2p spectrum, the main peak in O 1s is slightly displaced (0.9 eV), as a result of the interaction and consequent surface charge transfer between TiO_2 and MWCNTs [37]. The C 1s spectrum was adjusted to four peaks, representing the C=C binding energy of MWCNTs of 284.7 eV, as well as C-O, C=O, and COO-energies of 286.1, 287.9, and 290.1 eV, respectively. These results reinforce the results from FTIR analysis, evidencing the formation of functional groups on the MWCNT surface after acid functionalization. It is very important to note that, even after calcination at 1000°C, there is no evidence of carbon replacing oxygen atoms in the lattice of TiO_2, as the formation of O-Ti-C bonds commonly represented in the C 1s spectrum by a peak at 282.5 eV [34]. Additionally, Figure 5(d) shows the composite valence band (VB), where no additional electronic states are shown above the TiO_2 edge valence band, discarding any possible carbon doping in the TiO_2 network [38, 39].

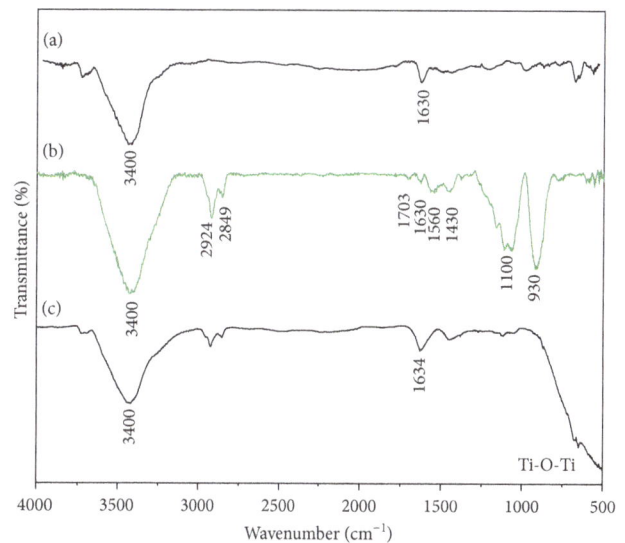

FIGURE 4: FTIR spectra of (a) pristine, (b) oxidized (functionalized), and (c) TiO_2-coated MWCNTs with I coating cycle after calcination at 1000°C.

3.4. X-Ray Diffraction.

After calcination at different temperatures, the presence of a crystalline TiO_2 layer was confirmed by X-ray diffraction, and the results are shown in Figure 6. Comparison of the diffraction patterns of MWCNT/TiO_2 core/shell nanocomposite samples treated at 500, 750, and 1000°C shows that the crystal form of the coating layer changes from totally anatase to totally rutile, as the calcination temperature increases. At the intermediate temperature of 750°C, a mixture of both crystalline structures is observed.

It is also important to note that no characteristic peaks of MWCNTs were found. The proximity of the most intense anatase (101) peak at $2\theta = 25.2°$ with the most intense (002) graphitic MWCNT peak at $2\theta = 25.9°$ impairs the nanotube identification by X-ray diffraction in the coated samples.

However, the presence of MWCNTs in the nanocomposites was further confirmed by Raman analysis as discussed later.

A rough approximation of the crystallite size of the samples was calculated by the well-known Scherrer equation: $D = k\lambda/(\beta\cos\theta)$, where D is the crystallite particle size, k is a constant of 0.9, λ is the wavelength (nm) of the X-rays used, β is the peak full width at half maximum (FWHM), and θ is the Bragg angle [34]. Calculations show that the crystallite size for the samples of II cycles calcined at 500°C was about 5 nm, increasing to about 87 nm in samples calcined at 750°C, and more significantly to about 118 nm after calcination at 1000°C. These results indicate that the phase transformation from anatase to rutile induced both grain growth and probably surface area reduction and are in agreement with the literature [28] and with scanning electron microscopy results.

TABLE 1: FTIR bands observed for pristine, acid-functionalized, and TiO_2-coated MWCNTs.

Band (cm^{-1})	Interpretation	Pristine	Acid functionalized	TiO_2 coated
3400	OH stretching vibrations of isolated surface –OH moieties and in sorbed water [31]	x	x	x
2924	Asymmetric stretching of CH bonds located on the MWCNT surface defects [32]	—	x	—
2849	Symmetric stretching of CH bonds located on the MWCNT surface defects [32]	—	x	—
1730	C=O bonds related to the carboxyl group COOH [33]	—	x	—
1630	OH stretch of adsorbed water and Ti-OH [34]	x	x	x
1560	C-O carboxylate anion stretch mode [33]	—	x	—
1430	O-H deformation vibrations [31]	—	x	—
1100	C-O stretch [31, 35]	—	x	—
930	O-H stretch [36]	—	x	—
500	Ti-O-Ti stretch [34]	—	—	x

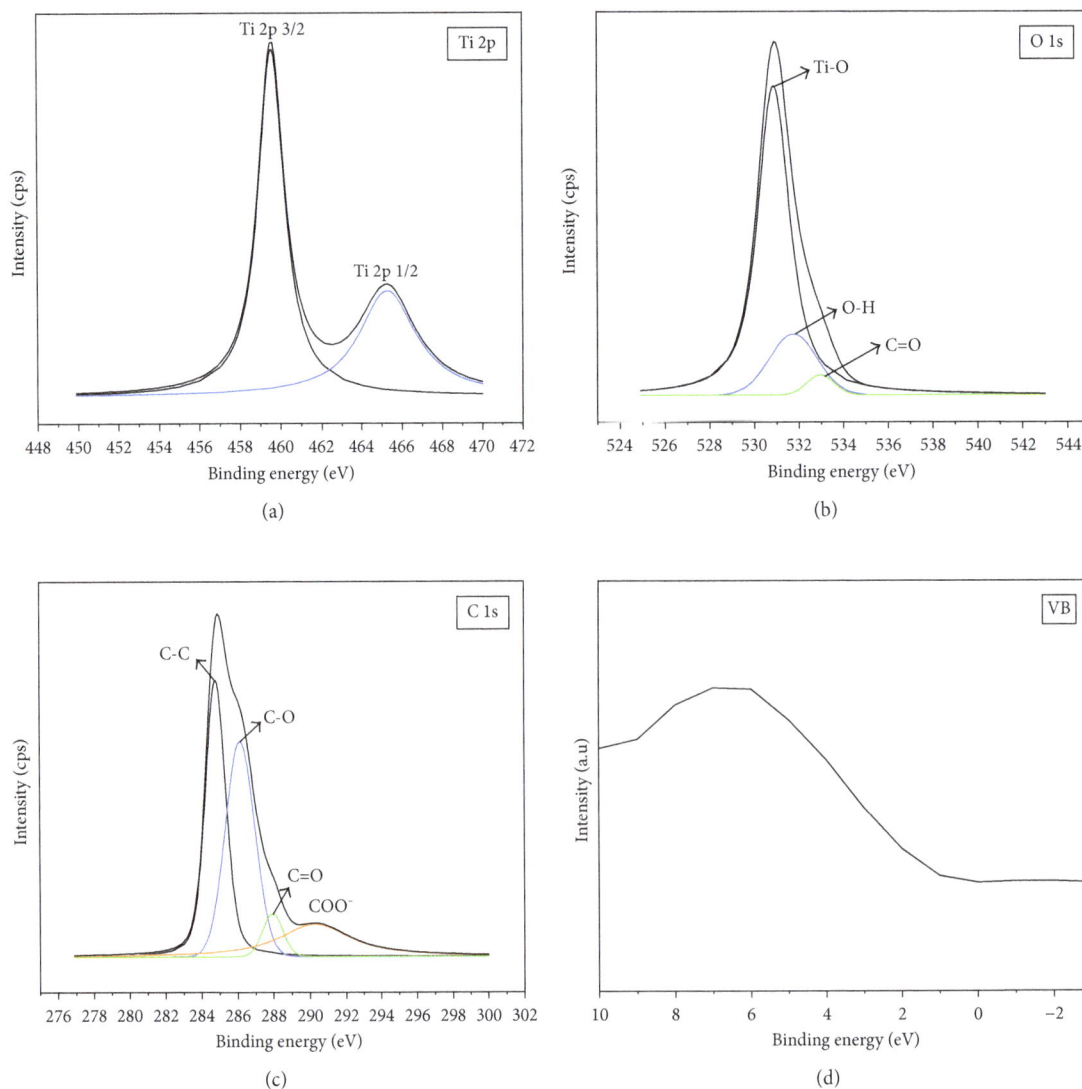

FIGURE 5: High-resolution XPS spectra of (a) Ti 2p, (b) O 1s, (c) C 1s, and (d) XPS valence band for TiO_2-coated MWCNTs with II coating cycles after calcination at 1000°C.

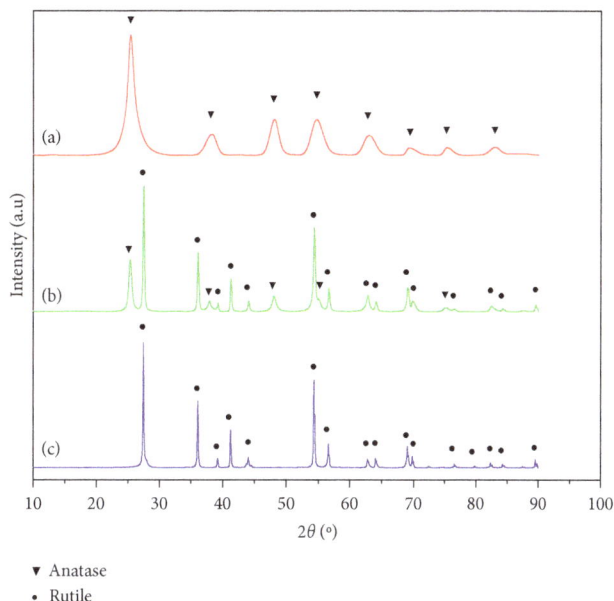

FIGURE 6: X-ray diffraction spectra for TiO$_2$-coated MWCNTs with II coating cycles and calcined at (a) 500°C, (b) 750°C, and (c) 1000°C.

3.5. Raman Spectroscopy. Figure 7(a) presents the Raman spectra obtained for the pristine MWCNTs. The *D* and *G* bands were identified at their typical positions: 1337.37 cm^{-1} and 1576.31 cm^{-1}, respectively. After acid functionalization, no shifting was observed on these *D* and *G* peak bands. However, their intensity and width changes slightly. Perfect highly ordered graphitic structures do not produce scatter *D*-band, which is associated with structural defects. Consequently, the *D/G* ratio is commonly used to assess the crystalline quality of carbon-based materials. In the present work, *D/G* ratio increases from 1.24 (pristine MWCNTs) to 1.36 after acid functionalization, indicating that the external wall of the nanotubes is damaged in a certain extension in order to introduce the functional groups. However, it is to be pointed out that this increase was not so intense.

Figure 7(b) shows the spectra for the MWCNT/TiO$_2$ nanocomposites calcined at different temperatures. Typical E_g (142 cm^{-1}), B_{1g} (399 cm^{-1}), A_{1g} (518 cm^{-1}), and E_g (641 cm^{-1}) scattering bands of anatase are present in samples treated at 500°C. At 750°C, scattering bands of B_{1g} (145 cm^{-1}), E_g (446 cm^{-1}), A_{1g} (612 cm^{-1}), and the multiphoton process (230 cm^{-1}) of rutile [40] start to be visible and are predominate after calcination at 1000°C. These results fully agree with the results from X-ray diffraction. Still referring to Figure 7(b), it is possible to observe the presence of the D and G bands from the functionalized MWCNT substrate, located between 1342–1359 cm^{-1} and 1583–1594 cm^{-1}, respectively. Both bands are slightly shifted to larger wavenumbers, when comparing to Figure 7(a). According to the literature [41, 42], upshift of the *D* band can be related to stresses induced by the presence of TiO$_2$ shell on the surface of the MWCNT core, and the upshift of the *G* band is due to a high-structural interaction (bonding) between the TiO$_2$ and the MWCNTs [43, 44].

Gui et al. [42] found in a MWCNT/TiO$_2$ composite that the intensity ratio of main anatase Eg mode related to the intensity of graphene band (I_A/I_G) reduces, as the ratio of TiO$_2$: MWCNTs increases. In the present work, in samples with pure anatase (calcined at 500°C), the I_A/I_G ratio obtained was observed to increase as the number of coating cycles increases. In samples subjected to a single coating cycle (TiO$_2$: MWCNT ratio of 1 : 0.4), I_A/I_G ratio obtained was 0.80. I_A/I_G ratio obtained after two and three cycles (TiO$_2$: MWCNT ratio of 1 : 0.115 and 1 : 0.0325) was found to be 6.32 and 9.35, respectively. These results are consistent with the results of Gui and et al. and reveal that the high intensity of the anatase E_g scattering band is related to the excess of TiO$_2$ related to MWCNTs [42].

As far as the I_D/I_G ratio is concerned, after calcination at 500°C, it was found to decrease to the range of 0.74–0.81, when comparing to pristine and acid-functionalized MWCNTs (1.24 and 1.36, resp.). This behavior can be attributed to the thermal-induced rearrangement of the carbon structure [45]. Based on this result, it would be expected that calcination at higher temperatures would continuously decrease the I_D/I_G ratio. However, an opposite effect was observed. After calcination at 750°C, I_D/I_G obtained was in the range of 0.83–0.95, and 0.98–1.11 after calcination at 1000°C. This behavior can be related to the anatase to rutile phase transformation, evidenced from X-ray diffraction and Raman spectroscopy. This transformation is reconstructive; that is, it requires rearrangement of the Ti-O atoms to fit to the new structure. Furthermore, significant grain growth and densification (reducing the specific surface area) are expected [28]. As Ti-O atoms are tightly bonded to the MWCNT surface, stresses associated with the TiO$_2$ reconstruction should be transferred to the nanotubes, increasing the MWCNT network distortion and consequently the intensity of the *D* scatter band.

3.6. Thermogravimetry. TGA analysis was performed in order to estimate the final amount of MWCNTs in the

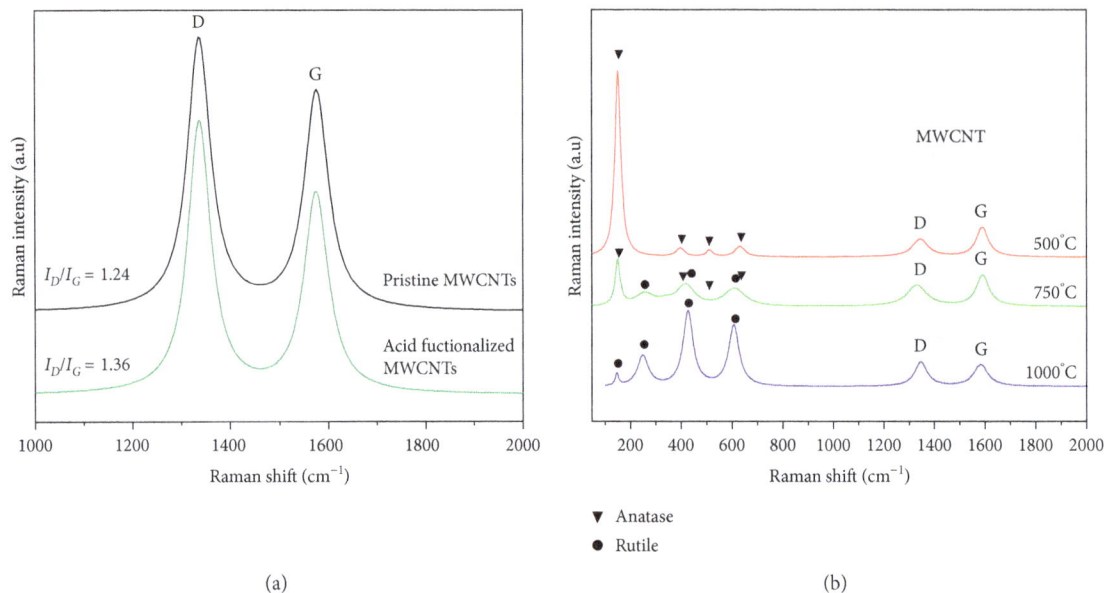

(a)   (b)

FIGURE 7: Raman spectra for pristine and acid functionalized MWCNTs. (b) TiO$_2$-coated MWCNTs with II coating cycles calcined at 500°C, 750°C, and 1000°C.

nanocomposite, and consequently the level of thermal protection offered by the layer of TiO$_2$ in different thicknesses and structure types. Eventual MWCNT oxidation is expected to form CO$_2$ gas, leaving the sample and resulting in composite weight loss as a function of temperature (Figure 8).

From the TGA results, several observations are worth to mention. All samples presented a mass loss up to about 400°C, which can be attributed to the decomposition of surfactant residues still present in the sample. This is more evident in the samples calcined at smaller temperatures: 500°C and 750°C. The thermal decomposition of SDS is very slow [46] and presents a mass loss also at about 750°C, due to the final decomposition of NaSO$_4$, which is very stable up to temperatures around 800°C [47]. After about 450°C, the samples start to present a significant mass loss, being more severe as the calcination temperature decreases. This mass loss is attributed to the decomposition of MWCNTs into CO$_2$ by the O$_2$ atmosphere. The oxidation peak, observed in DTG curves, occurs at about 550°C. The constant levels observed after 600°C in TGA curves should correspond to the complete oxidation of nanotubes. An estimative of the remaining carbon content in the samples can be made comparing this weight loss during TGA with the total weight of the MWCNT-coated TiO$_2$ nanocomposite (based on the mass balance of the precursors), assuming that loses are related to the MWCNTs. This hypothesis is feasible, as the TiO$_2$ coating is very stable, even after calcination at only 500°C, and the anatase to rutile phase transformation is not associated with changes in mass. This analysis, presented in Table 2, enables to quantitatively evaluate the protection level of the TiO$_2$ coating as a function of the layer thickness and calcination temperature. For the sample submitted to two coating cycles, calcined at 500 and 750°C, these ratios were 1.00 and 1.15, respectively, which are lower than the

values of the sample submitted to only one coating cycle and calcined at the same temperatures: 1.12 and 1.30, respectively. These results can be explained by the lack of homogeneity of the coatings obtained in Cycle II, as shown in Figure 3. However, it is possible to note that the protection level expressed by the ratio between initial weight and weight loss drastically increases from around 1.00 to 7.48, when comparing the sample coated by two coating cycles of TiO$_2$ calcined at 500°C with the sample coated by three coating cycles calcined at 1000°C. Similar results were found by Li. et al. [27] in TiO$_2$/MWCNT nanocomposites obtained by the sol-gel process using different titanium precursors and calcination at 500°C. However, the authors suggest that the remaining carbon after TGA analysis is related to elemental diffusion from the substrate into the TiO$_2$-coating structure during calcinations at 500°C, through a transfer-doping mechanism.

As shown in the FEG-SEM images and from XRD calculated crystallite size, the increase in the calcination temperature changes the morphology of the TiO$_2$-coating layer, making it denser and with lesser superficial area, favoring its growth. In fact, the TGA results showed that the higher the calcination temperature, the lower the MWCNT loss. Manivannan et al. [48] found that sintering MWCNTs coated with zirconia improved their resistance to oxidation, and this improvement is related to the proper selection of the sintering temperature. On the other hand, Inam et al. [30] found that the thicker the alumina layer deposited over MWCNTs, the better protection against oxidation it provides. However, it decreases the mechanical properties of the resulting composite by making the coated nanotubes more brittle. It is worth to reinforce that, contrarily to the work of Li et al. [27], in the present work the XPS analysis showed no evidences of carbon doping in the TiO$_2$ structure, strongly suggesting that the remaining carbon refers to intact, not

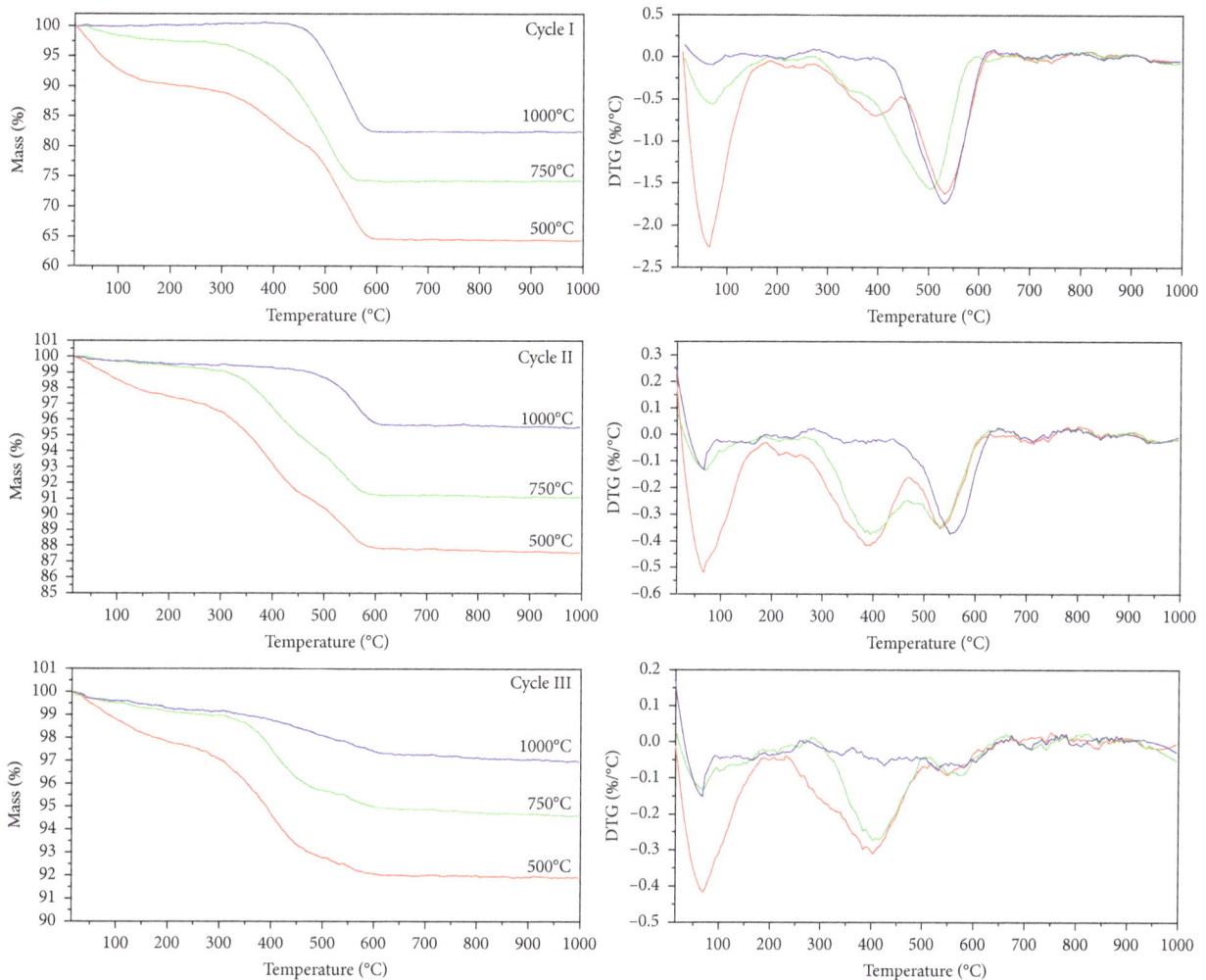

FIGURE 8: TGA and DTG curves of TiO_2-coated MWCNT composites prepared with I, II, and III coating cycles and calcined at 500°C, 750°C, and 1000°C.

TABLE 2: Carbon content in TiO_2-coated MWCNT composites.

Sample	Initial MWCNT %wt.[1]	Calcination temp. (°C)	TGA %wt. loss[2] (%wt[1]/%wt[2])
Cycle I	28.57	500	25.47 (1.12)
		750	21.88 (1.30)
		1000	17.22 (1.65)
Cycle II	8.23	500	8.21 (1.00)
		750	7.11 (1.15)
		1000	3.58 (2.29)
Cycle III	2.32	500	0.61 (3.80)
		750	0.40 (5.80)
		1000	0.31 (7.48)

Notes. The carbon contents presented here correspond to [1]the values before sample fabrication (percent of MWCNTs, related to the total weight of the MWCNT-coated TiO_2 nanocomposite) and [2]those determined by TGA analysis, respectively.

damaged MWCNTs, protected from oxidation by a dense and continuous layer of TiO_2 that completely coats the nanotube surface and acting as an efficient thermal shielding. Furthermore, as observed from the results depicted in Table 2, the MWCNT protection improves as the coating thickness increases. This result is similar to the results of Inam et al. [30], as already mentioned, and is

probably related to the fact that gaseous oxygen would take longer times to reach the MWCNT surface. Consequently, it seems to be feasible to coat MWCNTs with TiO_2 layers in order to protect them from oxidation during the high temperatures involved in MMC processing. Besides, it is expected that the TiO_2 coating layer formed over the MWCNT surface can improve their dispersion and adherence into the metallic matrix. With their integrity preserved, carbon nanotubes can fully contribute to strength in MMCs.

4. Conclusions

In this work, MWCNTs were uniformly coated with a TiO_2 continuous layer with core/shell structure, through the sol-gel method. The layer thickness formed could be controlled by the repetition of the coating cycles, and a stable and dense coating layer could be reached after calcination at 1000°C. Results from of X-ray diffraction and Raman spectroscopy confirm that the layer structure continuously changes from completely anatase to completely rutile, as the calcination temperature increases from 500 to 1000°C. TiO_2 coating formation was also confirmed by XPS, without any C doping in the TiO_2 network. The thermal protection effect of the TiO_2 layer against MWCNT oxidation was confirmed by TGA curves, showing that this protection is more efficient as the layer thickness increases, as well as the calcination temperature. The thermal protection of the MWCNTs is attributed to the fact that high-temperature calcination contributes to pore closure and densification of the TiO_2 layer, reducing the contact between oxygen and the substrate.

The results shown in this work can open new opportunities to fabricate metallic matrix composite materials reinforced by MWCNTs without damaging their structure during the high temperatures to which they are exposed during MMC processing.

Conflicts of Interest

The authors declare that they have no conflicts of interest regarding the publication of this paper.

Acknowledgments

The authors are thankful to the Associated Laboratory for Sensors and Materials (LAS-INPE–Brazil), for the FEG-SEM images, Raman spectroscopy, and XPS analysis and to CNPq for the financial support (Process no. 443395/2014-4).

References

[1] S. Iijima, "Helical microtubules of graphitic carbon," *Nature*, vol. 354, no. 6348, pp. 56–58, 1991.

[2] S. C. Tjong, "Recent progress in the development and properties of novel metal matrix nanocomposites reinforced with carbon nanotubes and graphene nanosheets," *Materials Science and Engineering: R: Reports*, vol. 74, no. 10, pp. 281–350, 2013.

[3] B. Boesl, D. Lahiri, S. Behdad, and A. Agarwal, "Direct observation of carbon nanotube induced strengthening in aluminum composite via in situ tensile tests," *Carbon*, vol. 69, pp. 79–85, 2014.

[4] H. Kwon, M. Estili, K. Takagi, T. Miyazaki, and A. Kawasaki, "Combination of hot extrusion and spark plasma sintering for producing carbon nanotube reinforced aluminum matrix composites," *Carbon*, vol. 47, no. 3, pp. 570–577, 2009.

[5] L. Ci, Z. Ryu, N. Y. Jin-Phillipp, and M. Rühle, "Investigation of the interfacial reaction between multi-walled carbon nanotubes and aluminum," *Acta Materialia*, vol. 54, no. 20, pp. 5367–5375, 2006.

[6] A. M. K. Esawi, K. Morsi, A. Sayed, A. A. Gawad, and P. Borah, "Fabrication and properties of dispersed carbon nanotube-aluminum composites," *Materials Science and Engineering: A*, vol. 508, no. 1-2, pp. 167–173, 2009.

[7] H. J. Choi, G. B. Kwon, G. Y. Lee, and D. H. Bae, "Reinforcement with carbon nanotubes in aluminum matrix composites," *Scripta Materialia*, vol. 59, no. 3, pp. 360–363, 2008.

[8] C. Deng, X. Zhang, D. Wang, Q. Lin, and A. Li, "Preparation and characterization of carbon nanotubes/aluminum matrix composites," *Materials Letters*, vol. 61, no. 8-9, pp. 1725–1728, 2007.

[9] C. F. Deng, D. Z. Wang, X. X. Zhang, and A. B. Li, "Processing and properties of carbon nanotubes reinforced aluminum composites," *Materials Science and Engineering: A*, vol. 444, no. 1-2, pp. 138–145, 2007.

[10] L. Wang, H. Choi, J. M. Myoung, and W. Lee, "Mechanical alloying of multi-walled carbon nanotubes and aluminium powders for the preparation of carbon/metal composites," *Carbon*, vol. 47, no. 15, pp. 3427–3433, 2009.

[11] R. Pérez-Bustamante, I. Estrada-Guel, P. Amézaga-Madrid, M. Miki-Yoshida, J. M. Herrera-Ramírez, and R. Martínez-Sánchez, "Microstructural characterization of Al-MWCNT composites produced by mechanical milling and hot extrusion," *Journal of Alloys and Compounds*, vol. 495, no. 2, pp. 399–402, 2010.

[12] A. M. K. Esawi, K. Morsi, A. Sayed, M. Taher, and S. Lanka, "Effect of carbon nanotube (CNT) content on the mechanical properties of CNT-reinforced aluminium composites," *Composites Science and Technology*, vol. 70, no. 16, pp. 2237–2241, 2010.

[13] A. Esawi and K. Morsi, "Dispersion of carbon nanotubes (CNTs) in aluminum powder," *Composites Part A: Applied Science and Manufacturing*, vol. 38, no. 2, pp. 646–650, 2007.

[14] A. M. K. Esawi and M. A. El Borady, "Carbon nanotube-reinforced aluminium strips," *Composites Science and Technology*, vol. 68, no. 2, pp. 486–492, 2008.

[15] D. N. Travessa and M. Lieblich, "Dispersion of carbon nanotubes in AA6061 aluminium alloy powder by the high energy ball milling process," *Materials Science Forum*, vol. 802, pp. 90–95, 2014.

[16] M. Rashad, F. Pan, J. Zhang, and M. Asif, "Use of high energy ball milling to study the role of graphene nanoplatelets and carbon nanotubes reinforced magnesium alloy," *Journal of Alloys and Compounds*, vol. 646, pp. 223–232, 2015.

[17] D. K. Lim, T. Shibayanagi, and A. P. Gerlich, "Synthesis of multi-walled CNT reinforced aluminium alloy composite via friction stir processing," *Materials Science and Engineering: A*, vol. 507, no. 1-2, pp. 194–199, 2009.

[18] H. Kurita, H. Kwon, M. Estili, and A. Kawasaki, "Multi-walled carbon nanotube-aluminum matrix composites prepared by combination of hetero-agglomeration method, spark plasma sintering and hot extrusion," *Materials Transactions*, vol. 52, no. 10, pp. 1960–1965, 2011.

[19] S. R. Bakshi, A. K. Keshri, and A. Agarwal, "A comparison of mechanical and wear properties of plasma sprayed carbon nanotube reinforced aluminum composites at nano and macro scale," *Materials Science and Engineering: A*, vol. 528, no. 9, pp. 3375–3384, 2011.

[20] T. Laha and A. Agarwal, "Effect of sintering on thermally sprayed carbon nanotube reinforced aluminum nanocomposite,"

Materials Science and Engineering: A, vol. 480, no. 1-2, pp. 323–332, 2008.

[21] T. Laha, Y. Chen, D. Lahiri, and A. Agarwal, "Tensile properties of carbon nanotube reinforced aluminum nanocomposite fabricated by plasma spray forming," *Composites Part A: Applied Science and Manufacturing*, vol. 40, no. 5, pp. 589–594, 2009.

[22] Y. Chen, F. Lu, K. Zhang et al., "Laser powder deposition of carbon nanotube reinforced nickel-based superalloy Inconel 718," *Carbon*, vol. 107, pp. 361–370, 2016.

[23] S. R. Bakshi, V. Singh, S. Seal, and A. Agarwal, "Aluminum composite reinforced with multiwalled carbon nanotubes from plasma spraying of spray dried powders," *Surface and Coatings Technology*, vol. 203, no. 10-11, pp. 1544–1554, 2009.

[24] J. K. Park and J. P. Lucas, "Moisture effect on SiCp/6061 Al MMC: dissolution of interfacial Al_4C_3," *Scripta Materialia*, vol. 37, no. 4, pp. 511–516, 1997.

[25] I. Jo, S. Cho, H. Kim, B. M. Jung, S.-K. Lee, and S.-B. Lee, "Titanium dioxide coated carbon nanofibers as a promising reinforcement in aluminum matrix composites fabricated by liquid pressing process," *Scripta Materialia*, vol. 112, pp. 87–91, 2016.

[26] B. Gao, C. Peng, G. Chen, and G. Lipuma, "Photo-electrocatalysis enhancement on carbon nanotubes/titanium dioxide ($CNTs/TiO_2$) composite prepared by a novel surfactant wrapping sol–gel method," *Applied Catalysis B: Environmental*, vol. 85, no. 1-2, pp. 17–23, 2008.

[27] Z. Li, B. Gao, G. Z. Chen, R. Mokaya, S. Sotiropoulos, and G. Li Puma, "Carbon nanotube/titanium dioxide (CNT/TiO_2) core-shell nanocomposites with tailored shell thickness, CNT content and photocatalytic/photoelectrocatalytic properties," *Applied Catalysis B: Environmental*, vol. 110, pp. 50–57, 2011.

[28] D. Eder and A. H. Windle, "Morphology control of CNT-TiO_2 hybrid materials and rutile nanotubes," *Journal of Materials Chemistry*, vol. 18, no. 17, pp. 2036–2043, 2008.

[29] S. Aksel and D. Eder, "Catalytic effect of metal oxides on the oxidation resistance in carbon nanotube–inorganic hybrids," *Journal of Materials Chemistry*, vol. 20, p. 9149, 2010.

[30] F. Inam, T. Vo, and S. Kumara, "Improving oxidation resistance of carbon nanotube nano-composites for aerospace applications," in *Proceedings of 2nd International Conference on Advanced Composite Materials and Technologies for Aerospace Applications*, pp. 1–6, Wrexham, North Wales, UK, June 2012.

[31] L. Stobinski, B. Lesiak, L. Kövér et al., "Multiwall carbon nanotubes purification and oxidation by nitric acid studied by the FTIR and electron spectroscopy methods," *Journal of Alloys and Compounds*, vol. 501, no. 1, pp. 77–84, 2010.

[32] B. Scheibe, E. Borowiak-Palen, and R. J. Kalenczuk, "Oxidation and reduction of multiwalled carbon nanotubes—preparation and characterization," *Materials Characterization*, vol. 61, no. 2, pp. 185–191, 2010.

[33] M. A. Atieh, O. Y. Bakather, B. Al-tawbini, A. A. Bukhari, F. A. Abuilaiwi, and M. B. Fettouhi, "Effect of carboxylic functional group functionalized on carbon nanotubes surface on the removal of lead from water," *Bioinorganic Chemistry and Applications*, vol. 2010, Article ID 603978, 10 pages, 2010.

[34] S. S. Mali, C. a. Betty, P. N. Bhosale, and P. S. Patil, "Synthesis, characterization of hydrothermally grown $MWCNT$-TiO_2 photoelectrodes and their visible light absorption properties," *ECS Journal of Solid State Science and Technology*, vol. 1, no. 2, pp. M15–M23, 2012.

[35] X. Yan, B. K. Tay, and Y. Yang, "Dispersing and functionalizing multiwalled carbon nanotubes in TiO_2 sol," *Journal of Physical Chemistry B*, vol. 110, no. 51, pp. 25844–25849, 2006.

[36] J. M. Silva-Jara, R. Manríquez-González, F. A. López-Dellamary, J. E. Puig, and S. M. Nuño-Donlucas, "Semi-continuous heterophase polymerization to synthesize nanocomposites of poly(acrylic acid)-functionalized carbon nanotubes," *Journal of Macromolecular Science, Part A*, vol. 52, no. 9, pp. 732–744, 2015.

[37] J. Li, S. Tang, L. Lu, and H. C. Zeng, "Preparation of nanocomposites of metals, metal oxides, and carbon nanotubes via self-assembly," *Journal of the American Chemical Society*, vol. 129, no. 30, pp. 9401–9409, 2007.

[38] S. Wang, L. Zhao, L. Bai, J. Yan, Q. Jiang, and J. Lian, "Enhancing photocatalytic activity of disorder-engineered C/TiO_2 and TiO_2 nanoparticles," *Journal of Materials Chemistry A*, vol. 2, no. 20, p. 7439, 2014.

[39] H. Tan, Z. Zhao, M. Niu et al., "A facile and versatile method for preparation of colored TiO_2 with enhanced solar-driven photocatalytic activity," *Nanoscale*, vol. 6, no. 17, pp. 10216–23, 2014.

[40] J. Yan, G. Wu, N. Guan, L. Li, Z. Li, and X. Cao, "Understanding the effect of surface/bulk defects on the photocatalytic activity of TiO_2: anatase versus rutile," *Physical Chemistry Chemical Physics*, vol. 15, no. 26, pp. 10978–88, 2013.

[41] C.-H. Wu, C.-Y. Kuo, and S.-T. Chen, "Synergistic effects between TiO_2 and carbon nanotubes (CNTs) in a $TiO_2/CNTs$ system under visible light irradiation," *Environmental Technology*, vol. 34, no. 17, pp. 2513–2519, 2013.

[42] M. M. Gui, S.-P. Chai, B.-Q. Xu, and A. R. Mohamed, "Visible-light-driven $MWCNT@TiO_2$ core–shell nanocomposites and the roles of MWCNTs on the surface chemistry, optical properties and reactivity in CO_2 photoreduction," *RSC Advances*, vol. 4, no. 46, p. 24007, 2014.

[43] W. Zhou, K. Pan, Y. Qu et al., "Photodegradation of organic contamination in wastewaters by bonding TiO_2/single-walled carbon nanotube composites with enhanced photocatalytic activity," *Chemosphere*, vol. 81, no. 5, pp. 555–561, 2010.

[44] S. B. A. Hamid, T. L. Tan, C. W. Lai, and E. M. Samsudin, "Multiwalled carbon nanotube/TiO_2 nanocomposite as a highly active photocatalyst for photodegradation of reactive black 5 dye," *Chinese Journal of Catalysis*, vol. 35, no. 12, pp. 2014–2019, 2014.

[45] Z. Peining, A. S. Nair, Y. Shengyuan, and S. Ramakrishna, "TiO_2–MWCNT rice grain-shaped nanocomposites: synthesis, characterization and photocatalysis," *Materials Research Bulletin*, vol. 46, no. 4, pp. 588–595, 2011.

[46] D. Ramimoghadam, M. Zobir, B. Hussein, and Y. H. Taufiq-yap, "The effect of sodium dodecyl sulfate (SDS) and cetyltrimethylammonium bromide (CTAB) on the properties of ZnO synthesized by hydrothermal method," *International Journal of Molecular Sciences*, vol. 13, no. 12, pp. 13275–13293, 2012.

[47] Y. Li, Y. Huo, C. Li, S. Xing, L. Liu, and G. Zou, "Thermal analysis of Cu-organic composite nanoparticles and fabrication of highly conductive copper films," *Journal of Alloys and Compounds*, vol. 649, pp. 1156–1163, 2015.

[48] R. Manivannan, A. Daniel, I. Srikanth et al., "Thermal stability of zirconia-coated multiwalled carbon nanotubes," *Defence Science Journal*, vol. 60, no. 3, pp. 337–342, 2010.

3

Synthesis of Hydrophilic Sulfur-Containing Adsorbents for Noble Metals Having Thiocarbonyl Group Based on a Methacrylate Bearing Dithiocarbonate Moieties

Haruki Kinemuchi and Bungo Ochiai ⓘ

Department of Chemistry and Chemical Engineering, Faculty of Engineering, Yamagata University, 4-3-16 Jonan, Yonezawa, Yamagata 992-8510, Japan

Correspondence should be addressed to Bungo Ochiai; ochiai@yz.yamagata-u.ac.jp

Academic Editor: Santiago Garcia-Granda

Novel hydrophilic sulfur-containing adsorbents for noble metals were prepared by the radical terpolymerization of a methacrylate bearing dithiocarbonate moieties (DTCMMA), hydrophilic monomers, and a cross-linker. The resulting adsorbents efficiently and selectively adsorbed noble metals (Au, Ag, and Pd) from various multielement aqueous solutions at room temperature owing to the thiocarbonyl group having high affinity toward noble metals. The metal adsorption by the adsorbents was proceeded by simple mixing followed by filtration. The noble metal selectivity of the adsorbent obtained from DTCMMA and N-isopropylacrylamide was higher than that of the adsorbent obtained from DTCMMA and N,N-dimethylacrylamide due to the lower nonspecific adsorption.

1. Introduction

Noble metals such as Au, Pd, and Ag are essential in the modern life and applied in various fields such as jewelry [1], synthetic catalysts [2–5], and materials for electronic industry [6]. The amounts of noble metals in mines are inherently very low, and the recent increase in their use is shortening their reserves to production ratios. Therefore, recovery and reuse of these metals from various used materials is very essential. For example, the recycling from industrial wastewater and urban mines is promising owing to the high amounts and contents.

Typical methods for collection of noble metals from water are solvent extraction [7–12], adsorption [13–31], and electrochemical processes [32]. The collection by adsorption is advantageous owing to its low cost, safety, and high efficiency. By contrast, solvent extraction requires organic solvents, lots of eluents, and in some cases supercritical eluents to attain high extraction efficiency. As a result, tremendous amounts of wastes are produced. Electrochemical processes require a troublesome procedure to scrape off the deposited metals on the electrode. Adsorbents bearing sulfur substituents are one

of the most examined adsorption agents for noble metals owing to the excellent affinity of sulfur toward soft cations due to the soft Lewis basicity and the facileness of its introduction into various organic structures [30, 33–37]. Dithiocarbamic acids are the most popular sulfur adsorbents for heavy metals [38]. Dithiocarbamic acids effectively adsorb various metals, but the selectivity of the metal adsorption is low. The emission of toxic carbon disulfide is also a severe problem, although the emission rate is slow [39]. Adsorbents based on thiols and thiocarbonyl moieties are advantageous by eliminating the possibility of the leakage of sulfur compounds such as carbon disulfide. Thiols are effective scavengers for various metals by forming stable thiolates [40–46]. However, thiol groups are susceptible to oxygen-converting thiols into disulfides via oxidative coupling [44]. Thiocarbonyl moieties such as thioamide [26–30, 47–53], thiourethane [31, 33, 35, 37], and thiourea [11, 25, 54–60] have also been examined as stable adsorbents. We also have found that a polymer bearing N,N-dialkylthiourethane moieties has high adsorption ability to Pd and Au, but the adsorption from aqueous solutions only proceeded in the presence of organic solvents such as chloroform due to the hydrophobic nature of the backbone [35]. Although these adsorbents bearing thiocarbonyl moieties certainly adsorb metals from aqueous or organic solvents, the effect of the structures adjacent to thiocarbonyl groups is still unclear.

In this study, we designed a hydrophilic adsorbent bearing thiocarbonyl moieties for selective adsorption of noble metals in an organic solvent free manner. As the source for thiocarbonyl groups, we focused on a methacrylate bearing dithiocarbonate moieties (5-(methacryloyloxy)methyl-1,3-oxathiolane-2-thione, DTCMMA) [61], which can be prepared via a simple reaction of glycidyl methacrylate with carbon disulfide. In order to adsorb noble metals from water in a facile manner, DTCMMA was copolymerized with hydrophilic monomers and a cross-linker to make the adsorbents swellable and insoluble. Selectivity of the adsorption was evaluated using three solutions containing multiple metal ions.

2. Materials and Methods

2.1. Materials. DTCMMA was prepared according to the reported procedure [61]. *N*-isopropylacrylamide (NIPAM) (Wako Pure Chemical, >98.0%) was purified by recrystallization from a mixed solvent of diethyl ether and *n*-hexane. *N,N*-dimethylacrylamide (DMAA) (Kanto Chemical, >97.0%) was dried over calcium hydride and distilled under reduced pressure. Dehydrated dimethyl sulfoxide (DMSO) (Wako Pure Chemical, >99.0%), *N,N'*-methylenebisacrylamide (MBAA) (Wako Pure Chemical, >97.0%), 2,2'-azobis(isobutyronitrile) (AIBN) (Tokyo Chemical Industry, >98.0%), palladium(II) chloride (Wako Pure Chemical, >99.0%), anhydrous copper(II) chloride (Wako Pure Chemical, >99.0%), nitric acid (Wako Pure Chemical, concentration: 60-61%), and hydrochloric acid (Kanto Chemical, concentration: 35.0–37.0%) were commercially available and used as received. Multielement solutions containing platinum elements (**A**) (10 mg/L of Sn, Ru, Rh, Pd, Sb, Te, Hf, Ir, Pt, and Au in 10% HCl/1% HNO$_3$ aq.), containing rare earth elements (**B**) (10 mg/L of Sc, Y, La, Ce, Pr, Nd, Sm, Eu, Gd, Tb, Dy, Ho, Er, Tm, Yb, Lu, and Th in 5% HNO$_3$ aq.), and containing alkali metals, alkali earth metals, typical metals in groups 12–16, Ag, and U (**C**) (10 mg/L of Li, Be, Na, Mg, Al, K, Ca, V, Cr, Mn, Fe, Co, Ni, Cu, Zn, Ga, As, Se, Rb, Sr, Ag, Cd, In, Cs, Ba, Tl, Pb, Bi, and U in 5% HNO$_3$ aq.) (PerkinElmer) were used for competitive adsorption experiments.

2.2. Instruments. IR spectra were recorded on a HORIBA FT-720 spectrometer. Inductive-coupled plasma mass spectrometry (ICP-MS) measurements were performed on a PerkinElmer ELAN DRC II spectrometer. Operating conditions were as follows: nebulizer gas flow, 0.91–1.01 mL/min; ICP RF power, 1.1 kW; lens voltage, 7.4 V; pulse stage voltage, 900 V; dwell time, 60 ns; sweeps, 3 times; readings per replicate, 3 times; and flow rate, 0.96 mL/min. The dissolution media for ICP-MS measurements of the samples after metal adsorption experiments were coincided with the multielement solutions, namely, 10% HCl aq. for **A**, 5% HNO$_3$ aq. for **B**, and 5% HNO$_3$ aq. for **C**.

2.3. Hydrophilic Sulfur-Containing Adsorbent from DTCMMA and NIPAM (1). A solution of DTCMMA (96.4 mg, 442 μmol), NIPAM (500 mg, 4.42 mmol), MBAA (6.8 mg, 4.4 μmol), and AIBN (21.8 mg, 133 μmol) in DMSO (3.0 mL) was stirred at 60°C for 24 h under a nitrogen atmosphere. The resulting precipitate was washed with a large amount of methanol and dried under reduced pressure at 50°C to obtain a hydrophilic sulfur-containing adsorbent (**1**) (586 mg, yield = 97.2%). IR spectra (KBr, cm^{-1}): 3438 (NH in C(=O)NHCH(CH$_3$)$_2$), 3303 (NH in C(=O)NHCH$_2$), 1730 (C=O in C(=O)OCH$_2$), 1657 (C=O in C(=O)NH), and 1192 (C=S).

2.4. Hydrophilic Sulfur-Containing Adsorbent from DTCMMA and DMAA (2). A solution of DTCMMA (96.4 mg, 442 μmol), DMAA (438 mg, 4.42 mmol), MBAA (6.8 mg, 4.4 μmol), and AIBN (21.8 mg, 133 μmol) in DMSO (3.0 mL) was stirred at 60°C for 24 h under a nitrogen atmosphere. The resulting precipitate was washed with a large amount of methanol and dried under reduced pressure at 50°C to obtain a hydrophilic sulfur-containing adsorbent (**2**) (524 mg, yield = 96.8%). IR spectra (KBr, cm^{-1}): 3473 (NH in C(=O)NH), 1732 (C=O in C(=O)OCH$_2$), 1631 (C=O in C(=O)N(CH$_3$)$_2$, C(=O)NH), and 1192 (C=S).

2.5. Metal Adsorption from Multielement Solution with Hydrophilic Sulfur-Containing Adsorbents (Typical Procedure). A multielement solution (2.0 mL, 10 mg/L of metal ions) and a hydrophilic sulfur-containing adsorbent ([C=S]/[total metal] = 4.5/1) were stirred in a glass vial at room temperature for 1.5 h. Then, the adsorbent was removed by filtration and washed 2 times with 100 mL amounts of water. The amounts of adsorbed metals were calculated from the concentrations of metals in the filtrate measured by ICP-MS.

2.6. Competitive Metal Adsorption from Pd/Cu Mixed Solution with 2 (Typical Procedure). CuCl$_2$ (54.9 mg/mL) in 1 N HCl solution (2.0 mL), PdCl$_2$ (0.36 mg/mL) in 1 N HCl solution (2.0 mL), and adsorbent **2** (50.0 mg) were added to a glass vial and stirred at room temperature for 1.5 h. Then, the adsorbent was removed by filtration and washed 2 times with 100 mL of water. The amounts of adsorbed metals were calculated from the concentrations of metals in the filtrate measured by ICP-MS.

3. Results and Discussion

3.1. Synthesis of Hydrophilic Sulfur-Containing Adsorbents. Hydrophilic sulfur-containing adsorbents were synthesized by the radical terpolymerization of DTCMMA, hydrophilic monomer NIPAM or DMAA, and a cross-linker MBAA (feed ratio, [DTCMMA]$_0$: [hydrophilic monomer]$_0$: [MBAA]$_0$ = 9.0 : 90.1 : 0.90) in DMSO at 60°C using AIBN (2.7 mol%) under a nitrogen atmosphere (Scheme 1). Products insoluble in common organic solvents and water were obtained in excellent yields. The products obtained using NIPAM and DMAA were denoted as **1** and **2**,

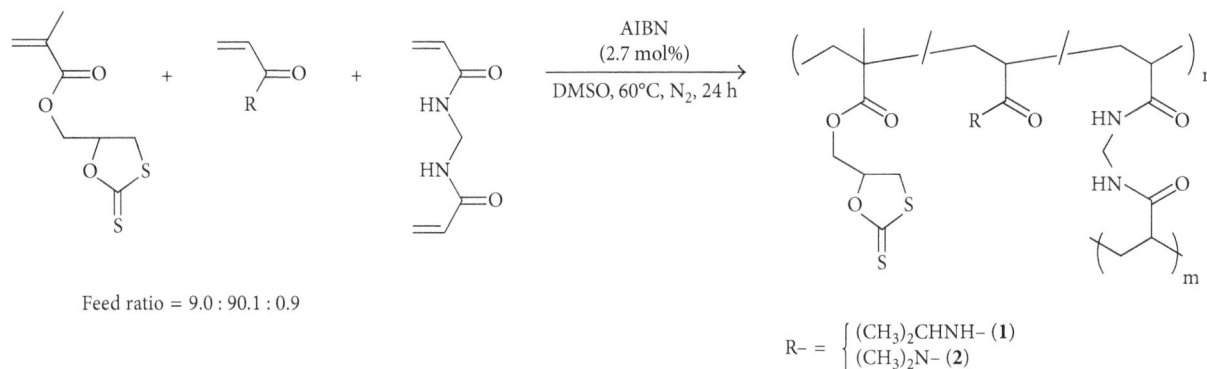

SCHEME 1: Synthesis of hydrophilic sulfur-containing adsorbents **1** and **2**.

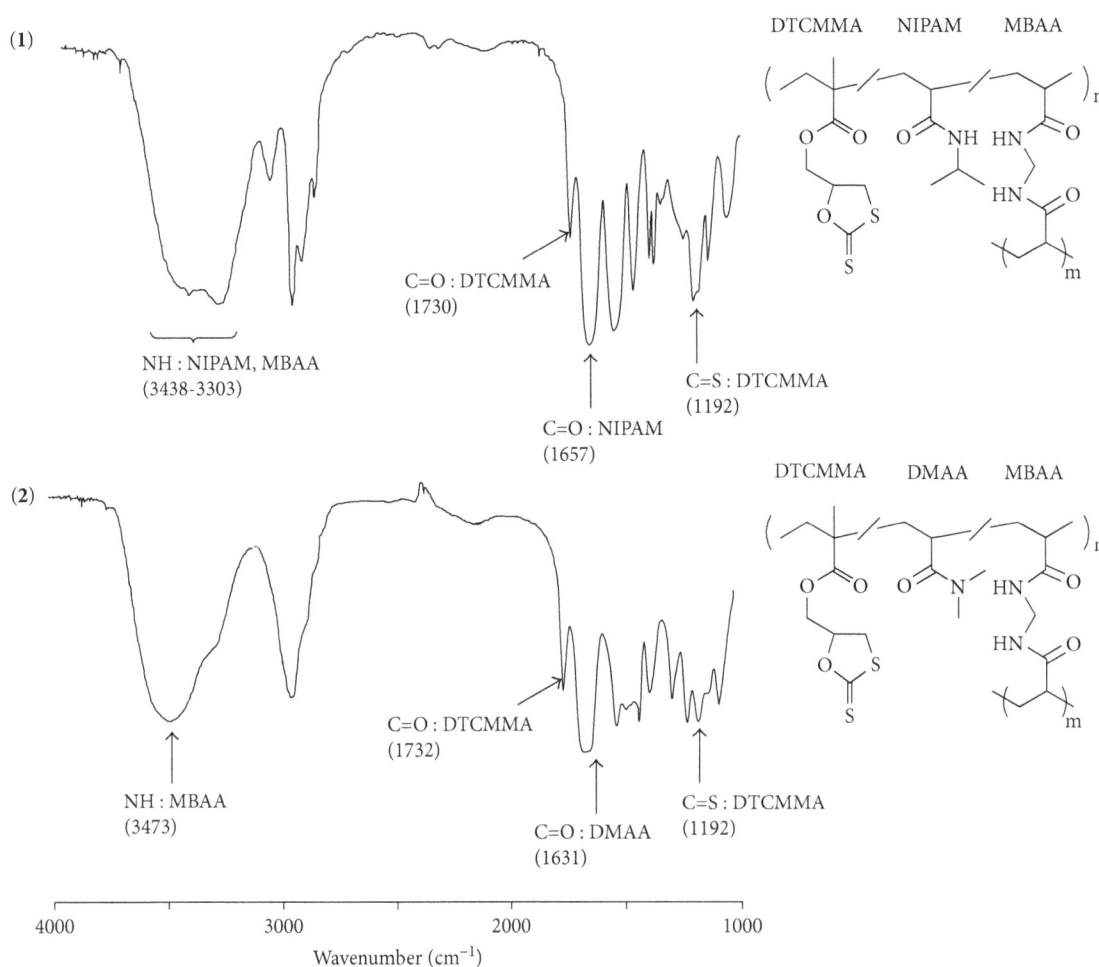

FIGURE 1: FT-IR spectra of hydrophilic sulfur-containing adsorbents (**1**) and (**2**).

respectively. The products could be swelled by water to yield soft gels. The structures were confirmed by FT-IR spectroscopic analysis (Figure 1). Characteristic peaks assignable to C=S moieties in the DTCMMA unit were observed at $1192 \, cm^{-1}$ [60]. Peaks assignable to the amide moieties and carbonyl group in the units originating from the hydrophilic monomers and MBAA were also observed (**1**: $3438 \, cm^{-1}$ for NH in NIPAM, $3303 \, cm^{-1}$ for NH in MBAA,

$1730 \, cm^{-1}$ for C=O in DTCMMA, and $1657 \, cm^{-1}$ for C=O in NIPAM; **2**: $3473 \, cm^{-1}$ for NH in MBAA, $1732 \, cm^{-1}$ for C=O in DTCMMA, and $1631 \, cm^{-1}$ for C=O in DMAA). The contents of the DTCMMA unit in the hydrophilic sulfur-containing adsorbents were estimated by elemental analysis. The contents of DTCMMA units in **1** and **2** were calculated to be $68.3 \, \mu mol$ and $73.0 \, \mu mol$ of DTCMMA units per 100 mg of the adsorbents, respectively, which almost agreed with the feed ratio (theoretical values for **1** and **2**: 73.3 and 81.6 μmol).

3.2. Metal Adsorption Experiments to Find the Metals Adsorbed by the Hydrophilic Metal Adsorbents Using Multi-element Solutions.

The hydrophilic metal adsorbents were used at the molar ratio of [C=S]/[total metal] = 4.5/1. The metal adsorption experiments were conducted by dispersing the adsorbents in multielement solutions, and the adsorbents adsorbing metals were separated by filtration. The amounts of the adsorbed metals were calculated by measuring the amounts of the metals in the filtrates by ICP-MS.

The metal adsorption ratios of some metals were calculated to be negative values, possibly because the apparent metal concentrations after the adsorption increased by uptake of water by the adsorbents resulting in increase in the concentration of metals.

First, a multielement solution **A** was employed for metal adsorption experiments at room temperature for 1.5 h and 24 h (Figure 2). As a result, Au was adsorbed almost quantitatively by both of the metal adsorbents in 1.5 h. Pd was also adsorbed quantitatively in 24 h, but the adsorption ratios at 1.5 h were differed to be 52.3 and 98.7% for **1** and **2**, respectively. The faster adsorption by **2** can be ascribed to the hydrophilicity of the DMAA unit higher than that of the NIPAM unit. Other metals were negligibly adsorbed. The selectivity of the metal adsorption by the sulfur-containing metal adsorbents can be ascribed to the very soft Lewis basicity of the thiocarbonyl group in the dithiocarbonate moieties having high affinity with the very soft Lewis acids Pd and Au.

Second, a multielement solution **B** was employed for the adsorption at room temperature for 24 h (Figure 3). As a result, all the adsorption ratios were below 10%. The low adsorption originated from the low affinity between the very soft thiocarbonyl group and the hard rare earth elements.

Third, a multielement solution **C** was employed for the adsorption at room temperature for 1.5 and 24 h (Figure 4). The exact metal concentrations of Na, Mg, Al, K, Ca, Fe, Zn, and Ba could not be obtained using the standard curves based on this multielement solution, probably due to the leakage of these metals from the vials used in this experiment. As a result, Ag was adsorbed almost quantitatively by both of the metal adsorbents within 1.5 h. Se, originating from SeO$_2$ and existing as selenous acid, was also adsorbed at 36.2 and 31.4% by **1** and **2**, respectively, and the adsorption ratios were increased to ca. 60% at 24 h. We presume that selenous acid coordinated to the Lewis basic carbonyl groups in **1** and **2** as a hard Brønsted acid. The selectivity of **1** was excellent by the negligible adsorption of other metals, but that of **2** was lower due to the slight adsorption of most of the elements, suggesting nonspecific adsorption described later. The experiments using these three multielement solutions proved that the hydrophilic sulfur-containing adsorbents, especially **1**, are effective adsorbents for noble metals with high selectivity and efficiency even in the presence of various elements.

We considered the excellent selectivity of the thiocarbonyl group in DTCMMA for noble metals. The

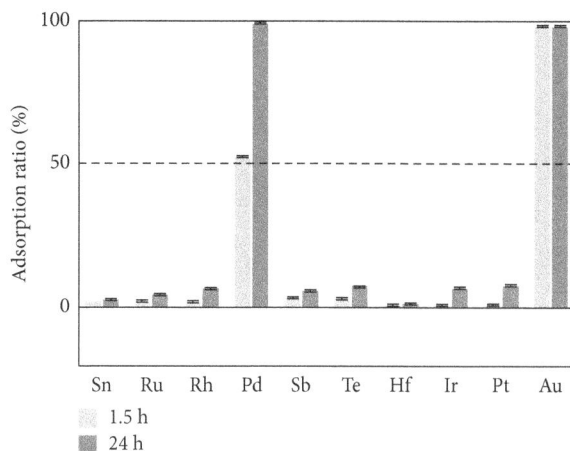

(a)

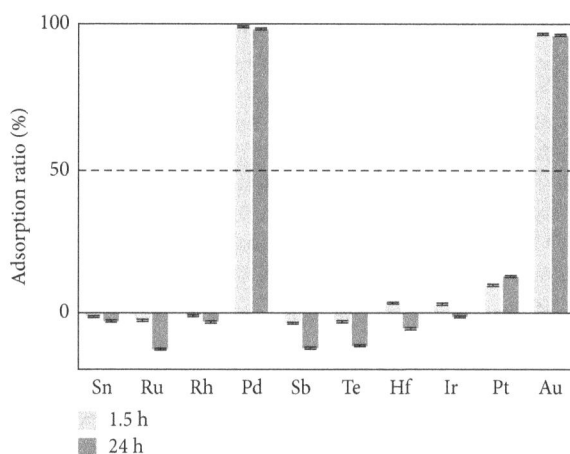

(b)

FIGURE 2: Metal adsorption from multielement solution **A** with hydrophilic sulfur-containing adsorbents **1** (a) and **2** (b) (feed molar ratio of [C=S]/[total metal] = 4.5/1).

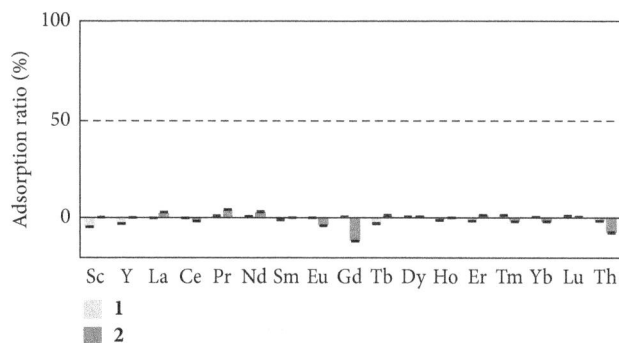

FIGURE 3: Metal adsorption from multielement solution **B** with hydrophilic sulfur-containing adsorbents **1** and **2** (feed molar ratio of [C=S]/[total metal] = 4.5/1).

selectivity of **1** and **2** toward Pd, Au, and Ag is higher than conventional adsorbents having thiocarbonyl groups such as thioamide [26–30, 46–52], thiourethane [32, 34, 36], and thiourea [11, 25, 53–59] including a commercial adsorbent QuadraPure™ TU (Aldrich) with the R–NH(C=S)NH$_2$

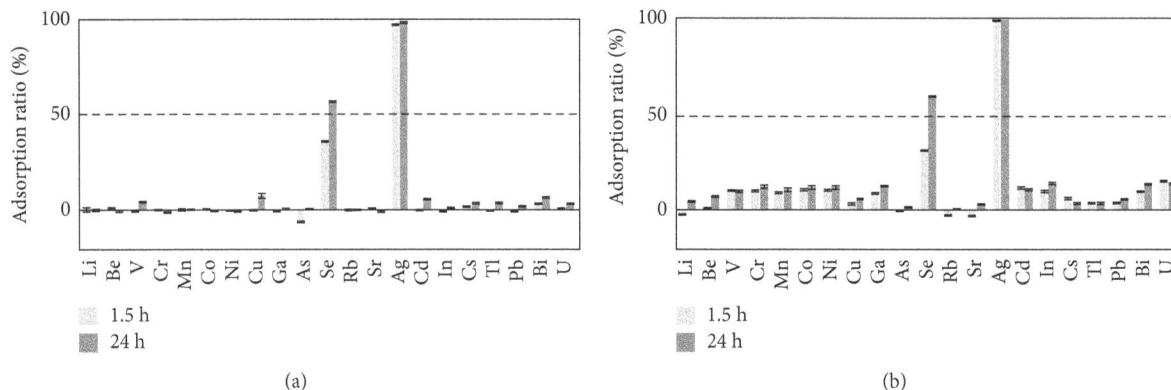

FIGURE 4: Metal adsorption from multielement solution **C** with hydrophilic sulfur-containing adsorbents **1** (a) and **2** (b) (feed molar ratio of [C=S]/[total metal] = 4.5/1).

structure [61]. The lower selectivity of these thiocarbonyl structures can be ascribed to the tautomerism between thiocarbonyl and thiol. For example, the thiourea structure in QuadraPure TU can take the tautomeric iminothiol structure (Scheme 2) [62]. The thiol group is a harder ligand than the thiocarbonyl group and has higher affinity with metals with harder Lewis acidity. By contrast, the selectivity of DTCMMA for the noble metals is higher by the dithiocarbonate moiety without taking harder tautomeric structures.

3.3. Control Experiment Using Cross-Linked Polymers without DTCMMA Unit.

We confirmed the effect of the DTCMMA unit using analogous cross-linked polymers without DTCMMA unit. The cross-linked polymers were synthesized by the radical copolymerization of the hydrophilic monomers NIPAM (**1**′) and DMAA (**2**′) with MBAA in DMSO at 60°C under a nitrogen atmosphere. Control experiments with **1**′ and **2**′ were conducted using the multielement solutions **A** and **C**, which contain the metals adsorbed by **1** and **2**. The metal adsorption experiments were conducted at room temperature for 24 h in a similar manner with the adsorption experiments using the sulfur-containing adsorbents. Most of the elements in **A** were hardly adsorbed as the cases of **1** and **2**, but the adsorption ratios of Au by **1**′ and **2**′ were slightly higher (19.0% and 16.7%, resp.) (Figure 5). A plausible reason for the adsorption of Au is the adsorption with the amide moieties as reported for poly(NIPAM), poly(DMAA), and poly(N,N-diethylacrylamide) gels [63], which should be occurred for the amide moieties in **1**′ and **2**′ leading to the nonspecific adsorption. In the adsorption from the multielement solution **C**, **1**′ negligibly adsorbed metals, indicating that nonspecific adsorption is ignorable (Figure 6). Contrary to **1**′, **2**′ adsorbed most of the metals though to lesser expects. These control experiments supported the excellent selectivity of **1** toward noble metals than **2**. A possible reason is the harder Lewis basicity of the carbonyl group in DMAA units that is likely to be responsible for nonspecific adsorption. The difference in the Lewis basicity was supported by FT-IR absorption of the carbonyl groups in **1**′ and **2**′ appeared at 1635 cm^{-1} and

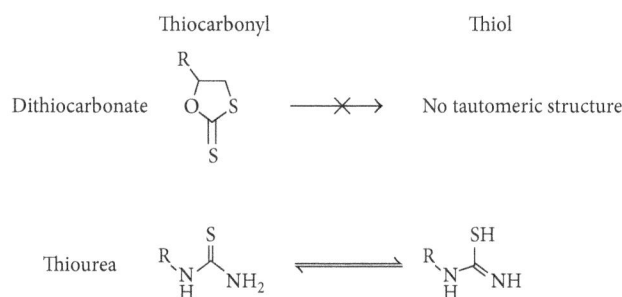

SCHEME 2: Tautomeric structures of thiocarbonyl and thiol.

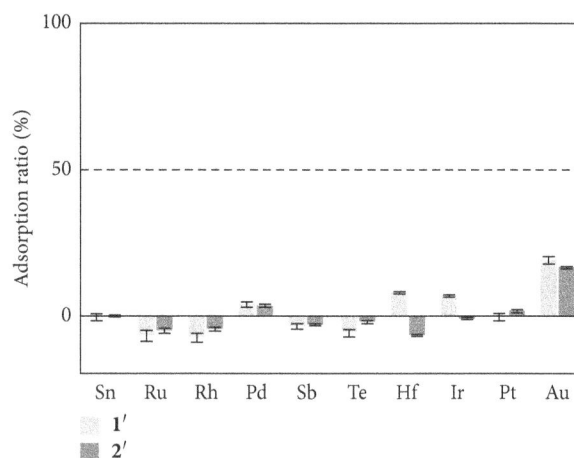

FIGURE 5: Metal adsorption from multielement solution **A** with hydrophilic gels **1**′ and **2**′.

1610 cm^{-1}, respectively. This result convinced us that the DTCMMA unit in the sulfur-containing adsorbents served as the important group for adsorption of noble metals.

3.4. Competitive Metal Adsorption from Mixed Solution of Pd and Excess Amount of Cu by 2.

Pd is a noble metal, and the concentration in nature is very low. In addition to the content in Pd mines, Pd is slightly contained in ores of other metals such as Cu. In industrial and catalytic uses of Pd, Pd are often

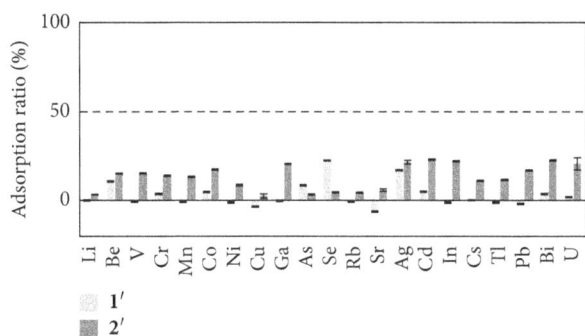

FIGURE 6: Metal adsorption from multielement solution **C** with hydrophilic gels **1′** and **2′**.

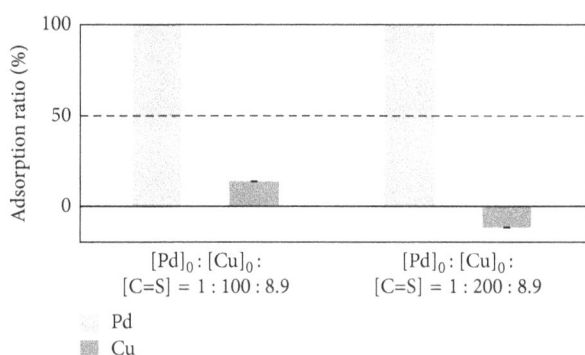

FIGURE 7: Competitive metal adsorption from mixed solutions of Pd and excess amount of Cu by **2**.

used with other metals such as catalysts for Sonogashira coupling reaction combined with copper halides [64]. For this reason, extraction of Pd from mixtures containing excess amounts of other metals such as Cu is an important subject. Accordingly, we conducted competitive adsorption experiments for Pd from mixtures containing excess amounts of Cu with **2**. The feed molar ratios of the C=S moieties in **2** and the metals examined were $[C=S]:[Pd]_0:[Cu]_0 = 8.9:1:100$ and $8.9:1:200$. The experiment was conducted for 1.5 h in a similar manner with the case of the multielement solutions. As a result, **2** adsorbed Pd in an effective and selective manner at both of the feed ratios in spite of the excess amounts of Cu (Figure 7). This result suggested the excellent selectivity of **2** as a Pd adsorption agent from solutions containing various base metals such as wastewater from Cu mines and waste solutions after Pd-catalyzed reactions [65].

4. Conclusions

We developed hydrophilic sulfur-containing adsorbents having thiocarbonyl groups based on a methacrylate bearing dithiocarbonate moieties and investigated the effect of the structure adjacent to the thiocarbonyl group. The hydrophilic sulfur-containing adsorbents **1** and **2** could selectively adsorb noble metals, Au, Ag, and Pd, from multielement aqueous solutions **A** and **B**. Au and Ag were adsorbed within 1.5 h by **1** and **2**. The adsorption of Pd from multielement solution **A** with **2** proceeded faster than **1** owing to the

higher hydrophilicity of DMAA than NIPAM but **1** also quantitatively adsorbed Pd for 24 h. Selective adsorption of Pd could be achieved with **2** in the presence of excess amounts of Cu by the high noble metal selectivity. The selectivity of **1** to noble metals was higher than **2** due to the lower nonspecific adsorption. The high selectivity of the hydrophilic sulfur-containing adsorbents is advantageous for practical recycling of noble metals from industrial and mining wastewater.

Conflicts of Interest

The authors declare that they have no conflicts of interest.

Acknowledgments

The authors are grateful for the financial supports from the Academia Showcase from Japan Chemical Innovation and Inspection Institute and the Grant-in-Aid for Scientific Research for Rare Metals Recycling from the Ministry of the Environment in Japan (K22030). The authors are also grateful to Dr. Atsushi Sasaki of Faculty of Engineering of Yamagata University for his kind suggestion and instruction on ICP-MS measurements.

References

[1] J. H. Potgieter, S. S. Potgieter, R. K. K. Mbaya, and A. Teodorovic, "Small scale recovery of noble metals from jewellery wastes," *Journal of the SA Institute of Mining & Metallurgy*, vol. 104, pp. 563–571, 2004.

[2] N. Miyaura and A. Suzuki, "Palladium-catalyzed cross-coupling reactions of organoboron compounds," *Chemical Reviews*, vol. 95, no. 7, pp. 2457–2483, 1995.

[3] M. S. Hegde, G. Madras, and K. C. Patil, "Noble metal ionic catalysts," *Accounts of Chemical Research*, vol. 42, no. 6, pp. 704–712, 2009.

[4] S. D. Lin and C. Song, "Noble metal catalysts for low-temperature naphthalene hydrogenation in the presence of benzothiophene," *Catalysis Today*, vol. 31, no. 1-2, pp. 93–104, 1996.

[5] P. Gelin and M. Primet, "Complete oxidation of methane at low temperature over noble metal based catalysts: a review," *Applied Catalysis B: Environmental*, vol. 39, no. 1, pp. 1–37, 2002.

[6] J. Cui and E. Forssberg, "Mechanical recycling of waste electric and electronic equipment: a review," *Journal of Hazardous Materials*, vol. 99, no. 3, pp. 243–263, 2003.

[7] M. Faisai, Y. Atsuta, H. Daimon, and K. Fujie, "Recovery of precious metals from spent automobile catalytic converters using supercritical carbon dioxide," *Asia-Pacific Journal of Chemical Engineering*, vol. 3, no. 4, pp. 364–367, 2008.

[8] N. T. Hung, M. Watanabe, and T. Kimura, "Solvent extraction of palladium(II) with various ketones from nitric acid medium," *Solvent Extraction and Ion Exchange*, vol. 25, no. 3, pp. 407–416, 2007.

[9] J. S. Wang and C. M. Wai, "Dissolution of precious metals in supercritical carbon dioxide," *Industrial & Engineering Chemistry Research*, vol. 44, no. 4, pp. 922–926, 2005.

[10] E. Lachowicz, B. Rozanska, F. Teixidor, H. Meliani, M. Barboiu, and N. Hovnanian, "Comparison of sulphur and sulphur–oxygen ligands as ionophores for liquid–liquid

extraction and facilitated transport of silver and palladium," *Journal of Membrane Science*, vol. 210, no. 2, pp. 279–290, 2002.

[11] T. Kai, T. Hagiwara, H. Haseba, and T. Takahashi, "Reduction of thiourea consumption in gold extraction by acid thiourea solutions," *Industrial & Engineering Chemistry Research*, vol. 36, no. 7, pp. 2757–2759, 1997.

[12] J. S. Preston and A. C. du Preez, "The separation of europium from a middle rare earth concentrate by combined chemical reduction, precipitation and solvent-extraction methods," *Journal of Chemical Technology & Biotechnology*, vol. 65, no. 1, pp. 93–101, 1996.

[13] C. S. Griffith, V. Luca, J. V. Hanna, K. J. Pike, M. K. Smith, and G. S. Thorogood, "Microcrystalline hexagonal tungsten bronze. 1. basis of ion exchange selectivity for cesium and strontium," *Inorganic Chemistry*, vol. 48, no. 13, pp. 5648–5662, 2009.

[14] D. Y. Zuo, B. K. Zhu, S. H. Wang, and Y. Y. Xu, "Membrane extraction for separation of copper cations from acid solution using polypropylene hollow fibre membrane," *Polymers for Advanced Technologies*, vol. 16, no. 10, pp. 738–743, 2005.

[15] A. Kargari, T. Kaghazchi, and M. Soleimani, "Extraction of gold (III) ions from aqueous solutions using polyamine type surfactant liquid membrane," *Canadian Journal of Chemical Engineering*, vol. 82, no. 6, pp. 1301–1306, 2004.

[16] W. S. W. Ho, B. Wang, T. E. Neumuller, and J. Roller, "Supported liquid membranes for removal and recovery of metals from waste waters and process streams," *Environmental Progress*, vol. 20, no. 2, pp. 117–121, 2001.

[17] E. R. Els, L. Lorenzen, and C. Aldrich, "The adsorption of precious metals and base metals on a quaternary ammonium group ion exchange resin," *Minerals Engineering*, vol. 13, no. 4, pp. 401–414, 2000.

[18] D. Clifford and W. J. Weber, "The determinants of divalent/monovalent selectivity in anion exchangers," *Reactive Polymers, Ion Exchangers, Sorbents*, vol. 1, no. 2, pp. 77–89, 1983.

[19] Q. Xu, P. Yin, G. Zhao, Y. Sun, and R. Qu, "Character of long-chain branching in highly purified natural rubber," *Journal of Applied Polymer Science*, vol. 117, no. 6, pp. 3645–3650, 2010.

[20] K. Deplanche, R. D. Woods, I. P. Mikheenko, R. E. Sockett, and L. E. Macaskie, "Manufacture of stable palladium and gold nanoparticles on native and genetically engineered flagella scaffolds," *Biotechnology and Bioengineering*, vol. 101, no. 5, pp. 873–880, 2008.

[21] C. R. Adhikari, D. Parajuli, H. Kawakita, K. Inoue, K. Ohto, and H. Harada, "Dimethylamine-modified waste paper for the recovery of precious metals," *Environmental Science & Technology*, vol. 42, no. 15, pp. 5486–5491, 2008.

[22] S. Kiyoyama, T. Maruyama, N. Kamiya, and M. Goto, "Immobilization of proteins into microcapsules and their adsorption properties with respect to precious-metal ions," *Industrial & Engineering Chemistry Research*, vol. 47, no. 5, pp. 1527–1532, 2008.

[23] L. Wan, Y. Wang, and S. Qian, "Study on the adsorption properties of novel crown ether crosslinked chitosan for metal ions," *Journal of Applied Polymer Science*, vol. 84, no. 1, pp. 29–34, 2002.

[24] J. M. Sanchez, M. Hidalgo, and V. Salvado, "The selective adsorption of gold (III) and palladium (II) on new phosphine sulphide-type chelating polymers bearing different spacer arms," *Reactive and Functional Polymers*, vol. 46, no. 3, pp. 283–291, 2001.

[25] G. Zuo and M. Muhammed, "Thiourea-based coordinating polymers: synthesis and binding to noble metals," *Reactive Polymers*, vol. 24, no. 3, pp. 165–181, 1995.

[26] K. Okamoto, T. Kanbara, T. Yamamoto, and A. Wada, "Preparation and characterization of luminescent SCS and NCN pincer platinum complexes derived from 3,5-bis(anilinothiocarbonyl)toluene," *Organometallics*, vol. 25, no. 16, pp. 4026–4029, 2006.

[27] M. Akaiwa, T. Kanbara, H. Fukumoto, and T. Yamamoto, "Luminescent palladium complexes containing thioamide-based SCS pincer ligands," *Journal of Organometallic Chemistry*, vol. 690, no. 18, pp. 4192–4196, 2005.

[28] T. Koizumi, T. Teratani, K. Okamoto, T. Yamamoto, Y. Shimoi, and T. Kanbara, "Nickel(II) complexes bearing a pincer ligand containing thioamide units: comparison between SNS- and SCS-pincer ligands," *Inorganica Chimica Acta*, vol. 363, no. 11, pp. 2474–2480, 2010.

[29] J. Kuwabara and T. Kanbara, "Synthesis and optical properties of pincer palladium and platinum complexes having thioamide units," *Journal of Photopolymer Science and Technology*, vol. 21, no. 3, pp. 349–353, 2008.

[30] S. Kagaya, E. Sato, I. Masore, K. Hasegawa, and T. Kanbara, "Polythioamide as a collector for valuable metals from aqueous and organic solutions," *Chemistry Letters*, vol. 32, no. 7, pp. 622–623, 2003.

[31] M. K. M. Z. Hyder and B. Ochiai, "Synthesis of a selective scavenger for Ag(I), Pd(II), and Au(III) based on cellulose filter paper grafted with polymer chains bearing thiocarbamate moieties," *Chemistry Letters*, vol. 46, no. 4, pp. 492–494, 2017.

[32] M. Spitzer and R. Bertazzoli, "Selective electrochemical recovery of gold and silver from cyanide aqueous effluents using titanium and vitreous carbon cathodes," *Hydrometallurgy*, vol. 74, no. 3-4, pp. 233–242, 2004.

[33] D. Nagai, T. Imazeki, H. Morinaga, H. Oku, and K. Kasuya, "Three-component polyaddition of diamines, carbon disulfide, and diacrylates in water," *Journal of Polymer Science Part A: Polymer Chemistry*, vol. 48, no. 4, pp. 845–851, 2010.

[34] S. Kagaya, H. Miyazaki, M. Ito, K. Tohda, and T. Kanbara, "Selective removal of mercury(II) from wastewater using polythioamides," *Journal of Hazardous Materials*, vol. 175, no. 1–3, pp. 1113–1115, 2010.

[35] B. Ochiai, T. Ogihara, M. Mashiko, and T. Endo, "Synthesis of rare-metal absorbing polymer by three-component polyaddition through combination of chemo-selective nucleophilic and radical additions," *Journal of the American Chemical Society*, vol. 131, no. 5, pp. 1636–1637, 2009.

[36] W. Yantasee, C. L. Warner, T. Sangvanich et al., "Removal of heavy metals from aqueous systems with thiol functionalized superparamagnetic nanoparticles," *Environmental Science & Technology*, vol. 41, no. 14, pp. 5114–5119, 2007.

[37] D. Nagai, T. Imazeki, H. Morinaga, and H. Nakabayashi, "Synthesis of a rare-metal adsorbing polymer by three-component polyaddition of diamines, carbon disulfide, and diacrylates in an aqueous/organic biphasic medium," *Journal of Polymer Science Part A: Polymer Chemistry*, vol. 48, no. 24, pp. 5968–5973, 2010.

[38] A. Ramesh, K. R. Mohan, and K. Seshaiah, "Announcement," *Talanta*, vol. 57, no. 2, pp. 243–252, 2002.

[39] G. D. Thorn and R. A. Ludwig, *The Dithiocarbamates and Related Compounds*, Elsevier Co., Amsterdam, Netherlands, 1962.

[40] B. Biannic, T. Ghebreghiorgis, and A. Aponick, "A comparative study of the Au-catalyzed cyclization of hydroxy-substituted allylic alcohols and ethers," *Beilstein Journal of Organic Chemistry*, vol. 7, pp. 802–807, 2011.

[41] G. Li, Z. Zhao, J. Liu, and G. Jiang, "Effective heavy metal removal from aqueous systems by thiol functionalized

magnetic mesoporous silica," *Journal of Hazardous Materials*, vol. 192, pp. 277–283, 2011.

[42] J. M. Richardson and C. W. Jones, "Strong evidence of solution-phase catalysis associated with palladium leaching from immobilized thiols during Heck and Suzuki coupling of aryl iodides, bromides, and chlorides," *Journal of Catalysis*, vol. 251, pp. 80–93, 2007.

[43] M. Yu, W. Tian, D. Sun, W. Shen, G. Wang, and N. Xu, "Systematic studies on adsorption of 11 trace heavy metals on thiol cotton fiber," *Analytica Chimica Acta*, vol. 428, no. 2, pp. 209–218, 2001.

[44] I. L. Lagadic, M. K. Mitchell, and B. D. Payne, "Highly effective adsorption of heavy metal ions by a thiol-functionalized magnesium phyllosilicate clay," *Environmental Science & Technology*, vol. 35, no. 5, pp. 984–990, 2001.

[45] G. Hernandez and R. Rodriguez, "Adsorption properties of silica sols modified with thiol groups," *Journal of Non-Crystalline Solids*, vol. 246, no. 3, pp. 209–215, 1999.

[46] A. Lezzi, S. Cobianco, and A. Roggero, "Synthesis of thiol chelating resins and their adsorption properties toward heavy metal ions," *Journal of Polymer Science Part A: Polymer Chemistry*, vol. 32, no. 10, pp. 1877–1883, 1994.

[47] P. D. Akrivos, "Recent studies in the coordination chemistry of heterocyclic thiones and thionates," *Coordination Chemistry Reviews*, vol. 213, no. 1, pp. 181–210, 2001.

[48] B. Slootmaekers, E. Manessi-Zoupa, S. P. Perlepes, and H. O. Desseyn, "The infrared spectra of complexes with planar dithiooxamides part VIII. The Au(III) complexes," *Spectrochimica Acta Part A: Molecular and Biomolecular Spectroscopy*, vol. 52, no. 10, pp. 1255–1273, 1996.

[49] S. H. J. De Beukeleer and H. O. Desseyn, "Vibrational analysis of some transition metal complexes with deprotonated and neutral malonamide," *Spectrochimica Acta Part A: Molecular Spectroscopy*, vol. 51, no. 14, pp. 1617–1633, 1995.

[50] E. S. Raper, "Copper complexes of heterocyclic thioamides and related ligands," *Coordination Chemistry Reviews*, vol. 129, no. 1-2, pp. 91–156, 1994.

[51] C.-Y. Liu, H.-T. Chang, and C.-C. Hu, "Complexation reactions in a heterogeneous system," *Inorganica Chimica Acta*, vol. 172, no. 2, pp. 151–158, 1990.

[52] R. L. Martin and A. F. Masters, "Thio derivatives of 1,3-diketones and their metal complexes. Dithiomalonamide and its nickel(II), palladium(II), and platinum(II) derivatives," *Inorganic Chemistry*, vol. 14, no. 4, pp. 885–892, 1975.

[53] R. W. Kluiber, "Inner complexes. V. copper(II) and nickel(II) chelates of N-alkylthiopicolinamides," *Inorganic Chemistry*, vol. 4, no. 6, pp. 829–833, 1965.

[54] Y. Jiang, H. Zhang, Q. He, and Z. Hu, "Selective solid-phase extraction of trace mercury(II) using a silica gel modified with diethylenetriamine and thiourea," *Microchimica Acta*, vol. 178, no. 3-4, pp. 421–428, 2012.

[55] M. J. Girgis, L. E. Kuczynski, S. M. Berberena et al., "Removal of soluble palladium complexes from reaction mixtures by fixed-bed adsorption," *Organic Process Research & Development*, vol. 12, no. 6, pp. 1209–1217, 2008.

[56] T. Mikysek, I. Svancara, K. Vytras, and F. G. Banica, "Functionalised resin-modified carbon paste sensor for the voltammetric determination of Pb(II) within a wide concentration range," *Electrochemistry Communications*, vol. 10, no. 2, pp. 242–245, 2008.

[57] A. Hinchcliffe, C. Hughes, D. A. Pears, and M. R. Pitts, "QuadraPure cartridges for removal of trace metal from reaction mixtures in flow," *Organic Process Research & Development*, vol. 11, no. 3, pp. 477–481, 2007.

[58] N. Nikbin, M. Ladlow, and S. V. Ley, "Continuous flow ligand-free heck reactions using monolithic Pd [0] nanoparticles," *Organic Process Research & Development*, vol. 11, no. 3, pp. 458–462, 2007.

[59] C. Ni, C. Yi, and Z. Feng, "Studies of syntheses and adsorption properties of chelating resin from thiourea and formaldehyde," *Journal of Applied Polymer Science*, vol. 82, no. 13, pp. 3127–3132, 2001.

[60] H. A. Abd El-Rehim, E. A. Hegazy, and A. El-Hag Ali, "Selective removal of some heavy metal ions from aqueous solution using treated polyethylene-g-styrene/maleic anhydride membranes," *Reactive and Functional Polymers*, vol. 43, no. 1-2, pp. 105–116, 2000.

[61] N. Kihara, H. Tochigi, and T. Endo, "Synthesis and reaction of polymers bearing 5-membered cyclic dithiocarbonate group," *Journal of Polymer Science Part A: Polymer Chemistry*, vol. 33, pp. 1005–1010, 1995.

[62] G. A. A. Al-Hazmi, A. A. El-Zahhar, K. S. Abou-Melha et al., "Elaborated spectral analysis and modeling calculations on Co(II), Ni(II), Cu(II), Pd(II), Pt(II), and Pt(IV) nanoparticles complexes with simple thiourea derivative," *Journal of Coordination Chemistry*, vol. 68, no. 6, pp. 993–1009, 2015.

[63] H. Rostkowska, L. Lapinski, A. Khvorostov, and M. J. Nowak, "Proton-transfer processes in thiourea: UV induced thione → thiol reaction and ground state thiol → thione tunneling," *Journal of Physical Chemistry A*, vol. 107, no. 33, pp. 6373–6380, 2003.

[64] H. Tokuyama and A. Kanehara, "Temperature swing adsorption of gold(III) ions on poly(N-isopropylacrylamide) gel," *Reactive and Functional Polymers*, vol. 67, no. 2, pp. 136–143, 2007.

[65] K. Sonogashira, Y. Tohda, and N. Hagihara, "A convenient synthesis of acetylenes: catalytic substitutions of acetylenic hydrogen with bromoalkenes, iodoarenes and bromopyridines," *Tetrahedron Letters*, vol. 16, no. 50, pp. 4467–4470, 1975.

4

Mode-I Metal-Composite Interface Fracture Testing for Fibre Metal Laminates

Periyasamy Manikandan(i)(d) **and Gin Boay Chai**(i)(d)

Aerospace Engineering Cluster, School of Mechanical and Aerospace Engineering, Nanyang Technological University, Singapore

Correspondence should be addressed to Gin Boay Chai; mgbchai@ntu.edu.sg

Academic Editor: Fabrizio Sarasini

The main contribution of the present paper is the determination of the mode-I fracture of metal-composite interface region for fibre metal laminates (FMLs). A hybrid DCB configuration is proposed to investigate the mode-I fracture between metal-composite interface using experimental and numerical approaches. A computationally efficient and reliable finite element model was developed to account for the influence of metal plasticity on the measured fracture energy. The results of the experimental and numerical studies showed that metal plasticity increases the fracture energy of the metal-composite interface as the fracture event progresses. The applied energy truly utilized to propagate metal-composite interface fracture was predicted numerically by extracting the elastic strain energy data. The predicted true fracture energy was found to be approximately 50% smaller than the experimentally measured average propagation energy. The study concluded that metal plasticity in hybrid DCB configuration overpredicted the experimentally measured fracture energy, and this can be alleviated through numerical methodology such as the finite element approach as presented in this paper.

1. Introduction

Hybridization of material not only makes it structurally efficient but also makes it multifunctional. With sustained research and advancements in adhesive bonding technology, designers are able to successfully develop a novel metal-bonded hybrid composite structure such as the fibre metal laminates (FMLs) in the aviation industry. Currently, FML structures are the third largest primary load-carrying aviation material by volume housing nearly 470 sq·m of the upper fuselage section of Airbus A380 passenger air carrier [1]. FML is a versatile and high performance hybrid material made up of alternating layers of thin metal sheets and thin fibre reinforce polymeric (FRP) composite layers. Compounding metal and composite makes FML stronger, lighter, better damage tolerant, good fatigue resistant, better corrosion, and impact resistant material which is not feasible through single material alone [2, 3].

The effectiveness of FML as a good load-carrying material relies heavily on the level of the bonding integrity between the metal and the FRP composite layers [4]. The bonding interface must possess sufficient interfacial strength to be able to distribute the load between two characteristically different materials. There are two different bonding interfaces in a traditional FML: (i) adhesive-metal and (ii) adhesive-composite. The level of bonding integrity is strictly dependent on the appropriate surface pretreatment of bonded metal to composite interfaces and the chemical compatibility of adhesive introduced. Apart from those, the potential bonding strength is also dependent on the stiffness imbalance and the thermal mismatch of dissimilar materials during testing and curing, respectively [5]. For the case of FML, the problem of thermal mismatch is a challenging one where the difference in thermal expansion coefficients between adhesive, metal, and composite layers needs to be resolved. A study [6] had shown that poststretching after curing helps to enhance delamination resistance in the FML panels.

Adhesive bonding between monolithic material substrates like metal-metal and composite-composite is fairly well understood, and the testing procedures to evaluate their fracture properties are fairly standardized. However, this is not the case for evaluating dissimilar material joints like metal-composite interface. Furthermore, plastic yielding is predominately sensitive in the DCB sample having thin metal adherent and may attenuate the crack propagation in fracture test thereafter [7]. Such plasticity can result in abnormally high values of fracture toughness, and it is necessary to account for its influence in order to obtain the true magnitude of the fracture energy. So far in literature, researchers had employed three different approaches to account for the plasticity effects: (i) experimental approach using thick metal doublers to attenuate plastic flow [6, 8], (ii) theoretical approach [9], and (iii) finite element approach [7]. In the last two approaches, efforts were made to estimate the fraction of plastically absorbed energy from the overall internal strain energy.

Reeder et al. [8] portrayed the method of bonding doublers to thin substrates in a manner to avoid plastic deformation or premature adherent failure due to bending before the interface delamination initiate. The standard Irwin–Kies fracture energy equation was revised for the samples having bonded doublers for all three modes of fracture. Vlot and Van Ingen [6] utilized a similar concept to evaluate the delamination fracture energy of metal-composite interface of FML. To achieve stable delamination propagation in the mode-I test using the double cantilever beam (DCB) geometry, the transverse displacement of loading platen was stopped at every interval when there exists a delamination growth and stayed until the load magnitude leveled off. Lawcock et al. [10] followed the standard testing procedure of metallic joints to evaluate mode-I fracture energy of FML assuming that the volume fraction of the composite in the considered DCB geometry is negligibly small compared to the metal volume.

Reyes et al. [11] evaluated the mode-I, mode-II, and mixed-mode fracture energy of metal-composite interface of FML using a single cantilever beam (SCB), a double end notched flexure, and a modified mixed-mode flexure geometry, respectively. Using the SCB geometry, stable interface crack propagation without causing plastic yielding of metal sheet would become feasible. However, if the fracture energy of the adhesive material is far larger than the yielding stress of the metal sheet, this method could enforce more challenge to initiate interface crack growth.

The aim of the present paper is to investigate the mode-I fracture energy evaluation of metal-composite joints in order to understand the interface crack growth in FML that involves the influence of the plastic yielding in thin metal sheets. Finite element approach was utilized to estimate the amount of energy dissipated through plastic deformation. The robustness of the developed numerical model was validated by comparing the fracture test results of monolithic composite-composite sample with those obtained from the numerical study. Such comparison also explores the influence of plasticity effects and allows one to quantitatively estimate how far the fracture energy magnitude has been altered due to such additional irreversible deformation.

2. Materials and Test Methodology

In this section, the mode-I fracture test geometry and the methodology used in the study are described. Subsequently, the investigation of mode-I fracture behavior of composite-composite and metal-composite joints are presented, and some conclusive remarks are narrated in the final section.

2.1. Materials. Aluminum alloy Al 2024-0, L-530 8-harness satin weave 7781 glass epoxy prepreg, and Redux 335K adhesive film are the materials used to fabricate the adhesively bonded fibre-metal laminated samples investigated in this study. Before bonding, the aluminum alloy bonded surfaces were abraded using 320 grit size sand paper, cleaned with acetone solvent, and appropriately surface treated in a motive to promote good chemical linkage between aluminum-adhesive-composite interface.

Metal-composite layers of designated stacking were laminated and bonded with the single layer of adhesive film. Stacked laminates were cured in accordance with the composite manufacturer data sheet using an autoclave chamber under 3 bar and 120°C in the presence of vacuum. The cured laminates were cut to the required dimension without disturbing the adhesively bonded interface using a high-speed water jet cutter, and quality checks were carefully carried out under the optical microscope.

2.2. Experimental Tests. In total, three different configurations of DCB tests were experimented to investigate the mode-I fracture behavior of metal-composite adhesive interface fracture growth. They are as follows:

(i) Thick metal doublers bonded with thin composite adherent, *M-C-M configuration*

(ii) Monolithic composite adherents with in situ epoxy matrix adhesive interface, *C-C configuration*

(iii) Hybrid geometry with firmly bonded metal-composite adherent as one arm and monolithic composite adherent as another arm, *C-M-C configuration*

Schematic representation of above listed configurations and relevant dimensions are shown in Figures 1(a)–1(c), respectively. Samples were tested at a rate of 3 mm/min using a universal testing machine. Magnitude of load P, cross head displacement δ, and crack length a were recorded during both crack initiation and instant propagation.

The fracture energy for M-C-M configuration is evaluated using the following guidelines of ASTM D3833 assuming that the composite layer is very thin compared to thick aluminum doublers [10]. The equation of mode-I fracture energy, G_I, is given by

$$G_I = \frac{4P^2}{EB^2 t_1^2}\left[3a_o^2 + t_1^2\right], \tag{1}$$

where E, B, and t_1 are Young's modulus, width, and thickness of the aluminum adherends.

Composite layer

Precrack

P

a_0

Metal adherent

L

P

t_1

t_1

B

$L = 160$ mm

$a_0 = 45$ mm

$B = 25$ mm

$t_1 = 10$ mm

(a)

Adhesive layer

Precrack

P

a_0

Composite adherent

L

P

t_1

t_1

B

$L = 135$ mm

$a_0 = 50$ mm

$B = 25$ mm

$t_1 = 1.5$ mm

(b)

Adhesive layer

Precrack

P

a_0

Composite adherent

Metal adherent

L

P

t_1

t_2

t_m

B

$L = 155$ mm

$a_0 = 45$ mm

$B = 25$ mm

$t_1 = 2.1$ mm

$t_m = 0.6$ mm

$t_1 = 1.2$ mm

(c)

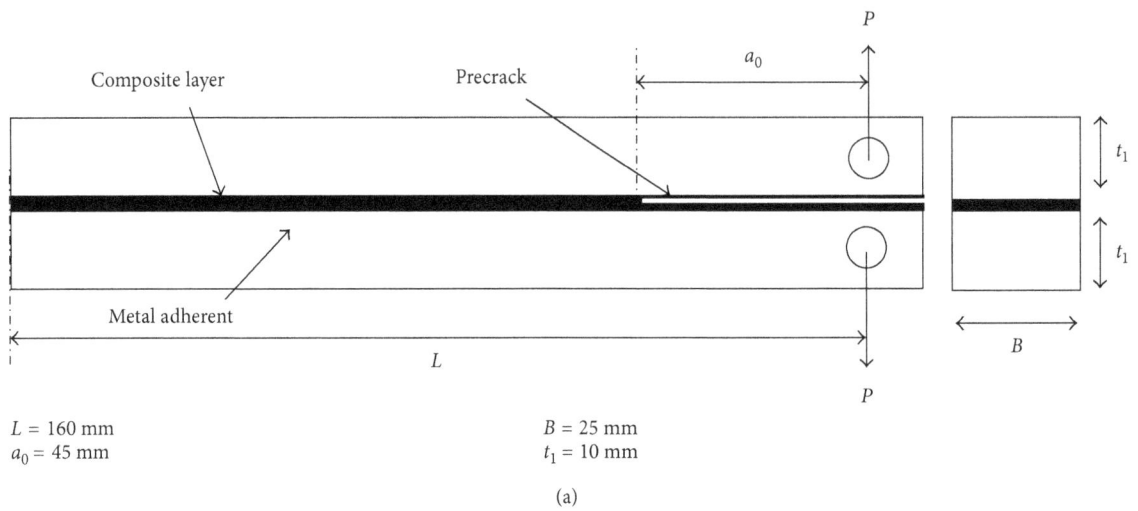

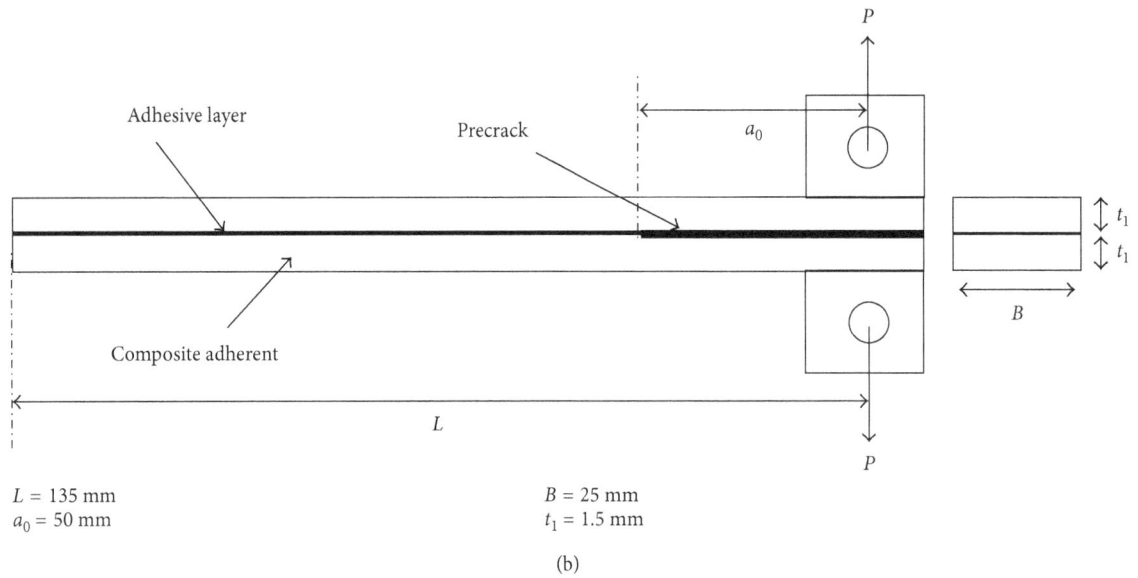

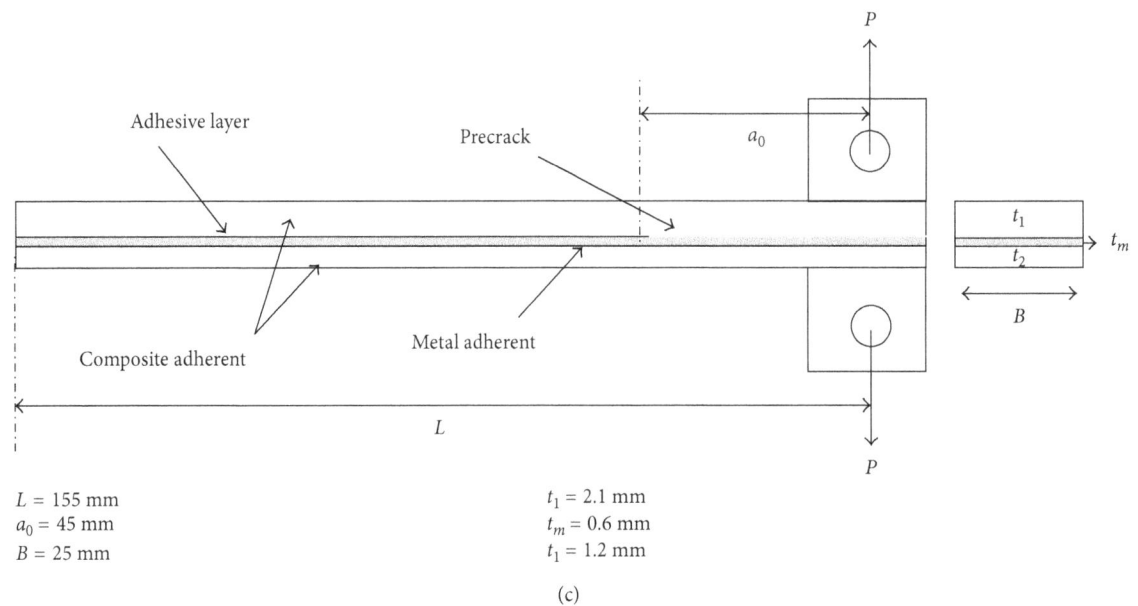

FIGURE 1: Schematic representation of DCB geometry: (a) M-C-M configuration, (b) C-C configuration, and (c) C-M-C configuration.

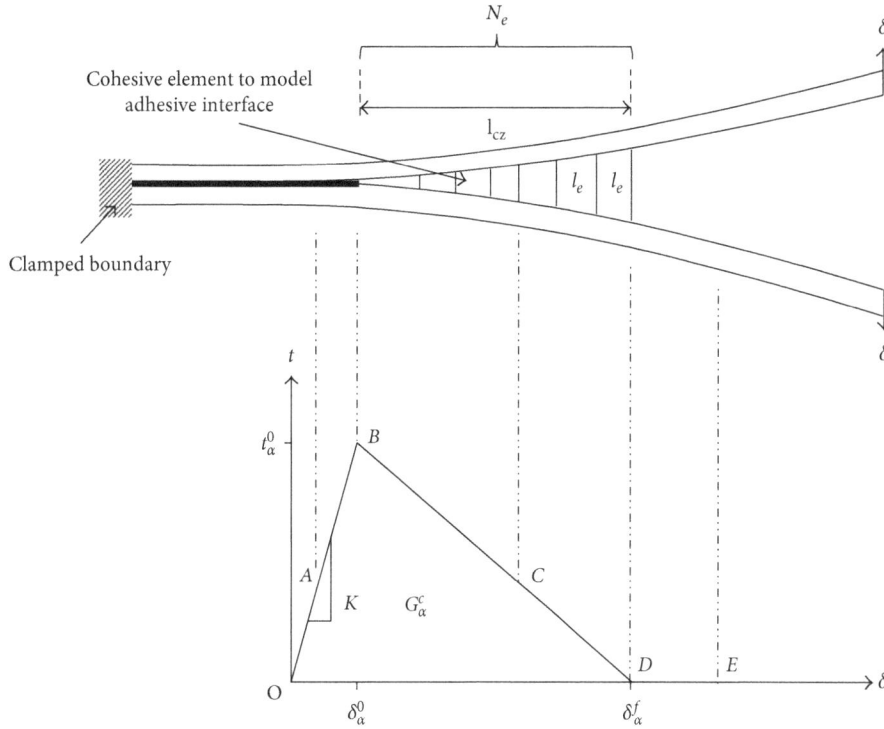

FIGURE 2: Bilinear traction-separation law for cohesive elements in the DCB model.

While for the C-C and C-M-C configurations, the value of fracture energy is ascertained using the Irwin–Kies equation [12] given by

$$G_I = \frac{P^2}{2B} \cdot \frac{dC}{da}, \tag{2}$$

where $C = \delta/P$ is the specimen compliance and a is the corresponding crack length for a given displacement of δ.

2.3. Finite Element Simulation. The simulation of interface crack initiation and propagation was performed using cohesive element in the commercial finite element platform, ABAQUS/Standard. The constitutive formulation of cohesive element is modeled based on the traction (t)-separation (δ) law that relates the crack opening displacement in the process zone to the resisting tractions [13]. The material model of cohesive element is defined by an initial elastic stiffness, K, peak traction which causes crack initiation, t_α^0, and critical energy release rate equal to area under traction-separation curve responsible to cause crack-propagation rate, G_α^c. The schematic of the bilinear traction-separation law used to simulate the progressive interface crack of DCB configuration in this paper is shown in Figure 2.

The input parameters required to model the cohesive elements are determined based on Turon's methodology [14]. The methodology provides a systematic procedure to calculate the length of the cohesive zone, l_{cz}, size of the cohesive element, l_e, and number of cohesive elements, N_e, required to model the fracture growth accurately. The methodology also allows us to relax the requirement of extremely fine meshes through artificially increasing the

length of cohesive zone by reducing the peak traction, t_α^0, magnitude. The mathematical derivations to obtain adjusted traction strength, t_α^a, and the optimum mesh parameters using Turon's methodology are summarized in the Appendix. Both metal and composite adherents are modeled using isotropic and orthotropic elastic material properties, respectively. These layers are discretized using an incompatible-mode eight-node brick element, C3D8I, and the interface region using three-dimensional cohesive element, COH3D8.

The material properties used in the finite element model for the substrates and interface region are provided in Table 1. The FE model was validated by comparing the predicted load-displacement results with the experimental results. Using this curve, fracture energy data are calculated by employing the area method given by [15]

$$G_I = \frac{\Delta U}{B \Delta a}, \tag{3}$$

where ΔU is the area under numerical P-δ curve between consecutive crack growth intervals and Δa is the corresponding crack length increment as shown in Figure 3.

3. Results and Discussion

3.1. Selection of DCB Configuration. The initial emphasis to determine the metal-composite fracture behavior was analyzed through the experimental study of the M-C-M configuration. Enhanced resistance to bending because of thick aluminum adherends resulting in multiple interfacial and interlaminar crack propagation was observed between M-C and C-C regions, respectively, as shown in Figure 4(a).

TABLE 1: Material properties used in the FE model.

Response	Property	Value
Metal	Density	$\rho = 2780 \, \text{kg/m}^3$
	Young's modulus	$E = 70 \, \text{GPa}$
	Poisson's ratio	$\nu = 0.3$
Composite	Density	$\rho = 1900 \, \text{kg/m}^3$
	In-plane tensile Young's modulus	$E_{1+} = E_{2+} = 24 \, \text{GPa}$
	In-plane compression Young's modulus	$E_{1-} = E_{2-} = 24 \, \text{GPa}$
	Shear modulus	$G_{12} = 3.6 \, \text{GPa}$
	Poisson's ratio	$\nu_{12} = 0.1$
Adhesive	Penalty stiffness	$K_{nn} = K_{ss} = K_{tt} = 1e^{15} \, \text{N/m}^3$
	Interface strength	$t_n^0 = t_s^0 = t_f^0 = 50 \, \text{MPa}$

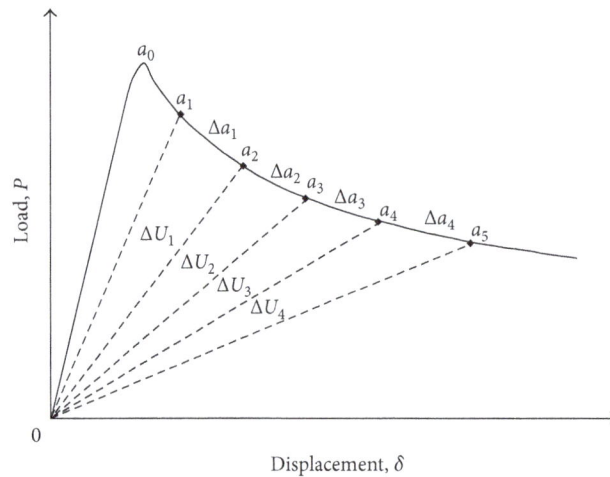

FIGURE 3: Area method of fracture energy determination.

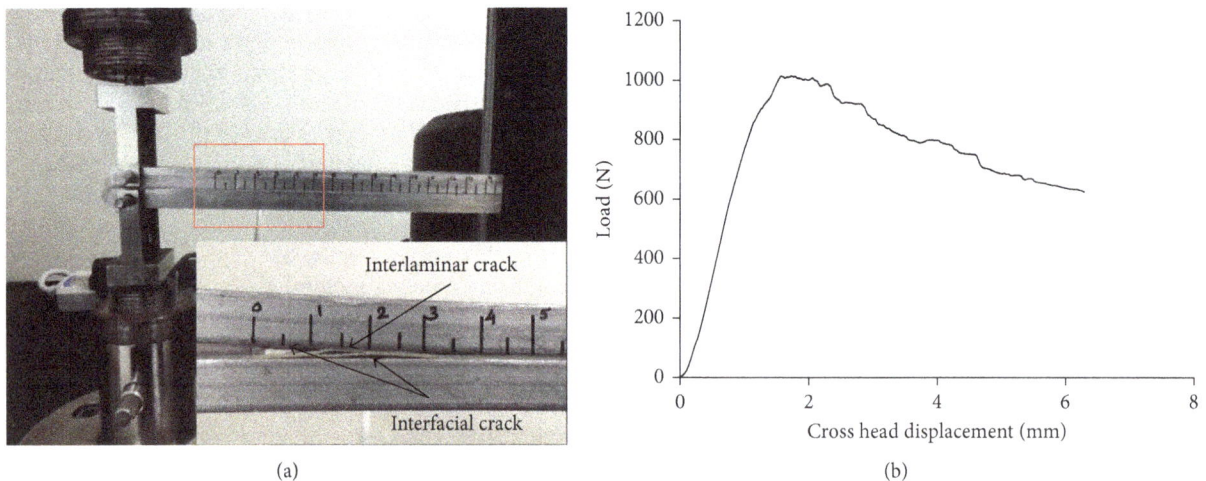

FIGURE 4: M-C-M DCB configuration: (a) multiple interface crack surfaces and (b) typical load-displacement curve.

Despite such multiple crack nucleation scenarios, the load variations associated with those multiple interface crack propagation do not show any discrete load drops as illustrated in Figure 4(b). This behavior indicates that the presence of thick aluminum adherends desensitized these multiple surface nucleation. Only the initiation fracture energy is feasible from this experiment using (1) which amounted to about 0.656 N/mm.

The DCB geometry with dissimilar adherends having bimaterial metal-composite adherent as one arm and monolithic composite as another arm (C-M-C configuration) was found to be a feasible design that allows the evaluation of the fracture energy not only in the crack initiation phase but also throughout crack propagation path. Thin adherends admit the arm to bend freely and facilitate the preimitated crack front propagation along metal-composite interface in stable manner even though there exist a significant amount of plastic deformation in the thin metal layer.

For the case of the hybrid configuration of DCB arms, the bending centroids of the two arms are generally different, and this difference results in unsymmetric flexure and inherently induces the mixed-mode loading state on the precrack front instead of pure mode-I loading state. Such mixed-mode inference was attenuated by selecting appropriate thickness for two DCB arms in a manner to have equal bending rigidity, EI, for both adherent arms. The derived geometric dimensions of the C-M-C configured DCB design is given in Figure 1(c).

3.2. DCB Results of C-C Configuration.

The DCB test using the C-C configuration is performed to obtain an accurate fracture energy of adhesive interface where the adherents undergo only elastic deformation. This study primarily serves as a benchmark that allows one to understand the metal sheet plastic deformation influence on G_I magnitude in the C-M-C configuration test. Moreover, it also provides opportunity to check the reliability of developed finite element model by ensuring that the internal energy utilized to delete the cohesive interface elements is the same as that of the experimentally predicted G_I.

The experimental and predicted mode-I fracture crack interface between composite adherends is shown in Figures 5(a) and 5(b), respectively. No visible fibre bridging patterns were found along the cracked interface. This is because the weave pattern of the woven composite adherends completely hindered the fibre pullout failure. Figure 6 illustrates the comparison plots of the load P versus deflection δ and the fracture energy G_I versus crack extension Δa. The smooth experimental load variation with no discrete load drops and change in fracture energy attributed in propagation phase confirms that the crack growth surfaces are pure cohesive in nature as shown in Figure 5(a). The average experimental fracture energy to cause crack initiation is found to be about 0.2 N/mm, and it increased to about 0.65 N/mm and maintained constant throughout the crack propagation phase. The latter propagation phase G_I is used as a damage evolution material data input for cohesive elements in finite element simulation.

3.3. Selection of Optimum Mesh Parameters.

The optimum mesh parameters required to simulate DCB mode-I fracture were found using the equations presented in the Appendix. The minimum cohesive zone length required to model mode-I fracture based on the experimental fracture energy for the C-C configuration is estimated about 0.29 mm using (A.1), and the material properties used are listed in Table 1. Based on that magnitude, parametric studies for three levels of mesh l_e ($\geq l_{cz}$) were conducted for four different N_e in the cohesive zone length. The corresponding magnitude of

adjusted traction strengths and artificially increased cohesive zone length for crack propagation was found using (A.3) and (A.1), respectively, and these are tabulated in Table 2.

The predicted load-deflection response for different levels of mesh refinement is shown in Figure 7. Irrespective of l_e, the response predicted by the adjusted interfacial strength $\bar{t}_\alpha^a$ with $N_e = 1$ did not converge and no numerical solution exists. For all other cases, the predicted softening crack propagation response is almost similar compared to the experimental measurement with some minor variations.

For the mesh size of $l_e = 1$ mm, the softening response shows spurious oscillations for smaller N_e which means the mesh size is too coarse to accurately predict the crack propagation. Take note that the amplitude of the spurious oscillation is found to be attenuated with increase in N_e. It can be seen that the responses obtained with the finer mesh size of $l_e = 0.5$ mm and $l_e = 0.3$ mm overpredicted the peak load. Meanwhile for this mesh size, the softening response exhibited smooth crack propagation. On comparing the number of iterations required to complete the solution, mesh sizes with more N_e in its artificially increased cohesive zone length l_{cz} have shown to be computationally efficient.

The above comparative study confirms that the interface strength $\bar{t}_\alpha^0$ required to initiate crack propagation does not have a significant influence on the overall response. Thus, the interface strength modification and the artificial cohesive zone length increment strategy initially proposed by Turon [14] can be used to model accurate crack propagation using cohesive elements at coarse mesh level. For the present DCB simulation, results obtained using lowest adjusted interfacial strength $\bar{t}_\alpha^a$ with larger N_e ($l_e = 1$ mm and $N_e = 8$) yield computationally efficient solution. All finite element simulations presented in the current paper are based on the abovementioned optimum mesh size.

It can be seen in Figure 6 that the numerical solution obtained for the C-C configuration model having artificially increased cohesive length and adjusted traction strength shows good agreement with the experimental results. Cohesive element deletion replicating mode-I crack growth happened approximately around the stated fracture energy material input of 0.65 N/mm.

3.4. DCB Results of C-M-C Configuration.

Having validated the reliability of the developed finite element model to simulate mode-I crack growth, the fracture of adhesive interface between plastically deformed metal adherent and composite adherent was investigated using numerical approach. Experimental observation confirmed that no stiffness imbalance exists between the two dissimilar DCB arms, and complete crack growth is caused strictly by the opening mode of fracture. Figure 8 shows the load-displacement plots and the final permanently deformed shape of the fractured C-M-C configuration sample obtained from the experiments and the finite element model. With an increase in crack opening displacement, the nature of crack growth in experiments was found to change from stable to unstable manner. The continuous cumulative effect of plastic flow in metal layers makes the crack growth in a *stick-slip manner* in

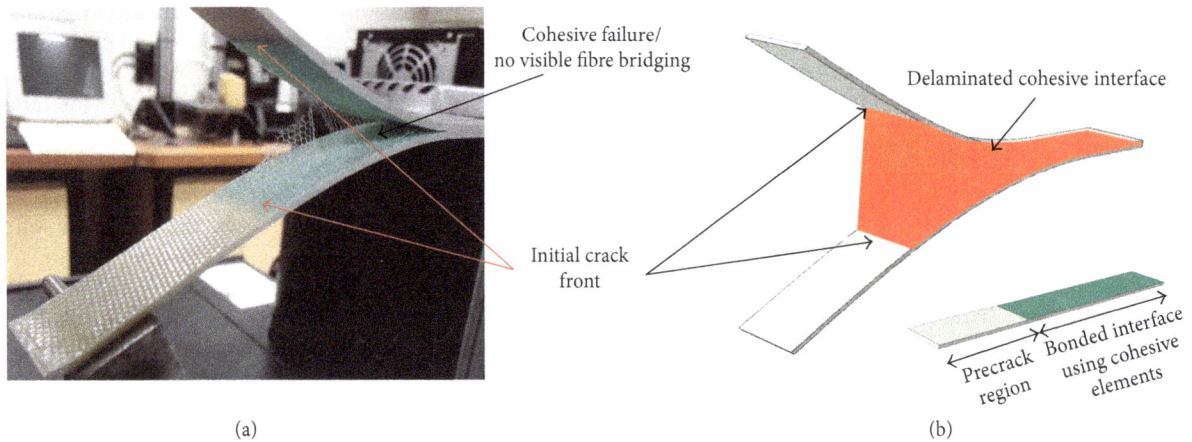

FIGURE 5: Mode-I fracture of C-C configuration: (a) experimental and (b) numerical.

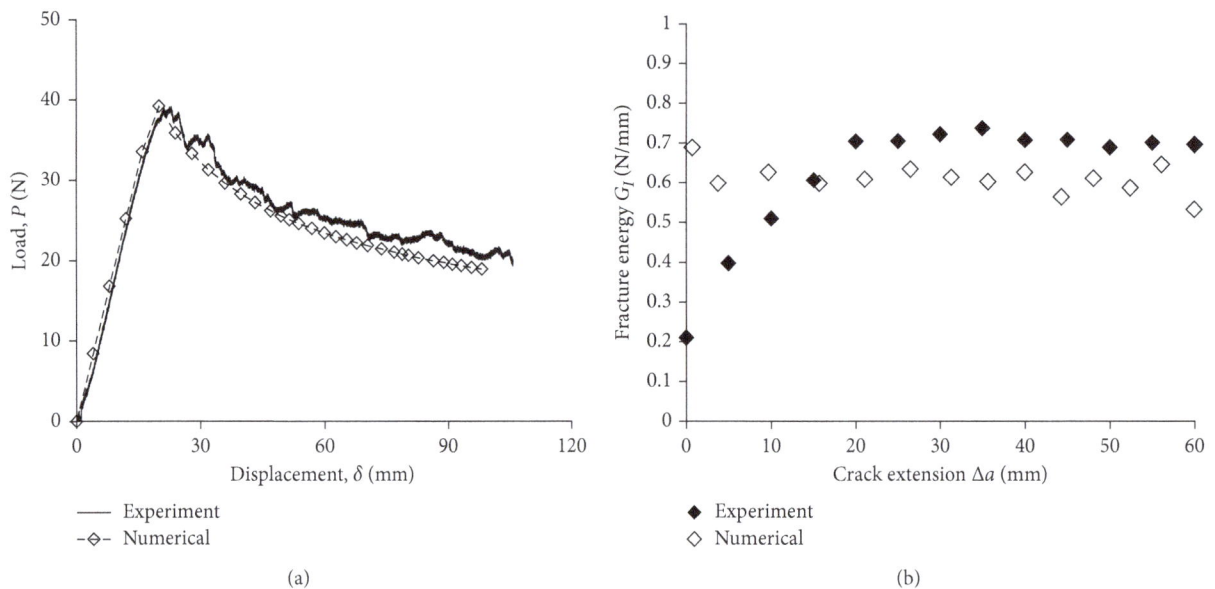

FIGURE 6: Finite element validation of the C-C configuration: (a) load-displacement curve and (b) fracture energy-crack extension.

TABLE 2: Adjusted normal interfacial strength with respect to different mesh sizes.

N_e	$l_e = 1$ mm		$l_e = 0.5$ mm		$l_e = 0.3$ mm	
	$\bar{t}_\alpha^a$ (MPa)	l_{cz} (mm)	$\bar{t}_\alpha^a$ (MPa)	l_{cz} (mm)	$\bar{t}_\alpha^a$ (MPa)	l_{cz} (mm)
1	25.5	1.0	36.1	0.5	45.8	0.3
3	14.7	3.0	20.8	1.5	26.4	0.9
5	11.4	5.0	16.1	2.5	20.5	1.5
8	9.0	8.0	12.8	4.0	16.2	2.4

the form of periodic short bursts and crack arrest. This scenario leaves a combined cohesive and adhesive kind of fracture surface along the bonded adherent surfaces. Figure 9(a) shows the variation of the measured fracture energy G_I using (2) against crack extension Δa; it can be seen that the plasticity effect causes G_I to increase with increase in crack length. It seems that the peak propagation fracture energy increases nearly by 3.5 times more than the initiation fracture energy (i.e., G_I at $\Delta a = 0$).

The absence of a single value of G_I from the experimental study makes the selection of an appropriate G_I input data required for numerical simulation a very difficult one. The fact that the magnitude of the fracture energy of the adhesives is constant and independent of the adherent material and geometry when their failure surface along the adherent interface is completely cohesive [16]. Assuming that the simulated fracture surface of the C-M-C model is presumably cohesive, a propagation G_I magnitude of

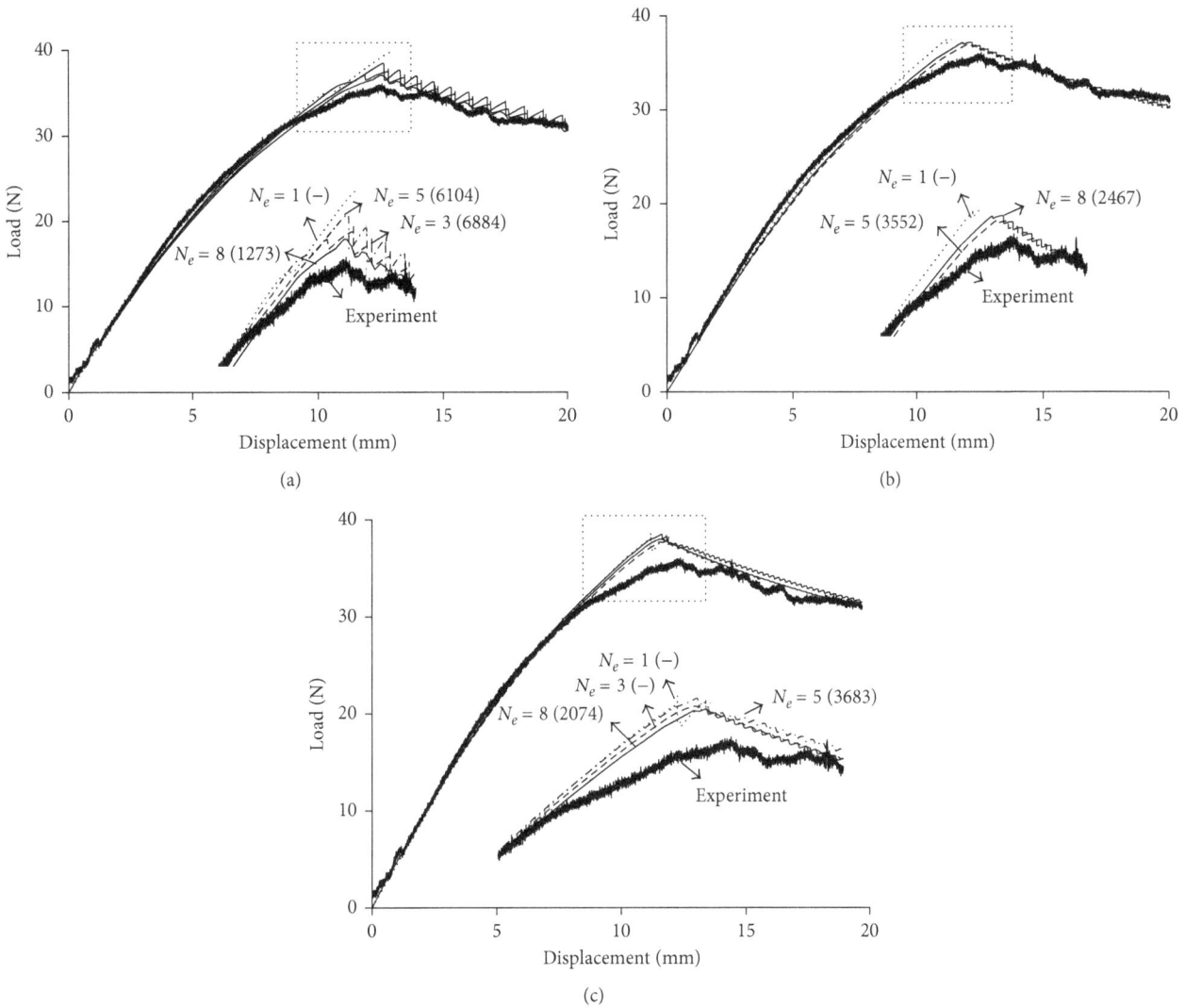

FIGURE 7: Load-displacement curve of C-C configuration for different cohesive element sizes: (a) $l_e = 1$ mm, (b) $l_e = 0.5$ mm, and (c) $l_e = 0.3$ mm.

0.65 N/mm obtained in the monolithic C-C DCB sample is used instead. Two different numerical cases were simulated with the distinct material model definition for thin metal sheets: (i) pure elastic (El) and (ii) combined elastic-plastic (El-Pl) to investigate the influence of plastic effects. The material data required to model plastic response of metal layer were obtained from the standard tensile test as given in [17].

From Figure 8(a), it appears that the predicted initial elastic responses of both cases are correlated well with the experimental load-displacement curve. After reaching the plastic state, only the case with metal plasticity inclusion coincides with the experimental crack propagation phase until the stable crack region. Numerically, the plasticity effects were found to increase the propagation fracture energy nearly by an average of 1.5 times compared to the pure elastic model as shown in Figure 9(a).

The true fracture energy value was calculated by extracting the elastic strain energy magnitude from the fractured elastic-plastic numerical model and substituted into (4) which is similar to (3):

$$(G_I)_{\text{Tr}} = \frac{\Delta U_R}{B\Delta a}, \tag{4}$$

where U_R is the recoverable (or elastic) strain energy obtained using ABAQUS output variable ALLSE in postprocessing.

By plotting U_R versus a as shown in Figure 9(b) and taking the derivative, the average true propagation fracture energy $(G_I)_{\text{Tr}}$ can be found using (4). The values of the fracture energy obtained from the different cases are summarized in Table 3. The obtained $(G_I)_{\text{Tr}}$ values are approximately equal to the initiation fracture energy $(G_I)_{\text{In}}$ predicted from experiment while it is 50% smaller than the average propagation fracture energy $(G_I)_{\text{Pr}}$. Such huge differences are ultimately caused by the influence of the plasticity effect in the thin metal layer and its induced unstable nature of crack propagation.

4. Conclusion

The effect of adhesion between metal and composite layers on the mode-I fracture energy G_I was investigated in the

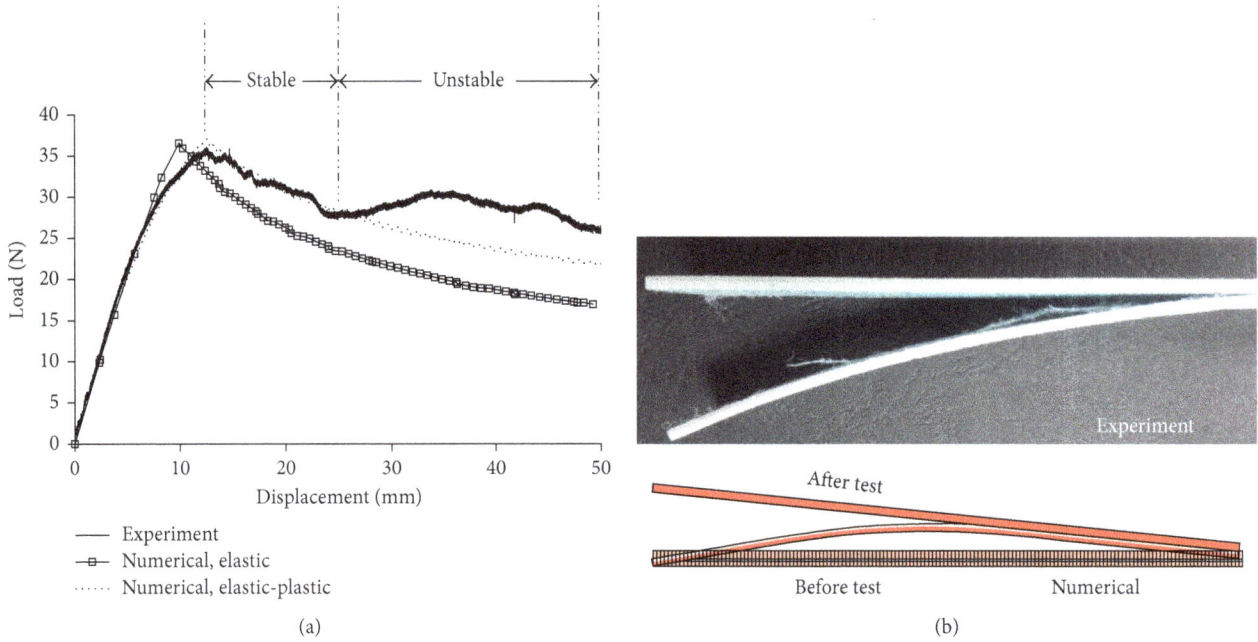

FIGURE 8: Mode-I fracture of the C-M-C configuration: (a) load-displacement curve and (b) permanent plastic deformation.

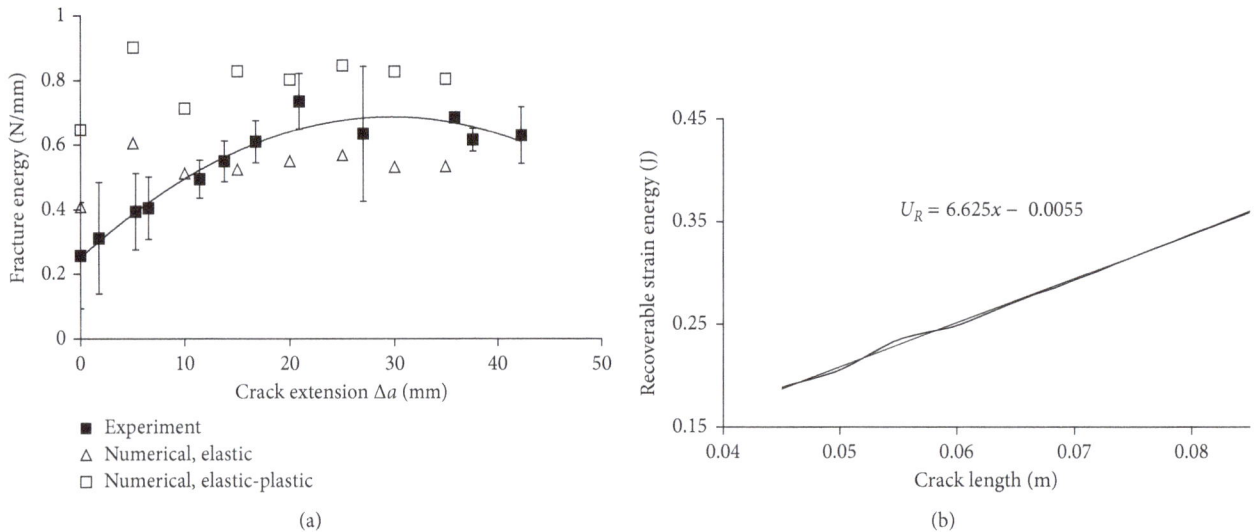

FIGURE 9: Mode-I fracture of C-M-C configuration: (a) fracture energy-crack extension and (b) recoverable elastic strain energy-crack length variation.

TABLE 3: Summary of fracture energy determined using C-M-C configuration.

Method	Experiment, using (1)	Numerical, elastic (3)	Numerical, elastic-plastic (3)	Numerical, elastic-plastic (4)
Fracture energy (N/mm)	$(G_I)_{In} = 0.257$ $(G_I)_{Pr} = 0.595$	$(G_I)_{El} = 0.529$	$(G_I)_{El-Pl} = 0.796$	$(G_I)_{Tr} = 0.265$

fibre metal laminates (FMLs) using experimental and finite element study. A robust numerical approach was proposed to calculate the true fracture energy of the interface crack propagation between the plastically deformed metal and elastic composite surfaces. Different hybrid DCB configurations based on the type of adherends were used in the experimental study to observe metal-composite interface fracture not only in initiation phase but also in the complete propagation phase. The M-C-M configuration containing thick metal adherent resulted in highly unstable crack growth via nucleating multiple interfacial and interlaminar cracks, and no reliable fracture data can be extracted.

An accurate fracture energy G_I of the adhesive used was found using monolithic woven composite C-C configuration. The crack propagation surfaces observed in the test were predominantly cohesive in nature, and no sign of fibre bridging was apparent. The obtained G_I was used as a material input data in the developed finite element model to simulate damage evolution of interface cohesive element. Good correlation was found with the developed finite element model, and this ensured the reliability in modeling mode-I fracture in the interface of dissimilar layers.

Finally, the influence of the plastic deformation on G_I was studied employing the C-M-C configuration where one arm of DCB adherent was designed to be metal-composite hybrid and the other with only composite material. This experimental study indicated that the plastic flow in thin metal layer continuously increases the measured G_I by up to 3.5 times as the metal-composite interface fracture progresses. Similar response was also predicted in finite element study where the inclusion of the plastic constitutive model for metal layer increases the average propagation energy by 1.5 times compared to the pure elastic numerical model. A true fracture energy $(G_I)_{\mathrm{Tr}}$ was obtained using the numerically predicted internal elastic energy in the finite element model. The predicted result indicated that $(G_I)_{\mathrm{Tr}}$ was nearly equal to experimentally measured initiation fracture energy $(G_I)_{\mathrm{In}}$ and 50% smaller than the average propagation fracture energy $(G_I)_{\mathrm{Pr}}$. This difference indicated the influence of metal plasticity effects on the measured G_I magnitude through experiments.

Appendix

Cohesive zone length, l_{cz}, defined as distance from crack front to the integration point of specified peak traction, t_α^0, is expressed as

$$l_{cz} = \mathrm{ME}\frac{G_\alpha^c}{(t_\alpha^0)^2}, \quad (A.1)$$

and the number of cohesive elements, N_e, corresponding to given l_{cz} is expressed as

$$N_e = \frac{l_{cz}}{l_e}, \quad (A.2)$$

where E is Young's modulus of the interface material, l_e is the cohesive element size, and M is the constant cohesive parameter ranges between 0.2 and 1. In the present work, the value of M has been taken as 1 [18].

The minimum requirement of N_e in l_{cz} in order to accurately simulate the interface crack growth is not well established and ranges widely from 1 to 8 depending on the kind of bonded substrate material (DaVilla, Falk). For the case of FRP composite substrates, typical range of l_{cz} is normally smaller than 1 mm and requires extremely fine mesh to simulate delamination. By combining (A.1) and (A.2), Turon [14] adjusted the magnitude of peak interface strength, t_α^0, by changing the size and number of cohesive element. The equation for adjusted interface strength, t_α^a, is shown in the following equation:

$$t_\alpha^a = \sqrt{\frac{EG_\alpha^c}{N_e l_e}}. \quad (A.3)$$

Conflicts of Interest

The authors declare that they have no conflicts of interest.

Acknowledgments

The financial support in the form of a research student scholarship provided by the School of Mechanical and Aerospace Engineering, Nanyang Technological University, and the permission to use the laboratory and computing facilities at the School of Mechanical and Aerospace Engineering are truly acknowledged. It must also be mentioned that the fabrication and testing of the specimens by final year undergraduate students and postgraduate interns are acknowledged.

References

[1] A. Vlot and J. W. Gunnink, *Fibre Metal Laminates: An Introduction*, Kluwer Academic Publishers, Norwell, USA, 2001.

[2] G. B. Chai and P. Manikandan, "Low velocity impact response of fibre-metal laminates—A review," *Composite Structures*, vol. 107, pp. 363–381, 2014.

[3] M. Sadighi, R. C. Alderliesten, and R. Benedictus, "Impact resistance of fiber-metal laminates: a review," *International Journal of Impact Engineering*, vol. 49, pp. 77–90, 2012.

[4] T. Sinmazçelik, E. Avcu, M. Ö. Bora, and O. Çoban, "A review: fibre metal laminates, background, bonding types and applied test methods," *Materials & Design*, vol. 32, no. 7, pp. 3671–3685, 2011.

[5] L. J. Hart Smith, "Adhesive joints for composites-phenomenological considerations," in *Proceedings of Conference on Composite Technology*, EI Segundo, CA, USA, September 1978.

[6] A. Vlot and J. W. Van Ingen, "Delamination resistance of post-stretched fibre metal laminates," *Journal of Composite Materials*, vol. 32, no. 19, pp. 1784–1805, 1998.

[7] B. A. Huppe, *High Reliability Adhesive Joining of Metal and Composite Components*, Department of Aeronautics and Astronautics, Massachusetts Institute of Technology, Cambridge, MA, USA, 2001.

[8] J. R. Reeder, K. Demarco, and K. S. Whitley, "The use of doubler reinforcement in delamination toughness testing," *Composites Part A: Applied Science and Manufacturing*, vol. 35, no. 11, pp. 1337–1344, 2004.

[9] M. D. Thouless, J. L. Adams, M. S. Kafkalidis, S. M. Ward, R. A. Dickie, and G. L. Westerbeek, "Determining the toughness of plastically deforming joints," *Journal of Materials Science*, vol. 33, no. 1, pp. 189–197, 1998.

[10] G. Lawcock, L. Ye, Y. W. Mai, and C. T. Sun, "The effect of adhesive bonding between aluminum and composite prepreg on the mechanical properties of carbon-fiber-reinforced metal laminates," *Composites Science and Technology*, vol. 57, no. 1, pp. 35–45, 1997.

[11] G. V. Reyes and W. J. Cantwell, "The mechanical properties of fiber-metal laminates glass fibre reinforced polypropylene," *Composites Science and Technology*, vol. 60, no. 7, pp. 1085–1094, 2000.

[12] S. Mostovoy, E. J. Ripling, and C. F. Bersch, "Fracture toughness of adhesive joints," *Journal of Adhesion*, vol. 3, no. 2, pp. 125–144, 1971.

[13] K. Song, K. C. G. Dávila, and C. A. Rose, "Guidelines and parameter selection for the simulation of progressive delamination," in *Proceedings of Abaqus User Conference*, Newport, RI, USA, May 2008.

[14] A. Turon, C. G. Dávila, P. P. Camanho, and J. Costa, "An engineering solution for mesh size effects in the simulation of delamination using cohesive zone models," *Engineering Fracture Mechanics*, vol. 74, no. 10, pp. 1665–1682, 2007.

[15] B. R. K Blackman, J. P. Dear, A. J. Kinloch, and S. Osiyemi, "The calculation of adhesive fracture energies from double-cantilever beam test specimens," *Journal of Materials Science Letters*, vol. 10, no. 5, pp. 253–256, 1991.

[16] B. R. K Blackman, A. J. Kinloch, F. S. Rodriguez-Sanchez, and W. S. Teo, "The fracture behaviour of adhesively-bonded composite joints: effects of rate of test and mode of loading," *International Journal of Solids and Structures*, vol. 49, no. 13, pp. 1434–1452, 2012.

[17] P. Manikandan and G. B. Chai, "A layer-wise behavioral study of metal based interply hybrid composites under low velocity impact load," *Composite Structures*, vol. 117, pp. 17–31, 2014.

[18] A. Hillerborg, M. Modéer, and P. E. Petersson, "Analysis of crack formation and crack growth in concrete by means of fracture mechanics and finite elements," *Cement and Concrete Research*, vol. 6, no. 6, pp. 773–781, 1976.

5

Preparation of Palladium-Impregnated Ceria by Metal Complex Decomposition for Methane Steam Reforming Catalysis

Worawat Wattanathana,[1] Suttipong Wannapaiboon,[2,3] Chatchai Veranitisagul,[4]
Navadol Laosiripojana,[5] Nattamon Koonsaeng,[6] and Apirat Laobuthee[1]

[1]Department of Materials Engineering, Faculty of Engineering, Kasetsart University, Chatuchak, Bangkok 10900, Thailand
[2]Synchrotron Light Research Institute, 111 University Avenue, Suranaree, Muang, Nakhon Ratchasima 30000, Thailand
[3]Chair of Inorganic and Metal-Organic Chemistry, Technical University of Munich, Lichtenbergstr. 4, 85748 Garching, Germany
[4]Department of Materials and Metallurgical Engineering, Faculty of Engineering,
 Rajamangala University of Technology Thanyaburi, Klong 6, Thanyaburi, Pathumthani 12110, Thailand
[5]The Joint Graduate School of Energy and Environment, CHE Center for Energy Technology and Environment,
 King Mongkut's University of Technology Thonburi, Bangkok 10140, Thailand
[6]Department of Chemistry, Faculty of Science, Kasetsart University, Chatuchak, Bangkok 10900, Thailand

Correspondence should be addressed to Apirat Laobuthee; fengapl@ku.ac.th

Academic Editor: Andres Sotelo

Palladium-impregnated ceria materials were successfully prepared via an integrated procedure between a metal complex decomposition method and a microwave-assisted wetness impregnation. Firstly, ceria (CeO_2) powders were synthesized by thermal decomposition of cerium(III) complexes prepared by using cerium(III) nitrate or cerium(III) chloride as a metal source to form a metal complex precursor with triethanolamine or benzoxazine dimer as an organic ligand. Palladium(II) nitrate was consequently introduced to the preformed ceria materials using wetness impregnation while applying microwave irradiation to assist dispersion of the dopant. The palladium-impregnated ceria materials were obtained by calcination under reduced atmosphere of 10% H_2 in He stream at 700°C for 2 h. Characterization of the palladium-impregnated ceria materials reveals the influences of the metal complex precursors on the properties of the obtained materials. Interestingly, the palladium-impregnated ceria prepared from the cerium(III)-benzoxazine dimer complex revealed significantly higher BET specific surface area and higher content of the more active $Pd^{\delta+}$ ($\delta > 2$) species than the materials prepared from cerium(III)-triethanolamine complexes. Consequently, it exhibited the most efficient catalytic activity in the methane steam reforming reaction. By optimization of the metal complex precursors, characteristics of the obtained palladium-impregnated ceria catalysts can be modified and hence influence the catalytic activity.

1. Introduction

Cerium oxide or so-called ceria (CeO_2) naturally contains a high concentration of mobile oxygen vacancy sites, which are created by replacement of a fraction of the Ce^{4+} occupancies within the ceria lattice by trivalent cerium ions (Ce^{3+}) [1]. These self-generated oxygen vacancy sites act as local sources or sinks for oxygen which take part in the reactions occurring at its surface and therefore enhance the interest in using ceria for a wide range of catalytic applications [2–9]. One of the promising applications of ceria is the use as catalyst support,

of which desired characteristics reveal significant influences on the impregnation process of such active catalytic species (e.g,. metal nanoparticles and metal oxide additives), catalytic activity, and stability of the embedded catalysts [10]. Hence, optimizing the preparation procedure of the catalytic support is crucial in order to achieve an effective performance of catalysts.

Among various synthetic procedures, thermal decomposition of metal complexes is one of the useful techniques for synthesizing ceramic materials. This method provides many advantages, that is, simple and straightforward procedure, low cost, high availability, and variability of complex formations with miscellaneous cations. Success is highlighted by our achievements within the past decades for preparations of various metal oxides by the thermal decomposition of metal-triethanolamine complexes [11–17]. In particular, ceria and rare-earth-doped ceria materials have been successfully prepared and examined for the applications as solid support catalysts [18–22] and solid electrolytes in solid oxide fuel cells [23, 24].

Apart from using cerium(III)-triethanolamine complexes as starting precursors for the thermal decomposition process, the alternative choices of chelating ligands have been investigated. Recently, the use of benzoxazine dimer derivatives as ligand for metal complex precursors has revealed a promising choice for synthesizing ceria through thermal decomposition of metal complex [25]. Therefore, in this work, we extend the state of the art toward the study of the influences of precursor choices used for the metal complex formations on characteristics of the obtained ceria materials and consequently on the catalytic activity after impregnation of palladium catalysts onto the ceria solid support. Specifically, three different combinations of metal sources and organic ligands have been used to prepare metal complex precursors, namely, (a) cerium(III) nitrate hexahydrate ($Ce(NO_3)_3 \cdot 6H_2O$) and triethanolamine ligand, (b) cerium(III) chloride heptahydrate ($CeCl_3 \cdot 7H_2O$) and triethanolamine ligand, and (c) cerium(III) nitrate hexahydrate and benzoxazine dimer, N,N-bis(5-methoxy-2-hydroxybenzyl) methyl amine (MeMD), of which the structure was reported previously [25]. Note that the metal complex precursor prepared by combining cerium(III) chloride heptahydrate and benzoxazine dimer MeMD ligand is excluded from this study due to the less stability of this metal complex.

Methane steam reforming is selected to be the proof-of-concept reaction for a catalytic activity test. Unlike the conventional homogeneous combustion of methane (known as one of the most powerful greenhouse gases) at high temperature which normally emits CO and NO_x as products [26], this reaction provides a useful outcome by conversion of methane to the more valuable fuel, hydrogen gas. Pd-based materials have been well known as an example of the most active catalysts for hydrogen production from natural gases such as by methane steam reformation and partial oxidation. However, the catalytic activity is not stable when directly used in the reactions. Therefore, impregnation and embedding of the Pd-based catalysts in metal oxide support are required to increase the activity and stability of the active Pd-based catalysts [10, 27]. Herein, an attempt to investigate the influences of different precursors used for preparation of ceria materials by thermal decomposition of metal complexes on the function as the catalytic support materials has been accomplished. Palladium catalyst is introduced onto the ceria solid support by employing microwave-assisted wetness impregnation of palladium(II) nitrate dihydrate with the different molar ratios. Calcination under reduced atmosphere was used to generate the catalytic-active palladium-impregnated ceria materials prior to the test of catalytic activity, which is examined by the percentages of methane conversion.

2. Experimental

2.1. Chemicals. Cerium(III) nitrate hexahydrate ($Ce(NO_3)_3 \cdot 6H_2O$) and cerium(III) chloride heptahydrate ($CeCl_3 \cdot 7H_2O$) were supplied from Acros Organics. Palladium(II) nitrate dihydrate ($Pd(NO_3)_2 \cdot 2H_2O$) was purchased from Sigma-Aldrich, while triethanolamine (TEA, $N(CH_2CH_2OH)_3$) was obtained from CARLO ERBA Reagents. Solvents (ethanol and 1-propanol) were bought from Merck. All chemicals were used without further purification.

2.2. Preparation of Cerium Complexes and Ceria. Preparation of cerium-triethanolamine complexes was carried out by the same procedure as our previous literatures [19, 22]. The cerium salt (cerium(III) nitrate hexahydrate or cerium(III) chloride heptahydrate, 150 mmol) was dissolved in 1-propanol in a round bottom flask. The equimolar amount of triethanolamine was then added to the solution of the cerium salt. The mixture was distilled for 5 h to eliminate water, a byproduct from complexation, and 1-propanol solvent. After the reaction completed, pale-yellow precipitates of the complexes were formed. The precipitates were dried using rotary evaporator.

One of the derivatives of benzoxazine dimers, so-called MeMD, was used as a ligand to form complex with cerium(III) nitrate. Hereinafter, we call the ligand "benzoxazine dimer." Cerium(III) nitrate hexahydrate and the benzoxazine dimer with the molar ratio of 1 : 6 were mixed together in ethanol in order to form cerium-benzoxazine dimer complex [25]. The complexation occurred very rapidly, since there was an immediate color change to dark purple after mixing two colorless substances together. The solution was heated and stirred for 30 min to ensure complete reaction.

All the obtained complexes were used as precursors for synthesizing ceria powders. According to the thermogravimetric analyses in the previous works [19, 22, 25], the calcination temperature was selected to be 600°C. Therefore, all the prepared cerium complexes were subjected to calcination at 600°C for 2 h under air atmosphere to turn into ceria powders via thermal decomposition. The ceria powders obtained after the heat treatment of the complexes were pale-yellow colored.

2.3. Preparation of Palladium-Impregnated Ceria Catalysts. The prepared ceria powders were intended to be used as catalyst support for palladium. The method used for introducing palladium onto the surface of ceria powders was microwave-assisted wetness impregnation, where the ceria support was soaked with palladium salt solution. Prior to impregnation, the obtained ceria powders were ground using mortar and pestle and then sieved with the size of 45 μm. Palladium(II) nitrate dihydrate with 1% and 3% by mole was dissolved in deionized water and then added to the prepared ceria. The mixtures were sonicated for 30 min and then heated with domestic microwave to help palladium particles distribute evenly on the surface of the ceria support. After that, the impregnated samples were calcined under 10% H_2/He atmosphere at 700°C for 2 h.

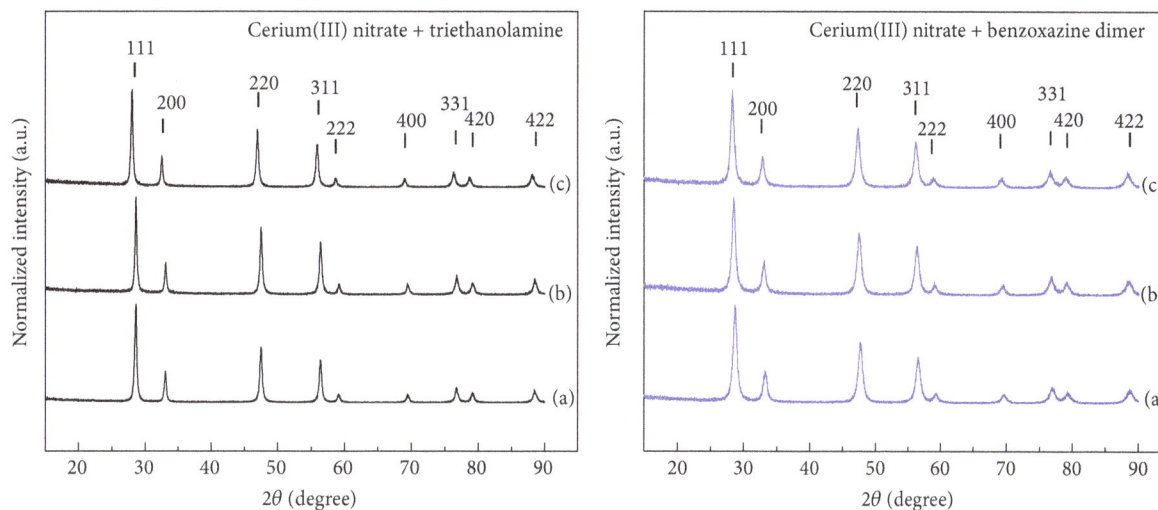

FIGURE 1: XRD patterns of the obtained products of (a) ceria after the thermal decomposition of the metal complex at 600°C for 2 h, further used as a catalytic support; (b) ceria support after microwave-assisted wetness impregnation of palladium(II) nitrate (3% mole); (c) Pd-impregnated ceria catalyst after calcination under reduced atmosphere of 10% H_2 in He stream at 700°C for 2 h, denoted as 3Pd-CeO2-T-N (red curves) and 3Pd-CeO2-B-N (blue curves) with regard to the use of cerium(III) nitrate to prepare metal complexes with triethanolamine and benzoxazine dimer ligand, respectively.

2.4. Materials Characterization. X-ray diffraction (XRD) analysis was carried out at room temperature on the prepared catalysts using an X-ray diffractometer (X'Pert PRO MPD diffractometer) operating at 40 kV and 30 mA using nickel-filtered Cu Kα radiation. Diffraction patterns were recorded in the 2θ range of 20–90° by step scanning with a step interval of 0.02° and a scanning time of 2 s for each step. The crystallite size (D_{XRD}) of the calcined powders was determined using the Scherrer equation: $D = 0.9\lambda/\beta\cos\theta$, where λ is the wavelength of the X-rays (1.5406 Å), θ is the scattering angle of the main reflection (111), and β is the corrected peak at full width at half maximum (FWHM) intensity. Morphology of the prepared catalysts was studied by scanning electron microscope (SEM, JEOL JSM5410) equipped with an energy dispersive X-ray analyzer (EDX). Prior to the study, the samples were mounted on stubs using carbon tape and then sputtered with Au to avoid particle charging. Moreover, transmission electron microscope (TEM, JEOL JEM-2100) equipped with an EDX unit was used for further investigation of microstructure of the materials. X-ray photoelectron spectroscopy (XPS) was operated to investigate the elemental species and their oxidation states in the obtained materials. The specific surface area (S_{BET}) of the obtained catalysts was measured from nitrogen adsorption isotherm at 77 K using a surface area analyzer (Micromeritics) based on the Brunauer-Emmett-Teller (BET) principle. The mesopore size distribution was further analyzed based on the Barrett-Joyner-Halenda (BJH) method.

2.5. Study on Catalytic Activity toward Methane Steam Reforming. Conversion of methane steam reforming reaction was studied in a temperature-programmed microflow reactor operated at ambient pressure. A catalyst with approximately 50 mg was packed in the quartz tube in the reactor. Reagent gases were flowed in the tube containing the catalyst with a flow rate in the range from 20 to 200 cm^3 min^{-1}. After the reactions, the exit gas mixture was transferred via trace-heated lines (100°C) to the analysis section, which consists of a Porapak Q column, Shimadzu 14B gas chromatograph (GC). The outlet of the GC column was directly connected to thermal conductivity detector (TCD) and flame ionization detector (FID). In order to satisfactorily separate all elements, the temperature setting inside the GC column was programmed varying with time. In the first 3 min, the column temperature was constant at 60°C; it was then increased steadily by the rate of 15°C min^{-1} until 120°C and lastly decreased to 60°C.

3. Results and Discussion

Ceria powders were synthesized by thermal decomposition at 600°C for 2 h of three different cerium(III) complexes formed by using different precursor sources as follows: (a) cerium(III) chloride heptahydrate as a metal source and triethanolamine as a ligand (coined as CeO2-T-Cl), (b) cerium(III) nitrate hexahydrate and triethanolamine ligand (CeO2-T-N), and (c) cerium(III) nitrate hexahydrate and benzoxazine dimer derivative ligand (CeO2-B-N). Note that structural characteristics of the three metal complex precursors have been thoroughly investigated and reported in our previous works [22, 25] and are not discussed further in this recent work. The obtained ceria powders exhibit high crystallinity indexed to be the face-centered cubic CeO$_2$ (JCPDS number 34-0394) and no crystalline impurity is observed in the XRD patterns (as demonstrated in Figure 1(a) in both red and blue curves).

TABLE 1: Crystallite size calculated by Scherrer equation according to the XRD patterns, percentage of different chemical states of the Pd-based species according to the Pd 3d XPS spectra, and specific surface area derived from BET calculation of N_2 adsorption isotherm at 77 K (S_{BET}) of the obtained Pd-impregnated ceria catalysts.

Pd-impregnated ceria catalysts	FWHM (degree)	Crystallite size (nm)	% of Pd-based species		S_{BET} (m^2/g)
			Pd^{2+}	$Pd^{\delta+}$ ($\delta > 2$)	
1Pd-CeO2-T-Cl	0.28328	30.20	—	—	17.22
3Pd-CeO2-T-Cl	0.28363	30.16	76.6	23.4	17.70
1Pd-CeO2-T-N	0.49834	17.17	—	—	14.30
3Pd-CeO2-T-N	0.39466	21.68	75.7	24.3	11.42
1Pd-CeO2-B-N	0.77191	11.10	—	—	62.89
3Pd-CeO2-B-N	0.70906	12.07	70.0	30.0	57.01

These obtained ceria powders were further used as catalytic support to impregnate palladium catalyst by using microwave-assisted wetness impregnation of palladium(II) nitrate dihydrate for 1% and 3% mole. For notation of the synthesized Pd-impregnated ceria catalysts reported herein, we coin the term **xPd-CeO2-a-b**, where **x** denotes percentages of **Pd** impregnating amount (**1** for 1% and **3** for 3% mole), **a** denotes type of organic ligand used for preparation of metal complex (**T** for triethanolamine and **B** for benzoxazine dimer), and **b** denotes type of cerium salts used for preparation of metal complex (**Cl** for cerium(III) chloride and **N** for cerium(III) nitrate). The materials were characterized by XRD after the wetness impregnation process, and there is no significant change of the XRD patterns (or crystallinity) of the ceria support before and after the impregnation (Figure 1(b)). In order to activate the Pd-impregnated ceria catalyst, the materials were further treated by calcination at 700°C for 2 h under reduced atmosphere of 10% H_2 in He stream. XRD patterns of the obtained Pd-impregnated ceria catalysts show no significant change from the crystallinity of the ceria support materials before impregnation (Figure 1(c)). Moreover, there are no additional diffraction peaks related to crystalline Pd-based materials, indicating the rather small particles as well as the lack of long-range crystallite ordering of the Pd-based species in the materials.

With closer inspection on the XRD patterns of the Pd-impregnated ceria catalysts derived from different metal complex precursors (Figure 2), it is noteworthy that a variation of metal complex precursors influences the crystallinity of the obtained materials. Specifically, crystallinity of ceria support is highest when using cerium(III) chloride and triethanolamine for preparing metal complex precursor (Figure 2, black curves), followed by the cases of using cerium(III) nitrate and triethanolamine (Figure 2, red curves), and the lowest crystallinity is in the cases of using cerium(III) nitrate and benzoxazine dimer (Figure 2, blue curves). Crystallite sizes calculated by Scherrer equation according to the XRD patterns are shown in Table 1, highlighting dependence on the metal complex precursors. The use of benzoxazine dimer as ligand for cerium complex formation notably reduces the crystallite size of the ceria product.

Scanning electron microscopic (SEM) images of the six different Pd-impregnated ceria catalysts are illustrated in

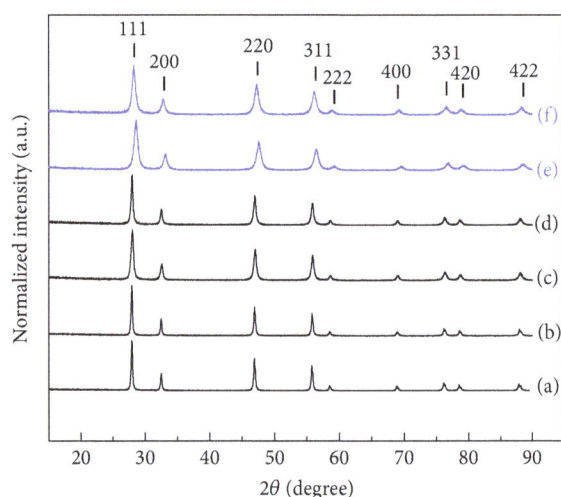

FIGURE 2: Comparison of XRD patterns of the Pd-impregnated ceria catalysts prepared by using ceria support derived from different metal complex precursors and impregnating Pd with two different concentrations: (a) 1Pd-CeO2-T-Cl, (b) 3Pd-CeO2-T-Cl, (c) 1Pd-CeO2-T-N, (d) 3Pd-CeO2-T-N, (e) 1Pd-CeO2-B-N, and (f) 3Pd-CeO2-B-N.

Figure 3. According to the overview of SEM images, the Pd-impregnated ceria materials are observed as aggregation of nanometer-sized particles, which cannot easily distinguish the influences of the different metal complex precursors and the impregnating concentrations on the particle size and morphology. However, the corresponding energy dispersive X-ray (EDX) spectrum and the EDX elemental mapping (shown in Figure 4 as an example) indicate that cerium, oxygen, and palladium are well distributed within the material matrix. This evidence reveals preliminary success of impregnation of Pd-based catalysts onto the ceria support.

To get insight into microstructures of the Pd-impregnated ceria catalysts, TEM is further used for microscopic characterization. According to the TEM images (Figure 5), the effect of different metal complex precursors on the particle size of Pd-impregnated ceria catalysts is clearly observed. Specifically, Pd-impregnated ceria prepared by cerium(III)-benzoxazine dimer complex shows significantly smaller particle size (approximately 10 nm) than the ones prepared

FIGURE 3: SEM images of the Pd-impregnated ceria catalysts. (a) 1Pd-CeO2-T-Cl, (b) 3Pd-CeO2-T-Cl, (c) 1Pd-CeO2-T-N, (d) 3Pd-CeO2-T-N, (e) 1Pd-CeO2-B-N, and (f) 3Pd-CeO2-B-N. Scale bar represents 10 μm.

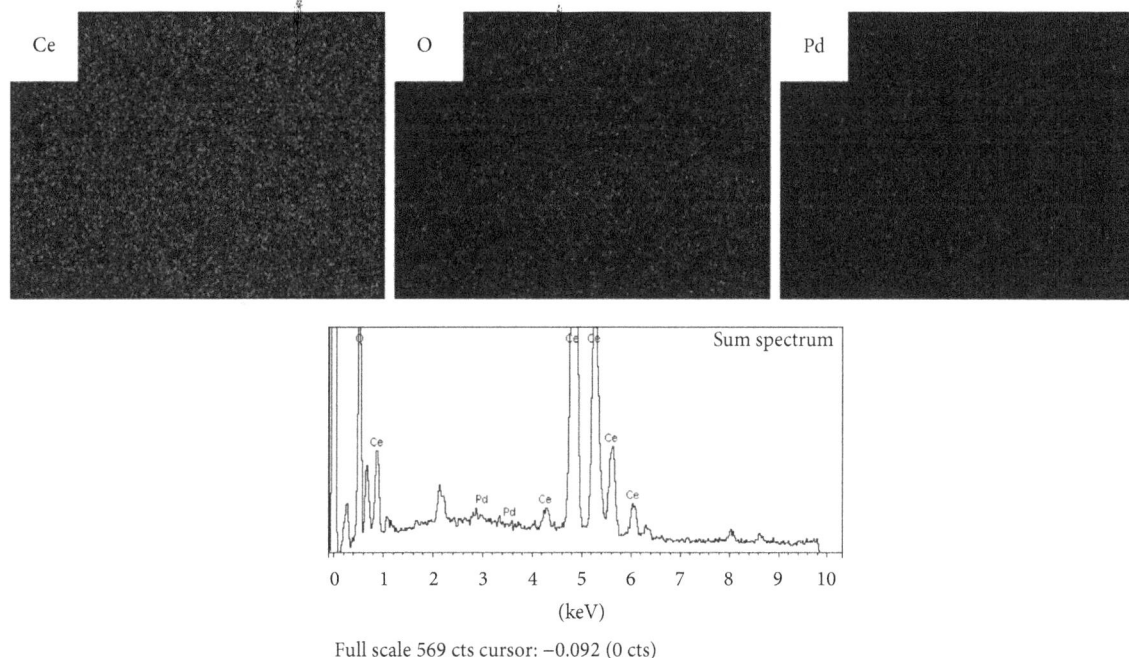

Full scale 569 cts cursor: −0.092 (0 cts)

FIGURE 4: EDX mapping of cerium, oxygen, and palladium within 1Pd-CeO2-B-N showing a good distribution of the three elements within the material matrix and its corresponding EDX spectrum based on SEM image.

from cerium(III)-triethanolamine complexes. High-resolution TEM image of the 3Pd-CeO2-B-N catalyst (Figure 5(d)) clearly indicates the crystallite planes (111) and (200) of CeO_2 with the crystallite plane size of 0.319 and 0.285 nm, respectively. Note that it is rather complicated to distinguish presence of Pd-based species by TEM due to a low content of Pd

doping amount. However, EDX spectrum (Figure 5(e)) indicates the presence of Pd-based species with a better signal of Pd content than the EDX spectra observed in the bulk materials according to SEM measurements (Figure 4).

To clearly elucidate the presence of Pd species impregnated in the materials, XPS spectra of the Pd-impregnated

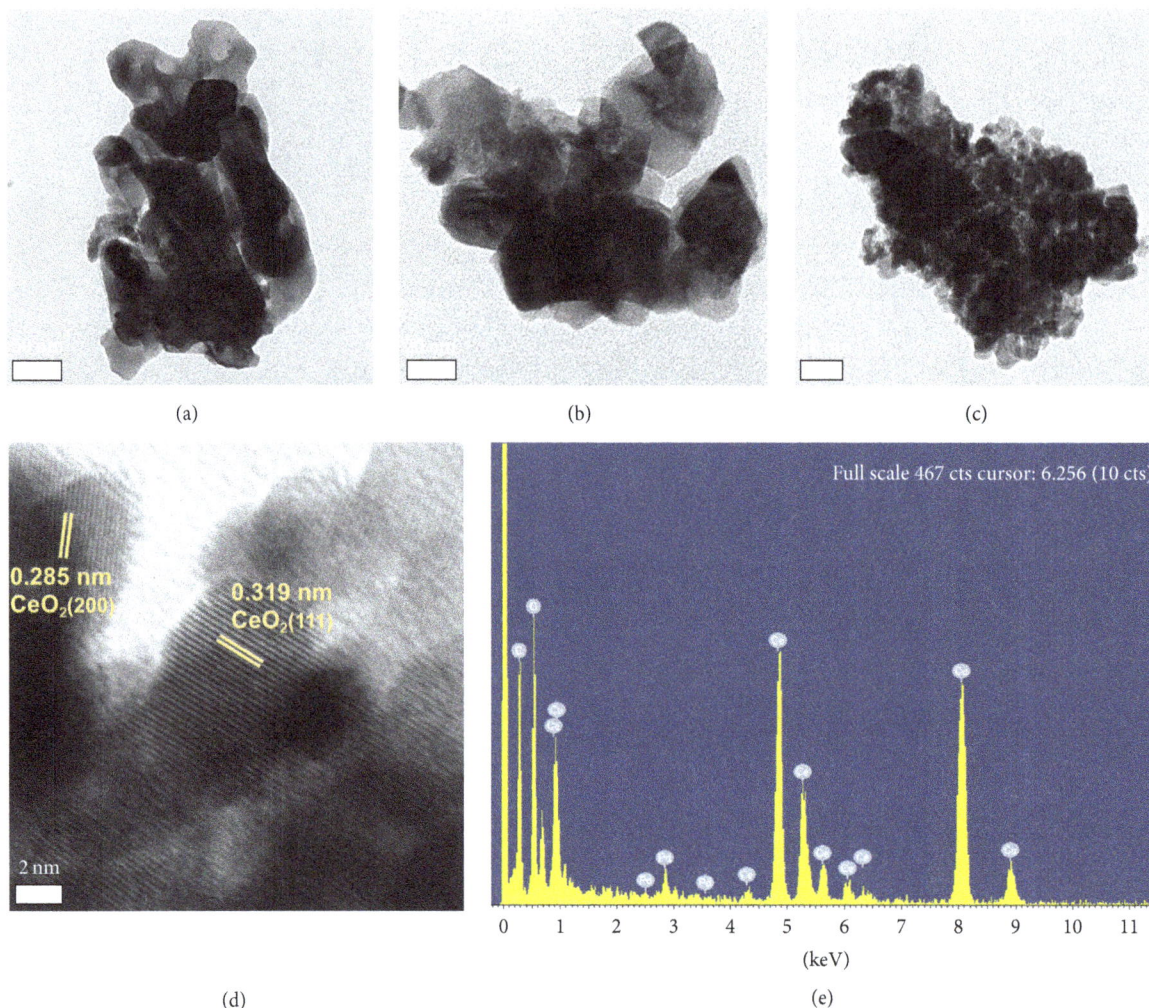

FIGURE 5: TEM images of the Pd-impregnated ceria catalysts. (a) 3Pd-CeO2-T-Cl, (b) 3Pd-CeO2-T-N, and (c) 3Pd-CeO2-B-N. (d) High-resolution TEM image of the 3Pd-CeO2-B-N catalyst indicating the crystallite planes of CeO_2; (e) the corresponding EDX spectrum of the 3Pd-CeO2-B-N catalyst based on TEM image in (c).

ceria catalysts are recorded (Figure 6). The XPS results of Pd $3d_{5/2}$ could be decomposed into two components at a binding energy of 337.0 and 339.0 eV, which could be assigned to oxidized Pd species of Pd^{2+} and $Pd^{\delta+}$ ($\delta > 2$), respectively. Moreover, the XPS results of Pd $3d_{3/2}$ could be also decomposed into two components of oxidized Pd species of Pd^{2+} and $Pd^{\delta+}$ ($\delta > 2$). Interestingly, the Pd-impregnated ceria prepared by cerium(III)-benzoxazine dimer complex exhibits the highest content of the high-oxidative $Pd^{\delta+}$ (30.0%, see Table 1). Herein, we confirm successful impregnation of oxidized Pd-based catalyst on the CeO_2 support.

Specific surface area (S_{BET} in Table 1) of the obtained Pd-impregnated ceria catalysts is derived from BET calculation of N_2 adsorption isotherm at 77 K (Figures 7(a) and 7(b)). The results clearly indicate that the use of benzoxazine dimer ligand leads to the formation of ceria materials with observably higher specific surface area (approximately 60 m^2/g) than

the cases of using triethanolamine ligand (approximately 15 m^2/g). This observation may be explained by referring to the particles sizes of ceria materials observed by TEM (Figure 5). The aggregation of these smaller particles creates a higher surface area of the ceria support (higher S_{BET} value), which is expected to distribute the Pd-impregnated nanoparticles on the ceria support in a better fashion. So the Pd-impregnated ceria catalyst derived from the use of cerium(III)-benzoxazine dimer complex as precursors shows the highest specific surface area among the other cases. Moreover, the Pd-impregnated ceria materials exhibit a hysteresis in the N_2 sorption isotherm curves, indicating the characteristics of mesopore within the materials. Their corresponding mesopore size distributions are calculated based on BJH method and illustrated in Figures 7(c) and 7(d). Again, the Pd-impregnated ceria prepared by the cerium(III)-benzoxazine dimer complex shows a relative smaller mesopore size distribution which is significantly different from

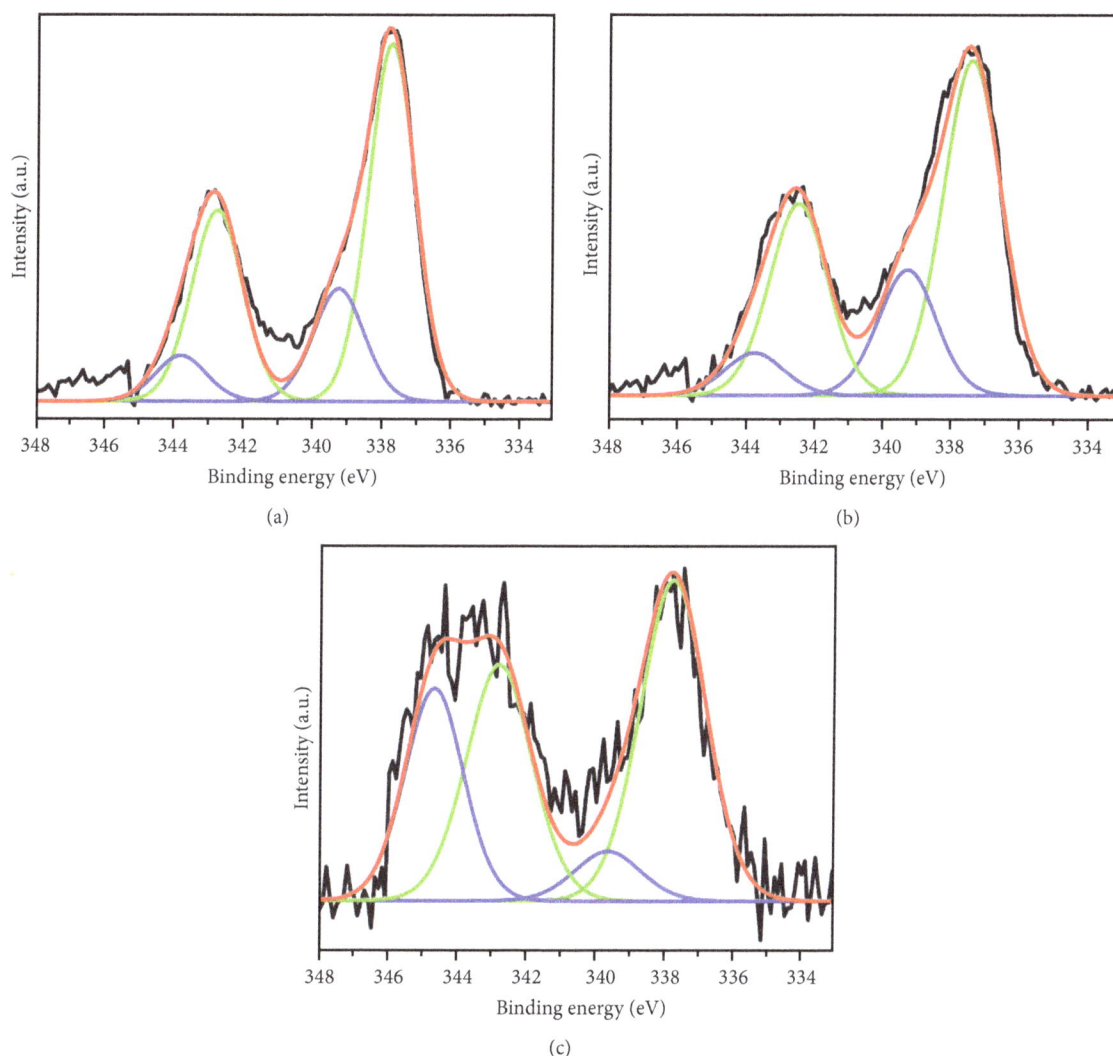

Figure 6: Pd 3d XPS profiles of the Pd-impregnated ceria catalysts. (a) 3Pd-CeO2-T-Cl, (b) 3Pd-CeO2-T-N, and (c) 3Pd-CeO2-B-N. The green and blue lines represent the decomposed spectra which could be assigned to the oxidized Pd species of Pd^{2+} and $Pd^{\delta+}$ ($\delta > 2$), respectively. The red line represents the combination of all decomposed spectra of Pd 3d.

the materials synthesized from cerium(III)-triethanolamine complexes, highlighting the influence of precursor choices on the characteristics of the obtained materials.

Catalytic activity of the Pd-impregnated ceria catalysts prepared by using ceria support derived from different metal complex precursors was tested in the methane steam reforming reaction operated at ambient pressure in a microflow reactor. Percentages of methane conversion as a function of reaction time using each Pd-impregnated ceria catalyst are shown in Figure 8. In each precursor type, it is clearly observed that the higher amount of Pd particles impregnated onto the ceria support is, the higher catalytic reactivity for the conversion of methane steam reforming is (the 3% Pd impregnation shows higher reactivity than the 1% Pd case with respect to the same metal complex precursors). Interestingly, a difference of catalytic activity as dependence of the metal complex precursors is notably observed. The use of Pd-impregnated ceria catalysts derived from the use of

cerium(III) nitrate-benzoxazine dimer complex (blue curves) shows the highest percentage of methane conversion of 75% after 1 h of reaction time, whereas the catalytic reactivity is gradually increased as a function of reaction time when using the Pd-impregnated ceria catalysts derived from the cerium(III) chloride-triethanolamine complex precursors (black curves), reaching 70% conversion after 8 h reaction time.

Referring to the XPS spectra and BET surface area analysis, the Pd-impregnated ceria catalysts derived from the use of cerium(III) nitrate-benzoxazine dimer complex contain higher amount of $Pd^{\delta+}$ ($\delta > 2$) species and higher BET surface area than the other cases. The $Pd^{\delta+}$ species reveal that more electrons transfer to the oxide support and are more oxidatively active than the metallic Pd and the Pd^{2+} species [28–32]. The higher oxidation state $Pd^{\delta+}$ should accelerate the C-H bond activation of the CH_4 molecule on the coordinatively unsaturated (CUS) Pd and consequently enhances the

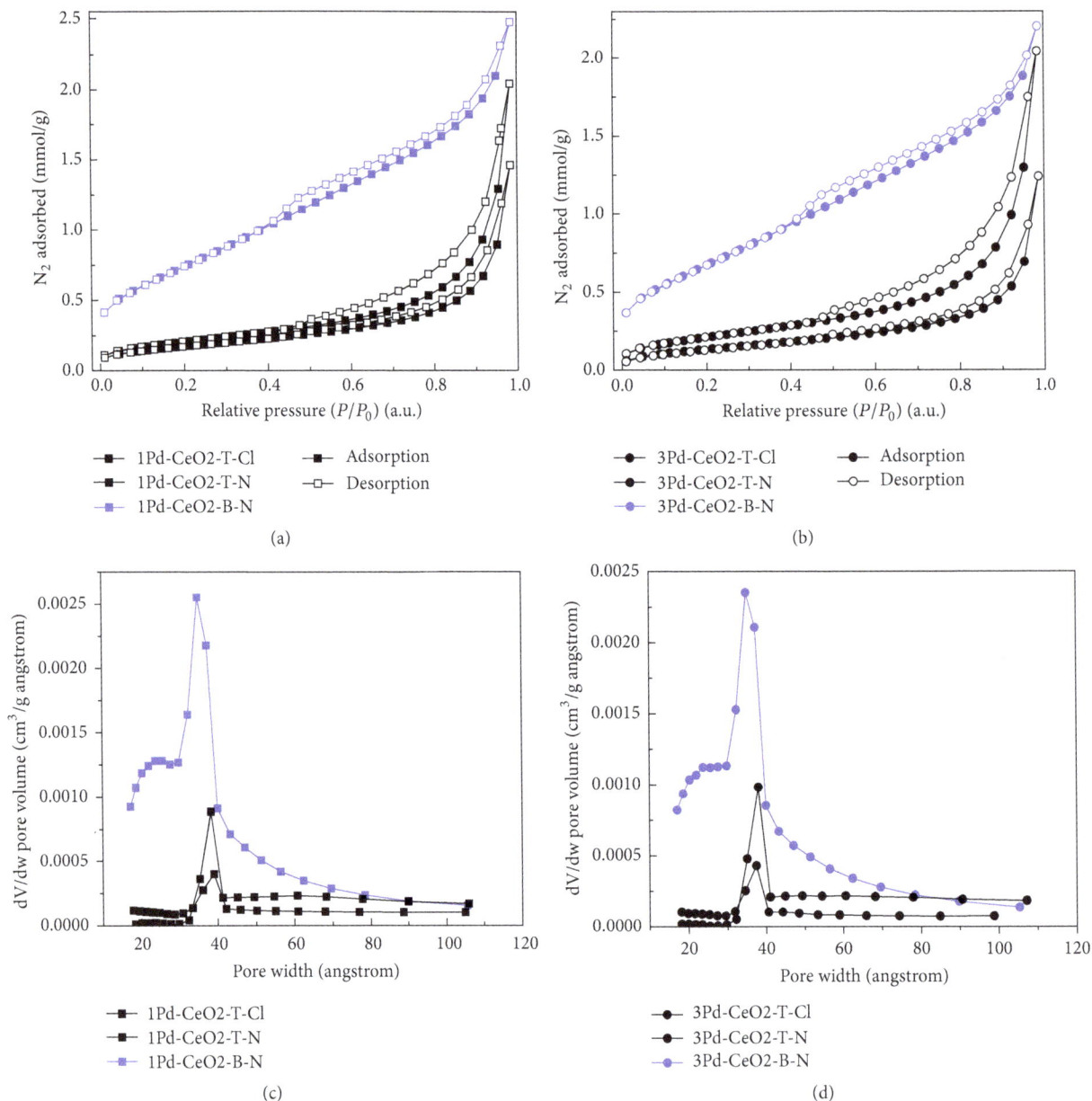

FIGURE 7: Nitrogen adsorption isotherms at 77 K of the (a) 1% and (b) 3% Pd-impregnated ceria catalysts; ((c) and (d)) their corresponding pore size distribution based on BJH method of the 1% and 3% Pd-impregnated ceria catalysts, respectively.

catalytic activity [31, 32]. Hence, it reveals a consequence of the higher catalytic activity of the Pd-impregnated ceria by means of the higher BET specific surface area and the higher content of the more active $Pd^{\delta+}$ ($\delta > 2$) species, which can be modified by the choice of metal complex precursors used for synthesis of the ceria support by the thermal decomposition method.

4. Conclusions

Characteristics of ceria support materials are strongly influenced by the choice of metal complex precursors used for the

thermal decomposition process. Optimizing the metal complex precursors leads to a systematic control of particles sizes and consequently specific surface area of the obtained ceria support. The higher surface area of ceria support, achieved by the use of cerium(III)-benzoxazine dimer complex as the thermal decomposition precursor, enhances the quality of impregnation of the catalytically active Pd-based species onto the ceria support. Moreover, the higher content of the more active $Pd^{\delta+}$ ($\delta > 2$) species within the Pd-impregnated ceria catalysts derived from the cerium(III)-benzoxazine dimer complex creates a higher catalytic performance for the methane steam reforming reaction. By optimization of the

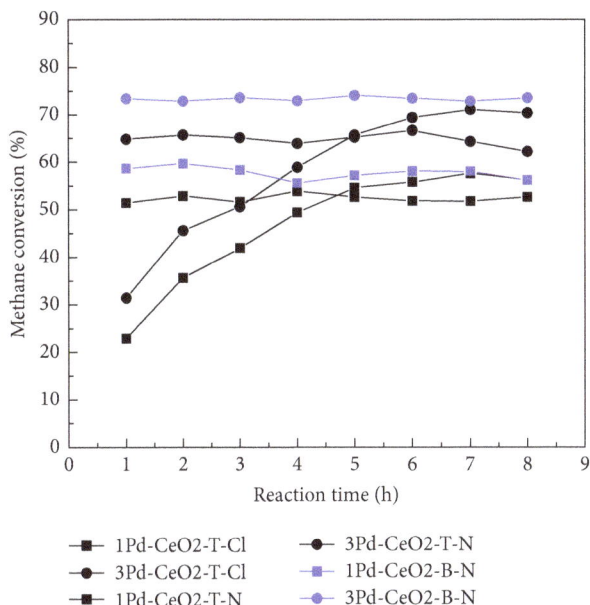

FIGURE 8: Catalytic activity of the obtained Pd-impregnated ceria catalysts according to percentages of methane conversion during the methane steam reforming reaction for hydrogen gas production.

metal complex precursors, characteristics and catalytic activity of the obtained Pd-impregnated ceria catalysts can be diligently modified and enhanced.

Conflicts of Interest

The authors declare that there are no conflicts of interest regarding the publication of this paper.

Acknowledgments

This work was supported by Kasetsart University Research and Development Institute (KURDI); Department of Chemistry, Faculty of Science, Kasetsart University; Department of Materials Engineering, Faculty of Engineering, Kasetsart University; and the Development and Promotion of Science and Technology Talents Project (DPST).

References

[1] K. C. Anjaneya, J. Manjanna, G. P. Nayaka, V. M. Ashwin Kumar, G. Govindaraj, and K. N. Ganesha, "Citrate-complexation synthesized $Ce_{0.85}Gd_{0.15}O_{2-\delta}$ (GDC15) as solid electrolyte for intermediate temperature SOFC," *Physica B: Condensed Matter*, vol. 447, pp. 51–55, 2014.

[2] A. Trovarelli, "Catalytic properties of ceria and CeO_2-containing materials," *Catalysis Reviews-Science and Engineering*, vol. 38, no. 4, pp. 439–520, 1996.

[3] P. Fornasiero, G. Balducci, R. Di Monte et al., "Modification of the redox behaviour of CeO_2 induced by structural doping with ZrO_2," *Journal of Catalysis*, vol. 164, no. 1, pp. 173–183, 1996.

[4] T. Miki, T. Ogawa, M. Haneda et al., "Enhanced oxygen storage capacity of cerium oxides in cerium dioxide/lanthanum sesquioxide/alumina containing precious metals," *The Journal of Physical Chemistry*, vol. 94, no. 16, pp. 6464–6467, 1990.

[5] C. Padeste, N. W. Cant, and D. L. Trimm, "The influence of water on the reduction and reoxidation of ceria," *Catalysis Letters*, vol. 18, no. 3, pp. 305–316, 1993.

[6] S. Kacimi, J. Barbier Jr., R. Taha, and D. Duperz, "Oxygen storage capacity of promoted Rh/CeC_2 catalysts. Exceptional behavior of $RhCu/CeO_2$," *Catalysis Letters*, vol. 22, no. 4, pp. 343–350, 1993.

[7] G. S. Zafiris and R. J. Gorte, "Evidence for a second CO oxidation mechanism on Rh/Ceria," *Journal of Catalysis*, vol. 143, no. 1, pp. 86–91, 1993.

[8] G. S. Zafiris and R. J. Gorte, "Evidence for low-temperature oxygen migration from ceria to Rh," *Journal of Catalysis*, vol. 139, no. 2, pp. 561–567, 1993.

[9] S. Imamura, M. Shono, N. Okamoto, A. Hamada, and S. Ishida, "Effect of cerium on the mobility of oxygen on manganese oxides," *Applied Catalysis A: General*, vol. 142, no. 2, pp. 279–288, 1996.

[10] M. Cargnello, J. J. Delgado Jaen, J. C. Hernandez Garrido et al., "Exceptional activity for methane combustion over modular $Pd@CeO_2$ subunits on functionalized Al_2O_3," *Science*, vol. 337, no. 6095, pp. 713–717, 2012.

[11] A. Laobuthee, S. Wongkasemjit, E. Traversa, and R. M. Laine, "$MgAl_2O_4$ spinel powders from oxide one pot synthesis (OOPS) process for ceramic humidity sensors," *Journal of the European Ceramic Society*, vol. 20, no. 2, pp. 91–97, 2000.

[12] A. Laobuthee, N. Koonsaeng, B. Ksapabutr, M. Panapoy, and C. Veranitisagul, "Doped $MgAl_2O_4$ spinel screen print thick film as sensing material for humidity measurement," *International Journal of Materials Structure Reliability*, vol. 3, no. 2, pp. 95–103, 2005.

[13] P. Hasin, N. Koonsaeng, and A. Laobuthee, "Nickel-aluminium complex: a simple and effective precursor for nickel aluminate ($NiAl_2O_4$) spinel," *Maejo International Journal of Science and Technology*, vol. 2, no. 1, pp. 140–149, 2008.

[14] S. Ummartyotin, S. Sangngern, N. Koonsaeng, N. Yoswath-ananont, M. Sato, and A. Laobuthee, "Preliminary study of nickel aluminate prepared from nickel complex as a solid support for hydrogenation reaction in a continuous-flow microreactor," *Journal of Research in Engineering and Technology*, vol. 5, pp. 375–391, 2008.

[15] S. Ummartyotin, S. Sangngern, A. Kaewvilai, N. Koonsaeng, H. Manuspiya, and A. Laobuthee, "Cobalt aluminate ($CoAl_2O_4$) derived from Co-Al-TEA complex and its dielectric behaviors," *Journal of Sustainable Energy & Environment*, vol. 1, pp. 31–37, 2010.

[16] T. Thaweechai, A. Wisitsoraat, A. Laobuthee, and N. Koonsaeng, "Ethanol sensing of $La_{1-x}Sr_xFeO_3$ (x = 0, 0.1, and 0.3) prepared by metal-organic complex decomposition," *Kasetsart Journal- Natural Science*, vol. 43, no. 5, pp. 218–223, 2009.

[17] W. Nantharak, W. Wattanathana, W. Klysubun et al., "Effect of local structure of Sm^{3+} in $MgAl_2O_4:Sm^{3+}$ phosphors prepared by thermal decomposition of triethanolamine complexes on their luminescence property," *Journal of Alloys and Compounds*, vol. 701, pp. 1019–1026, 2017.

[18] A. Laobuthee, C. Veranitisagul, N. Koonsaeng, V. Bhavakul, and N. Laosiripojana, "Catalytic activity of ultrafine $Ce_xGd_ySm_zO_2$ synthesized by metal organic complex method toward steam reforming of methane," *Catalysis Communications*, vol. 12, no. 1, pp. 25–29, 2010.

[19] W. Wattanathana, A. Lakkham, A. Kaewvilai, N. Koonsaeng, A. Laobuthee, and C. Veranitisagul, "Preliminary study of Pd/CeO$_2$ derived from cerium complexes as solid support catalysts for hydrogenation reaction in a micro-reactor," *Energy Procedia*, vol. 9, pp. 568–574, 2011.

[20] C. Veranitisagul, N. Koonsaeng, N. Laosiripojana, and A. Laobuthee, "Preparation of gadolinia doped ceria via metal complex decomposition method: Its application as catalyst for the steam reforming of ethane," *Journal of Industrial and Engineering Chemistry*, vol. 18, no. 3, pp. 898–903, 2012.

[21] A. Laobuthee, C. Veranitisagul, W. Wattanathana, N. Koonsaeng, and N. Laosiripojana, "Activity of Fe supported by Ce$_{1-x}$Sm$_x$O$_{2-\delta}$ derived from metal complex decomposition toward the steam reforming of toluene as biomass tar model compound," *Renewable Energy*, vol. 74, pp. 133–138, 2015.

[22] W. Wattanathana, N. Nootsuwan, C. Veranitisagul, N. Koonsaeng, N. Laosiripojana, and A. Laobuthee, "Simple cerium-triethanolamine complex: Synthesis, characterization, thermal decomposition and its application to prepare ceria support for platinum catalysts used in methane steam reforming," *Journal of Molecular Structure*, vol. 1089, pp. 9–15, 2015.

[23] C. Veranitisagul, A. Kaewvilai, W. Wattanathana, N. Koonsaeng, E. Traversa, and A. Laobuthee, "Electrolyte materials for solid oxide fuel cells derived from metal complexes: Gadolinia-doped ceria," *Ceramics International*, vol. 38, no. 3, pp. 2403–2409, 2012.

[24] W. Wattanathana, C. Veranitisagul, S. Wannapaiboon, W. Klysubun, N. Koonsaeng, and A. Laobuthee, "Samarium doped ceria (SDC) synthesized by a metal triethanolamine complex decomposition method: Characterization and an ionic conductivity study," *Ceramics International*, vol. 43, no. 13, pp. 9823–9830, 2017.

[25] C. Veranitisagul, A. Kaewvilai, S. Sangngern et al., "Novel recovery of nano-structured ceria (CeO$_2$) from Ce(III)-benzoxazine dimer complexes via thermal decomposition," *International Journal of Molecular Sciences*, vol. 12, no. 7, pp. 4365–4377, 2011.

[26] D. Ciuparu, M. R. Lyubovsky, E. Altman, L. D. Pfefferle, and A. Datye, "Catalytic combustion of methane over palladium-based catalysts," *Catalysis Reviews: Science and Engineering*, vol. 44, no. 4, pp. 593–649, 2002.

[27] T. M. Onn, S. Zhang, L. Arroyo-Ramirez et al., "Improved thermal stability and methane-oxidation activity of Pd/Al$_2$O$_3$ catalysts by atomic layer deposition of ZrO$_2$," *ACS Catalysis*, vol. 5, no. 10, pp. 5696–5701, 2015.

[28] P. Gélin and M. Primet, "Complete oxidation of methane at low temperature over noble metal based catalysts: a review," *Applied Catalysis B: Environmental*, vol. 39, no. 1, pp. 1–37, 2002.

[29] Y.-H. Chin, C. Buda, M. Neurock, and E. Iglesia, "Consequences of metal-oxide interconversion for C-H bond activation during CH$_4$ reactions on Pd catalysts," *Journal of the American Chemical Society*, vol. 135, no. 41, pp. 15425–15442, 2013.

[30] F. Yin, S. Ji, P. Wu, F. Zhao, and C. Li, "Deactivation behavior of Pd-based SBA-15 mesoporous silica catalysts for the catalytic combustion of methane," *Journal of Catalysis*, vol. 257, no. 1, pp. 108–116, 2008.

[31] N. M. Martin, M. Van den Bossche, A. Hellman et al., "Intrinsic ligand effect governing the catalytic activity of Pd oxide thin films," *ACS Catalysis*, vol. 4, no. 10, pp. 3330–3334, 2014.

[32] N. Yang, J. Liu, Y. Sun, and Y. Zhu, "Au@PdO$_x$ with a PdO$_x$-rich shell and Au-rich core embedded in Co$_3$O$_4$ nanorods for catalytic combustion of methane," *Nanoscale*, vol. 9, no. 6, pp. 2123–2128, 2017.

6

Metallurgical and Mechanical Research on Dissimilar Electron Beam Welding of AISI 316L and AISI 4340

A. R. Sufizadeh and S. A. A. Akbari Mousavi

School of Metallurgy and Materials Engineering, College of Engineering, University of Tehran, P.O. Box 11155-4563, Tehran, Iran

Correspondence should be addressed to A. R. Sufizadeh; sufizadeh@ut.ac.ir

Academic Editor: Jörg M. K. Wiezorek

Dissimilar electron beam welding of 316L austenitic stainless steel and AISI 4340 low alloy high strength steel has been studied. Studies are focused on effect of beam current on weld geometry, optical and scanning electron microscopy, X-ray diffraction of the weld microstructures, and heat affected zone. The results showed that the increase of beam current led to increasing depths and widths of the welds. The optimum beam current was 2.8 mA which shows full penetration with minimum width. The cooling rates were calculated for optimum sample by measuring secondary dendrite arm space and the results show that high cooling rates lead to austenitic microstructure. Moreover, the metallography result shows the columnar and equiaxed austenitic microstructures in weld zone. A comparison of HAZ widths depicts the wider HAZ in the 316L side. The tensile tests results showed that the optimum sample fractured from base metal in AISI 316L side with the UTS values is much greater than the other samples. Moreover, the fractography study presents the weld cross sections with dimples resembling ductile fracture. The hardness results showed that the increase of the beam current led to the formation of a wide softening zone as HAZ in AISI 4340 side.

1. Introduction

It is evident that deployment of dissimilar metals plays the most important part of construction processes due to the technological advances of production in the industry. Dissimilar joints would improve not only products quality in terms of cost but also design and enhancement of the structures' stability. The intrinsic properties of the materials used in dissimilar welding play a major role in determining the weld quality [1].

Electron beam welding is one of the widely used welding methods for joining dissimilar materials in industries. The primary advantage of electron beam welding is its high depth to width ratio which results in a very strong weld. The quality of weld depends upon the parameters, namely, accelerating voltage, beam current, welding speed, focus current, and vacuum level [2].

Electron beam welding has high power density and consequently led to small heat affected zone and high heating and cooling rates. The spot size of the EBW can vary between 0.2 mm and 13 mm, though only smaller sizes are used for welding. The depth of penetration is proportional to the amount of power supplied but is also dependent on the location of the focal point. Penetration is maximized when the focal point is slightly below the surface of the work piece [3].

The AISI 4340 stainless steel is of those high strength low alloy steels that their strengths are achieved with appropriate heat treatments [4]. It has good weld ability and is usually welded by the GTAW process [5].

Resistance welding and many other arc welding processes are used to weld austenitic stainless steels. But the resistance to corrosion and cracking in austenitic stainless steels are the major problems where careful selection on welding process or any other further solution is required. The corrosion properties of AISI 316L stainless steel are enhanced by addition of molybdenum. The AISI 316L stainless steel microstructure contains austenite with a little ferrite [6].

Basically, in dissimilar welding, the degree of dilution of each material in the weld bead is very important. This matter is also one of the influential parameters in the weld microstructures in the laser welding process [7]. The degree of dilution of the weld can be achieved by adjusting beam-focal point position relative to the joint interface of the weld. There is a possibility that the beam diameter is deviated away from the interface during laser welding in practice. Positioning of the focused beam by CNC facilitates location control and chemical compositions of the weld. This procedure requires close tolerances during the preparation of the joint [8].

Arivazhagan et al. studied the microstructure and mechanical properties of AISI 304 stainless steel and AISI 4140 low alloy steel joints by electron beam welding (EBW). The yield strengths of weldments performed by EBW were higher than those carried out by GTAW. In the weldments performed by EBW, the failures occurred on the HAZ of the AISI 4140 steels [9].

The fatigue performance of an electron-beam welded joint between AISI 4140 and AISI 316L steel has been investigated by Çalik et al. Results indicated that a good strength weld can be achieved between the two dissimilar steels by electron beam welding with a fatigue limit approaching 190 MPa, which is a value between the fatigue limits of the base materials. The microstructure of the electron-beam welded dissimilar AISI 4140-AISI 316L joint consisted of both columnar and equiaxed austenite grains in the weld interface [10].

2. Research Method

In this study, welding was performed with the electron beam welding machine. Table 1 shows the chemical compositions of the AISI 4340 low alloy high strength steel and AISI 316L austenitic stainless steel obtained by wave dispersive X-ray fluorescence spectroscopy. Table 2 presents the mechanical properties of base metals.

Figures 1(a) and 1(b) depict the microstructures of the AISI 316L stainless steel and the AISI 4340 steel. The typical picture of the welded sample is shown in Figure 1(c) in which a butt square weld joint was used. The thickness of the base metals at the joint cross section was 0.5 mm. No filler metal was used for electron beam welding and the welding was performed autogenously.

In order to investigate the effects of EBW parameters on the weld microstructures, the welding tests were (conducted) carried out with selected welding parameters according to Table 3.

The chemical etching was performed for optical examinations with the solution containing 40 mL methanol, 20 mL choleric acid, and 15 mL HNO_3 for 40 seconds. The electrochemical etching was carried out for SEM examinations with 6 V voltage and NaOH solution in 20 seconds. Also in order to reveal the microstructure and phases, the optical and scanning electron microscopy and X-ray diffraction were implemented.

The EDS analysis was carried out to investigate the variations (in weight%) of the main chemical elements across the weld cross section.

The tensile tests were carried out according to ASTM E8 for subsize specimen [11]. Figures 2(a) and 2(b) show the schematic dimensions of tensile test specimen and typical tensile test welded sample, respectively.

The Vickers microhardness was carried out with applying 100 gr load in 15 seconds. The microhardness profiles were obtained across the 316L stainless steel, the weld zone, and the 4340 steel. Figure 3 shows the schematic of the microhardness points investigated across the welding zone.

3. Results and Discussion

3.1. The Effects of Beam Current on Weld Properties

3.1.1. The Effects of Beam Current on Weld Geometry. Beam current is one of the most vital parameters in the electron beam welding process. Experiments have been conducted to note its effects on the depth of penetration.

Trials were conducted, to determine the range of beam current that can be used for a thickness of 0.5 mm. Hence, the range of beam currents is determined to be in between 1.6 mA and 4 mA. At constant voltage, weld focus value, speed of 0.2 cm/min, duration, and vacuum, beam current was varied as tabulated in Table 3 and observations were made.

According to previous investigation [12], the heat input was directly increased with beam current. The results were shown in Figure 4. The depth and width of the weld were increased with beam current. The width of the weld is more influential to the beam current than the depth of the weld (see Figure 4). The maximum depth was obtained for 2.8, 3.4, and 4 mA and was full penetration weld with 500 μm depths. The maximum width for the 4 mA beam current was 1950 μm. In equal weld depths, the beam current increase could cause the excessive increase of weld widths which is not desirable in such low heat input process. Therefore, sample number 3 (2.8 mA) was selected as the optimum one, since the penetration of all three samples (2.8, 3.4, and 4 mA) is the same.

3.1.2. The Effects of Beam Current on Weld Microstructure. Based on Figure 5, fine grained austenitic microstructures are seen as the dissimilar weld microstructure. The electron beam welding parameters play important role on weld microstructure and mechanical properties due to their effect on heat input. The effects of beam current on weld grain size were investigated.

Figure 5 shows that the weld grain size increases with beam current. The reason can be attributed to the increase of power and heat input [13] which reduces the cooling rate. Therefore, there is enough time for excessive grain growth in the weldment solidification period. Moreover, according to Figure 6, the observation showed that increasing the beam current values more than 2.8 mA led to significant grain growth rate but in lower beam current values the gradual growth was accrued. The reason can be attributed to the excessive increase of beam current only which led to increase in weld volume and decrease in cooling rate. At this condition, there is much more time for growth stage during solidification.

3.2. The Microhardness Tests of the Welded Specimen.
Figure 7 shows the microhardness profiles. The x-axis in Figure 7 shows the locations where microhardness tests were carried out. These locations are shown in Figure 3.

TABLE 1: The chemical composition of the base metals.

Material elements	Si	P	S	C	Mo	Mn	Ni	Cr	Fe	Cr_{eq}	Ni_{eq}
AISI 316L	0.39	0.031	0.005	0.027	2.89	1.84	10.65	16.62	68.95	19.9	12.38
AISI 4340	0.35	0.010	0.001	0.391	0.21	0.79	1.90	0.88	74.79	1.44	13.99

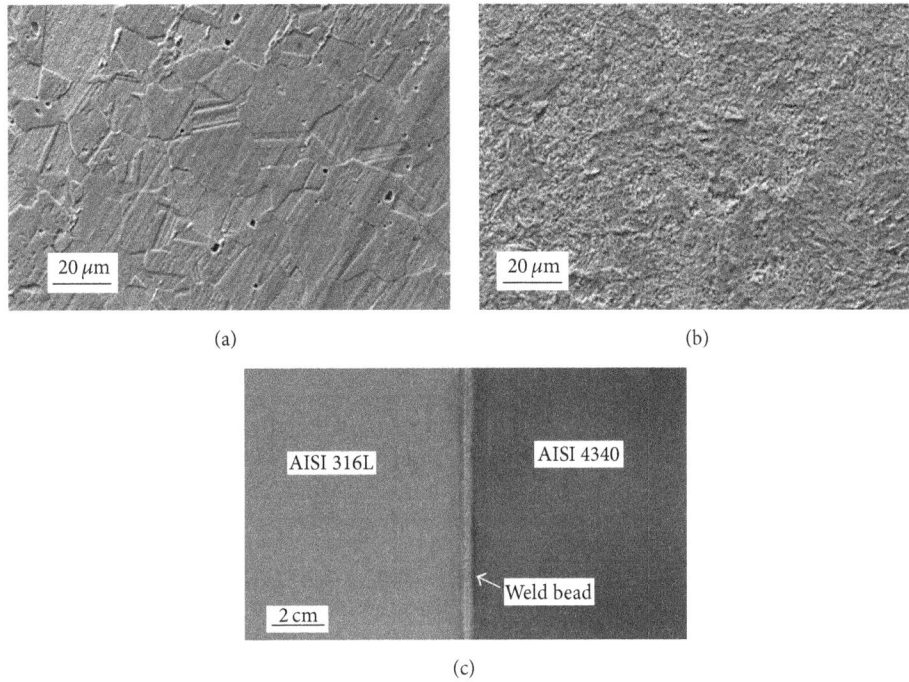

(a)

(b)

(c)

FIGURE 1: (a) Microstructures of AISI 316L. (b) Microstructures of AISI 4340 steel. (c) The welded joint macrograph.

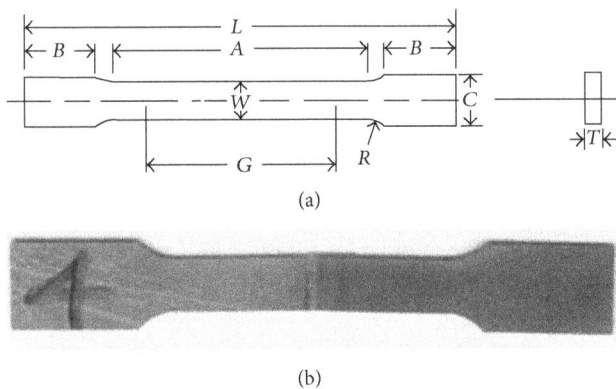

(a)

(b)

FIGURE 2: (a) Schematic of tensile test sample. (b) Tensile test sample [11]. G: gage length, W: width, T: thickness, R: radius of fillet, min, L: overall length, min, A: length of reduced section, min, B: length of grip section, min, and C: width of grip section, approximate.

TABLE 2: The mechanical properties of the base metals.

Alloy	Ultimate tensile strength (MPa)	Yield strength (MPa)	Elongation%
AISI 316L	515	190	45
AISI 4340	744	475	22

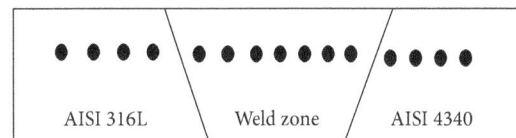

FIGURE 3: The schematic microhardness profiles.

According to Figure 7, if the austenitic weld microstructure is formed, the magnitude of the weld microhardness will be between those of base metals (450–550 HV). On the other hand, the microhardness of the weld zone is greater than that of the AISI 316L stainless steel base metal due to producing fine grain austenitic weld microstructure. The microhardness of the weld zone is lower than that of the AISI 4340 steel base metal because the austenitic weld microstructure is softer than the martensitic structure of the AISI 4340 steel base metal. The fluctuation of hardness profile in weld center line is the result of the change of solidification mode from columnar dendrite to equiaxed dendrite by transition from fusion line to center line according to Figure 16.

In the HAZ near AISI 316L stainless steel side, the hardness is similar to base metal. However, on the other side (near the AISI 4340 steel) the softened zone is observed. The

TABLE 3: The parameters for electron beam welding process.

Sample number	Voltage, KV	Weld focus value, Amps	Beam current, I, mA	Vacuum, Torr	Speed, cm/s
1	45 KV	2.2 A	1.6 mA	5×10^{-4}	0.2 cm/s
2	45 KV	2.2 A	2.2 mA	5×10^{-4}	0.2 cm/s
3	45 KV	2.2 A	2.8 mA	5×10^{-4}	0.2 cm/s
4	45 KV	2.2 A	3.4 mA	5×10^{-4}	0.2 cm/s
5	45 KV	2.2 A	4.0 mA	5×10^{-4}	0.2 cm/s

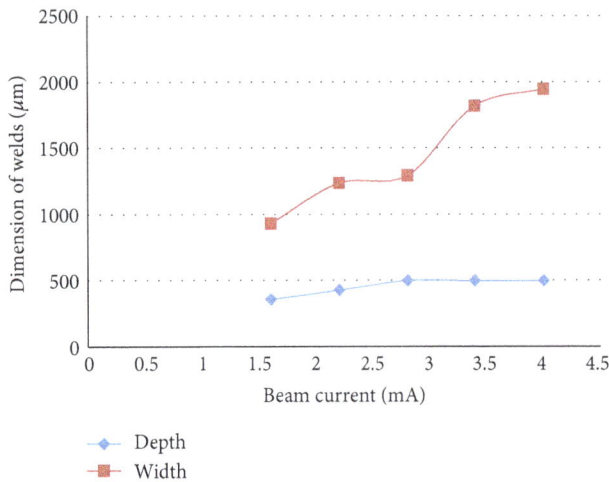

FIGURE 4: The effect of beam current on welds geometry.

reason can be attributed to producing coarse grained softening zone in the HAZ near the fusion boundary, according to Figure 17(b). The minimum hardness occurred on HAZ sample number 5. This is attributed to the beam current value in this sample which is the highest among all samples. So the heat affected zone was much longer exposed to high temperature and wider softening zones were created.

3.3. The Tensile Tests of the Welded Samples.

The engineering stress-engineering strain curves obtained by the tensile tests of the transverse welded full penetration samples (numbers 3, 4, and 5) are shown in Figure 8. According to the results of the tensile tests, samples number 3 (optimum depth/width ratio) and number 4 fractured above 515 MPa (the ultimate tensile strength of the AISI 316L stainless steel). On the other hand, it was observed that the strength magnitude of EBW-welded samples was lower than that of the AISI 4340 steel and higher than that of the AISI 316L stainless steel. These results were confirmed with the effect of producing fine grained microstructure in the austenitic weld zone. Figure 9 shows the fracture locations in the samples. Samples numbers 3 and 4 fractured in the AISI 316L stainless steel sides while sample number 5 was fractured on HAZ in side of AISI 4340. The reason for low tensile strength for sample number 5 can be described. The beam current of sample number 5 is the highest value so that the wide softening zone in HAZ was created. The failure occurred in the softening zone in heat affected zone. These results are in good agreement with the hardness results of the HAZ in AISI 4340 side, shown in Figure 7.

Figure 10 shows the fractography analysis obtained by SEM. The dimples in Figures 10(a) and 10(b) verified the ductile fracture behavior of AISI 316L as the base metal in samples numbers 3 and 4. Moreover, the dimples on Figure 10(c) shows ductile fracture behavior of AISI 4340 as the HAZ in sample number 5.

3.4. The Microstructure Investigations.

Considering the results obtained at previous sections, one may conclude that the optimum parameters were applied on sample number 3 from mechanical properties and weld geometry point of view. Therefore, in this section the metallurgical properties, solidification patterns, and the microstructure of sample number 3 were studied.

3.4.1. The Microstructure Prediction by Schaeffler Diagram.

The dissimilar weld microstructures prediction of Schaeffler diagram (Figure 11) with those achieved by the experiments had been investigated by the EDS analyses of Cr, Ni, Mn, C, Si, and Mo elements for the points shown in Figure 12. The EDS results were used to calculate Cr and Ni equivalent content for all three regions considered (Table 4).

$$Cr_{eq} = Cr\% + Mo\% + 1.5Si\% + 0.5Cb\% \qquad (1)$$

$$Ni_{eq} = Ni\% + 30C\% + 0.5Mn\% \qquad (2)$$

According to the Schaeffler diagram (see Figure 11), the austenite + martensite microstructures of are predicted; however, Figure 12 shows that the full austenitic microstructure is produced in the weld microstructures. The reason can be attributed to high cooling rate in EBW process which causes changes in the solidification mode from ferritic to austenitic mode according to Figure 13(a) [14]. The results of Figure 12 are in good agreement with previous investigations [15, 16].

3.4.2. The Calculation of Welding Cooling Rate.

In order to study the cooling rate of sample number 3, the measurement of secondary dendritic arm space (SDAS) was carried out. According to the investigation of [17], the following formula was explained for the relationship between SDAS and cooling rate:

$$SDAS = K \cdot \dot{T}^{-n}, \qquad (3)$$

where $\dot{T}$ is cooling rate (K/s) and K and n are material constants which are approximately 100 and 0.5 for austenitic microstructure [17]. Table 5 shows the results of SDAS and cooling rate measurements. The results reveal that, based on the high cooling rate (816 K/s), referring to Figure 13(a), the

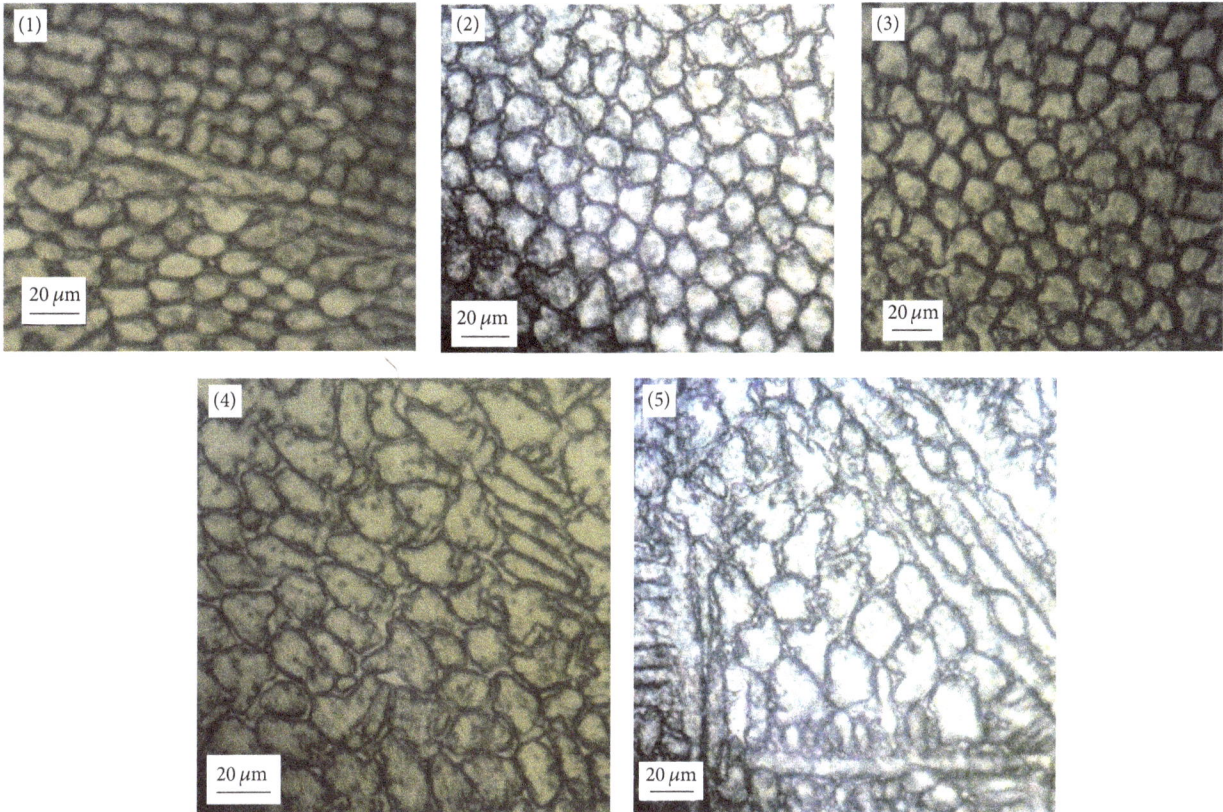

FIGURE 5: The grain size from welded microstructure.

TABLE 4: The calculation based on Schaeffler diagram for prediction of welds microstructures for sample number 3.

Sample location	Cr_{eq}	Ni_{eq}	Predicted microstructure	Actual microstructure
Assumption dilution (50%)	11.2	13.2	A + M	A
Figure 12(d) (near 4340)	8.1	12.8	M	A
Figure 12(b) (near 316L)	17.8	10.6	A + M + F	A
Figure 12(c) (weld center)	15.5	11.3	A + M	A

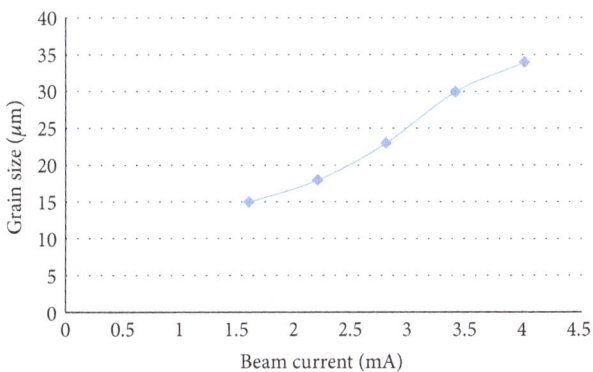

FIGURE 6: The effect of beam current on weld zone grain size.

TABLE 5: The results from SDAS and cooling rate measurements for sample number 3.

DAS (μm)	SDAS (μm)	SDA thickness (μm)	Cooling rate (K/s)
5.99	3.5	0.8	816

austenite dendrite tip temperature is higher than ferrite and this attributed to change solidification mode from austenite + martensite to fully austenite. High cooling rate effect is in accordance with formation of austenitic microstructures in Figure 13(b), which is also in accordance with the pervious investigations [15, 16].

3.4.3. The Microsegregation in Dendritic Microstructure. In the next step, the EDS analyses for sample number 3 were carried out in order to investigate the microsegregation during solidification and differences of percent of alloys element in subgrain, dendrites, and interdendritic spaces were done. The results were shown in Figure 14 and Table 6. According to Table 6 at point A (subgrain boundary), the percent of ferrite promotion such as Cr and Mo was 13.39% and 3.23%,

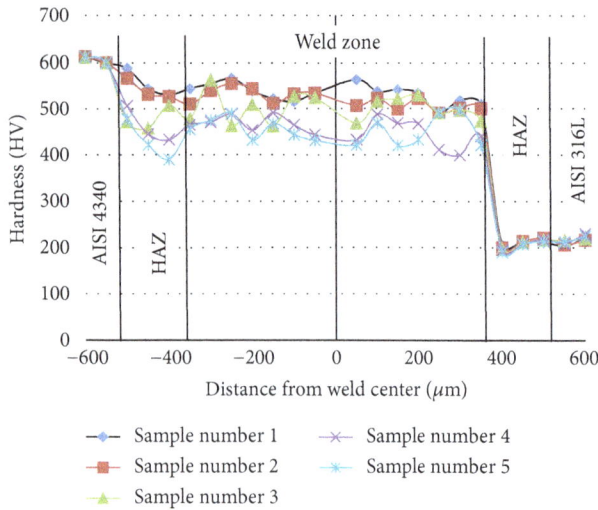

FIGURE 7: The profile from hardness tests for EBW weld samples.

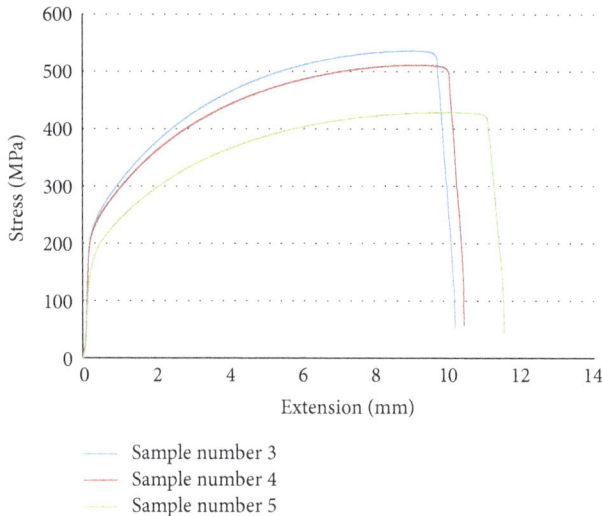

FIGURE 8: The tensile tests curves for EBW full penetration weld samples.

TABLE 6: The EDS analyses from subgrain boundary and subgrain spaces in the weld zone (sample number 3).

Element	Weight percent of elements (%)			
	A	B	C	D
Silicon	0.44	0.66	0.3	0.6
Chromium	13.39	10.38	11.33	12.26
Manganese	0.77	0.11	0.01	0.01
Nickel	7.25	5.92	7.61	6.03
Molybdenum	3.23	1.75	1.84	2.55

respectively, while in the matrix and point B that element percent was reduced to 10.38% and 1.75%, respectively.

These characterizations are the consequence of the austenite dendrites growth during solidification and also the reduction of the solubility of ferrite promotions elements

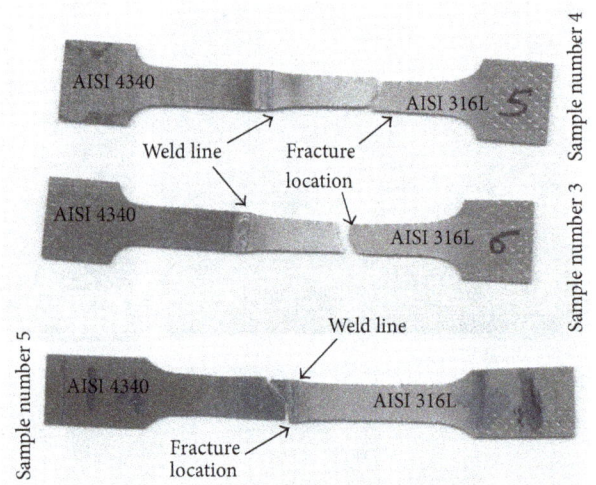

FIGURE 9: The fracture locations of EBW-welded specimens after tensile tests.

TABLE 7: The EDS analyses from matrix and grain boundary in AISI 316L heat affected zone side.

Element	Weight percent (%) at point E	Weight percent (%) at point F
Silicon	0.4	0.5
Chromium	19.5	16.4
Manganese	0.8	0.9
Nickel	7.1	10.5
Molybdenum	2.9	2.2

such as Cr and Mo. In this process, the austenitic former elements are concentrating on the dendritic structures and the ferritic former elements are traveling to interdendritic structures. In order to investigate ferrite formation in interdendritic spaces, the XRD test (Figure 15) was carried out. The results show that no ferrite phases were created and the microstructure is completely austenite. This result is in accordance with Figures 12 and 13, which is also conforming to the previous investigations [15, 16].

According to the EDS results (Table 7), the secondary dendritic arm space (SDAS) shows that the microsegregation was less than dendritic arm space (DAS) and the percent of Cr and Mo was 12.24% and 2.54%, respectively. The smaller diffusion zone, solute redistribution of elements in dendrite tips, and very high cooling rate are the major factors resulting in formation of SDAS and so the reduction of microsegregation in SDAS.

3.4.4. The Solidification Mode in the Weld Zone. According to Figure 16, transitions from fusion line to center line lead to changing the microstructure of weld from columnar dendrites to equiaxed dendrites. These results were attributed to the low growth rate (R) in the fusion boundary and it increases from the fusion boundary to the weld centerline. On the other hand, the thermal gradient (G) is minimum in the weld center and maximum in the fusion boundary. Therefore,

(a)

(b)

(c)

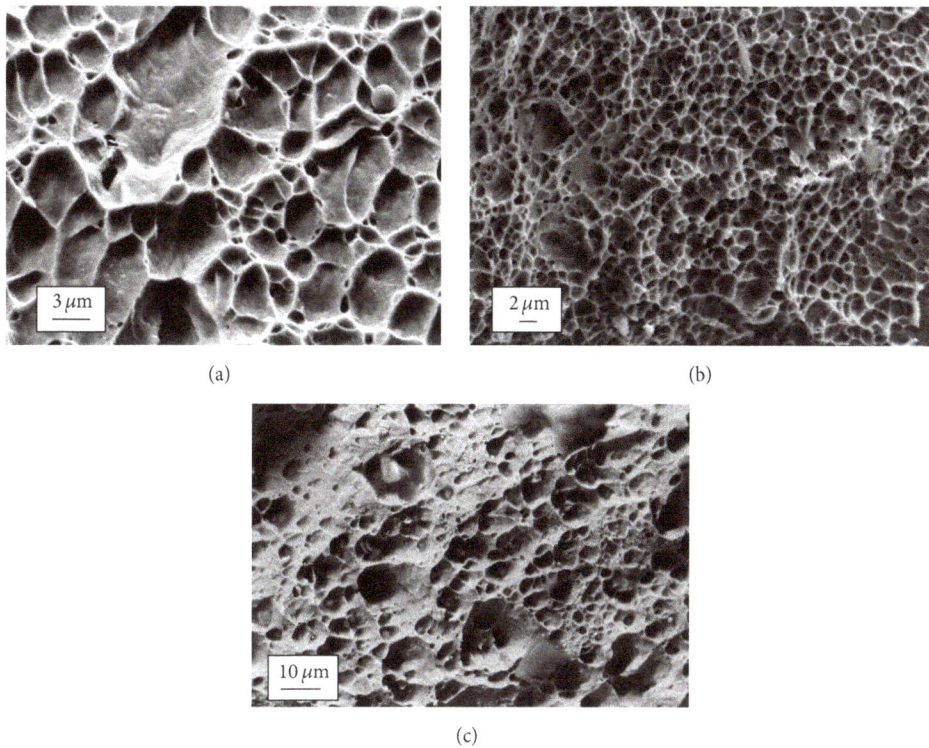

FIGURE 10: The fractography of welded samples: (a) sample 3; (b) sample 4; (c) sample 5.

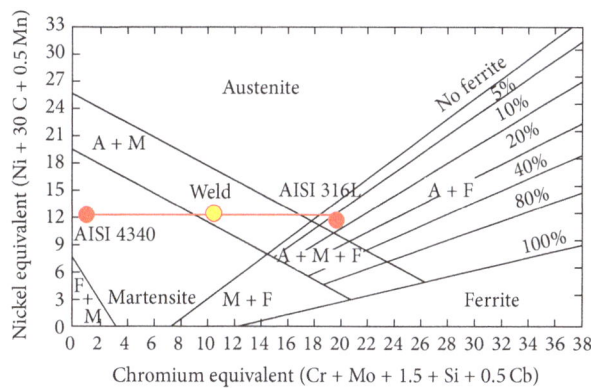

FIGURE 11: The predicted weld microstructure by Schaeffler diagram [14].

the G/R ratio which determines the solidification mode of the microstructure is maximum in the fusion boundary and minimum in the weld center line. Therefore, the microstructure is columnar dendrite in the fusion boundary which changes to equiaxed dendrite structure by reducing the G/R ratio in the center of the weld. This change attributed to fluctuation of hardness in weld center line on the microhardness diagram in Figure 7.

3.5. The Heat Affected Zone (HAZ) Microstructure. According to Figure 17(a), the close zone near fusion line formation of new phase was detected in HAZ area on AISI 316L side. Two phenomena would probably occur: the formation of Cr

carbide and ferrite. In this zone attributed to low percent of carbon, the probability of Cr carbide formation was too low which is in accordance with the results of EDS analyzes depicted in Table 7 with no formation of Cr carbide. However, according to the EDS results in Figure 18 and Table 7, the increase in amount of Cr in white zone comparison with matrix (in Figure 18) will increase the probability of formation of ferrite phase in grain zone. On the other hand, the absence of grain growth in HAZ (the lock austenite grain boundaries by ferrite) makes this phenomenon more feasible and probable.

In order to investigate formation of ferrite phase, the XRD test was done (Figure 19). The results show little percent

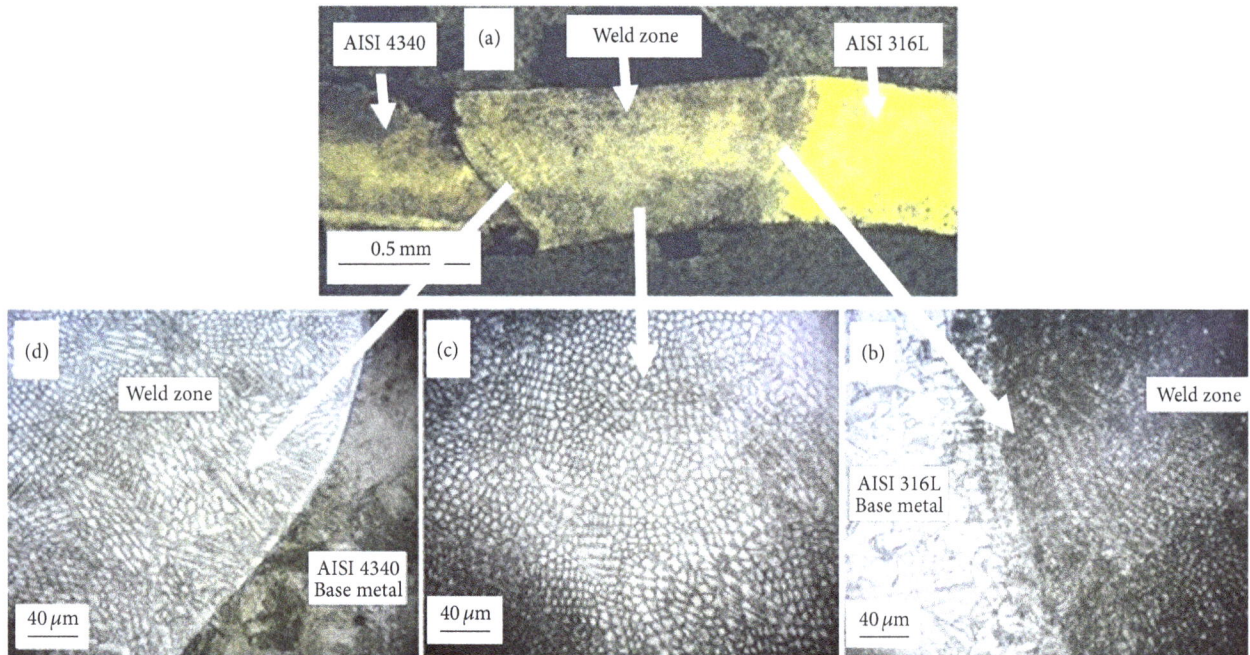

FIGURE 12: The weld microstructure for sample number 3: (a) weld macrograph; (b) near AISI 316L; (c) center of weld; (d) near AISI 4340.

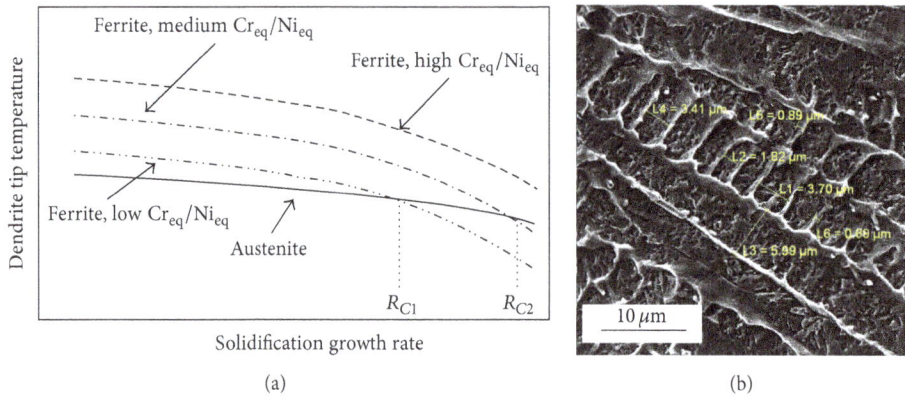

FIGURE 13: The SDAS and DAS sizes for EBW weld zone for sample number 3.

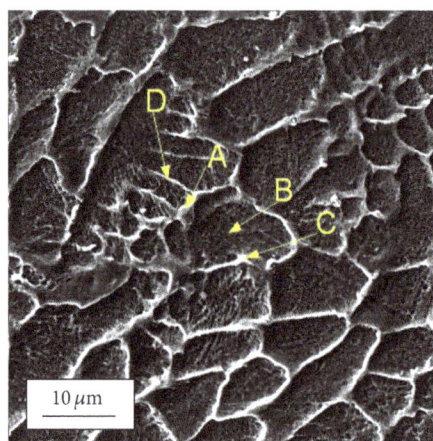

FIGURE 14: The EDS points to investigate microsegregation in subgrain boundary and subgrain spaces in the weld zone (sample number 3).

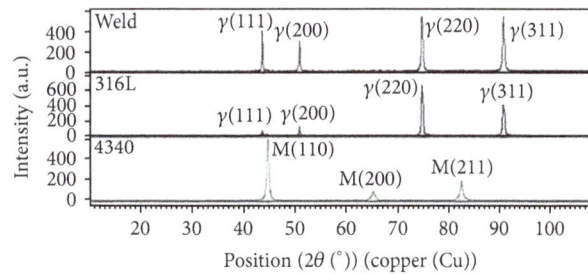

FIGURE 15: The XRD pattern from weld zone (sample number 3) and base metals.

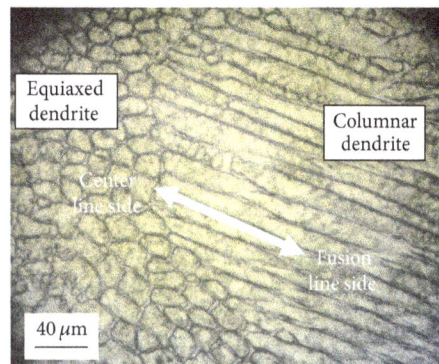

FIGURE 16: The change solidification mode by transition from fusion line to center line.

ferrite was created. This result strengthens the possibility of the formation of ferrite phase in white zone according to Figure 18.

According to Figure 17, the HAZ widths are different in the AISI 316L and AISI 4340 sides, 50 and 10 μm, respectively. The reason was due to using different materials with different heat conduction coefficients and also due to performing various heat inputs. Based on Figure 17, the HAZ width in AISI 316L stainless steel side was 3 or 4 times greater than that of AISI 4340 steel side. The major reason is attributed to the different heat conduction coefficients of AISI 316L stainless steel and AISI 4340 steel. The heat conduction coefficients were 44.5 and 16.3 W/m·K for AISI 4340 steel and AISI 316L stainless steel, respectively. In fact, the heat conduction coefficient of AISI 4340 is approximately 3 times greater than that of AISI 316L stainless steel. Therefore, the heat input induced in the materials leads to producing smaller HAZ size in the AISI 4340 steel side than the HAZ size in the AISI 316 stainless steel side.

Figure 17(b) shows the heat affected zone on AISI 4340 base metal side. The small narrow zone with approximately 10 μm width containing large grains was detected close to fusion line. This attributed to low heat input in electron beam welding process. The softening zone in this sample was too small and can be ignored. These results were in accordance with microhardness results in Figure 7. Moreover according to Figure 17(c), the HAZ width in AISI 4340 side for sample number 5 was 45 μm. The softening zone with coarse microstructure was detected which is in accordance with microhardness results (Figure 7) and tensile test results

(Figures 8, 9, and 10). This could be attributed to high beam current value in sample number 5 as was explained in Sections 3.1.1 and 3.1.2.

4. Conclusions

This paper presents electron beam welding of AISI 316L stainless steel and AISI 4340 low alloy steel. The welding process was carried out by changing the beam current values from 1.6 mA to 4 mA. The depth and width of the weld were increased with beam current. The full penetration weld with 500 μm depth was obtained at 2.8, 3.4, and 4 mA. The microhardness values of weld zone were greater than that in the AISI 316L stainless steel due to fine grain austenitic formation in weld microstructure. In the HAZ near the AISI 4340 steel, the softening zone was observed. The tensile tests results showed that samples 3 and 4 fractured from base metal in AISI 316L side with the UTS values much greater than the UTS values of sample number 5 which fractured from HAZ in AISI 4340 side due to high beam current value and wide softening zone occurred in sample number 5. The full austenitic microstructure was produced in the weld zone. The SDAS measurements showed the high cooling rate during the process. Such cooling rate is in accordance with the formation of austenitic microstructure. The low heat input resulted in formation of small HAZ and the absence of Cr carbide. The study showed that the HAZ width in the AISI 316L stainless steel side was 3 or 4 times greater than that of the AISI 4340 side. The optimum beam current was obtained at 2.8 mA.

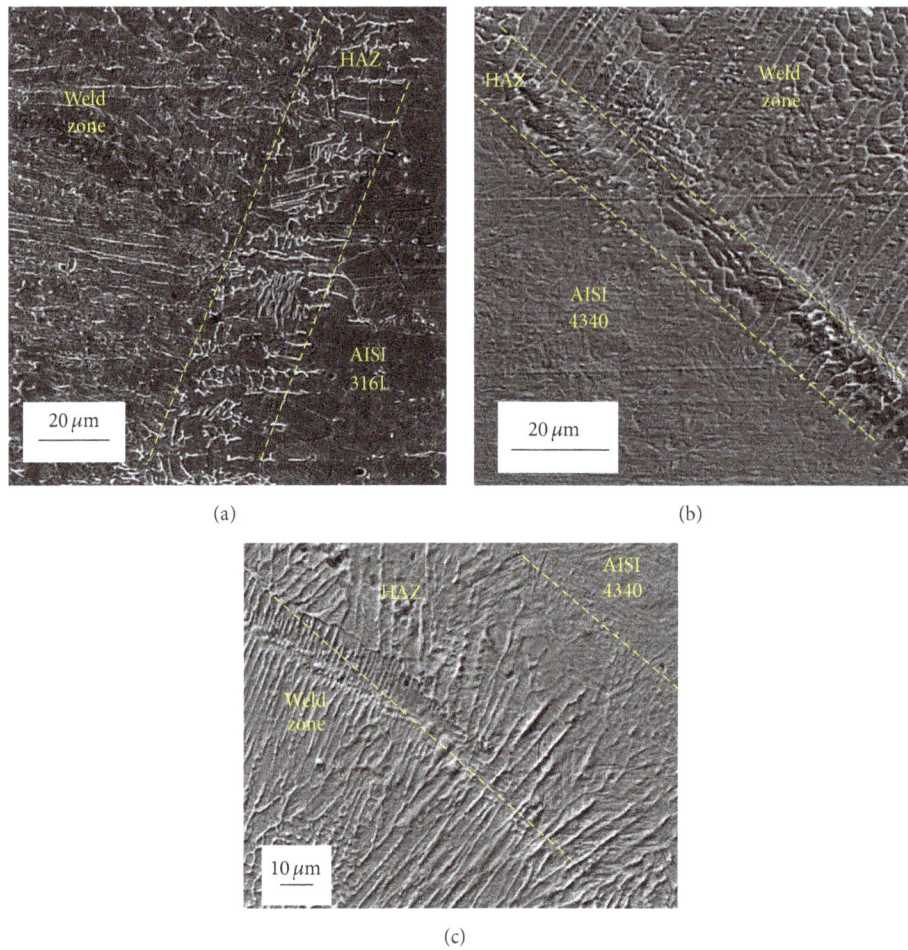

FIGURE 17: The HAZ of welded specimens: (a) AISI 316L side (sample number 3), (b) AISI 4340 side (sample number 3), and (c) AISI 4340 side (sample number 5).

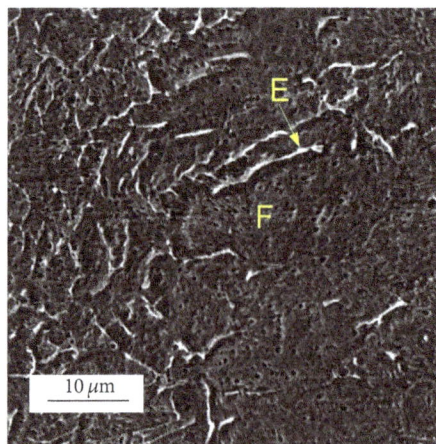

FIGURE 18: The selected region to detect Cr carbide in HAZ near AISI 316L base metal.

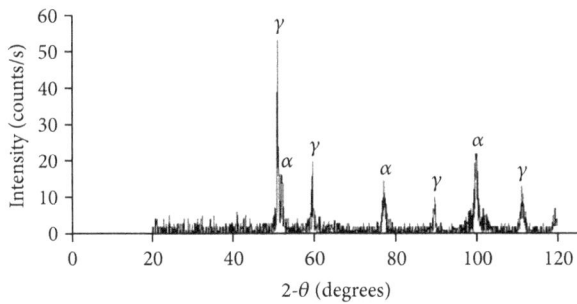

FIGURE 19: The XRD pattern from HAZ in 316L side for sample number 3.

Competing Interests

The authors declare that there are no competing interests regarding the publication of this paper.

References

[1] Z. Sun and R. Karppi, "The application of electron beam welding for the joining of dissimilar metals: an overview," *Journal of Materials Processing Technology*, vol. 59, no. 3, pp. 257–267, 1996.

[2] J. M. Fragomeni and A. C. Nunes Jr., "A study of the effects of welding parameters on electron beam welding in the space environment," *Aerospace Science and Technology*, vol. 7, no. 5, pp. 373–384, 2003.

[3] C. Y. Ho, "Fusion zone during focused electron-beam welding," *Journal of Materials Processing Technology*, vol. 167, no. 2-3, pp. 265–272, 2005.

[4] N. Özdemir, F. Sarsılmaz, and A. Hasçalık, "Effect of rotational speed on the interface properties of friction-welded AISI 304L to 4340 steel," *Materials & Design*, vol. 28, no. 1, pp. 301–307, 2007.

[5] A. Hasçalik, E. Ünal, and N. Özdemir, "Fatigue behaviour of AISI 304 steel to AISI 4340 steel welded by friction welding," *Journal of Materials Science*, vol. 41, no. 11, pp. 3233–3239, 2006.

[6] V. A. Ventrella, J. R. Berretta, and W. De Rossi, "Pulsed Nd:YAG laser seam welding of AISI 316L stainless steel thin foils," *Journal of Materials Processing Technology*, vol. 210, no. 14, pp. 1838–1843, 2010.

[7] B. Kurt, "The interface morphology of diffusion bonded dissimilar stainless steel and medium carbon steel couples," *Journal of Materials Processing Technology*, vol. 190, no. 1–3, pp. 138–141, 2007.

[8] M. Vedani, "Microstructural evolution of tool steels after Nd-YAG laser repair welding," *Journal of Materials Science*, vol. 39, no. 1, pp. 241–249, 2004.

[9] N. Arivazhagan, S. Singh, S. Prakash, and G. M. Reddy, "Investigation on AISI 304 austenitic stainless steel to AISI 4140 low alloy steel dissimilar joints by gas tungsten arc, electron beam and friction welding," *Materials and Design*, vol. 32, no. 5, pp. 3036–3050, 2011.

[10] A. Çalik, M. Karakas, and R. Varol, "Fatigue behavior of electron beam welded dissimilar metal joints," *Welding Research*, vol. 91, pp. 50–52, 2012.

[11] ASTM E8/E8M-08: Standard Test Techniques for Tension Testing of Metallic Materials, 2010.

[12] A. Krishnan, A. R. Poduri, and S. A. Reddy, "Effect of beam current, weld speed and dissolution on mechanical and microstructural properties in electron beam welding," *International Journal of Research in Engineering and Technology*, vol. 2, no. 6, pp. 1120–1129, 2013.

[13] M. Agilan, T. Venkateswaran, and D. Sivakumar, "Effect of heat input on microstructure and mechanical properties of inconel-718 EB Welds," *Procedia Materials Science*, vol. 5, pp. 656–662, 2014.

[14] J. C. Lippold and D. J. Kotecki, *Welding Metallurgy and Weldability of Stainless Steels*, John Wiley & Sons, Hoboken, NJ, USA, 2011.

[15] S. Kou, *Welding Metallurgy*, John Wiley & Sons, New Jersey, NJ, USA, 2nd edition, 2003.

[16] J. M. Vitek, S. A. David, and C. R. Hinman, "Improved ferrite number prediction model that accounts for cooling rate effects—part 1: model development," *Welding Journal*, vol. 82, no. 1, 2003.

[17] J. W. Elmer, S. M. Allen, and T. W. Eagar, "Microstructural development during solidification of stainless steel alloys," *Metallurgical Transactions A*, vol. 20, no. 10, pp. 2117–2131, 1989.

7

The Standard Fire Testing and Numerical Modelling of the Behavior of Calcium Silicate Board Metallic-Framework Drywall Assembly with Junction Box

Yinuo Wang [ID],[1] Ying-Ji Chuang,[2] Ching-Yuan Lin,[2] and Hao Zhang[1,3]

[1]*China Academy of Building Research, Beijing 100013, China*
[2]*Department of Architecture, National Taiwan University of Science and Technology, Taipei, Taiwan*
[3]*University of Science and Technology, Beijing, China*

Correspondence should be addressed to Yinuo Wang; wyn_up@foxmail.com

Academic Editor: Ana S. Guimarães

The metallic-framework drywall is used as the specimens in this research. The standard fire test and finite element simulation were performed once on 300 cm × 300 cm area specimen and twice on 100 cm × 100 cm area specimens, to quantify and evaluate the effect of the junction boxes on the fireproof property after being embedded into the metallic-framework drywall. The results of the experiment show that the temperature of unexposed surface rises faster due to the higher thermal conductivity of the internal metal junction box. The general junction box whose material is PVC can be softened off when heated, affecting the integrity of the firewall and also leading to rapid transfer to the unexposed surface. The prediction of finite element simulation temperature is highly correlated with the results of the real experiment. It is effective to strengthen the original weaknesses by adding a calcium silicate board behind the junction box and using metal panels instead of PVC. The temperature of the temperature junction box surface which is the highest temperature point of unexposed surface decreased most significantly at 72.9°C after the reinforcement. In addition, after reinforcement, the fire resistance time can reach to 1 hour by inserting the junction box into the metallic-framework drywall.

1. Introduction

With the development of architectural technology and fire engineering during the last decade, the construction project tends to develop high-rise and giant buildings. To adapt the tendency, a method of reducing the weight of the building, avoiding the construction risks and shortening the constructional duration, becomes an important issue for architectural engineering. In addition, the usage of thick and heavy building materials such as traditional brick wall or concrete wall must be reduced. For example, different dry metallic-framework wall systems which are expected to replace the traditional thick and heavy building materials appear constantly. These systems have features of optimization of construction methods, short constructional duration, and various constructional methods and light materials whose quality is more stable than concrete. With the gradual popularity of new materials and new constructional methods, such as dry metallic-framework wall system, whether they can achieve a certain time of fire resistance and can be applied in the fire division become more and more important. Whether the building components have the appropriate fire safety should be detected by the standard fire test [1–6]. In addition, they should be applied to the buildings after they are equipped with the capacity of thermal resistance or fire integrity [7]. Based on the test time, these products can be classified by the fire resistance. For example, the firewalls can be classified into 1-hour, 2-hour, or 3-hour fire resistance.

There are many investigated studies on the performance issues of the dry metallic-framework wall partitioning system. Lin et al. [8] investigated the combination of metallic framework and calcium silicate board. Ho and Tsai [9]

proposed that the quality of boards had a great effect on the fire resistance of the wall. Nithyadharan and Kalyanaraman [10] researched on the strength of the connection of screw and calcium silicate board. Chuang et al. [11] came up with the conclusion that the room temperature had a direct influence on the surface temperature when the specimen was tested for fire resistance. Maruyama et al. [12] researched the aging of calcium silicate board and found that its strength weakened as time goes by.

The above research studies on the fire resistance of dry metallic-framework wall are based on the standard fire test experiments. With CAE (computer-aided engineering) increasingly being applied in various engineering fields, as an important part of CAE, CFD (computational fluid dynamics) has been developed rapidly during the last two decades. The principle of CFD is to solve the differential equations of nonlinear simultaneous quality, energy, component, momentum, and scalar with numerical methods. The results of solutions are able to predict the details of movement, heat transmission, mass transmission, and burning, becoming an efficient tool to optimize process equipment and enlarge quantitative design. The basic features of CFD are numerical modelling and computer experiment. Beginning from the fundamental physical theorems, to a large extent, they replaced the expensive equipment for fluid dynamic experiments, greatly influencing the scientific researches and engineering technology.

CFD is mainly applied in cutting-edge designs, such as aerospace design, automobile design, and turbine design. In addition, more and more numerical simultaneous aided researches in building field are processed by making use of CFD. For example, Collier and Buchanan [13] presented the prediction model for fire resistance of drywall by the finite element method; Do et al. [14] came up with that the thermal conductivity of porous material is mainly related to the thermal conductivity of its components and spatial arrangement of its complex structure by formula, microstructure observation, and experiments; Nassif et al. [15] presented the comparison of thermal conductivity of a dry gypsum board wall by the standard fire experiment and by numerical modelling.

According to regulations in different countries or the above researches which focus on the standard fire test experiment or computer simulation, they are only studying and discussing focusing on the wall. Wang et al. [16] once proposed that installing devices, such as the embedded junction box in the wall, could influence the fire resistance of the wall. However, this research focused on the quality control of the board with standard fire test experiment. It does not consider numerical modelling and corresponding quantitative research.

Based on the above foundation, this research takes the dry metallic-framework wall with embedded junction box as the experimental specimens. The quantitative analysis of its fire resistance through a physical experiment and CFD numerical modelling gives improvement measures for its destruction behavior. Aiming at disruptive behaviors, filling the existing gaps, and supplementing the fields are not involved in regulations of various countries at present.

The research conducts a total of 3 standard fire resistance tests and numerical modelling simulations. Test 1 uses the standard of ISO 834-1 [2] to perform the test on a test specimen with a of size 300 cm (width) * 300 cm (height), proposes the numerical model to simulate the process of transient heat transmission, compares the results of computer simulation and the test results, and properly optimizes the digital model parameters. In Test 2, the fire area of the specimen is 100 cm (width) * 100 cm (height), and an embedded junction box in the testing specimen was used to compare the numerical modelling of CFD models. By quantitative analysis, the weakness of the embedded junction box in the wall can be analyzed, which influences the fire resistance, and the reinforcement scheme was proposed. Test 3 simulates the feasibility of the reinforcement scheme by CFD numerical modelling, tests on the specimen of which fire area is 100 cm (width) * 100 cm (height), and verifies the feasibility of numerical modelling and reinforcement scheme by the use of the standard fire test.

2. Experimental Details

2.1. Fire Test Furnaces. In this research, two sets of test equipment are applied, and both can conduct material testing horizontally or vertically. The large test furnace of the first equipment is 300 cm (width) * 300 cm (height) * 240 cm (depth) (Figure 1). The small test furnace of the second equipment is 120 cm (width) * 120 cm (height) * 120 cm (depth) (Figure 2). They both adopt the electronic ignition, and the control system used is the computer PID temperature controller. There are 8 burners in the large test furnace among which only 4 are switched on for the wall test. In the furnace, there are two thermocouples to, respectively, control the operation of 2 high-speed burners on two sides and other 7 thermocouples are to measure the temperature in the furnace. All thermocouples are inserted from the top of the test furnace. There are 4 burners in the small test furnace. When the wall is tested, only 2 burners which are close to the wall are opened. In the furnace, there are two thermocouples to, respectively, control the operation of 1 high-speed burner on two sides and other 2 thermocouples are to measure the temperature in the furnace. All thermocouples are inserted from the left side to right side of the test furnace. The ceramic wool is paved around and the top of the furnace wall of which maximum temperature is 1400°C and density is 240 kg/m^3. The furnace bottom is made by the adiabatic board of which thermal resistance is 1400°C and density is 1140 kg/m^3. The refractory mortar is applied in the gap and connection of adiabatic boards. The exterior body of the test furnace is made by steel plate and steel frame. At the back of the test furnace, there is an air outlet for exhaust air, and it is connected to the outdoor chimney. All the thermocouples are 10 cm away from the fire testing area of the specimen. The temperature in the furnace is measured by a K-type thermocouple of which specification conforms to the regulation of CNS 5534 [17] that the thermocouple shall possess property above 0.75 Grade. The thermocouple wire is covered by the heat-resistant stainless steel pipe of which the diameter is 6.35 mm

(a)   (b)

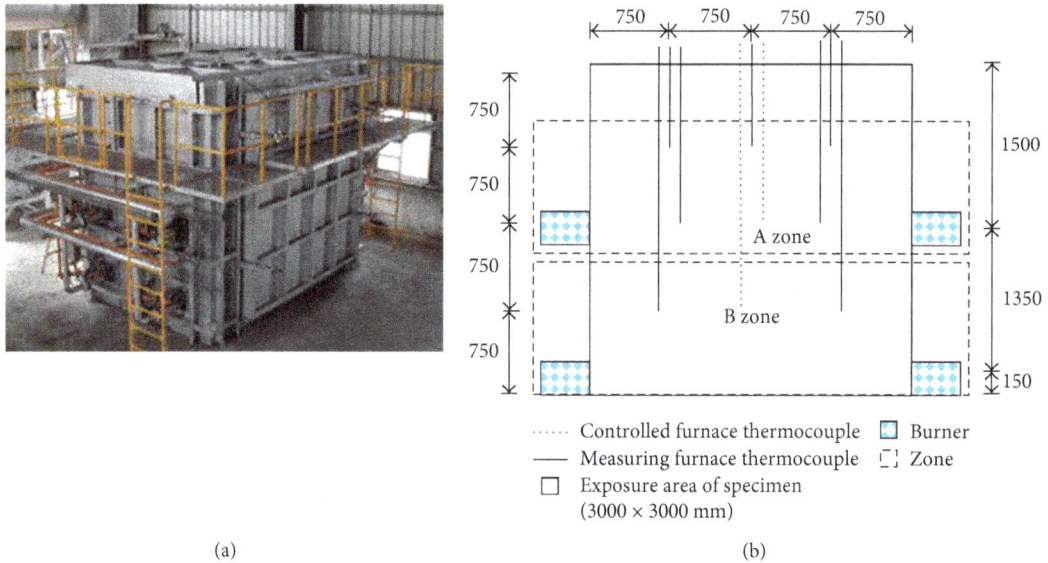

FIGURE 1: Full-size high-temperature furnace (inner size: 300 cm in width, 300 cm in height, and 240 cm in depth).

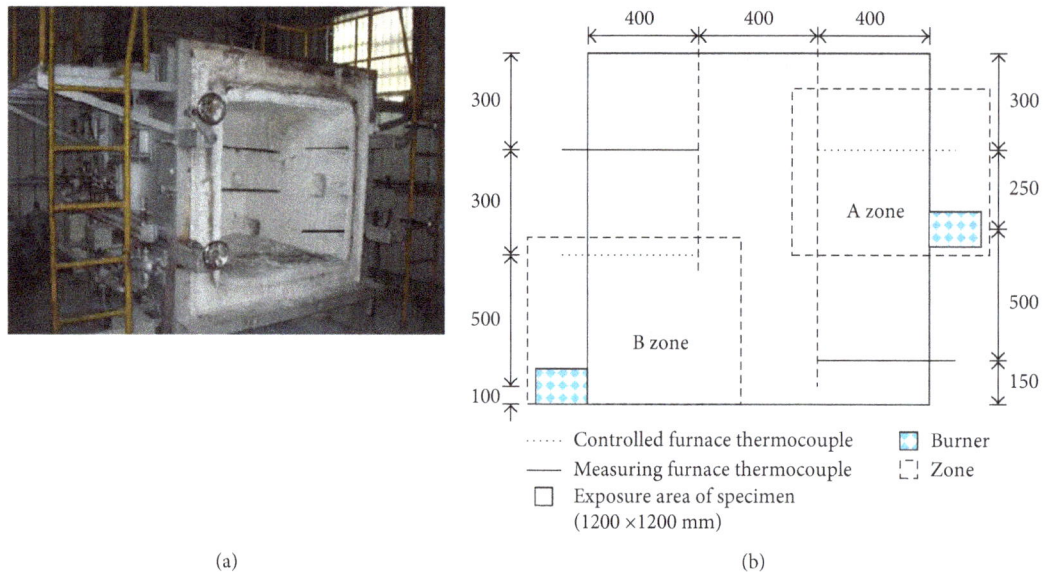

(a)   (b)

FIGURE 2: Small-size high-temperature furnace (inner size: 120 cm in width, 120 cm in height, and 120 cm in depth).

(16 gauge). In addition, the heat-resistant stainless steel pipe is placed in the insulated stainless steel pipe with an inner diameter of 14 mm and the front end is open and the hot junction of front end extrudes 25 mm. All the thermocouples in the furnace have been placed in the environment at a temperature of 1000°C for 1 hour [2] before the first use to increase the sensitivity of measuring the temperature, and the accuracy requirement is ±3%. All instrument signals are connected to the DS600 data recorder first, and DS600 processes and converses the signal to DC 100. At last, the data capture recorder converses the signal and outputs to the ThinkPad W540 laptop by a network cable. The data capture recorder is set to record once in every 6 seconds.

2.2. Test Specimens. The material used in the research is 9 mm calcium silicate board which is an erect blanking plate and fixed by a self-tapping screw. The self-tapping screw's diameter, length, and distance are 3.5 mm, 25.4 mm, and 250 mm, respectively. Its column is $65 \times 35 \times 0.6$ mm C-shaped steel, and the upper and down channels are $67 \times 25 \times 0.6$ mm C-shaped steel. The distance of the intermediate column is 406 mm, and the distance of the column away from two sides is 297 mm. The thickness of mineral wool is 50 mm, and a density of 60 kg/m^3 is applied to the material. For the size of the embedded socket, the external switch panel is 120 mm × 70 mm, and the internal junction box is $101 \times 55 \times 36$ mm. There are two kinds of external switches. In Test 2, the material of the switch panel

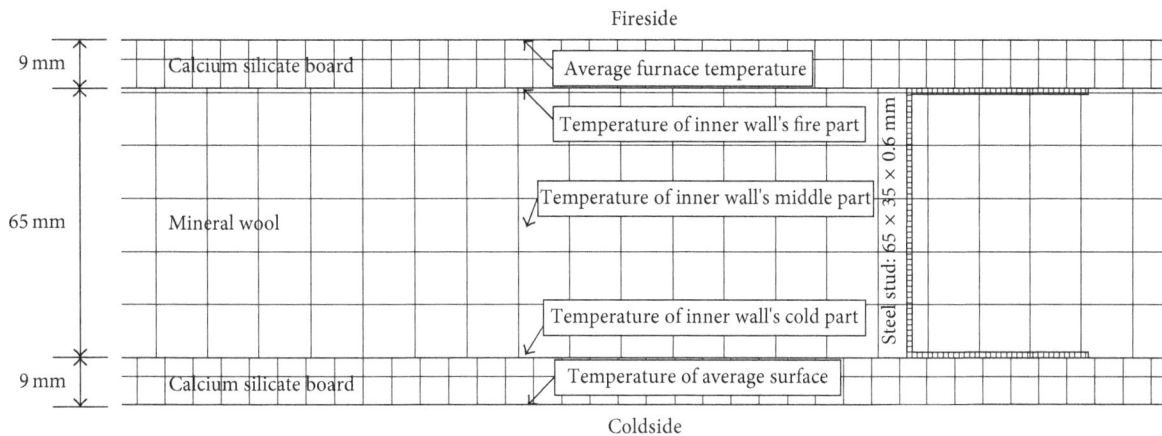

FIGURE 3: Cross section of the wall with indication of numerical modelling and the thermocouple position in Test 1.

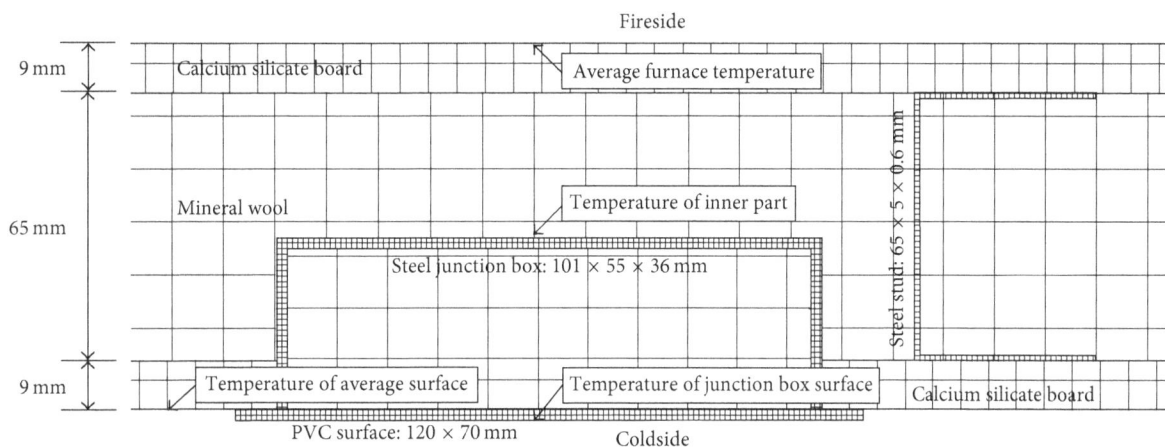

FIGURE 4: Cross section of the wall with indication of numerical modelling and the thermocouple position in Test 2.

is PVC (polyvinyl chloride), and the internal material is galvanizing steel box. In Test 3, the material of the external switch panel is steel, and the internal material is the galvanizing steel box. The junction box in Test 2 is not equipped with any reinforcement measures. Besides that the steel is selected in Test 3 as the material of the external panel, 9 mm calcium silicate board is added behind the junction box which is close to the fire source.

Three 60-minute standard heating tests were performed in the research. Test 1 is a standard test in which the specimen size is 3 m (height) × 3 m (width) and the density of fireproof cotton is 60 kg/m³ as shown in Figure 3. Test 2 is a standard test with small high-temperature furnace in which the specimen size is 1 m (height) × 1 m (width) and the density of fireproof cotton is 60 kg/m³. The socket junction box is embedded in the unexposed surface of the specimen. In addition, the material of the switch panel is PVC (polyvinyl chloride), and the internal material is the galvanizing steel box as shown in Figure 4. Test 3 is a test with small high-temperature furnace in which the specimen size is 1 m (height) × 1 m (width) and the density of fireproof cotton is 60 kg/m³. The socket junction box is embedded in the unexposed surface of the specimen. In addition, the material of the switch panel is steel, and the internal material

is the galvanizing steel box. A 9 mm calcium silicate board is added at the back of the box, as shown in Figure 5. Because there is no limit for the height that the socket junction box should be placed in the wall, this research expects to observe and simulate the most typical model in reality. In addition, according to the regulations of ISO 834-1 [2] that the weakness of the specimen shall be located in the center, the specimen of this research is 1 m (height) × 1 m (width) and the socket junction box is placed in the position 55 cm away from the ground. The furnace pressure is lower when it is more close to the bottom of the furnace. In short, the furnace pressure increases linearly as the height of the specimen increases. However, the furnace pressure is the negative pressure when it is 50 cm below the ground. As a result, the socket junction box is placed in the positive pressure position. As the test expects to verify the similarity of results of numerical modelling and test by Test 1, it takes the full-scale standard test of 3 m (height) × 3 m (width). The research expects to discuss the devastation that the embedded electronic junction box affects the components of wall, propose the reinforcement measures combining with the result of numerical modelling, and verify these measures through Test 2 and Test 3. As a result, it selects the test with small high-temperature furnace, and the size of the specimen is 1 m (height) × 1 m (width).

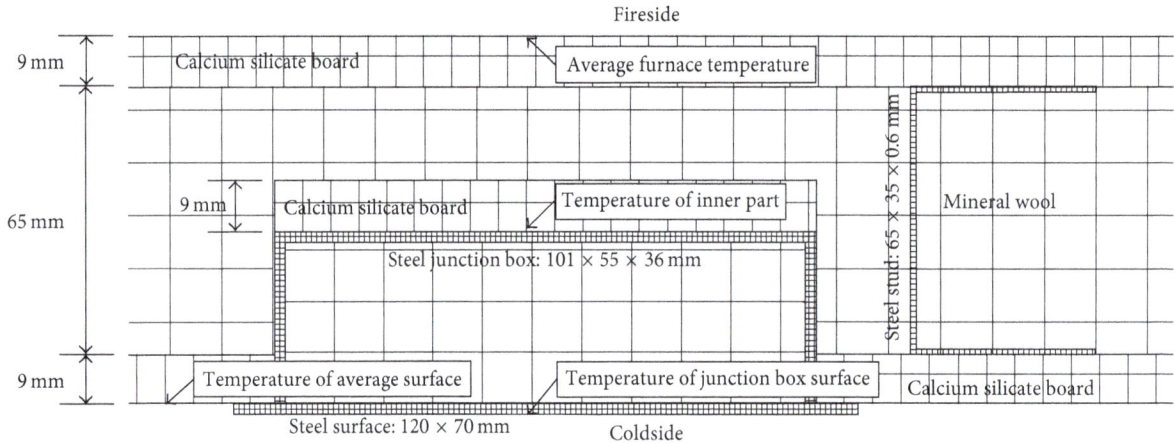

FIGURE 5: Cross section of the wall with indication of numerical modelling and the thermocouple position in Test 3.

2.3. Test Conditions. Test 1 follows the ISO 834-1 [2] standard. The size of the fire testing specimen is 3 m (height) × 3 m (width), and the zero pressure of the test furnace is at the height of 50 cm from the furnace bottom. As a result, according to the regulation of ISO 834-1 [2], 8 Pa should increase as the height increases every 1 m and the furnace pressure on the top of the specimen should not exceed 20 Pa; the standard heating curve of the test furnace is calculated using the following equation, and the furnace pressure measurement is recorded by a computer every 6 seconds:

$$T = 20 + 345 \times \log_{10}(8t + 1), \tag{1}$$

where T: average standard furnace temperature (°C) and t: time (min).

The heating temperature of Test 2 and Test 3 is adopted from the ISO 834-1 [2] standard heating curve, and the furnace pressure of the specimen is also set that the zero pressure is at the height of 50 cm from the furnace bottom. As a result, according to the regulation of ISO 834-1 [2], 8 Pa should increase as the height increases every 1 m and the furnace pressure on the top of the specimen is 4 Pa, and the pressure of the junction box is about 0.8 Pa.

2.4. Measurement and Recording of the Temperature of Standard Fire Test of Specimens. Test 1 sets 8 thermocouples on the unexposed surface of the specimen, as shown in Figure 6. According to the requirements of ISO 834-1 [2], temperature distribution is observed. Because it is required to compare the similarity of results of numerical modelling and tests and optimize the computer model, three measuring points are set in the middle layer of the wall and the measuring points are, respectively, in the position of 9 mm, 41 mm, and 74 mm, as shown in Figure 3. In Test 2 and Test 3, they, respectively, install the thermocouples on the unexposed surface of the specimens, as shown in Figure 7; four thermocouples are, respectively, located in the center of unexposed surface of the specimen, one in the center of wall, one on the panel of junction box, and another one in the center of mineral wool. The temperature measurement is recorded once in

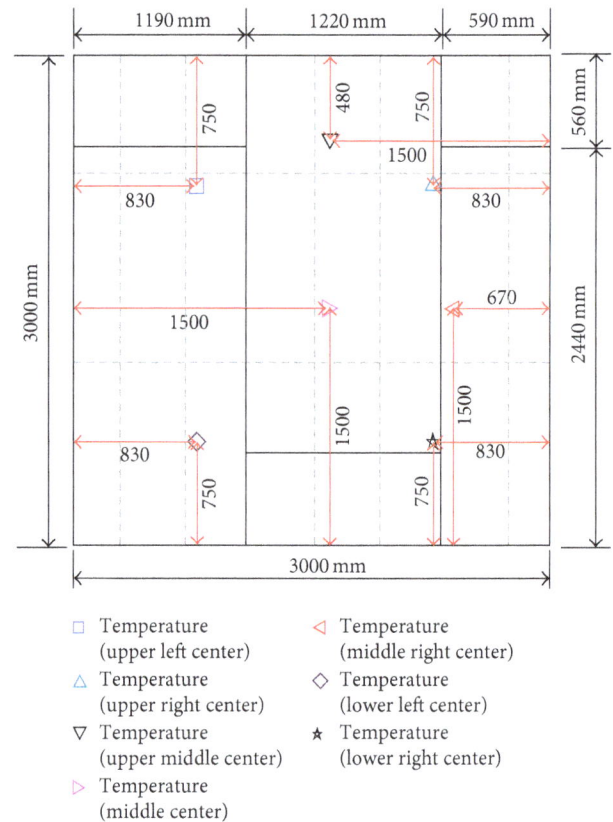

FIGURE 6: Test 1: the unexposed surface of the specimen and geometry of the thermocouple.

every 6 seconds by a computer, and it is recorded by photographs during the test.

3. Numerical Modelling

3.1. General Modelling. In this study, numerical modelling is based on CFD technology for a series of computer simulation analysis, the use of software fluent to solve [18]. It can be roughly divided into three parts: preprocessing, solve, and postprocessing. Preprocessing mainly focuses on how to build the geometric model and mesh. This research needs to

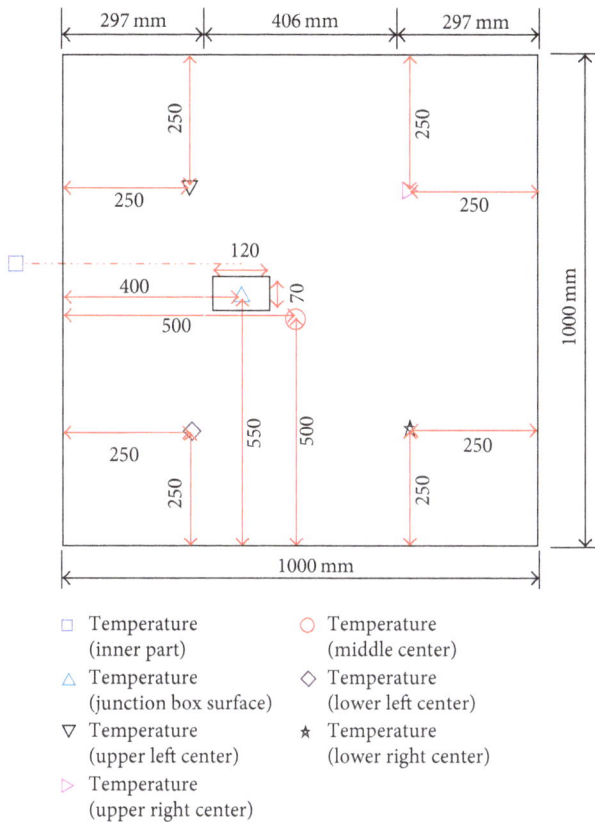

FIGURE 7: Test 2 and Test 3: the unexposed surface of the specimen and geometry of the thermocouple.

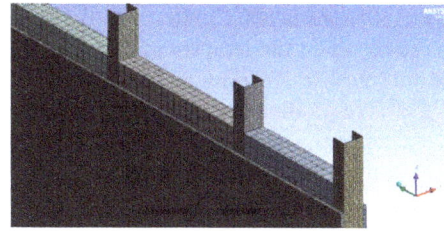

FIGURE 8: Numerical model of meshing.

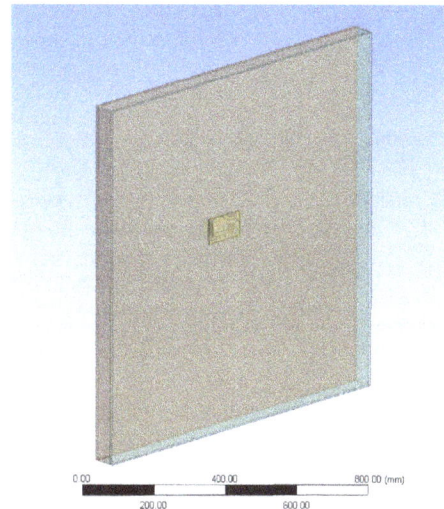

FIGURE 9: Numerical geometric model of wall embedded with the junction box.

get the computer model which matches the standard fire test result, so the geometric model is designed corresponding to the standard fire test of the specimen. The numerical modelling takes the finite element numerical analysis in which the principle is to divide the solution domains into several interrelated subdomain units, assume a proper approximate solution to every unit, deduce the general satisfied conditions of the domain, and work out the solution of the question. Therefore, it is necessary to build the geometric model and mesh it into several units. Based on the shape and size of building components, Test 1 meshes it by hexahedron, as shown in Figure 8. The meshing sizes of the calcium silicate board, steel stud, and fireproof cotton are $9 \times 9 \times 9$ mm, $0.6 \times 0.6 \times 5$ mm, and $25 \times 25 \times 25$ mm, respectively. The geometric model in Test 2 and Test 3 is added with the embedded junction box, as shown in Figure 9. In addition, this part is added to the meshing. The meshing size is $0.5 \times 0.5 \times 0.5$ mm, and the other part is the same as Test 1. Solve mainly includes how to set up the related material parameters, set boundary conditions, and select the mathematics model and calculation methods. Postprocessing aims to analyze the data solved by modelling.

3.2. Parameter Setting. The materials used in the test include steel, calcium silicate board, PVC, and mineral wool. The parameters involved in the modelling material are specific heat (J/kg°C), thermal conductivity (W/m°C), and density

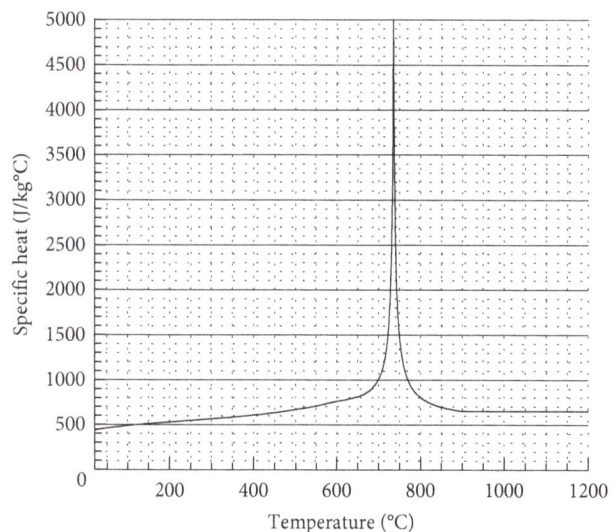

FIGURE 10: Specific heat for steel at elevated temperatures (BS EN 1993).

(kg/m^3). The parameters used in steel refer to the regulations of BS EN 1993 [19]. In addition, the specific heat of steel linearly increases as the temperature rises, as shown in Figure 10. The thermal conductivity of steel is 53.3 W/m°C when the temperature is 20°C, and it rises to 27.4 W/m°C when the temperature is 800°C. However, the thermal conductivity is stable when the temperature is more than 800°C. The density of steel is 7850 kg/m³. Walker and Pavía [20] performed research on the thermal performance of a series of insulation materials. According to their related data, the specific heat of the calcium silicate board is 819.4 J/kg°C. The

regulation of GB/T 10699-1998 [21] stipulates the thermal conductivities of the calcium silicate board at various temperatures and other parameters. Its thermal conductivity is shown in the following equation, and the thermal conductivity of the calcium silicate board linearly increases as the temperature rises, as shown in Figure 11:

$$\lambda = \begin{cases} 0.0564 + 7.786 \times 10^{-5} \times t + 7.8571 \times 10^{-8} \times t^2 & (t \leq 500°C) \\ 0.0937 + 1.67397 \times 10^{-10} \times t^3 & (500 < t \leq 800°C), \\ 0.179 & (t > 800°C) \end{cases} \quad (2)$$

where λ: thermal conductivity (W/m°C) and t: temperature (°C).

In this research, the PVC (polyvinyl chloride) panel of the junction box used in Test 2 conforms to the regulations of CNS 3142 [22]. According to the related researches made by Mansour et al. [23], the values of the specific heat, thermal conductivity, and density of PVC are 900 J/kg°C, 0.16 W/m°C, and 1380 kg/m³, respectively. In this research made by Nassif et al. [15], it presents some material property parameters related to mineral wool. Combining the specifications of mineral wool used in this research, the specific heat, respectively, is 840 J/kg°C and thermal conductivity increases as the temperature rises [24], as shown in Table 1. The density of mineral wool is, respectively, 60 kg/m³. The specific setting is shown as Table 2.

4. Results and Discussion

4.1. Test Results. The test time for Test 1 lasted 60 minutes. After 9 minutes of the test, trace smokes with abnormal smell burst out above the unexposed surface of the specimen and the seams of framework. At this time, all temperatures of measuring points obviously tend to rise (Figure 12). Until the 14th minute, the temperature of the measuring point on unexposed surface tends to decline until the 35th minute. From the 35th minute to the end, the temperatures rise all the time. The temperature of the inner wall's fire part rises rapidly after 9 minutes, and it slowly rises to the end after 22 minutes. At the end of test time, the temperature of the measuring point is 738.1°C. The temperature of the inner wall's middle part rises rapidly after the test begins till 9 minutes, and the rise begins to slow down towards the end after 38 minutes. At the end of test time, the temperature of the measuring point is 487.8°C. The situation of temperature of the inner wall's cold part is generally the same to that of the temperature of the side wall within the first 18 minutes. After that, the temperature gradually goes up to the end, and the final temperature is 316.5°C. At the 21st minute, a transverse crack appears on the upward side of the left board of the unexposed surface, and the crack extends to the center at 38th minute. When the test time is over, the temperature in the upper left center is the highest one (104.7°C) among temperatures on the unexposed surfaces, and the highest average temperature was 97.5°C (Figure 13), which does not exceed the stipulated fire

resistance given in regulations of ISO 834-1 [2]. After the test, the integrity of the unexposed surface of the specimen is still good (Figure 14). Therefore, the specimen meets the demand of 60-minute fire resistance.

The test time of Test 2 lasted 60 minutes. After 6 minutes of the test, some smokes burst out and all temperatures of measuring points obviously tend to rise. Until the 37th minute, the temperatures of the thermocouple steadily increase, and the temperature junction box surface rises most sharply. It can be found that the embedded junction box has a great effect on the fire resistance of the metallic wall system. After that, the temperature significantly decreases and maintains around 40°C until the end of test. This is because the external panel of the junction box is PVC. According to the regulation of CNS 3142 [22], the softening temperature of PVC is not less than 73°C. The temperature of the thermocouple is higher than 100°C at the 28th minute, and the external panel of the junction box softens comprehensively (Figure 15) and falls out. This thermocouple is placed on the panel of the junction box, so the recorded temperature tends to plunge. Other temperatures of measuring points increase steadily. Among them, the temperature of the upper left center rises most significantly. This is because the furnace pressure is in a rising trend, and the temperature of the upper specimen is higher than that of the below specimen. The junction box is placed left-to-center, so the hot gas is released to the unexposed side from the weak surface after the weak surface is destroyed. After that, the hot gas rises rapidly making the temperature of the thermocouple higher than other temperatures. Until the 57.6th minute, the temperature of the upper left center is 207.8°C (Figure 16) and its initial temperature is 25°C, increasing by 182.8°C. According to ISO 834-1 [2], when the highest temperature of the unexposed surface is higher than the initial temperature by 180°C, it can be judged that the fire resistance is destroyed. After the test, the junction box panel on the unexposed surface of the specimen falls out, so the integrity is destroyed (Figure 17). Therefore, this specimen does not meet the demand of original 60 min fire resistance.

The test time of Test 3 lasted 60 minutes. After 6 minutes of the test, some smokes burst out and all temperatures of measuring points obviously tend to rise. After 15 minutes, the temperature of the inner part increases more obviously and the temperature is 463.4°C at the end of the test. All

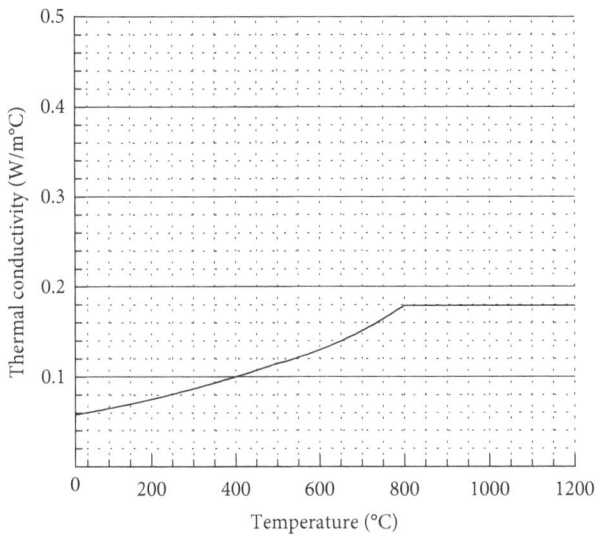

FIGURE 11: Thermal conductivity for the calcium silicate board at elevated temperatures.

TABLE 1: Thermal conductivity of the mineral wool.

Mineral wool					
Temperature (°C)	23.9	93	149	260	371
Thermal conductivity (W/m°C)	0.038	0.045	0.057	0.069	0.082

TABLE 2: Thermal properties.

Material	Specific heat (J/kg°C)	Thermal conductivity (W/m°C)	Density (kg/m³)
Steel	According to Figure 10	20°C, 53.3 $T \geq 800$°C, 27.4	7850
Calcium silicate	819.4	According to (2)	1350
PVC	900	0.16	1380
Mineral wool	840	According to Table 1	60

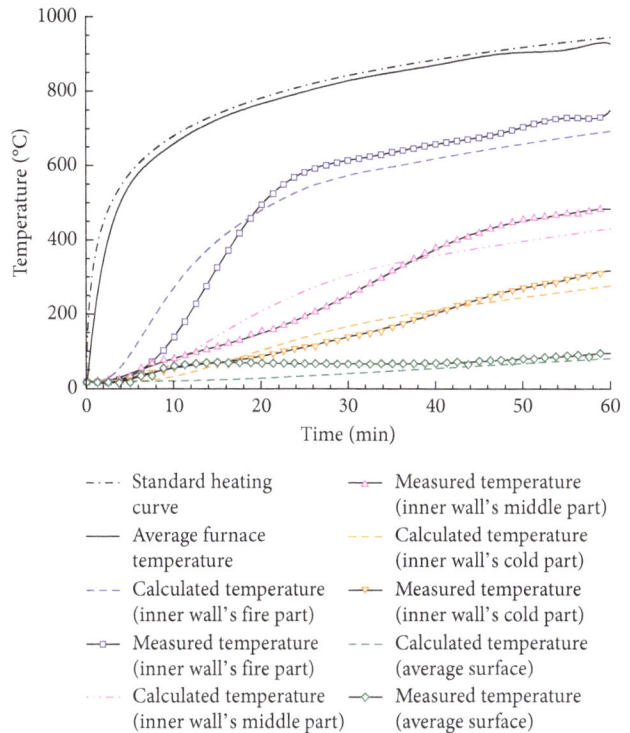

FIGURE 12: Measured temperatures compared to the calculated values in Test 1.

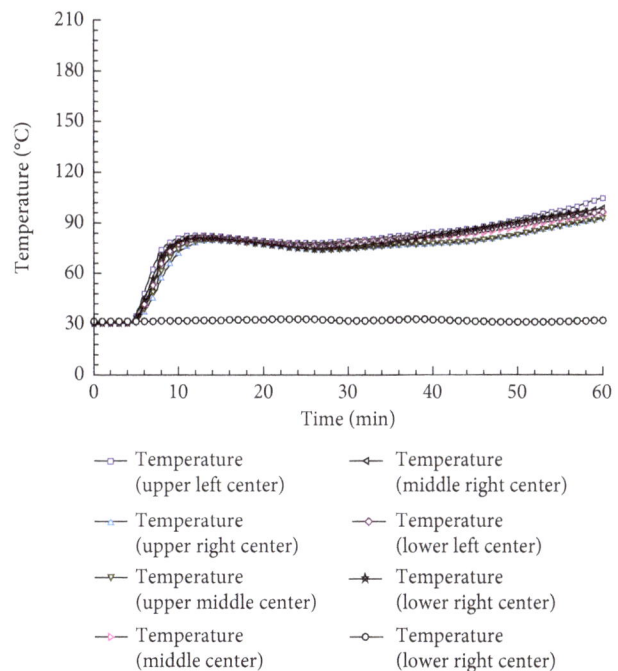

FIGURE 13: Time-temperature chart for the specimen in Test 1.

temperatures steadily rise. From 20th to 30th minute, the temperatures rise more sharply. After that, they tend to be stable until the 48th minute. From the 48th minute, the rising range of the temperature becomes greater towards the end. Among the temperatures, the temperature of the junction box surface rises most obviously, followed by the temperature of the upper left center. It is because the junction box fire resistance performance is still weaker than the surrounding integrated wall although it has been strengthened. Therefore, the junction box temperature is higher than other places. At the end of test, the temperature of the junction box surface is highest on the unexposed surface which is 198.2°C (Figure 18). The initial temperature is 25°C, increasing by 173.2°C. The highest average temperature of all the measuring points is 136.5°C, increasing by 111.5°C. According to ISO 834-1 [2], the highest temperature rise of unexposed surfaces should not be higher than 180°C from the initial temperature and the

average temperature rise should not be higher than 140°C. During the test, these temperatures did not exceed the value given in the regulation of ISO 834-1 [2]. After the test, the integrity of unexposed surface of the specimen is still good (Figure 19). Therefore, the specimen meets the demand of 60-minute fire resistance.

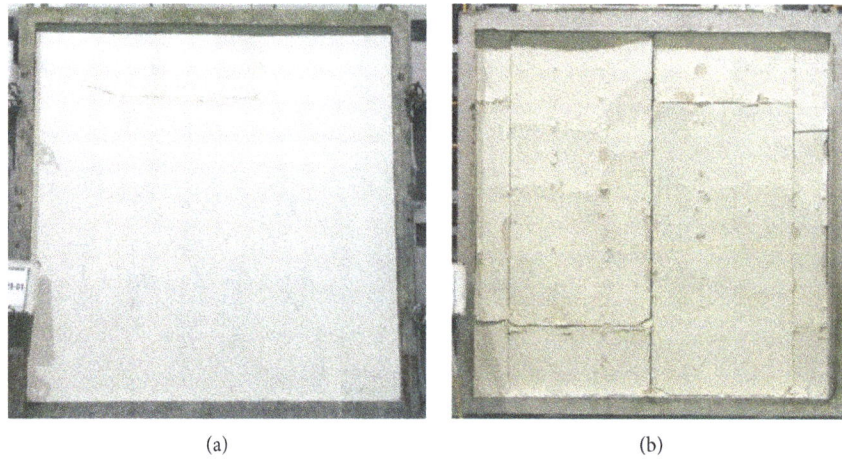

(a)   (b)

FIGURE 14: The result of the specimen after 60 min standard fire test in Test 1. (a) Unexposed surface. (b) Exposed surface.

4.2. General Discussion. Test 1 is the full-scale standard test with 3 m (height) × 3 m (width), and its standard fire test specimen and measuring temperatures of CFD finite element numerical modelling are in accordance with the test. The results obtained by comparing numerical modelling and standard fire test are shown in Figure 12. The heating curve of the test furnace is in accordance with the modelling standard heating curve which shows that the standard fire test temperature confirms with the regulation of ISO 834-1 [2]. In the standard fire test, the temperature of the inner wall's fire part rises faster in the first 22 minutes of the test and rises slower from 22nd minute to 60th minute. Similarly, the heating curve of numerical modelling on this measuring point is relatively ideal, which is basically coincident with the escalating trend of this measuring point in the standard fire test. In the first 20 minutes, the temperature of the inner wall's middle part rises slower, and after that, it rises faster. It is because at the 21st minute, a transverse crack appears on the upward side of the left board of unexposed surface and causes the temperature to rise faster. Because the numerical modelling is in an ideal condition, the numerical modelling temperature on this point rises slowly and continuously, so there will be a curve alternating situation of temperature slope and solid test temperature slope. However, the overall escalating trends are highly coincident. The temperature of the inner wall's cold part is away from the fire source, so its temperature heating curve is gentler than previous two temperatures, and it is highly correlated to the numerical modelling temperature. In the standard fire test, trace smokes with abnormal smell burst out above the unexposed surface of the specimen and the seams of framework at the 9th minute, so the average temperature on the unexposed surface greatly rises but tends to rise slower, corresponding with the numerical modelling temperature. Through the comparison, it can be found that the temperature ascending curves of different measuring points of numerically predicted values are highly correlated with the data of all temperature measuring points in the standard fire test in 60 minutes.

From Test 1, it can be found that the specimen without the junction box can meet the demand of 1-hour fire

FIGURE 15: The external panel of the junction box softens at 28th minute in Test 2.

resistance and the CFD numerical modelling result is highly correlated with the standard fire test. Test 2 expects to focus on exploring the effect of the specimen with the embedded junction box on the fire resistance and whether its related numerical modelling has correlation with the standard fire test result. Therefore, in Test 2, the specimen size is designed to be 1.0 m (height) × 1.0 m (width) and the junction box is embedded into the unexposed surface. The external panel of the junction box is PVC (polyvinyl chloride). In addition, the numerical model is added to the junction box, including the construction of the internal box and the external panel. The comparison of the standard fire test result and the numerical modelling result is shown in Figure 20. In the figure, it can be seen that the temperature heating curve of the test high-temperature furnace is very close to the standard heating curve in 60 minutes. It shows that the temperature of the burning furnace in the standard fire test meets the demand of ISO 834-1 [2]. The temperature value of the inner part in the standard fire test has limited difference from the numerical modelling temperature value on

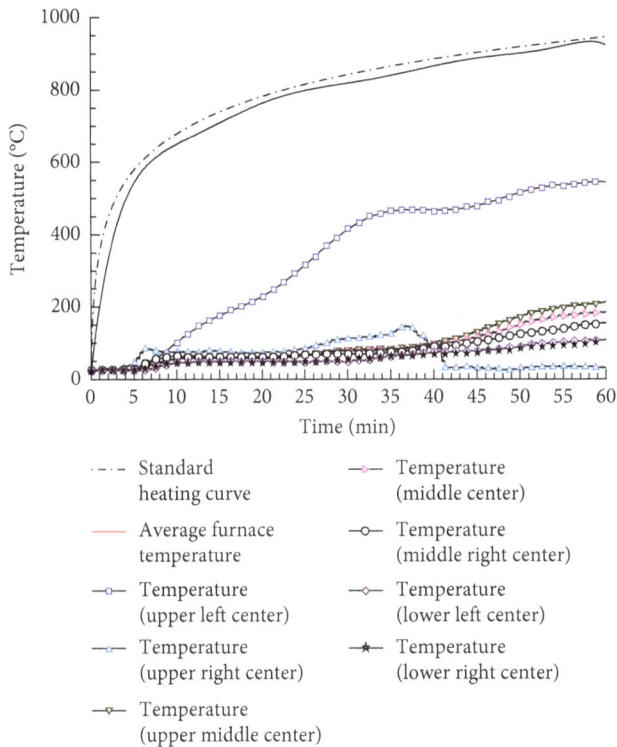

FIGURE 16: Time-temperature chart for the specimen in Test 2.

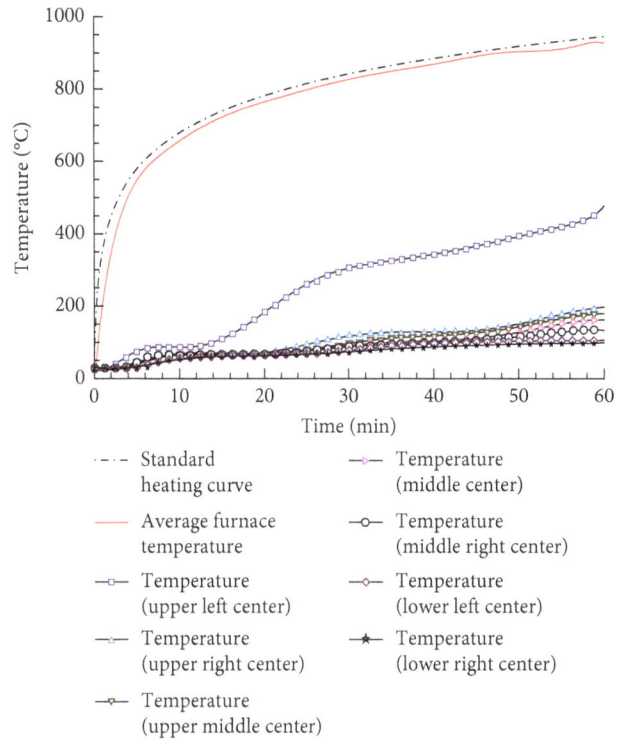

FIGURE 18: Time-temperature chart for the specimen in Test 3.

FIGURE 17: The result of the specimen after the 60 min standard fire test in Test 2.

FIGURE 19: The result of the specimen after the 60 min standard fire test in Test 3.

this point, and their escalating trends are basically corresponding and highly correlated. The actual temperature of the junction box surface is highly correlated with the temperature of the calculated junction box surface from test beginning to 37th minutes. After that, the test recorded temperatures dramatically reduce until the end of the test. This is because the external panel of the junction box is PVC. According to the regulation of CNS 3142 [22] that the softening temperature of PVC is not less than 73°C, the temperature of the thermocouple is more than 100°C at 28th minutes and the external panel of the junction box softens

comprehensively and falls out at 37th minute. As a result, the test temperature dramatically reduces after that, which is obviously different from the calculated temperature after 37 minutes. The average surface temperature and calculated average surface temperature's escalating trends are highly correlated. By comparison, it can be found that the standard fire test result is in accordance with the numerical modelling result in Test 2. Through the observation of the standard fire test and prediction of numerical modelling, it can be seen that after the junction box is embedded in the wall member, it will cause damage in two aspects, resulting in the

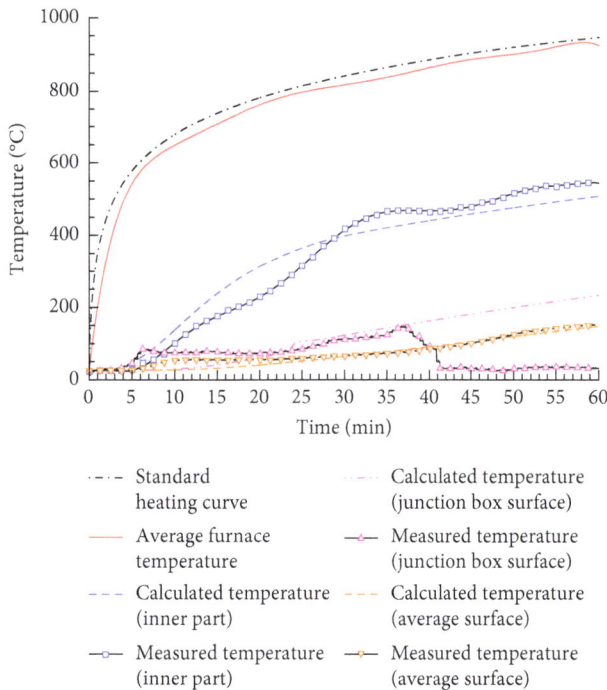

FIGURE 20: Measured temperatures compared to the calculated values in Test 2.

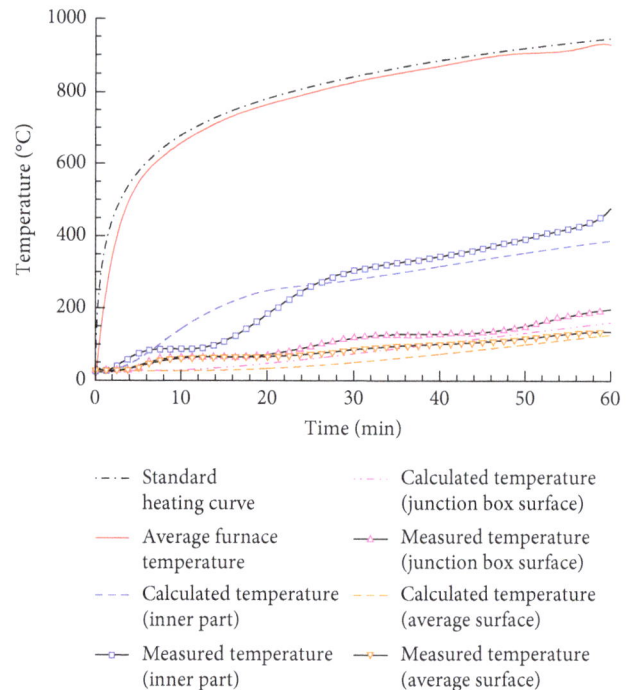

FIGURE 21: Measured temperatures compared to the calculated values in Test 3.

component failing to reach the original 1-hour fire resistance limitation. First, the material of the junction box is metal and its thermal conductivity is much higher than that of mineral wool and calcium silicate board, so the heat transmits to the unexposed surface faster. Second, the external panel material of the junction box is PVC and it begins to soften when the temperature is at 80~100°C. When the temperature rises, the panel falls out and the external junction box exposes. The material of the junction box is metal and it has high thermal conductivity, so it destroys the fire resistance of component.

The result of Test 2 shows that the position of the embedded junction box has a higher temperature than other parts and that the PVC junction box panel softens and falls out in heat which destroys the integrity of wall. According to above weaknesses, Test 3 proposes related reinforcement measures. Considering the density of the calcium silicate board as 1350 kg/m^3 and the mineral wool as 60 kg/m^3, the fire resistance performance of the calcium silicate board is much better than that of the mineral wool under the same thickness, although its thermal conductivity is slightly higher than the mineral wool. Therefore, a reasonable reinforcement measure is adding the calcium silicate board to the back of junction box and changing the PVC material to metal material. After these adjustments, modify the numerical models and parameters. When the modelling result meets the demand of 1-hour fire resistance, the standard fire test can be processed to verify. The final comparison of the standard fire test result and numerical modelling result is shown in Figure 21. According to the regulation of GB/T 10699-1998 [21], the water content of the calcium silicate board is not more than

10%. Therefore, the board contains certain water. The specific heat of water is higher and the heating temperature variation is not obvious, so the temperature of the inner part of the standard fire test at the beginning of heating is lower than the numerical modelling temperature. However, after the water evaporates completely, the temperature rapidly rises, which is highly correlated with the numerical modelling temperature. As shown in Figure 21, it can be seen that the temperature of the junction box surface is still higher than the temperature of the average surface. Although two aspects on the weakness are reinforced, it is still limited by the material and it cannot be completely identical with the original wall. The temperature of the junction box surface is the highest temperature on unexposed surface, but it can meet the demand of 1-hour fire resistance. The final result of the standard fire test in Test 2 shows that the temperature of the average surface is 153.0°C. When the standard fire test of Test 3 is over, this temperature is 138.2°C, reducing by 14.8°C. Because the panel falls out at after 37 minutes, the measurement of the temperature of the junction box surface in Test 2 distorts. The final temperature of the junction box surface of the standard fire test in Test 3 is 186.2°C, conforming to the standard of ISO 834-1 [2]. When the numerical modelling of Test 2 is over, the temperature of the average surface is 148.6°C and temperature of the junction box surface is 234.2°C. When the numerical modelling of Test 3 is over, the temperatures are, respectively, 127.5°C and 161.3°C, reducing by 21.1°C and 72.9°C, respectively. It can be seen that compared with Test 2, the temperatures on unexposed surface obviously reduce after reinforcement in Test 3 and the temperature of the junction box surface reduces most significantly. However, in numerical modelling and standard fire test, the metallic-framework walls with embedded junction boxes after reinforcement in Test 3 can meet the demand of 1-hour fire resistance.

This is an innovative research. It is the first time to put forward the quantitative research on influence of embedded junction box on fire resistance of metal frame walls and analyze the weakness. In addition, it proposes effective reinforcement aiming at the weakness. After the reinforcement, the wall meets the demand of 1-hour fire resistance when the metallic-framework wall is embedded with the junction box. During the process, it successfully builds the CFD numerical models which is corresponding with the specimens, successfully predicts the effect of reinforcement by modified model parameters, and verifies it in the following standard fire test. This research not only systematically analyzes the metallic-framework wall with the embedded junction box and proposes effective reinforcement measures but also predicts them by CFD numerical modelling and verifies them successfully in the test. It is proved that the pattern of numerical modelling before the standard fire test is effective. Similar patterns can be applied to other researches on the wall systems and that can greatly save the cost of the test. Before the test, the numerical modelling can be processed to work out the predicted results. When the predicted result is satisfying, the standard fire test can be processed to verify.

5. Conclusions

(1) When the junction box is embedded into the metallic wall, the fire resistance of wall may be damaged because the metal junction box has a larger thermal conductivity and transfers the heat faster, while the PVC panel of the junction box softens in heat.

(2) The result of numerical modelling predicting the temperature by finite element is highly correlated to the result of standard fire test.

(3) After reinforcement, the temperature of the junction box surface decreases most significantly, and in Test 3, it reduces by 72.9°C when compared to Test 2.

(4) Adding the calcium silicate board to the back of the junction box and using the metal panel instead of the PVC panel can reinforce the original weakness effectively and help the metallic wall with the embedded junction box to meet the demand of 1-hour fire resistance.

(5) Making use of the finite element numerical modelling method to predict the test result is an effective way. It can be applied to other related fire prevention research and product development.

Conflicts of Interest

The authors declare that there are no conflicts of interest regarding the publication of this paper.

Acknowledgments

This work was supported by the CABR Application Technology Research Project: Key Technology of Smoke Control for Ultra Thin and Tall Atriums (no. 20150111330730049).

References

[1] NFPA 251, *Standard Methods of Tests of Fire Resistance of Building Construction and Material*, National Fire Protection Association, Quincy, MA, USA, 2006.

[2] ISO 834-1, *Fire-Resistance Tests-Elements of Building Construction: Part 1: General Requirements*, International Organization for Standardization, Geneva, Switzerland, 1999.

[3] JIS A 1304, *Method of Fire Resistance Test for Structural Parts of Buildings*, Japan Standards Association, Tokyo, Japan, 1994.

[4] CNS 12514, *Method of Fire Resistance Test for Structural Parts of Building*, Taiwan Standard, 2005.

[5] UL 263, *Fire Tests of Building Construction and Materials*, UL, Northbrook, IL, USA, 1997.

[6] BS 476-22, *Fire Tests on Building Materials and Structures. Methods for Determination of the Fire Resistance of Non-Loadbearing Elements of Construction*, BSI Group, London, UK, 1987.

[7] CNS 14652, *Glossary of Terms Used for Fire Protection in Building-Fire Test*, Taiwan Standard, 2008.

[8] S.-H. Lin, C.-L. Pan, and W.-T. Hsu, "Monotonic and cyclic loading tests for cold-formed steel wall frames sheathed with calcium silicate board," *Thin-Walled Structures*, vol. 74, pp. 49–58, 2014.

[9] M.-C. Ho and M.-J. Tsai, "Relative importance of fire resistance performance of partition walls," *Journal of Marine Science and Technology*, vol. 18, no. 3, pp. 430–434, 2010.

[10] M. Nithyadharan and V. Kalyanaraman, "Experimental study of screw connections in CFS-calcium silicate board wall panels," *Thin-Walled Structures*, vol. 49, no. 6, pp. 724–731, 2011.

[11] Y.-J. Chuang, W.-T. Wu, H.-Y. Chen, C.-H. Tang, and C.-Y. Lin, "Experimental investigation of fire wall insulation during a standard furnace fire test with different initial ambient air temperatures," *Journal of Applied Fire Science*, vol. 15, no. 1, pp. 41–55, 2006.

[12] I. Maruyama, G. Igarashi, Y. Nishioka, and Y. Tanigawa, "Carbonation and fastening strength deterioration of calcium silicate board: analysis of materials collected from a ceiling collapse accident," *Journal of Structural and Construction Engineering*, vol. 78, no. 689, pp. 1203–1208, 2013.

[13] P. C. R. Collier and A. H. Buchanan, "Fire resistance of lightweight timber framed walls," *Fire Technology*, vol. 38, no. 2, pp. 125–145, 2002.

[14] C. T. Do, D. P. Bentz, and P. E. Stutzman, "Microstructure and thermal conductivity of hydrated calcium silicate board materials," *Journal of Building Physics*, vol. 31, no. 1, pp. 55–67, 2007.

[15] A. Y. Nassif, I. Yoshitake, and A. Allam, "Full-scale fire testing and numerical modelling of the transient thermo-mechanical behaviour of steel-stud gypsum board partitionwalls," *Construction and Building Materials*, vol. 59, pp. 51–61, 2014.

[16] Y. Wang, Y.-J. Chuang, and C.-Y. Lin, "The performance of calcium silicate board partition fireproof drywall assembly with junction box under fire," *Advances in Materials Science and Engineering*, vol. 2015, Article ID 642061, 12 pages, 2015.

[17] CNS 5534, "*Thermocuples*," Part 4: *General Structural Matters*, Taiwan Standard, 1982.

[18] W. Yinuo, *The Standard Fire Testing and Numerical Modelling of the Behavior of Calcium Silicate Board Light Partition Wall Assembly with Junction Box*, Ph.D. thesis, National Taiwan University of Science and Technology, Taipei, Taiwan, 2018.

[19] BS EN 1993-1-2, *Eurocode 3: Design of Steel Structures—Part 1-2: General Rules Structural Fire Design*, BSI Group, London, UK, 2005.

[20] R. Walker and S. Pavía, "Thermal performance of a selection of insulation materials suitable for historic buildings," *Building and Environment*, vol. 94, pp. 155–165, 2015.

[21] GB/T 10699-1998, *Calcium Silicate Insulation, Part 5: Technical Requirements for Physical Performance*, China Standard, 1998.

[22] CNS 3142, *Polyvinyl Chloride Board, Part 3: General Quality*, Taiwan Standard, 2013.

[23] S. A. Mansour, M. E. Al-Ghoury, E. Shalaan, M. H. I. El Eraki, and E. M. Abdel-Bary, "Thermal properties of graphite-loaded nitrile rubber/poly(vinyl chloride) blends," *Journal of Applied Polymer Science*, vol. 116, no. 6, pp. 3171–3177, 2010.

[24] H. Kim, D. Park, E. S. Park, and H. M. Kim, "Numerical modeling and optimization of an insulation system for underground thermal energy storage," *Applied Thermal Engineering*, vol. 91, pp. 687–693, 2015.

8

Effect of Partial Cladding Pattern of Aluminum 7075 T651 on Corrosion and Mechanical Properties

E. Rendell,[1] A. Hsiao,[1,2] and J. Shirokoff[3]

[1]*Mechanical Engineering Department, Faculty of Engineering and Applied Science, Memorial University of Newfoundland, St. John's, NL, Canada A1B 3X5*

[2]*University of Prince Edward Island, Charlottetown, PEI, Canada C1A 4P3*

[3]*Process Engineering Department, Faculty of Engineering and Applied Science, Memorial University of Newfoundland, St. John's, NL, Canada A1B 3X5*

Correspondence should be addressed to J. Shirokoff; shirokof@mun.ca

Academic Editor: Alicia E. Ares

The corrosion resistance of aluminum 7075 T651 in full clad (Alclad), partial clad, and bare (unclad) forms was compared after 300 hours of corrosion exposure in an acidic salt spray cabinet test at 36°C. After corrosion exposure, severe to moderate exfoliation corrosion was observed on the unprotected medium sized test panel, light general corrosion was observed on the partially clad panel, and patches of corrosion not penetrating the clad layer were observed on the fully clad panel. After corrosion tests, the tensile strength of partially clad, fully clad, and unprotected panels decreased by 3.4%, 4.0%, and 5.3%, respectively.

1. Introduction

One of the primary drivers for materials selection in the aerospace industry is to maximize the economic efficiency of aircraft [1]. Commercial aircraft of the past utilized the high specific strength of the high strength aluminum alloys of the 2000 and 7000 series almost exclusively for the construction of structural components [2]. These materials were selected in an effort to minimize aircraft weight resulting in maximized aircraft payloads and reduce fuel consumption. High strength aluminum alloys are susceptible to localized corrosion such as pitting, exfoliation, and stress corrosion cracking. Presently, corrosion resistance of materials is also of great concern because of the cost of corrosion inspections and the high cost of unscheduled maintenance and aircraft downtime associated with replacing corroded aircraft components [3]. Corrosion as well as the ability to construct larger structural components that are more easily joined has resulted in a shift in preferred construction materials from high strength aluminum alloys to composites for new aircraft [1]. In spite of this, aluminum alloys will remain an important structural material for aircraft components, especially in compression where composites are less suitable. This shift in materials selection has caused the need for aluminum alloy innovation to maximize specific strength while maintaining acceptable corrosion resistance for aluminum to remain a competitive material choice [2]. The susceptibility of high strength aluminum alloys to corrosion requires them to be protected from corrosive environment using an anodic coating for components with complex geometries or by using Alclad products with an aerospace coating system on sheet and plate components. Alclad products are produced by the metallurgical bonding of a high strength aluminum alloy core sandwiched between two layers of a more electronegative aluminum alloy. The outer aluminum cladding layers corrode preferentially when exposed to a corrosive environment and prevent corrosion to the core by cathodic protection. The Alclad layers can comprise up to 4% of the total sheet or plate thickness [4] and are assumed to carry no load, reducing the overall material specific strength by increasing the weight without contributing to the strength.

Petroyiannis et al. have suggested in previous works that a continuous cladding layer may be excessive and that the application of a partial cladding pattern may provide equivalent corrosion protection for aluminum 2024 T3 alloys [5, 6]. They have found that a partial cladding layer covering only 7% of the core aluminum substrate provides equivalent corrosion protection to the mechanical properties of Al 2024 T3 when compared to Alclad products after a 300-hour immersion exposure in a neutral 3.5% NaCl solution. However, they have also concluded that this accelerated corrosion test is likely too mild to accurately represent the corrosion environment experienced by in service aircraft [6].

In this work, an appropriate partial cladding geometry for Al 7075 T651 is estimated by exposing an aluminum panel with a single clad spot to an acidic salt fog accelerated corrosion environment. The area of the panel protected from corrosion by the clad spot is then estimated and used to determine dimensions for a two-dimensional array of clad spots applied to a medium scale test panel. For comparison, medium scale test panels in both the as-received Alclad state and having the entire cladding layer removed have also been produced. These three test panels were exposed to the acidic salt fog accelerated corrosion environment and the resulting corrosion was compared through visually rating and by producing characteristic cross sections. Tensile specimens were then machined from the three corroded panels as well as tensile specimens having undergone the same machining processes as the three panels unexposed to the corrosion environment. All of the tensile specimens were then tested and compared.

2. Experimental

2.1. Corrosion Environment.
The corrosion environment selected for all of the experiments presented in this work was a 300-hour continuous acidic salt fog cabinet test in accordance with ASTM G85 Annex A1 [7]. The tests were conducted in a 120-liter benchtop Ascott S120ip salt spray chamber maintained at 36°C. The solution used to produce the acidic salt fog was composed of deionized water prepared with a Purite DC9 deionizing cylinder. The pH was then adjusted to a value between 3.1 and 3.3 by the addition of 99.7+% ACS reagent grade acetic acid purchased from Alfa Aesar. A solution salt concentration of 4–6 weight percent sodium chloride was produced by the addition of "Corro-Salt" purchased from Ascott. "Corro-Salt" meets the strict salt purity requirements detailed in ASTM B117 [8], namely, total impurities less than 0.3%, total halide (excluding chloride) composition of less than 0.1%, and a copper content of less than 0.3 ppm. The fog fallout rate was set between 1.0 and 2.0 mL per hour per 80 cm^2 horizontal area.

2.2. Single Central Clad Spot Panel.
Initially a single clad spot test panel was produced from 6.35 mm thick Alclad Al 7075 T651 plate having a total area exposed to the corrosion environment of 10 × 10 cm^2 with a square 1 cm^2 clad spot located in the panel center. The panel surface geometry was produced by mechanically milling approximately 0.19 mm over the panel surface removing the Alclad layer except at the clad spot. The cut edges and back surface of the test panel were protected from the corrosion environment with corrosion resistant polyvinyl chloride tape with the seams

sealed with super glue gel. After corrosion exposure, the panel was cleaned in accordance with ASTM G1 [9]. The panel was cleaned with water and stiff brush to remove deposited salt and some of the bulk corrosion products. This was followed by a four-minute immersion in nitric acid at room temperature to remove the remainder of the corrosion products. The exfoliation damage to the panel was characterized by applying a 2.5 × 2.5 mm grid to the surface and estimating the percentage of the panel surface affected by corrosion in each grid section. This data was then used to produce a surface plot showing the distribution of corrosion damage relative to the protective clad spot. To characterize the penetration of the corrosion damage, the machining method for determining pit depth described in ASTM G46 [10] was adapted to better describe exfoliation penetration. Approximately 0.02 mm and 0.17 mm were milled from the corroded panel surface and surface plots were produced characterizing the extent of exfoliation corrosion penetrating to these depths. The surface plots of the initial single clad spot test panel were used to estimate the area of the test panel effectively protected by the clad spot and used to determine appropriate dimensions for a partial cladding pattern to be applied to a medium scale test panel.

2.3. Medium Scale Test Panels.
A series of medium scale test panels were produced again from 6.35 mm thick Alclad Al 7075 T651 plate to investigate if a partial cladding pattern provides equivalent corrosion resistance when compared to Alclad in the selected corrosion environment. The medium scale test panels had a surface exposed to the corrosion environment of approximately 30 × 25 cm. The back surface and cut edges of the panels were protected for the corrosion environment using the same method described above for the initial single clad spot panel. Three medium scale test panels were produced to compare the relative corrosion resistance of fully clad aluminum plate, aluminum plate with a partial cladding pattern, and aluminum plate with no cladding. The fully clad test panel remains in the as-received Alclad form. The partially clad test panel was produced by mechanically milling the Alclad layer over the majority of the panel surface producing 1 × 1 cm clad spots. The clad spots were arranged in a two-dimensional array spaced 2 cm apart. During the machining process of the partially clad test panel, it was desired that only the Alclad layer be removed by milling approximately 0.19 mm from the panel surface to produce the clad spots. Due to difficulties in the machining process, such as the panel not being perfectly flat and issues with affixing the aluminum panel to the CNC milling machine, an average thickness of 0.69 mm was removed from the panel thickness to produce the clad spots. This resulted in the clad spots being comprised of both the cladding layer and a significant thickness of the high strength aluminum 7075 core. The presence of the layer of Al 7075 within the clad spots is not expected to significantly affect the corrosion behavior of the partially clad panel but may influence the mechanical testing results. The test panel with no cladding was produced by removing the entire Alclad layer through mechanical milling.

The experimental approach is highlighted in Figure 1 indicating schematic cross sections of the three medium sized

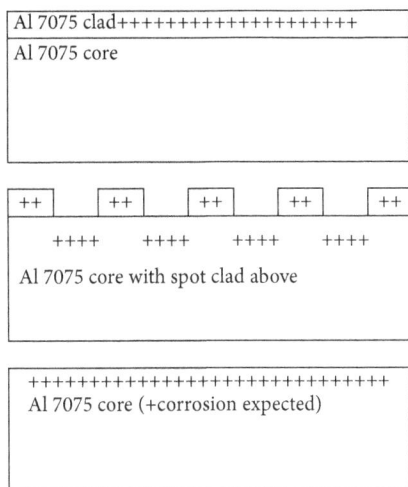

FIGURE 1: Schematic cross sections of the three medium sized test panels indicating the location (+) of corrosion products expected after corrosion testing. All test panels have visual exposed areas of 30 × 25 cm to salt spray. Alclad means full clad, while spot clad (partial clad) and unclad were machined to remove part or all of the clad layer.

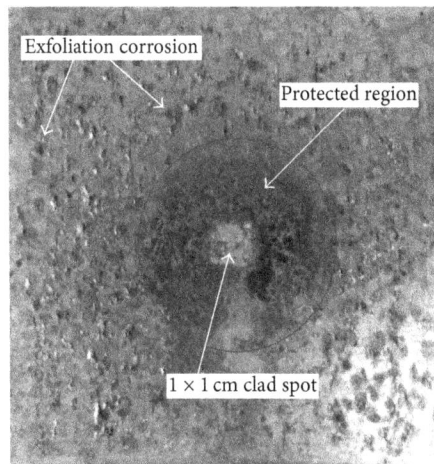

FIGURE 2: Optical microscopy of a cleaned central clad spot panel, 1 square cm clad spot.

test panels with a plus symbol (+) to indicate the expected development of corrosion products starting at the surface. All test panels have visual exposed surface areas of 30 × 25 cm to salt spray corresponding to Figure 1. Alclad means full clad, while spot clad (partial clad) and unclad were machined to remove part or all of the clad layer.

After exposure to the corrosion environment, the medium scale test panels were cleaned in the same method as the initial single clad spot test panel described above. The corrosion damage resulting from exposure to the acidic salt fog environment on the medium scale panels was qualitatively compared through visual observations and through cutting representative cross sections from each of the test panels. The cross sections were cut from various locations on the medium scale test panels with and Isomet 11-1180 low speed saw and then mounted and polished in SamplKwick fast dry acrylic. The cross sections were viewed at 100x magnification with a Nikon Eclipse 50i optical microscope and images were captured with an Infinity 1 microscopy camera.

Rectangular tensile specimens were cut from each of the corroded medium scale test panels in accordance with ASTM B557 [11]. For comparison, tensile specimens were prepared from uncorroded Al 7075 T651 that has undergone the same milling processes as the medium scale test panels, two tensile specimens for each of the panel surface geometries. All of the tensile samples were tested in accordance with ASTM B557 with an Instron 5585H load frame using wedge type grips. The crosshead speed of all the tensile tests was 5 mm/minute and the strain during the elastic section of the tests was measured using an Instron 2630-106 clip on extensometer with a gauge length of 25 mm.

Materials characterization by scanning electron microscopy-energy dispersive spectroscopy (SEM-EDS) was performed using a JEOL-JSM-6000 operating at 15 kV in order to verify the Alclad 7075 cladding, core, and corrosion product elemental analysis (i.e., red corrosion product containing 7.71 mass% Cu). The SEM-EDS results verified the alloying elements of the Alclad layers (cladding, core) as specified by the North American supplier of Alclad in their Al 7075 technical specification, expected reactions, and corrosion products. The Al-alloy clad layer reacts to form Al_2O_3 with Al dissolution; and Al-alloy core reacts to form Al-Cu corrosion product via the elemental Cu level increase from 1.50 to 7.71 mass% to be discussed further in the results and discussion sections. The Al_2O_3 was confirmed by X-ray diffraction using a Rigaku Ultima IV operating with Cu K-alpha radiation at 40 kV and 44 mA. Image analysis was also employed using a PAX system in order to estimate the percentage of surface area that corroded.

3. Results

3.1. Single Central Clad Spot Panel. After exposure to the acidic salt fog corrosion environment and cleaning, it can be seen in Figure 2 that there are two regions of interest on the central clad spot panel. A roughly circular region adjacent to the clad spot has been protected from the corrosion environment due to the clad spot acting as a sacrificial anode. The remainder of the panel shows corrosion that could be described as moderate exfoliation [12] or as poorly defined pitting with a large degree of horizontal propagation and delamination [9].

The optical image analysis system application of a grid on the corroded surface was used for an estimation of the percentage of the area of each grid section showing signs of corrosion on the three panels (central clad spot, 0.02 mm and 0.17 mm milled). From these surface plots, the area of the central clad spot test panel effectively protected by the clad spot was estimated as a 3 × 3 cm square. This dimension was used to produce the medium sized test panel with a clad spot pattern. It was estimated that a two-dimensional array of

1 cm^2 clad spots spaced 2 cm apart is sufficient to protect the panel surface from corrosion after a 300-hour exposure in the acidic salt fog environment.

3.2. Medium Scale Test Panels

3.2.1. Initial Visual Observations of the Medium Scale Test Panels. Upon visual inspection of the Alclad medium scale test panel, shallow irregularly shaped corrosion patches are fairly evenly distributed over the panel surface. Smaller, pit-like corrosion is also evident in the cladding layer. It appears that pitting has initiated in the cladding layer and progressed until reaching the aluminum core where the pits widened laterally, combining to form larger corrosion spots. From these observations it appears that the Alclad layer has successfully protected the high strength aluminum core by corroding preferentially in the acidic salt spray environment.

After corrosion exposure, exfoliation corrosion was observed over the entire surface of the panel with the cladding removed. Using the exfoliation corrosion rating system described in ASTM G34, the observed corrosion could be described between pitting and moderate exfoliation. Distinct pit-blisters were observed but with more lifting of the aluminum along the pit edges than is described in the standard. The blistering and lifting of slivers of uncorroded aluminum at the pit edges more closely resembled moderate exfoliation but there is less layering than described in the standard. Overall the observed corrosion can best be described as pit-blisters that have exposed grain boundaries allowing intergranular corrosion in the form of exfoliation to occur in tandem. Where multiple pit-blisters interacted to lift continuous sheets of uncorroded aluminum, the corrosion began to resemble moderate exfoliation. An unexpected observation is the "red colored" corrosion products presumably due to copper since the elemental Cu level increase from 1.50 to 7.71 mass% as measured by energy dispersive spectroscopy over the majority of the panel surface.

The corrosion observed over the partially clad medium scale test panel after acidic salt spray exposure was not uniform. Five distinct regions of corrosion behavior were observed: (1) The surface of the clad spots experienced general corrosion, where the Alclad layer was depleted as a sacrificial anode, protecting the overall panel surface. (2) The majority of the panel surface was successfully protected by the partial cladding pattern. This is evident where the surface retained the shiny appearance and milling marks from the machining process. (3) General corrosion with red color resembled what was observed on the unprotected test panel. (4) Light general corrosion located between the red colored region and the shiny noncorroded region. (5) Finally, there are small patches of corrosion distributed throughout the noncorroded region of the panel. These patches are generally located adjacent and below the clad spots. These patches of corrosion can possibly be explained by the acidic salt solution stagnating in the raised edges of the clad spots or within milling marks and leading to crevice corrosion.

3.2.2. Representative Cross Sections of the Medium Scale Test Panels. Many cross sections cut from the medium scale test

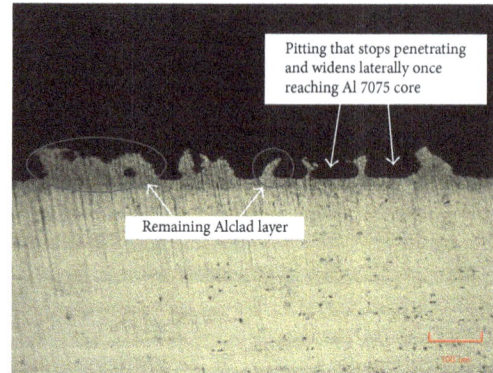

FIGURE 3: Clad panel and region after corrosion, 100x magnification and 100-micron bar.

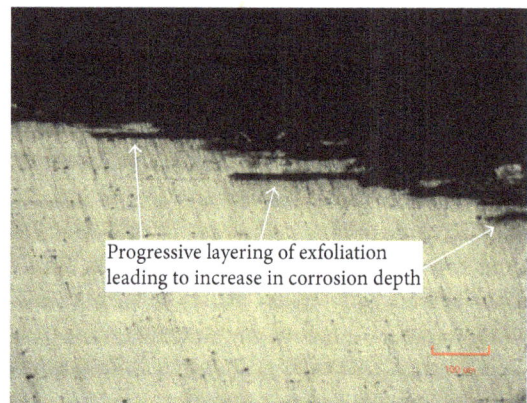

FIGURE 4: Unclad panel deep exfoliation damage, 100x magnification and 10-micron bar.

panels largely confirm the initial visual observations made of the corroded panels. Figure 3 shows one of the large corrosion patches distributed over the Alclad test panel surface. Dense pitting penetrating the Alclad layer until reaching the high strength aluminum core and widening laterally, combining to form larger shallow pits, is evident. Alclad that was representative of the majority of the test panel exhibited a smooth surface and intact cladding layer, indicating that these regions were unaffected by the corrosion environment.

Figures 4 and 5 are cross sections representing the corrosion observed on the bare test panel. Figure 4 shows successive layers of exfoliation resulting in increasing corrosion penetration depth after corrosion product expansion. As slivers of uncorroded aluminum are lifted from the substrate, additional grain boundaries are exposed to the corrosion environment and a new layer of exfoliation corrosion is initiated. Figure 5 shows one of the few examples of a well-defined pit in the unclad test panel. The absence of many examples of pitting can be explained by the large number of grain boundaries exposed by the pit. The pit initiates intergranular corrosion that progresses into exfoliation and obliterates evidence of the pit [13, 14].

TABLE 1: Mechanical properties summary.

| Property/material | Mechanical property and change (Δ) due to corrosion | | | | | |
| | No cladding | | Spot cladding | | Full clad | |
	Corrosion	No corrosion	Corrosion	No corrosion	Corrosion	No corrosion
Tensile strength (MPa)	550 Δ-5.3%	581	561 Δ-3.4%	581	551 Δ-4.0%	574
Yield strength (MPa)	520 Δ2.2%	509	528 Δ3.9%	508	520 Δ3.4%	503
Young's modulus (GPa)	66.5 Δ-7.4%	71.8	70.5 Δ1.0%	69.8	69.1 Δ3.6%	66.7
Elongation (%)	13.7 Δ-18.4%	16.2	13.5 Δ-8.7%	14.8	17.3 Δ-4.2%	18.1

FIGURE 5: Unclad panel and evidence of pitting, 100x magnification and 100-micron bar.

3.3. Mechanical Properties. Table 1 shows the mechanical properties summary of the tensile specimens cut from the corroded medium scale test panels and the tensile specimens that have undergone the same mechanical processes but have not been exposed to the corrosion environment. When comparing the corroded and the uncorroded tensile specimens, there is a decrease in tensile strength in each case. The partially clad test panel exhibits the smallest decrease in tensile strength with a percent decrease of 3.4%. This was followed by the fully clad test panel with a tensile strength percentage decrease of 4.0%. Finally, the cladding removed panel showed a percentage decrease in tensile strength of 5.3%.

In all cases, the yield strength of the tensile specimens increased after corrosion exposure. The spot clad, fully clad, and cladding removed test specimens showed a percent increase in yield strength of 3.9%, 3.4%, and 2.2%, respectively. This trend is interesting, as it is expected that the yield strength would decrease with corrosion exposure.

Young's modulus of the spot clad and fully clad test panels increased after corrosion exposure while there was a decrease for the bare test panel. The percent increase in Young's modulus for the spot clad and fully clad test panels was 1.0% and 3.6%, respectively. The percentage decrease in Young's modulus for the unprotected panel was 7.4%.

In all cases, there was a decrease in percent elongation after corrosion exposure. The percent decrease in percent elongation for the cladding removed, spot clad, and fully clad test specimens was 18.4%, 8.7%, and 4.2%, respectively.

4. Discussion

The single central clad spot panel was used to evaluate the region of corrosion surrounding the surface containing a 1×1 cm spot clad. Optical image analysis tools can provide an estimate of corroded surface area and from this approach the spot clad spacing of 2 cm was determined to be sufficient for adequate corrosion protection. For more quantitative analysis additional work would be required and it should therefore be considered by characterization methodologies such as scanning electron microscopy with energy dispersive spectroscopy capable of X-ray mapping chemical elements and also comparison of one spot to perhaps an array of four spots. From the visual observations and representative cross sections of the medium scale test panels, it is apparent that the cladding layer of the Alclad test panel successfully protected the high strength aluminum core from the corrosion environment. Conversely, the absence of the cladding layer was detrimental to the corrosion resistance of the bare test panel. Uniform exfoliation corrosion was observed over the entire panel surface after corrosion exposure. The corrosion to the high strength aluminum core observed on the partially clad test panel remained mild when compared to the bare panel but was somewhat more severe than the Alclad panel.

From the visual observations and representative cross sections of the medium scale test panels, it is apparent that the cladding layer of the Alclad test panel successfully protected the high strength aluminum core from the corrosion environment. Conversely, the absence of the cladding layer was detrimental to the corrosion resistance of the bare test panel. Uniform exfoliation corrosion was observed over the entire panel surface after corrosion exposure. The corrosion to the high strength aluminum core observed on the partially clad test panel remained mild when compared to the bare panel but was somewhat more severe than the Alclad panel.

An unexpected result was the red coloring of corrosion products on the bare test panel. It is believed that the red colored corrosion products are the result of alloyed copper dissolving into the corrosive electrolyte and plating back onto the aluminum alloy substrate.

During the corrosion of aluminum alloys containing copper as a major alloying element, in the presence of chloride ions, copper containing intermetallic particles found in Al 7075 T651 such as $MgCu_2$ or Al_2CuMg can be subjected to dealloying where magnesium and aluminum are selectively dissolved, leaving behind a microporous region consisting primarily of elemental copper. These high surface area regions of copper can then act as a local cathode and cause pitting and trenching to the alloy adjacent to the copper. It is then possible for the copper region to detach from the aluminum alloy, dissolve into the corrosive electrolyte, and electrically be plated back onto the alloy surface [15, 16]. The presence of the plated copper could then cause galvanic corrosion, accelerating the overall corrosion rate of the bare test panel.

The red corrosion products believed to be copper were also present on the partially clad test panel. The location of the red corrosion products on the top edge of the panel surface where there are not yet any clad spots may suggest that the ability of the clad spots to protect the aluminum substrate from corrosion exposure may be influenced by the downward flow of the corrosive electrolyte as the acidic salt fog is deposited on the surface. If the red coloring is plated copper, as is believed, the area of light general corrosion on the partially clad panel may be explained by galvanic corrosion caused by the copper accelerating the corrosion rate in this area.

From the results of the tensile testing of the medium scale test panels, the tensile strength and percent elongation show the most conclusive results. The results for the yield strength and Young's modulus remain fairly ambiguous, possibly because the aluminum plate selected for the experiment was rather thick (6.35 mm) and it is possible that the corrosion damage on all three test panels was superficial when compared to the plate thickness.

From the tensile strength results, the partially clad test panel had the smallest reduction in tensile strength after corrosion exposure with a percent decrease of 3.4% while the Alclad and bare panels had percent decreases of 4.0% and 5.3%, respectively. This suggests that the partial cladding pattern successfully protected the test panel from the corrosion environment when compared to the Alclad and bare panels. The fact that the tensile strength of the bare test panel was not more severely affected by the corrosion environment may suggest that the selected corrosion environment was too mild for the corrosion damage to significantly affect the tensile strength for 6.35 mm aluminum plate.

The results for the degradation in percent elongation of the test panels in this experiment suggest that the ductility of the aluminum was significantly affected by both the machining processes conducted on the aluminum plate and the corrosion environment exposure. The percent decrease in percent elongation for the bare, spot clad, and fully clad test specimens after corrosion exposure was 18.4%, 8.7%, and 4.2%, respectively. In this instance, the Alclad test panel outperformed the partially clad test panel but the partially clad panel performed much better than the bare panel. The percent elongation of the uncorroded spot clad, bare, and Alclad test panels was 14.8%, 16.2%, and 18.1%, respectively. These values differ significantly and imply that the machining processes subjected to each of the test panels had an effect on ductility. It is possible that the milling of the surface of the bare panel had a work hardening effect that reduced overall panel ductility. For the partially clad test panel, it is believed that the presence of the clad spots comprising not only the soft Alclad layer but also a thickness of the high strength aluminum core reduced overall ductility by reinforcing the sections of the test specimens with clad spots. As a result, the majority of the deformation of the tensile specimens occurred only in locations without the clad spot and the specimens fractured with an overall lower percent elongation.

When correlating mechanical properties results to corrosion test data and microstructure the issue of uncertainties deserves some discussion. Since these mechanical properties results are presented after several times of testing it is important to include the following caveat (i.e., the results fall within the uncertainties of experimentation that includes sample homogeneity, aging, microstructure, machining, anisotropy, residual stress, equipment calibration, grip forces, and work hardening); otherwise conclusions can be misinterpreted, misused, and misleading.

Therefore additional research study inclusive of detailed uncertainties analysis is recommended in order to verify the general trends in results that only approximate spot clad and fully clad mechanical properties. In general, the overall performance of full clad (Alclad) to corrosion should favor the clad versions (full clad = Alclad, partial clad) over unclad because aircraft alloys in service are normally held to a high level of standard in terms of performance, risk, and safety.

5. Conclusions

(1) The spot clad spacing of 2 cm was determined to be sufficient for adequate corrosion protection.

(2) After corrosion tests, the tensile strength of the partially clad, fully clad (Alclad), and unprotected test panels decreased by 3.4%, 4.0%, and 5.3%, respectively.

Conflicts of Interest

The authors declare that they have no conflicts of interest.

Authors' Contributions

The authors involved are the graduate student Evan Rendell who performed the original research and master's thesis under supervision of Dr. Amy Hsiao and Dr. John Shirokoff the latter of whom created the paper from the thesis and performed some additional research and writing to clarify and validate specific results.

Acknowledgments

The research in this paper is supported by an industrial regional benefit program of the Government of Newfoundland at Memorial University of Newfoundland.

References

[1] E. A. Starke Jr. and J. T. Staley, "Application of modern aluminum alloys to aircraft," *Progress in Aerospace Sciences*, vol. 32, no. 2-3, pp. 131–172, 1996.

[2] A. Merati, "Materials replacement for aging aircraft," in *Corrosion Fatigueand Environmentally Assisted Crackingin Aging Military Vehicles*, NATO Science and Technology Organization, 2011.

[3] W. Tsai, "The economics of maintenance," in *The Standard Handbook for Aeronautical and Astronautical Engineers*, McGraw-Hill, New York, NY, USA, 2003.

[4] ALCOA Mill Products Inc., Alloy 7075 sheet and plate, 2001.

[5] P. V. Petroyiannis, S. G. Pantelakis, and G. N. Haidemenopoulos, "Protective role of local Al cladding against corrosion damage and hydrogen embrittlement of 2024 aluminum alloy specimens," *Theoretical and Applied Fracture Mechanics*, vol. 44, no. 1, pp. 70–81, 2005.

[6] S. G. Pantelakis, A. N. Chamos, and D. Setsika, "Tolerable corrosion damage on aircraft aluminum structures: Local cladding patterns," *Theoretical and Applied Fracture Mechanics*, vol. 58, no. 1, pp. 55–64, 2012.

[7] "Standard Practice for Modified Salt Spray (Fog) Testing," *ASTM*, no. G85-11, 2011.

[8] "Standard Practice for Operating Salt Spray (Fog) Apparatus," ASTM B117-11, 2011.

[9] "Standard Practice for Preparing, Cleaning and Evaluating Corrosion Test Specimens," ASTM G1-03(2011), 2011.

[10] "Standard Guide for Examination and Evaluation of Pitting Corrosion," *ASTM*, no. G46-94(2013), 2013.

[11] "Standard Test Methods for Tension Testing Wrought and Cast Aluminum-and Magnesium-Alloy Products," ASTM B557-14, 2014.

[12] "Standard Test Method for Exfoliation Corrosion Susceptibility 2XXX and 7XXX Series Aluminum Alloys (EXCO Test)," ASTM G34-01(2013), 2013.

[13] X. Zhao, *Exfoliation corrosion kinetics of high strength aluminum alloys [Ph.D. thesis]*, The Ohio State University, 2006.

[14] D. W. Hoeppner and C. A. Arriscorreta, "Exfoliation corrosion and pitting corrosion and their role in fatigue predictive modeling: State-of-the-art review," *International Journal of Aerospace Engineering*, Article ID 191879, 2012.

[15] M. B. Vukmirovic, N. Dimitrov, and K. Sieradski, "Dealloying and corrosion of Al alloy 2024-T3," *Journal of the Electrochemical Society*, vol. 149, no. 9, pp. B428–B439, 2002.

[16] H. M. Obispo, L. E. Murr, R. M. Arrowood, and E. A. Trillo, "Copper deposition during the corrosion of aluminum alloy 2024 in sodium chloride solutions," *Journal of Materials Science*, vol. 35, no. 14, pp. 3479–3495, 2000.

9

Multiaxial Cycle Deformation and Low-Cycle Fatigue Behavior of Mild Carbon Steel and Related Welded-Metal Specimen

Weilian Qu, Ernian Zhao, and Qiang Zhou

Hubei Key Laboratory of Roadway Bridge & Structure Engineering, Wuhan University of Technology, Wuhan, Hubei 430070, China

Correspondence should be addressed to Ernian Zhao; zhaoern@126.com

Academic Editor: Luciano Lamberti

The low-cycle fatigue experiments of mild carbon Q235B steel and its related welded-metal specimens are performed under uniaxial, in-phase, and 90° out-of-phase loading conditions. Significant additional cyclic hardening for 90° out-of-phase loading conditions is observed for both base metal and its related weldment. Besides, welding process produces extra additional hardening under the same loading conditions compared with the base metal. Multiaxial low-cycle fatigue strength under 90° out-of-phase loading conditions is significantly reduced for both base-metal and welded-metal specimens. The weldment has lower fatigue life than the base metal under the given loading conditions, and the fatigue life reduction of weldment increases with the increasing strain amplitude. The KBM, FS, and MKBM critical plane parameters are evaluated for the fatigue data obtained. The FS and MKBM parameters are found to show better correlation with fatigue lives for both base-metal and welded-metal specimens.

1. Introduction

Engineering components are always subjected to complex cycle loading during the service period, and the failure eventually occurs due to the accumulated fatigue damage [1]. In engineering applications, fatigue failures occur in local regions, where stress concentrations generate multiaxial stress/strain states [2]. The multiaxial stress/strain states commonly arise from multidirectional external loads, notch effects, and complex geometric features, which indeed influence the fatigue strength of engineering components [3].

Understanding of multiaxial fatigue strength of metallic materials is always based upon the experimental observations from thin-walled tubular specimens under axial-torsional loading. For a long period, the multiaxial fatigue tests are mainly focused on the base material of metals. A review of the multiaxial fatigue experiments for metallic materials can be found in [4, 5]. In the last decade, Chen et al. [6] studied the multiaxial strength of type 304 stainless steel under sequential biaxial loading. Gao et al. [7, 8] tested the multiaxial fatigue strength of 16MnR steel and 7075-T651 aluminum alloy under various multiaxial loading paths. Shang and Wang [9] conducted the fatigue tests on hot-rolled medium-carbon 45 steel under the axial-torsional loading using sinusoidal wave forms. The multiaxial cycle deformation and fatigue behaviors of type 304 stainless steel and medium-carbon 1050 steel are studied by Shamsaei [10]. Gladskyi and Shukaev [11] conducted the contrastive analysis of the uniaxial and multiaxial low-cycle fatigue strength of type BT1-0 titanium alloy.

For welded steel structures, residual stresses, welding defects, material inhomogeneity in the weld zone, and so forth caused by the welding process can significantly reduce the fatigue strength of the welded joints [12–14]. At present, the fatigue design of welded joints is mainly based on the fatigue resistance *S-N* curves, which are obtained from the statistical analysis of the uniaxial fatigue test results of classified welding structural details [15–17]. However, multiaxial fatigue of welded joints is rarely investigated. Chen et al. [18] conducted the low-cycle fatigue experiments on 1Cr-18Ni-9Ti stainless steel and related weld metal under axial, torsional, and 90° out-of-phase loading. Fatigue of welded components under bending-torsion proportion and nonproportion loading was studied in [19].

The multiaxial fatigue of metallic materials has attracted a widespread attention in the past decades. Investigations on multiaxial fatigue are often conducted by using an equivalent parameter which makes it possible to compare the multiaxial

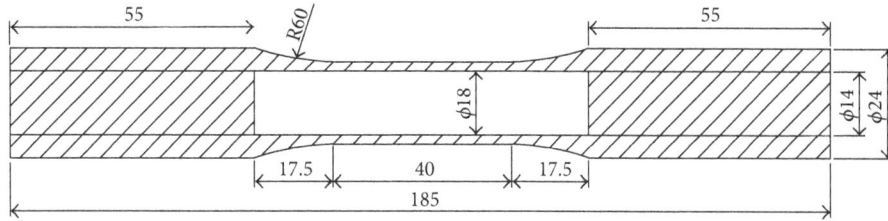

FIGURE 1: Geometry of the base-metal specimen.

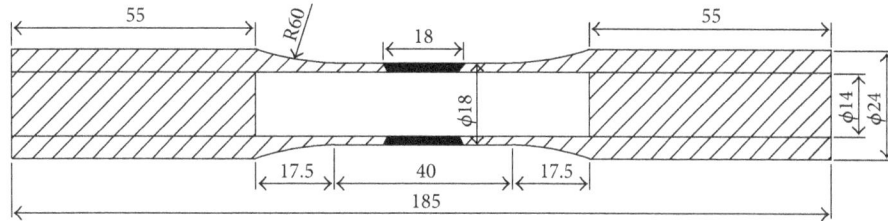

FIGURE 2: Geometry of the welded-metal specimen.

TABLE 1: Chemical composition of Q235B steel (wt%).

C	Mn	Si	P	S	V	Alt	Nb
0.16	0.45	0.26	0.021	0.025	0.006	0.004	0.003

TABLE 2: Chemical composition of welding wire (wt%).

C	Mn	Si	P	S	V	Ni	Cr	Mo	Cu
0.077	1.54	0.92	0.011	0.012	0.002	0.006	0.023	0.004	0.126

loading with uniaxial one [20]. Then the fatigue analysis methods developed for the uniaxial case can be employed to solve the multiaxial fatigue problems. The well-known Manson-Coffin criterion, which is widely used in the uniaxial low-cycle fatigue analysis, is modified for the multiaxial loading condition. In the recent decades, the Manson-Coffin criterion in terms of critical plane parameters, such as KBM and FS parameters, plays an important role in the multiaxial fatigue damage evaluation (more details are given in [21, 22]).

To study the multiaxial cycle deformation and low-cycle fatigue behaviors of mild carbon Q235B welded joints, which are more and more widely used in the steel constructions in China, the fatigue experiments are conducted on Q235B steel and its weldment by using thin-walled tubular specimens under fully reversed strain-controlled loading conditions with uniaxial, in-phase, and 90° out-of-phase loading. The KBM, FS, and MKBM critical plane parameters are evaluated for the experimental data gathered in this study.

2. Experimental Procedure

2.1. Materials and Specimens.
The investigated material in the present study is mild carbon Q235B structural steel, which is widely used in China's steel constructions. The investigated Q235B steel has Young's modulus of 204 GPa, yield strength of 270 MPa, ultimate strength of 390 MPa, Poisson's ratio of 0.3, and elongation of 36.9%. The chemical composition of Q235B steel is presented in Table 1.

The base-metal specimen has a tubular geometry with the outside and inside diameters of 18 mm and 14 mm, respectively. The wall thickness in the gage section is 2 mm.

The geometry of the base-metal specimen is displayed in Figure 1.

The welded specimen is made by the manual CO_2 gas-shielded welding process. The welding wire of MG70S-6 with a diameter of 2 mm is used. The chemical composition of welding wire is presented in Table 2.

The manufacture of the welded metal specimen is followed by [18]. A well-designed notch is first machined at the center of the base metal bar, and the notch is then filled with weld metal. Finally, the welded-metal specimen is machined to the shape in accordance with the base-metal specimen. The manufactured welded-metal specimen has an 18 mm long welded zone at the center of the gauge length. The geometry of the welded thin-walled tubular specimen tested is identical to the base one (see Figure 2).

The monotonic mechanical properties for the base-metal and welded-metal specimens are listed in Table 3.

2.2. Fatigue Tests.
Fatigue tests under uniaxial, in-phase, and 90° out-of-phase loading conditions are conducted under the fully reversed strain-controlled loading at constant amplitudes. The applied waveforms for both base-metal and welded-metal specimens are sinusoidal. The three test strain paths are displayed in Figure 3. The horizontal axis is the term of axial strain, ε, and the vertical axis is the term of shear strain, $\gamma/\sqrt{3}$. The correlation of the horizontal axis and the vertical axis is derived from the von Mises criterion of $\bar{\varepsilon} = \sqrt{\varepsilon^2 + (1/3)\gamma^2}$ in which $\bar{\varepsilon}$ is the equivalent von Mises strain.

TABLE 3: Mechanical properties of Q235B base and welded metal.

	Young's modulus E (GPa)	Shear modulus G (GPa)	Yield strength σ_y (MPa)	Tensile strength σ_u (MPa)	Ultimate strain ε_u (%)	Elongation δ (%)
Base metal	204	81.4	269	390.9	15.2	36.9
Welded metal	198	76.3	265	371.3	4.39	14.3

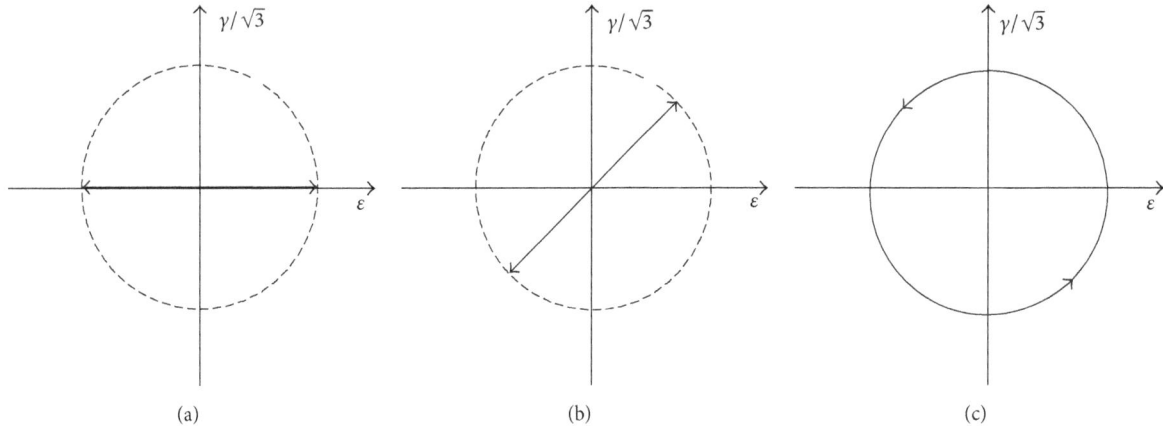

(a)   (b)   (c)

FIGURE 3: Fatigue test loading paths: (a) uniaxial (UA), (b) in-phase (IP), and (c) 90° out-of-phase (OP).

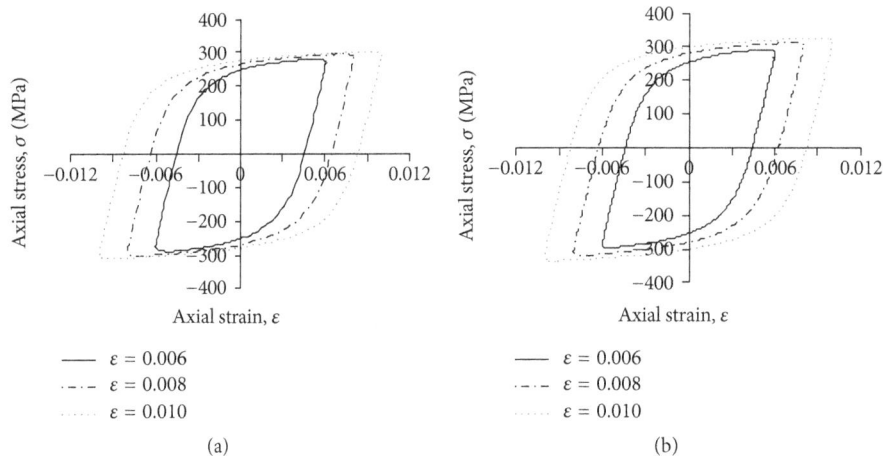

(a)   (b)

FIGURE 4: Axial hysteresis loops for uniaxial loading: (a) base metal and (b) welded metal.

Fatigue tests were conducted on an MTS tension-torsion machine under strain-controlled loading using a tension-torsion strain extensometer with the gauge length of 25 mm, which is mounted at the center of the outside of the specimen gauge section to measure the strain responses. The loading frequency for constant-amplitude tests is 1.0 HZ. Fatigue life is assumed as the number of cycles for which there is 30% reduction with respect to the maximum tensile or shear stress of the uniaxial test.

3. Results and Discussion

The stable hysteresis loops of base-metal and welded-metal specimens under uniaxial, in-phase, and 90° out-of-phase

loading at different strain amplitude are presented in Figures 4–6, respectively. It can be observed that the multiaxial cycle deformation behavior for in-phase loading condition is basically the same as the uniaxial one, while the multiaxial cycle deformation behaviors for out-of-phase loading condition significantly changed. The maximum shear and axial stress responses as well as shear and axial strains are simultaneous under in-phase loading conditions for both base-metal and welded-metal specimen. However, the maximum values of cyclic stress and strain responses do not always occur at the same time under the 90° out-of-phase loading, which indicates that the metal's plastic yield flow for 90° out-of-phase loading condition is different from that under the uniaxial one.

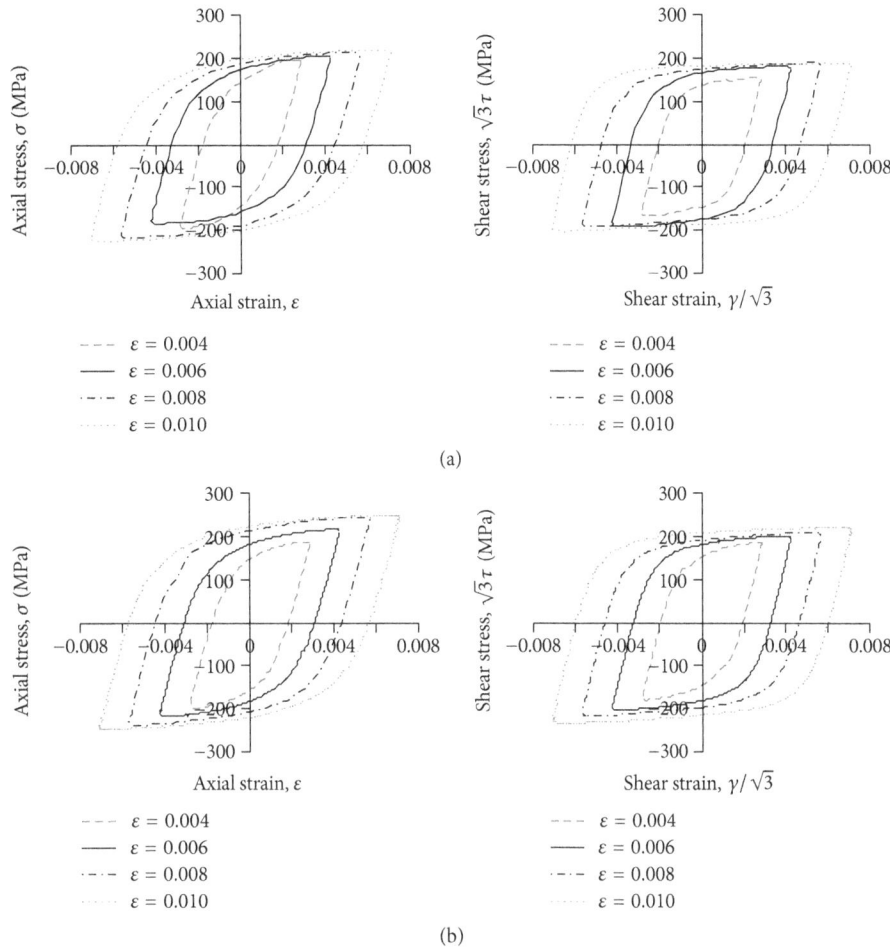

FIGURE 5: Axial (left) and shear (right) hysteresis loops for in-phase loading: (a) base metal and (b) welded metal.

Crack initiation under uniaxial fatigue test is in the circumferential direction as usual. In the in-phase fatigue test, the patterns of the macroscopic crack are observed similar to that of uniaxial fatigue test. For the 90° out-of-phase fatigue test, the crack direction is irregular because the maximum principal stress direction changed with respect to the nonproportionality loading condition, and the fracture shape of the macrocrack for 90° out-of-phase fatigue test was jagged. The crack patterns for base metal and welded metal under in-phase and out-of-phase loading conditions are shown in Figures 7 and 8, respectively.

The fatigue experimental and analytical results for the base-metal and welded-metal specimens under uniaxial, in-phase, and 90° out-of-phase loading are presented in Table 4. The table includes axial and shear stress and strain amplitudes, maximum shear strain amplitude acting on the maximum shear plane (critical plane), $\Delta\gamma_{max}/2$, strain ration, the normal strain range acting on the critical plane, $\Delta\varepsilon_n$, maximum normal stress acting on the critical plane, $\sigma_{n,max}$, and strain ration parameter, λ, for each fatigue test.

The stabilized cycle stress-strain relationship can be represented by the Ramberg-Osgood equation [23, 24].

$$\frac{\Delta\bar{\varepsilon}}{2} = \frac{\Delta\bar{\varepsilon}_e}{2} + \frac{\Delta\bar{\varepsilon}_p}{2} = \frac{\Delta\bar{\sigma}}{2E} + \left(\frac{\Delta\bar{\sigma}}{2K'}\right)^{1/n'}, \quad (1)$$

where K' is the cyclic hardening coefficient and n' is the cyclic hardening exponent, E is Young's modulus of the investigated material, and $\Delta\bar{\varepsilon}$ and $\Delta\bar{\sigma}$ are the equivalent strain range and the equivalent stress range, respectively. $\Delta\bar{\varepsilon}_e$ and $\Delta\bar{\varepsilon}_p$ are the equivalent elastic strain range and the equivalent plastic strain range, respectively. The equivalent elastic strain can be calculated based on Hooke's law. The equivalent plastic strain range can be then calculated by

$$\frac{\Delta\bar{\varepsilon}_p}{2} = \frac{\Delta\bar{\varepsilon}}{2} - \frac{\Delta\bar{\varepsilon}_e}{2} = \frac{\Delta\bar{\varepsilon}}{2} - \frac{\Delta\bar{\sigma}}{2E}. \quad (2)$$

Comparison of cyclic stress-strain curves between base metal and welded metal for uniaxial, in-phase, and out-of-phase loading conditions is presented in Figure 9, in which the solid lines and the dotted lines are the fitted cyclic stress-strain curves for base-metal specimens and welded-metal

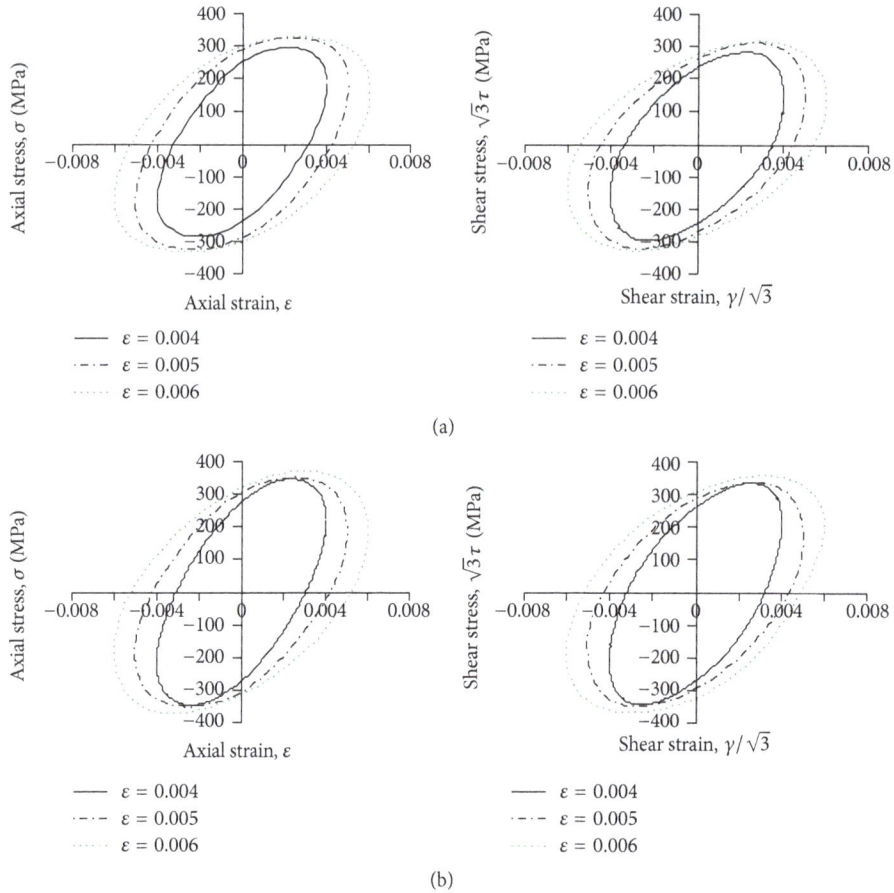

FIGURE 6: Axial (left) and shear (right) hysteresis loops for out-of-phase loading: (a) base metal and (b) welded metal.

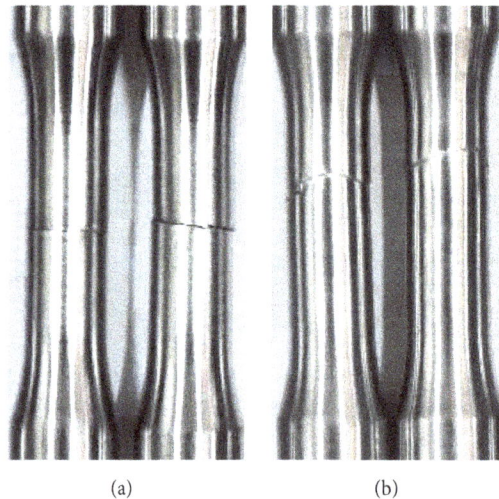

FIGURE 7: Patterns of crack growth for in-phase test: (a) base metal and (b) welded metal.

specimens, respectively. It can be seen that the fitted cyclic stress-strain curves for both base metal and welded metal under in-phase loading are similar to those obtained for uniaxial loading, while the fitted cyclic stress-strain curves for the 90° out-of-phase loading conditions are above those relative to the uniaxial loading case, which indicates that a significant additional cyclic hardening effect occurs for both base metal and welded metal under the out-of-phase loading

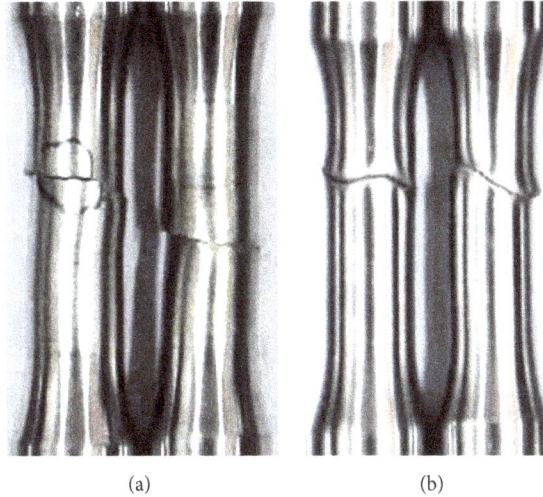

FIGURE 8: Patterns of crack growth for out-of-phase test: (a) base metal and (b) welded metal.

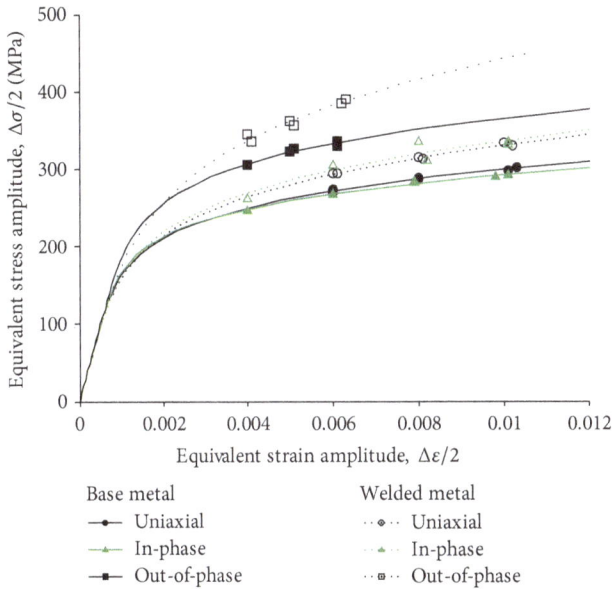

Base metal
- ━●━ Uniaxial
- ━▲━ In-phase
- ━■━ Out-of-phase

Welded metal
- ··◎·· Uniaxial
- ··▲·· In-phase
- ··▣·· Out-of-phase

FIGURE 9: Comparison of cyclic stress-strain curves between base metal and welded metal for uniaxial, in-phase, and out-of-phase loading conditions.

conditions. It can be also seen that the stabilized cyclic stress-strain curves of the welded metal are over those of the base metal under the same loading condition. It can be concluded that the welding process produces extra additional hardening for the weldment compared with the base metal.

4. Fatigue Life Analysis

The Manson-Coffin equation in terms of the equivalent strain parameter for fatigue evaluation under uniaxial loading can be written as [20–22]

$$\frac{\Delta \bar{\varepsilon}}{2} = \frac{\sigma'_f}{E} \left(2N_f\right)^b + \varepsilon'_f \left(2N_f\right)^c, \qquad (3)$$

where $\Delta \bar{\varepsilon}$ is the equivalent von Mises strain range, N_f is the cycle numbers to fatigue failure, and σ'_f, ε'_f, b, and c are the fatigue strength coefficient, the fatigue ductility coefficient, the fatigue strength exponent, and the fatigue ductility exponent, respectively. The fatigue properties fitted by (3) for the base metal and welded metal under the uniaxial, in-phase, and 90° out-of-phase loading are listed in Table 5.

The relationships of equivalent strain parameters versus fatigue life under the investigated loading conditions are shown in Figure 10(a) for the base-metal specimens and in Figure 10(b) for the welded-metal specimens, respectively. It can be seen that the fatigue life under in-phase loading conditions is slightly longer than that under uniaxial loading conditions for both base metal and welded metal, while the fatigue life for 90° out-of-phase loading conditions is significantly reduced for both base metal and welded metal compared with the uniaxial case. Therefore, the following conclusion can be drawn: the equivalent strain parameter is not well correlated with fatigue life in the case of out-of-phase loading.

Fatigue lives of base-metal and welded-metal specimen for the three investigated loading conditions are compared in Figure 11. It can be observed that the fatigue life of welded-metal specimen is greatly reduced compared with that of base metal under the same loading conditions, and the fatigue life reduction of weldment increases with increasing strain amplitude.

5. Critical Plane Parameters

Three strain-based critical plane parameters, which are adopted to correlate the fatigue life for the different loading conditions, are investigated in this section.

Kandil-Brown-Miller [22] takes the linear combination of the shear and the normal strain ranges acting on the critical plane as the fatigue parameter. The correlation of the KBM parameters and fatigue lives can be obtained by employing

TABLE 4: Fatigue test and analysis results for base-metal and welded-metal specimens.

Strain paths	$\dfrac{\Delta\varepsilon}{2}$ (%)	$\dfrac{\Delta\gamma}{2}$ (%)	λ	$\dfrac{\Delta\sigma}{2}$ (MPa)	$\dfrac{\Delta\tau}{2}$ (MPa)	$\dfrac{\Delta\gamma_{\max}}{2}$ (%)	$\Delta\varepsilon_n$ (%)	$\sigma_{n,\max}$ (MPa)	N_f
Base-metal fatigue test									
UA	0.60	0	0	270	0	0.87	0.16	135	3443
UA	0.60	0	0	273	0	0.87	0.16	137	4012
UA	0.80	0	0	289	0	1.17	0.21	145	1643
UA	0.80	0	0	288	0	1.17	0.21	144	1817
UA	1.03	0	0	302	0	1.51	0.27	151	1257
UA	1.01	0	0	299	0	1.49	0.27	149	1254
IP	0.27	0.50	1.852	194	88	0.63	0.08	119	16269
IP	0.42	0.74	1.762	196	105	0.96	0.11	107	4443
IP	0.42	0.74	1.762	195	107	0.96	0.11	102	4535
IP	0.56	0.96	1.714	214	108	1.26	0.15	119	2929
IP	0.56	0.97	1.732	213	108	1.27	0.15	118	2971
IP	0.71	1.25	1.761	219	111	1.63	0.19	123	1475
IP	0.71	1.18	1.662	218	110	1.58	0.19	117	2053
OP	0.40	0.69	1.725	305	176	0.69	0.17	313	1494
OP	0.40	0.70	1.750	306	170	0.70	0.17	312	1519
OP	0.51	0.85	1.667	326	182	0.81	0.22	322	1002
OP	0.50	0.81	1.620	321	181	0.85	0.22	326	992
OP	0.61	1.01	1.656	336	189	1.01	0.27	336	703
OP	0.61	0.99	1.623	329	186	0.99	0.27	329	642
Welded-metal fatigue test									
UA	0.61	0	0	295	0	0.89	0.17	147	289
UA	0.60	0	0	293	0	0.87	0.16	146	432
UA	0.81	0	0	313	0	1.18	0.22	156	118
UA	0.80	0	0	315	0	1.17	0.22	157	177
UA	1.02	0	0	329	0	1.50	0.27	164	70
UA	1.00	0	0	334	0	1.47	0.27	167	60
IP	0.28	0.49	1.750	189	104	0.63	0.08	102	2697
IP	0.42	0.75	1.786	221	121	0.96	0.12	118	1262
IP	0.42	0.74	1.762	220	122	0.97	0.12	120	701
IP	0.57	1.01	1.772	230	131	1.31	0.15	127	384
IP	0.56	1.00	1.786	241	134	1.29	0.15	129	393
IP	0.71	1.25	1.761	253	128	1.63	0.19	142	160
IP	0.70	1.27	1.814	243	129	1.63	0.19	135	145
OP	0.40	0.69	1.725	345	195	0.69	0.18	345	359
OP	0.41	0.68	1.659	335	190	0.68	0.19	335	388
OP	0.49	0.88	1.796	350	198	0.91	0.23	362	220
OP	0.50	0.91	1.820	362	207	0.88	0.24	356	172
OP	0.63	1.03	1.635	405	234	0.88	0.23	390	98
OP	0.56	1.07	1.911	410	237	1.03	0.30	383	76

the Manson-Coffin equation. Thus, the Kandil-Brown-Miller (KBM) model can be given by

$$\frac{\Delta\gamma_{\max}}{2} + S\Delta\varepsilon_n = \left[1 + \nu_e + (1 - \nu_e)\,S\right] \frac{\sigma_f'}{E} \left(2N_f\right)^b \\ + \left[1 + \nu_p + \left(1 - \nu_p\right)S\right] \varepsilon_f' \left(2N_f\right)^c, \quad (4)$$

where $\Delta\gamma_{\max}$ and $\Delta\varepsilon_n$ are the maximum shear strain range and the normal strain range acting on the critical plane, respectively. ν_e and ν_p are the elastic and plastic Poisson's ratios, respectively. Consistency of volume requires the elastic Poisson's ratio to typically equal 0.3 and the plastic Poisson's ratio to be 0.5. S is an experimental coefficient of the investigated materials.

TABLE 5: Fatigue properties of base metal and welded metal.

Fatigue properties	Uniaxial tests	In-phase tests	90° out-of-phase tests
		Base-metal specimen	
σ'_f (MPa)	407.59	438.40	535.09
ε'_f	0.8091	0.6196	0.5630
b	−0.0424	−0.0520	−0.0702
c	−0.5827	−0.5277	−0.6723
		Welded-metal specimen	
σ'_f (MPa)	481.14	508.27	519.35
ε'_f	0.0375	0.0654	0.0202
b	−0.0755	−0.0708	−0.3555
c	−0.3154	−0.0624	−0.3181

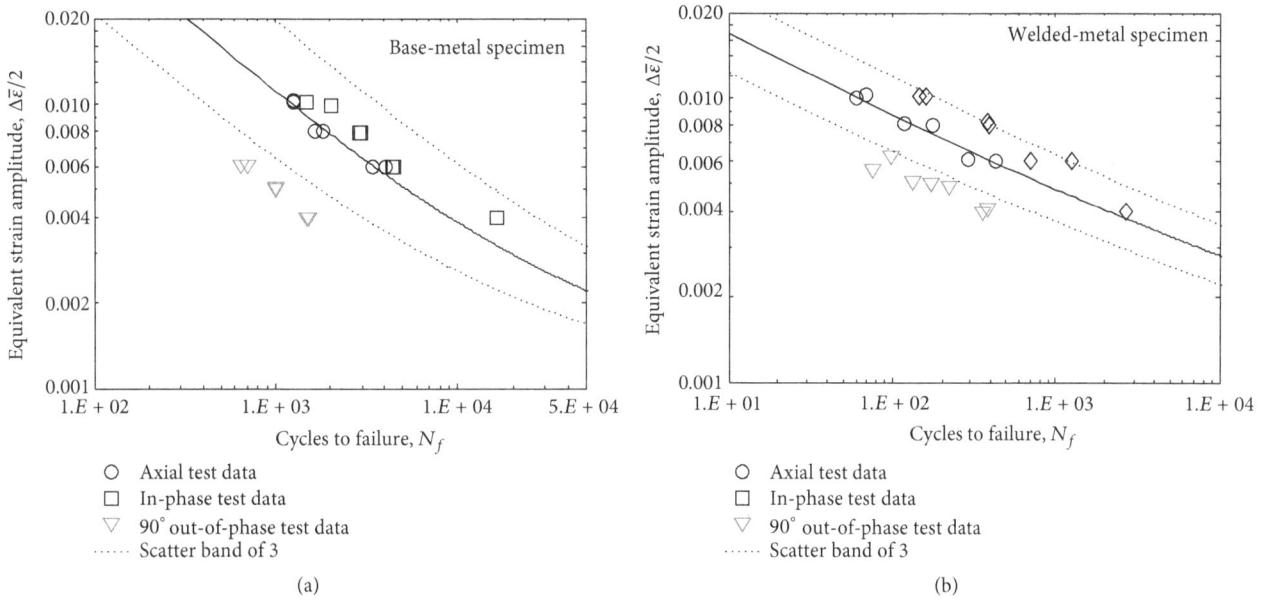

(a)

(b)

FIGURE 10: Relationship of equivalent strain versus fatigue life: (a) base metal and (b) welded metal.

Fatemi and Socie [25] proposed a widely accepted critical plane concept using the maximum normal stress, which can reflect the effect of nonproportional cyclic additional hardening on multiaxial fatigue damage to replace the normal strain in KBM parameter. The Fatemi-Socie (FS) model can be written as

$$\frac{\Delta\gamma_{\max}}{2}\left(1 + k\frac{\sigma_{n,\max}}{\sigma_y}\right)$$
$$= \left[\left(1 + \nu_e\right)\frac{\sigma'_f}{E}\left(2N_f\right)^b + \left(1 + \nu_p\right)\varepsilon'_f\left(2N_f\right)^c\right] \quad (5)$$
$$\cdot\left(1 + k\frac{\sigma'_f}{2\sigma_y}\left(2N_f\right)^b\right),$$

where $\sigma_{n,\max}$ is the maximum normal stress acting on the critical plane. σ_y is the yield strength for the investigated

materials. k is an experimental coefficient. As an approximation, one may simply assume the experimental coefficient in FS model to be 1.0 [10].

Based on the FS critical plane concept, Li and the coauthors [4] developed a stress-correlated factor to consider the effect of the nonproportional additional hardening on multiaxial fatigue damage, which can be used to modify the KBM parameter. The Modified KBM (MKBM) model can be rewritten as

$$\frac{\Delta\gamma_{\max}}{2} + \left(1 + \frac{\sigma_{n,\max}}{\sigma_y}\right)\frac{\Delta\varepsilon_n}{2}$$
$$= \left[\frac{\sigma'_f}{E}\left(2N_f\right)^b + \varepsilon'_f\left(2N_f\right)^c\right]\left(1 + \frac{\sigma'_f}{\sigma_y}\left(2N_f\right)^b\right). \quad (6)$$

Figures 12–14 present the correlation of the KBM, FS, and MKBM parameters with the observed fatigue life. It can be seen that KBM parameters fit well fatigue life data of both base-metal and welded-metal specimens for the uniaxial and

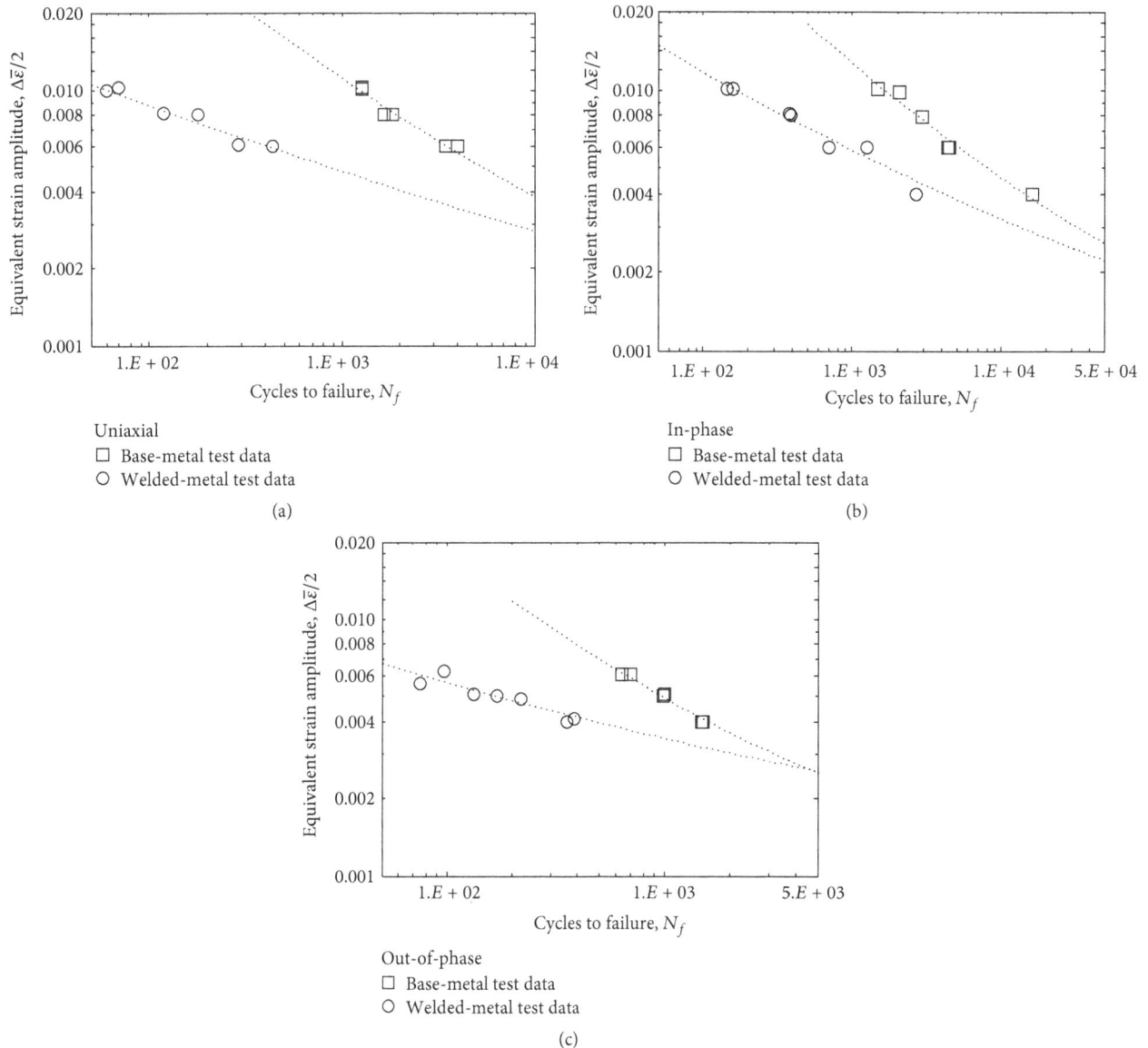

FIGURE 11: Comparison of fatigue life between base-metal and welded-metal specimen: (a) uniaxial loading, (b) in-phase loading, and (c) 90° out-of-phase loading.

in-phase tests. No correlation could be found for the 90° out-of-phase loading condition. It should be noted that the experimental coefficient S in KBM model is taken as 0.8 for the base metal and 1.0 for the welded metal, respectively, which are the averaged values over the fatigue test data.

FS parameters and fatigue life are well correlated for base metal and welded metal for all loading conditions. The correlation of the MKBM parameters with the fatigue life for the out-of-phase loading is greatly improved with respect to KBM, especially for the welded-metal specimens under 90° out-of-phase loading. The improving predictions of the FS and MKBM parameters for out-of-phase loading conditions can be attributed to the introduced stress term $\sigma_{n,\max}$, which reflects the effect of the material additional cyclic hardening due to the nonproportionality of the cyclic loading on multiaxial fatigue damage.

A critical plane approach should be able to predict both the fatigue life and the critical planes, where cracks are predicted to initiate [26, 27]. Comparison of theoretical predictions with experimental data allows evaluating the validity of a proposed critical plane approach. However, due to the difficulties in defining cracking direction observed experimentally due to the roughness of the crack surface, limited work has been done in the evaluation of a critical plane approach for predicting the cracking directions. Jiang [26] investigated the cracking behavior predictions of FS model by using 35 thin-walled tubular specimens of S460N steel and found that only about 20% specimens are predicted correctly by the FS model for the cracking orientations, which reveals that the predictions for the cracking directions are far less desirable than the fatigue life predictions. The time-consuming cracking behavior observation may have

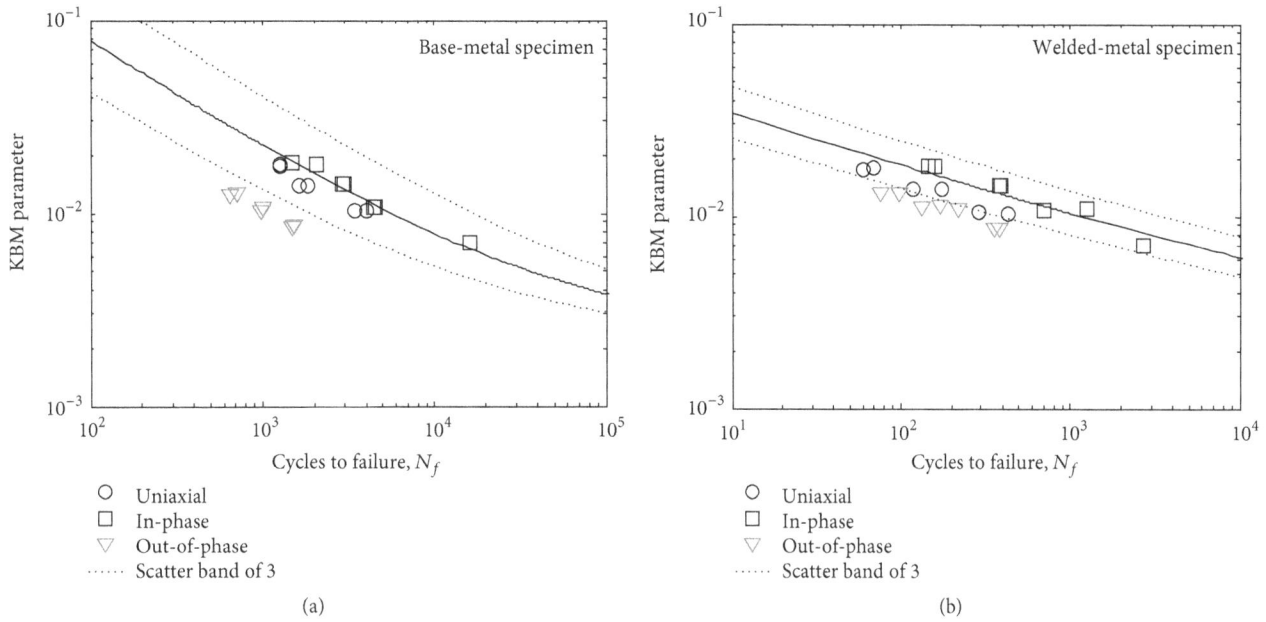

FIGURE 12: Correlation of KBM parameters with the fatigue life: (a) base metal and (b) welded metal.

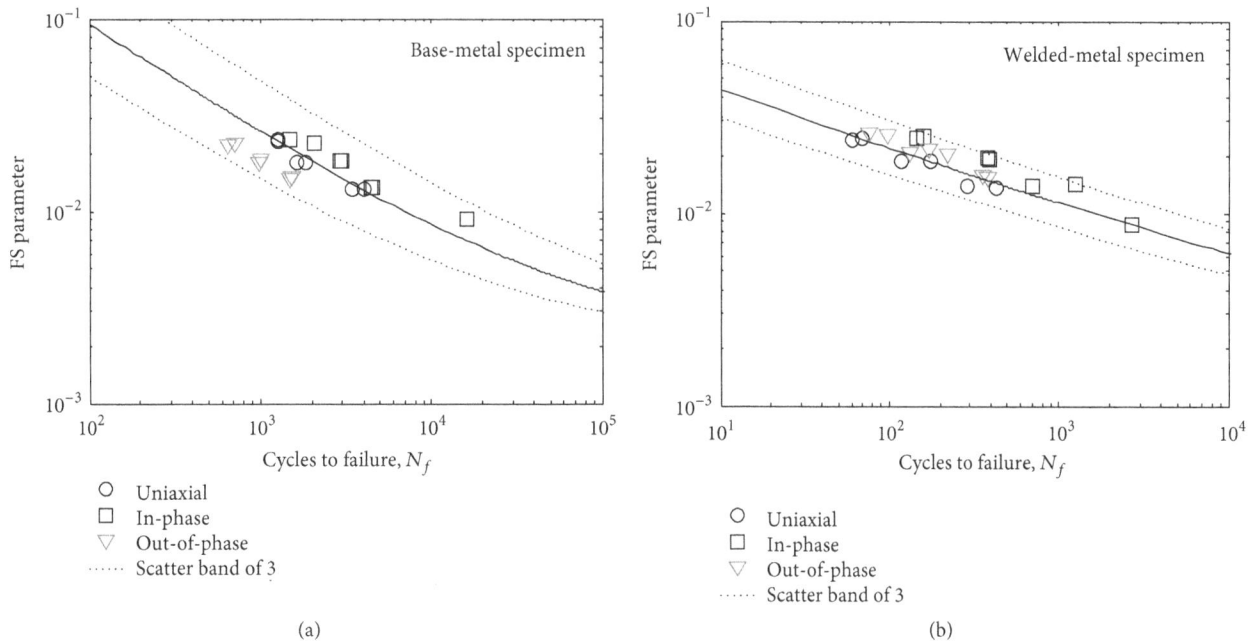

FIGURE 13: Correlation of FS parameters with the fatigue life: (a) base metal and (b) welded metal.

prevented more detailed work on the cracking direction examinations [26]. Accordingly, it is worth pointing out that the predictions for cracking directions using different critical plane models certainly warrant investigations in the future.

6. Conclusions

The multiaxial cycle deformation and fatigue behavior of Q235B mild carbon steel and its related welded metal are experimentally investigated in the present paper. The following conclusions can be drawn:

(1) Significant additional cyclic hardening effect is observed for both base steel and welded metal under out-of-phase loading conditions. Besides, welding proc-ess produces extra additional hardening compared with the base metal.

(2) Fatigue strength under in-phase loading is slightly higher than that under uniaxial loading for both base

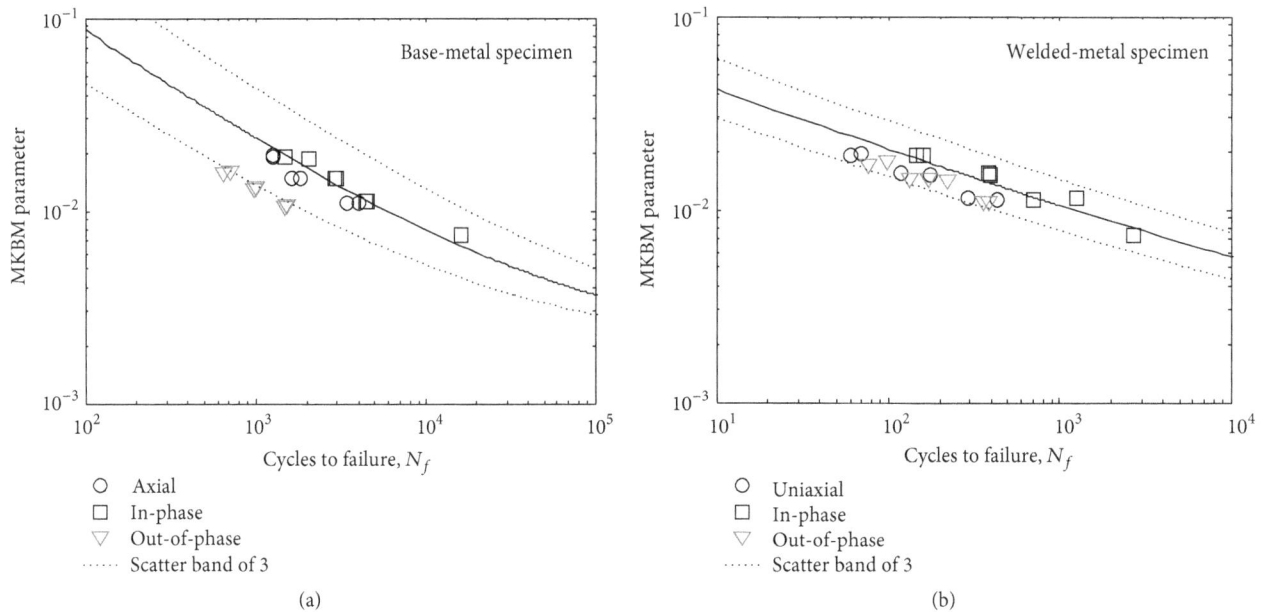

FIGURE 14: Correlation of MKBM parameters with the fatigue life: (a) base metal and (b) welded metal.

metal and welded metal, while the fatigue strength under 90° out-of-phase loading is significantly reduced.

(3) Fatigue strength for welded-metal specimens is greatly reduced compared with the base metal under the same loading conditions, and the fatigue life reduction of the weldment increases with increasing applied strain amplitudes.

(4) The FS and the MKBM parameters show better correlation with fatigue life for both base-metal and welded-metal specimens. The good correlation can be attributed to the introduction of maximum normal stress acting on the critical plane, which can reflect the influence of the nonproportional hardening on multiaxial fatigue damage.

Competing Interests

The authors declare that there are no competing interests regarding the publication of this paper.

Acknowledgments

The authors would like to thank Professor Keshi Zhang in Guangxi University for providing the fatigue test equipment and invaluable comments on the experimental study. The authors also gratefully acknowledge the financial support for this work from National Natural Science Foundation of China (nos. 51378409 and 51438002).

References

[1] A. Karolczuk and E. Macha, "A review of critical plane orientations in multiaxial fatigue failure criteria of metallic materials," *International Journal of Fracture*, vol. 134, no. 3-4, pp. 267–304, 2005.

[2] J. Das and S. M. Sivakumar, "An evaluation of multiaxial fatigue life assessment methods for engineering components," *International Journal of Pressure Vessels and Piping*, vol. 76, no. 10, pp. 741–746, 1999.

[3] C. M. Sonsino, "Effect of residual stresses on the fatigue behaviour of welded joints depending on loading conditions and weld geometry," *International Journal of Fatigue*, vol. 31, no. 1, pp. 88–101, 2009.

[4] J. Li, Z.-P. Zhang, Q. Sun, and C.-W. Li, "Multiaxial fatigue life prediction for various metallic materials based on the critical plane approach," *International Journal of Fatigue*, vol. 33, no. 2, pp. 90–101, 2011.

[5] B. C. Li, C. Jiang, X. Han, and Y. Li, "A new path-dependent multiaxial fatigue model for metals under different paths," *Fatigue & Fracture of Engineering Materials & Structures*, vol. 37, no. 2, pp. 206–218, 2014.

[6] X. Chen, D. Jin, and K. S. Kim, "Fatigue life prediction of type 304 stainless steel under sequential biaxial loading," *International Journal of Fatigue*, vol. 28, no. 3, pp. 289–299, 2006.

[7] Z. Gao, T. Zhao, X. Wang, and Y. Jiang, "Multiaxial fatigue of 16MnR steel," *Journal of Pressure Vessel Technology*, vol. 131, no. 2, pp. 73–80, 2009.

[8] T. Zhao and Y. Jiang, "Fatigue of 7075-T651 aluminum alloy," *International Journal of Fatigue*, vol. 30, no. 5, pp. 834–849, 2008.

[9] D. G. Shang and D. J. Wang, *Multiaxial Fatigue Strength*, Science Press, Beijing, China, 2007 (Chinese).

[10] N. Shamsaei, *Multiaxial Fatigue and Deformation Including Non-Proportional Hardening and Variable Amplitude Loading Effects*, The University of Toledo Dissertations, 2010.

[11] M. Gladskyi and S. Shukaev, "A new model for low cycle fatigue of metal alloys under non-proportional loading," *International Journal of Fatigue*, vol. 32, no. 10, pp. 1568–1572, 2010.

[12] D. Radaj, C. M. Sonsino, and W. Fricke, *Fatigue Assessment of Welded Joints by Local Approaches*, Wood Publishing Limited, Cambridge, UK, 2006.

[13] V. Caccese, P. A. Blomquist, K. A. Berube, S. R. Webber, and N. J. Orozco, "Effect of weld geometric profile on fatigue life of cruciform welds made by laser/GMAW processes," *Marine Structures*, vol. 19, no. 1, pp. 1–22, 2006.

[14] S. J. Kim, R. T. Dewa, W. G. Kim, and M. H. Kim, "Cyclic stress response and fracture behaviors of Alloy 617 base metal and weld joints under LCF loading," *Advances in Materials Science and Engineering*, vol. 2015, Article ID 207497, 11 pages, 2015.

[15] A. Hobbacher, *Recommendations for Fatigue Design of Welded Joints and Components*, International Institute of Welding, XIII-1539-96/XV-845-96, 2007.

[16] Eurocode 3, *Design of Steel Structures. Part 1-9: Fatigue*, European Standard EN 1993-1-9, 2005.

[17] BS5400, *Steel, Concrete and Composite Bridges. Part 10: Code of Practice for Fatigue*, BS5400, 1980.

[18] X. Chen, K. An, and K. S. Kim, "Low-cycle fatigue of 1Cr-18Ni-9Ti stainless steel and related weld metal under axial, torsional and 90° out-of-phase loading," *Fatigue & Fracture of Engineering Materials & Structures*, vol. 27, no. 6, pp. 439–448, 2004.

[19] M. Bäckström and G. Marquis, "A review of multiaxial fatigue of weldments: experimental results, design code and critical plane approaches," *Fatigue & Fracture of Engineering Materials & Structures*, vol. 24, no. 5, pp. 279–291, 2001.

[20] J. Szusta and A. Seweryn, "Low-cycle fatigue model of damage accumulation—the strain approach," *Engineering Fracture Mechanics*, vol. 77, no. 10, pp. 1604–1616, 2010.

[21] A. Karolczuk and E. Macha, "A review of critical plane orientations in multiaxial fatigue failure criteria of metallic materials," *International Journal of Fracture*, vol. 134, no. 3-4, pp. 267–304, 2005.

[22] D. F. Socie and G. B. Marquis, *Multiaxial Fatigue*, Society of Automotive Engineers, Warrendale, Pa, USA, 2000.

[23] N. Shamsaei and A. Fatemi, "Effect of microstructure and hardness on non-proportional cyclic hardening coefficient and predictions," *Materials Science and Engineering A*, vol. 527, no. 12, pp. 3015–3024, 2010.

[24] Y. Wang, X. Liu, G. Dai, and Y. Shi, "Experimental study on constitutive relation of steel SN490B under cyclic loading," *Journal of Building Structures*, vol. 35, no. 4, pp. 142–148, 2014.

[25] A. Fatemi and D. F. Socie, "A critical plane approach to multiaxial fatigue damage including out-of-phase loading," *Fatigue & Fracture of Engineering Materials & Structures*, vol. 11, no. 3, pp. 149–165, 1988.

[26] Y. Jiang, "Fatigue criterion for general multiaxial loading," *Fatigue and Fracture of Engineering Materials and Structures*, vol. 23, no. 1, pp. 19–32, 2000.

[27] D. McClaflin and A. Fatemi, "Torsional deformation and fatigue of hardened steel including mean stress and stress gradient effects," *International Journal of Fatigue*, vol. 26, no. 7, pp. 773–784, 2004.

10

Controllable Synthesis of Silver Nanoparticles Using Three-Phase Flow Pulsating Mixing Microfluidic Chip

Guojun Liu ⓘ,[1] **Xiang Ma** ⓘ,[1] **Xiaodong Sun** ⓘ,[2] **Yanhui Jia** ⓘ,[1] and **Tengfei Wang** ⓘ[1]

[1]*College of Mechanical Science and Engineering, Jilin University, Changchun 130025, China*
[2]*College of Communication Engineering, Jilin University, Changchun 130025, China*

Correspondence should be addressed to Yanhui Jia; jiayh@jlu.edu.cn

Academic Editor: Fernando Lusquiños

On the basis of liquid-phase reduction mechanism, a novel synthesis method to prepare silver nanoparticles (AgNPs) is proposed, which uses piezoelectric-actuated three-phase flow pulsating mixing microfluidic chip. In order to study and explore the influence of different factors on the synthesis of AgNPs, a series of related synthesis experiments were carried out. The corresponding experimental conditions include the concentration of sodium hydroxide and reducing agent solution, polyvinylpyrrolidone (PVP) dosage, inlet flow rate, and synthesis temperature. The synthesized AgNPs were characterized by the UV-Vis absorption spectrophotometer and transmission electron microscopy. The effects of different experimental conditions on the controllable synthesis of AgNPs were analyzed, and the optimum synthesis conditions of AgNPs were obtained. Experimental results show that the spherical AgNPs with an average particle diameter of about 29 nm, high yield, fine morphology, and good monodispersity were synthesized using the microfluidic chip under the conditions of the working frequency (200 Hz), the initial concentration of silver nitrate (1 mM), the synthesis temperature (80°C), the concentration ratio of sodium hydroxide to silver nitrate (2 : 1), the concentration ratio of glucose to silver nitrate (4 : 1), the inlet flow rate (3.5 ml/min), and the quality ratio of PVP to silver (more than 1 : 1). The related research shows that it is an efficient synthesis method to develop the controllable synthesis experiments of AgNPs under multifactors using the three-phase pulsating mixing microfluidic chip.

1. Introduction

Silver nanoparticles (AgNPs), due to the volume effect, surface effect, quantum size effect, tunneling effect, and some new unique properties, are ideal candidates for biological [1], catalytic [2, 3], agricultural [4], chemical sensor [5], chemical probe [6], and medical [7, 8] applications. However, the properties and applications of these nanoparticles mainly depend upon the shape, size, and distribution of nanoparticles [9]. Hence, how to prepare AgNPs with good performance is of vital importance.

Since the 21st century, there have been many new methods of preparing AgNPs in the world [10–14]. For example, Agrawal et al. [10] synthesized Ag and Al nanoparticles using an innovative approach of ultrasonic dissociation of thin films. Küünal et al. [11] reported on the study of surfactant-free silver nanoparticles synthesized using nonhydrolytic sol-gel methods. Roopan et al. [12] used mesocarp layer extract of *Cocos nucifera* as a reduction agent to synthesize silver nanoparticles. However, in recent years with the rapid development of microfluidic technology, the microfluidic chips based on microfluidic technology have been widely used in the synthesis of AgNPs due to their characteristics of low consumption of reagent, high reaction efficiency, controllable reaction process, and safe and reliable [15–18].

Compared to other synthetic methods, the synthesis of AgNPs by microfluidic chip effectively shortens the synthesis time, simplifies the preparation process, and enables the controllable synthesis of AgNPs. But the controllable synthesis of AgNPs is affected by many factors [19–23], and how to synthesize AgNPs with good performance by adjusting their influencing factors in microfluidic system is attracting more and more attention. Baber et al. [19] synthesized AgNPs with an average diameter of 3.1~9.3 nm and narrow size distribution by varying the concentration and

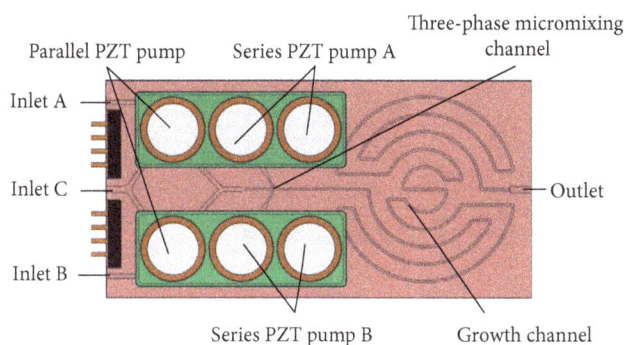

(a)

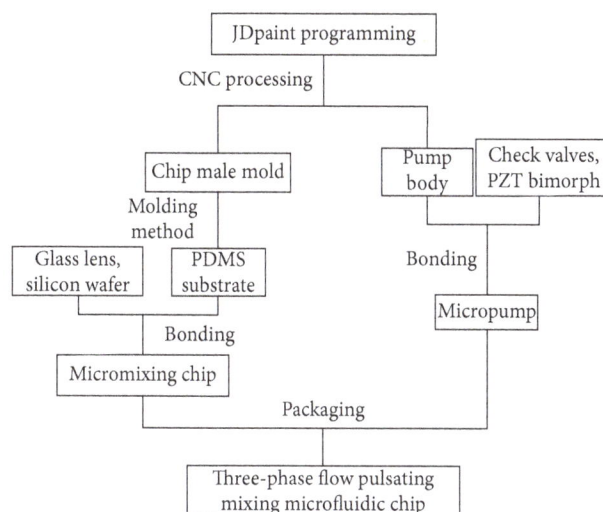

(b)

(c)

FIGURE 1: (a) The prototype schematic of the microfluidic chip. (b) The fabrication flowchart of microfluidic chip for three-phase flow pulsating mixing. (c) The prototype photo of the microfluidic chip.

flow rate of the reaction reagents in a coaxial flow reactor. They found that higher flow rates resulted in the appearance of larger AgNPs (as well as increased polydispersity), and higher silver nitrate concentrations will be increasing the likelihood of reactions occurring in silver nitrate-rich zones, which leads to an increased formation of larger nanoparticles and polydispersity. He et al. [20] synthesized AgNPs with different particle size and monodispersity by adjusting the reaction temperature and the flow rate of the precursor solution in the microfluidic device. Lazarus et al. [21] analyzed various flow conditions to determine optimal flow rates for producing gold and silver nanoparticles. They demonstrated that the combination of fast and controlled microfluidic mixing with ionic liquid solvents allows for the synthesis of high-quality, small, and monodisperse gold and silver nanoparticles in the microfluidic system. Yang et al. [22] synthesized AgNPs with uniform particle size by controlling the flow rate of the solution and chitosan concentrations in the microfluidic chip. They found that reducing the flow rate of the continuous phase or increasing the flow rate of the dispersed phase achieves a narrow size distribution of AgNPs, and they also revealed that the particle size increases with increasing chitosan concentration. Xu et al. [23] observed that an increase in the mean diameter of AgNPs with increasing

fluid flow rates. However, AgNPs formed clusters at higher flow rates. In addition, they found that EDTA as a complexant presented significant effect on the morphology and particle size control of spherical AgNPs with good dispersion. Through experiments, they also proved that the controllable synthesis of AgNPs can be realized by using the microfluidic device.

It is clear from our literature survey that, on the basis of the microfluidic chip, the AgNPs with good performance can be synthesized controllably by changing the synthesis conditions of AgNPs. However, the current studies on the influencing factors of controllable synthesis of AgNPs by microfluidic devices are relatively simple. Besides, the microfluidic devices are mostly Y- or T-type which are driven by a syringe pump, and the operation of microfluidic devices are also more complicated. Based on the current situation, this paper presented a piezoelectric driven integrated three-phase pulsating mixing microfluidic chip, and the synthesis of silver nanoparticles were carried out by changing the synthesis parameters of silver nanoparticles. Analyzing the test results of UV-Vis absorption spectrophotometer (UV-Vis) and transmission electron microscopy (TEM), we discussed the effects of the concentration of sodium hydroxide (NaOH) and glucose ($C_6H_{12}O_6$) solutions, polyvinylpyrrolidone (PVP) dosage, inlet flow rate

(a)

(b)

(c)

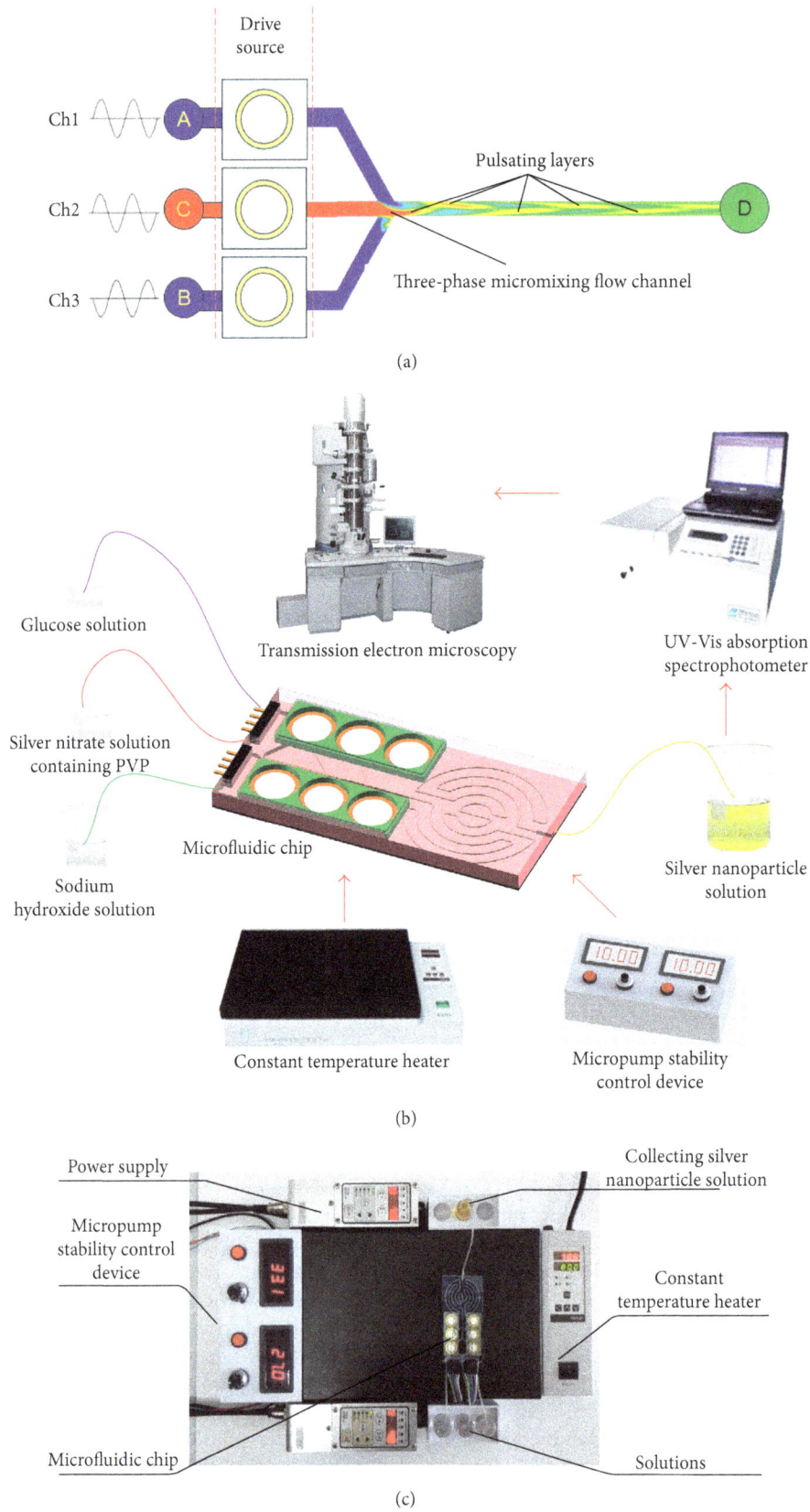

FIGURE 2: (a) Schematic diagram of the three-phase flow pulsating mixing mechanism. (b) Schematic diagram of experimental scheme. (c) The experimental platform photo.

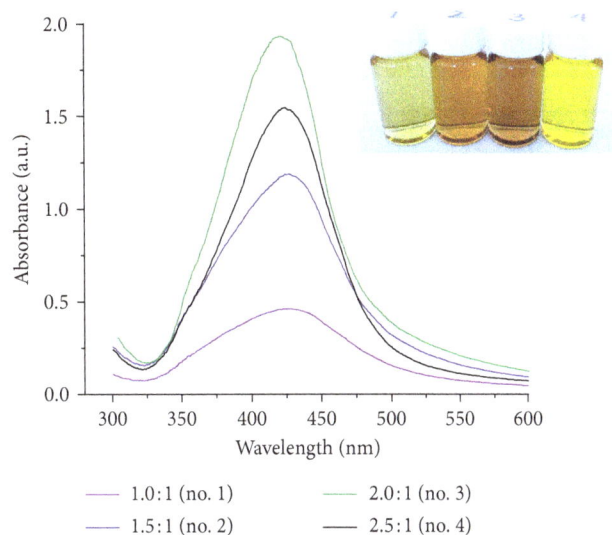

(a)

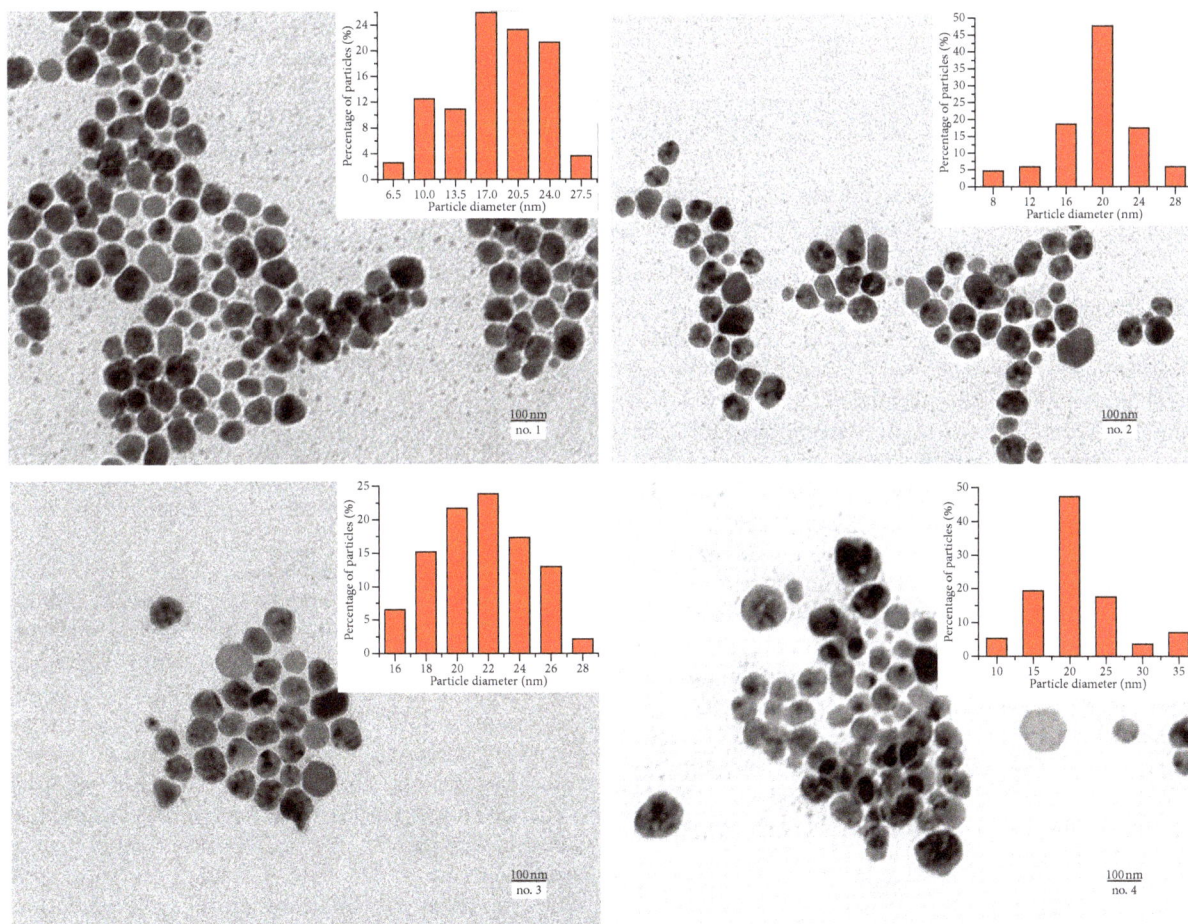

(b)

FIGURE 3: UV-Vis absorption spectra (a) and TEM images (b) of four groups of AgNPs synthesized under different NaOH concentrations.

and synthesis temperature on the particle size, morphology, homogeneity, monodispersity, and yield of AgNPs and then determined the optimum synthesis conditions of AgNPs.

2. Experiment

2.1. Reagents. AgNO$_3$ (precursor), C$_6$H$_{12}$O$_6$ (reducing agent), NaOH (accelerating agent), and PVP (stabilizer) were purchased

TABLE 1: Test results of AgNPs synthesized under different NaOH concentrations.

Number	$C_{\text{NaOH}} : C_{\text{AgNO}_3}$	λ_{m} (nm)	α_{m}	PWHM (nm)
1	1.0 : 1	425	0.462	115.53
2	1.5 : 1	427	1.187	99.47
3	2.0 : 1	421	1.943	91.63
4	2.5 : 1	424	1.544	93.47

from Sinopharm Chemical Reagent Corporation (Shanghai, China). The water used in the experiments was of Milli-Q grade.

2.2. Apparatus. A UV-Vis absorption spectrophotometer (UV-2501 PC, SHIMADZU, Japan) was used to characterize the characteristic absorption spectrum of silver colloid. The size and morphology of synthesized AgNPs were characterized by high-resolution transmission electron microscopy (TEM TECNAI-12, Philips, Holland). The micropumps and microfluidic chip used in experiment were manufactured by mature MEMS process in lab. In addition, a flat panel temperature control heater (BHW-05A Shanghai Broadcom Chemical Technology Co. Ltd.), precision electronic balance (Sartorius Scientific Instrument Co. Ltd.), and PZT actuator control power supply (Nanjing CUH Science & Technology Co. Ltd.), and so on were used in specific experiment test.

2.3. Fabrication of the Microfluidic Chip. The three-phase flow pulsating mixing microfluidic chip is the core apparatus for the controllable synthesis of AgNPs, which was designed and manufactured by MEMS technology and processes in laboratory. The prototype schematic of the microfluidic chip is shown in Figure 1(a), which consists of two series PZT pumps (pump A and B), one parallel PZT pump, three-phase micromixing channel, growth channel, fluid inlet, and fluid outlet. The key structural parameters of mixing channel have been optimized by the CFD simulation software.

Figure 1(b) is the fabrication flowchart of the microfluidic chip for three-phase flow pulsating mixing. In order to ensure the microfluidic chip having good processing technology on the aspects of chip integration and bonding, and having good working performance in high temperature and corrosive environment, we selected PDMS, silicon, and FR-4 epoxy glass fiber board as material of chip, substrate, and pump body, respectively. The finished prototype has an overall dimension: 100 mm × 50 mm × 5 mm, as shown in Figure 1(c).

2.4. Controllable Synthesis of AgNPs. Based on chemical reduction mechanism, the liquid-phase synthesis method of AgNPs includes three processes, namely, precipitation of silver atoms, nucleation, and growth of crystal nuclei. In the initial stage of reaction of the reducing agent and precursor, the high efficiency mixing is of significant importance. Therefore, it is essential to know how to enhance the mixing characteristic of three-phase microfluid for the controllable synthesis of AgNPs in a microfluidic chip. Figure 2(a) is the

schematic diagram of the three-phase flow pulsating mixing mechanism. In the experiment, the working frequency of PZT micropumps is chosen to be 200 Hz, namely, the number of mixing per second reaches 200. Two series PZT micropumps are applied with sine wave electrical signal with a phase difference of 180 degrees. They drive A and B solutions alternately through the micromixing channel in pulsating form. While, for the parallel PZT micropump applied with sine signal, the C solution is pumped into the mixing channel in a continuous pulse form. Under the reasonable structural parameters and control parameter condition, a thinner pulsating layer can be formed in the micromixing channel. It can effectively increase the contact area and contact opportunity between the solutions, shorten the mixing time, and greatly improve the mixing effect between three solutions [17, 24, 25]. Figure 2(b) is the schematic diagram of experimental scheme, A represents the glucose ($C_6H_{12}O_6$) solution, C represents the silver nitrate ($AgNO_3$) solution containing PVP, and B represents the sodium hydroxide (NaOH) solution. The above three-phase solutions are mixed intensively and react simultaneously, then entering the annular channel for the growth of crystal nuclei. Finally, the reaction fluid is collected at the outlet and will be inspected after 15 minutes. In the experiment testing, the self-made stability control device is responsible for the stable output of three-phase flow rate. The temperature control heater is responsible to provide with the appropriate reaction temperature. And the quality of the synthesized AgNPs is mainly analyzed and characterized by the TEM and UV-Vis. The experimental platform photo is shown as Figure 2(c).

3. Results and Discussion

3.1. Related Experimental Parameters. During the controllable synthesis of AgNPs using the microfluidic chip and characterization of AgNPs using the related apparatuses, some important experimental parameters are involved. Among them, C_{NaOH}, C_{AgNO_3}, and $C_{C_6H_{12}O_6}$ represent the concentration of NaOH, $AgNO_3$, and $C_6H_{12}O_6$ solution, respectively. The m_{PVP} and m_{Ag}, respectively, indicate the mass of PVP and the mass of Ag in $AgNO_3$ solution. Q is the inlet flow rate, and T is the temperature; where λ_{m} is the wavelength of absorption maximum, α_{m} is the absorption maximum, and PWHM is the peak width at half maximum. Generally, the position of the absorption peak, namely, λ_{m}, can reflect the total situation about the size of nanoparticles. Generally, the larger the size of nanoparticles, the greater the value of λ_{m}, namely, the peak position shifts to a longer wavelength range [26–28]. Normally, the value of α_{m} is proportional to the concentration of AgNPs' solution. The larger the value of α_{m}, the higher the concentration of the solution and the higher the yield of AgNPs [29]. Besides, PWHM mainly reflects the particle size deviation of AgNPs; in general, the smaller the value of PWHM, the smaller the particle size deviation, the better the uniformity of nanoparticles appearance [30].

3.2. Test Results and Analysis. In the lab, several groups of controllable synthesis of AgNPs were completed using the

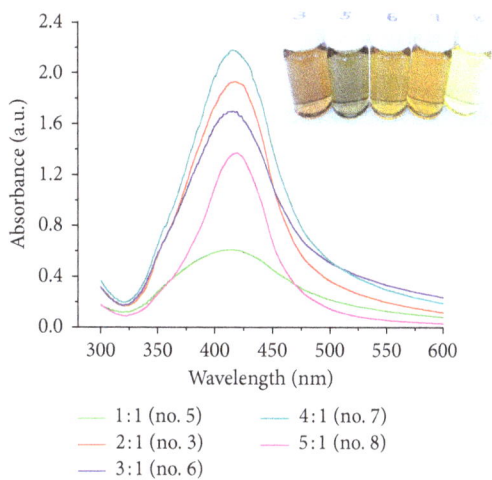

(a)

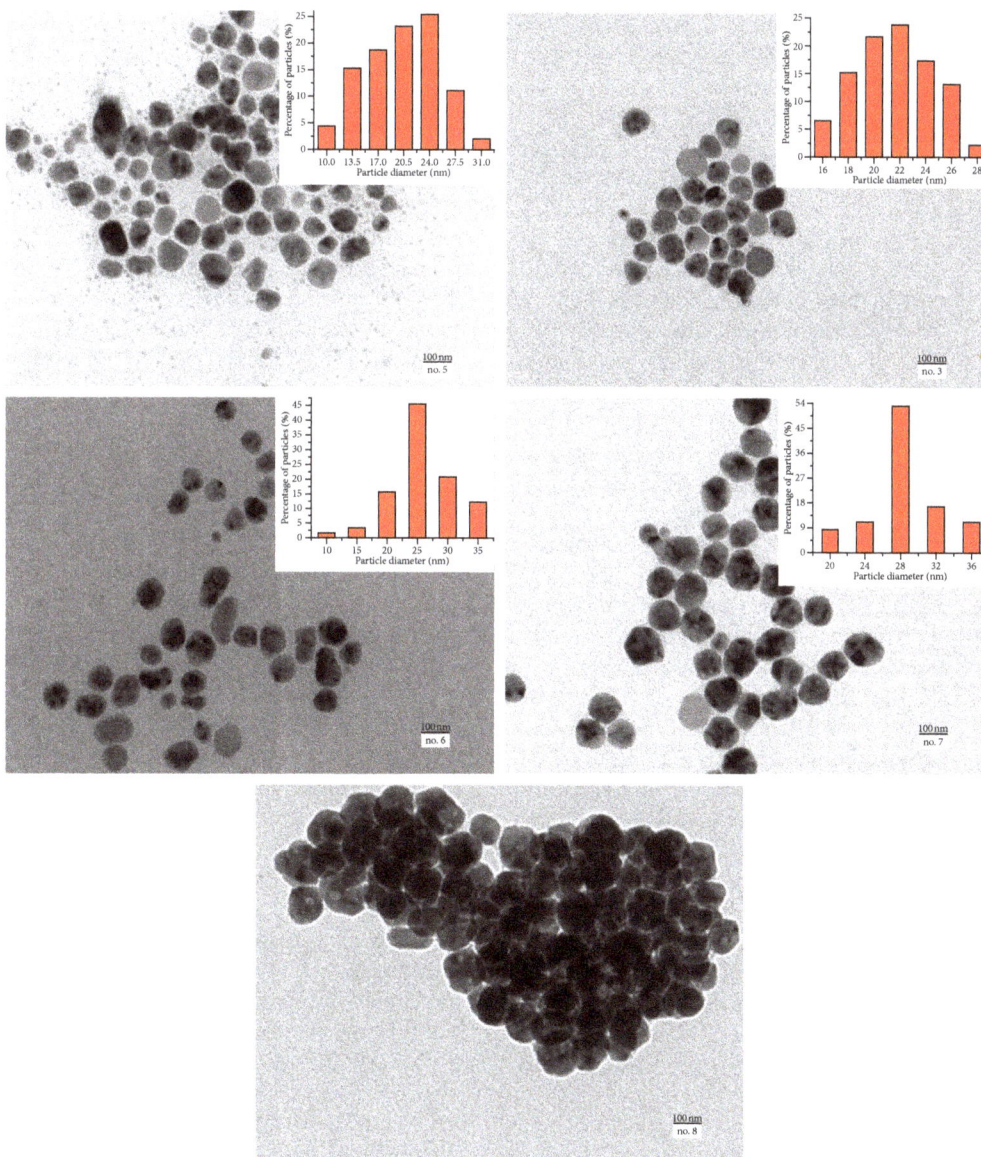

(b)

FIGURE 4: UV-Vis absorption spectra (a) and TEM images (b) of five groups of AgNPs synthesized under different glucose concentrations.

TABLE 2: Test results of AgNPs synthesized under different glucose concentrations.

Number	$C_{C_6H_{12}O_6} : C_{AgNO_3}$	λ_m (nm)	α_m	PWHM (nm)
5	1.0 : 1	414	0.614	118.42
3	2.0 : 1	421	1.943	91.63
6	3.0 : 1	419	1.704	96.58
7	4.0 : 1	421	2.194	91.84
8	5.0 : 1	419	1.379	65.22

self-made microfluidic chip, based on the liquid-phase reduction method. Comprehensively analyzing the effects of different experimental factors on the synthesis results of AgNPs, the optimum synthesis conditions of AgNPs were obtained.

3.2.1. Effect of NaOH Concentration on the Synthesis of AgNPs.

Under the different NaOH concentrations, four groups of synthesis experiments have been carried out. And other experimental conditions were set as follows: the working frequency of PZT micropump was 200 Hz, the initial concentration of AgNO$_3$ solution was 1 mM, the temperature was 80°C, the inlet flow rate Q was set as 3.5 ml·min^{-1}, $C_{C_6H_{12}O_6} : C_{AgNO_3} = 2 : 1$, and $m_{PVP} : m_{Ag} = 1.5 : 1$. Figure 3 shows the UV-Vis absorption spectra and TEM images of four groups of AgNPs synthesized under different NaOH concentrations. The detailed test results are shown in Table 1.

Analyzing the UV-Vis absorption spectra of Figure 3(a) and tests data in Table 1, we can find that the maximum absorption wavelengths of four groups of sample solutions are in the range of 400~430 nm, which all accord with the typical optical characteristics of AgNP colloidal system. From the approximately same value of λ_m of four groups of samples, it can be seen that the concentration of NaOH has little influence on the particle size of AgNPs. Observing the change of the value of α_m, we can see that with the increasing of NaOH concentration, the maximum absorbance of AgNP sol increases first and then decreases. Among them, sample no. 3 has the largest value of the absorption maximum α_m, 1.943. It also indicates that the concentration (i.e., yield) of AgNPs similarly increases first and then decreases. The changes of test data directly coincide with the colour changes of samples. Accordingly, it first turns darker and then becomes shallower, and sample no. 3 has the darkest colour. Analyzing the values of PWHM, we can find that sample no. 1 has the largest value, while no. 3 has the smallest one. It means that the synthesized AgNPs of no. 3 has the smallest deviation of particle size and more uniform particle distribution. Thus, the synthesis conditions of sample no. 3 are relatively better. The TEM photographs of Figure 3(b) show that the AgNP morphologies of other three groups are approximately spherical and monodisperse, with the exception of no. 4. Sample no. 4 shows the undesirable particle aggregation and poor morphology. For the reason of aggregation, we think it may be the excessive NaOH concentration ($C_{NaOH} : C_{AgNO_3} = 2.5 : 1$) which gives rising to a rapid increase of the silver atom concentration in solution. While the concentration of reduced silver atoms is too large,

the probability of nuclei collision will increase during the formation process of ArgentCrystal. The rapid aggregation of nuclei will lead to the formation of irregular nanostructures and even cause partial agglomeration of nanoparticles. Hence, the synthesis effect under this condition is not ideal. In addition, by analyzing the spectrogram of Figure 3(a), we can also find that the maximum absorbance does not maintain a growth trend with the increase of NaOH concentration. This phenomenon still may be due to the higher NaOH concentration, which results in the precipitation reaction of silver hydroxide because of the remaining excessive OH$^-$ and Ag$^+$ in reaction solution.

Comprehensively considering the test results and discussion, it can be found that the concentration of NaOH solution has a great influence on the yield, morphology, monodispersity, and particle size uniformity of the synthesized AgNPs. So, as the important accelerating agent, a suitable optimum NaOH concentration is required. In the actual experiment, when $C_{NaOH} : C_{AgNO_3} = 2 : 1$, the spherical AgNPs with higher concentration, smaller particle size deviation, and better dispersion were synthesized.

3.2.2. Effect of Glucose Concentration on the Synthesis of AgNPs.

In order to study the effect of C$_6$H$_{12}$O$_6$ concentration on the synthesis of AgNPs, five groups of synthesis experiments were carried out using the self-made microfluidic chip under different concentrations of glucose ($C_{C_6H_{12}O_6}$). The basic experimental conditions are the same as the research on effect of NaOH concentration, except of a concentration ratio ($C_{NaOH} : C_{AgNO_3} = 2 : 1$). The UV-Vis absorption spectra and TEM images of five groups of AgNPs synthesized under different glucose concentrations are shown in Figure 4. The detailed test results are listed in Table 2.

Analyzing the UV-Vis absorption spectra of Figure 4(a) and tests data in Table 2, we can find that the maximum absorption wavelengths of five groups of sample solutions are in the range of 410~430 nm, and when $C_{C_6H_{12}O_6} : C_{AgNO_3} = 1 : 1$, the maximum absorption wavelengths λ_m has a little small value. After the concentration ratio of glucose to precursor exceeds 2, the maximum absorption wavelength of the sample shows obvious red shift, but the value of λ_m changes little, which means the synthesized nanoparticle sizes are less affected. Comparing the changes in the maximum absorption peaks (the values of α_m), it can be concluded that with the increasing of glucose concentration, the maximum absorbance of AgNP sol increases first and then decreases. Among them, sample no. 7 has the largest value of the absorption maximum α_m (2.194) which means sample no. 7 has the highest yield of AgNPs. Based on the values of PWHM, sample no. 8 will have the smallest deviation of nanoparticle size and then followed by no. 3 and no. 7 which have approximately the same deviation in nanoparticle size. But comprehensively considering the TEM images of Figure 4(b), we will find that AgNPs of sample no. 8 have poor monodispersity and morphology. For sample no. 8, not only the maximum absorbance is lower, but the more severe agglomeration occurs. As for the possible reasons, we believe that when the glucose concentration of reaction solution is

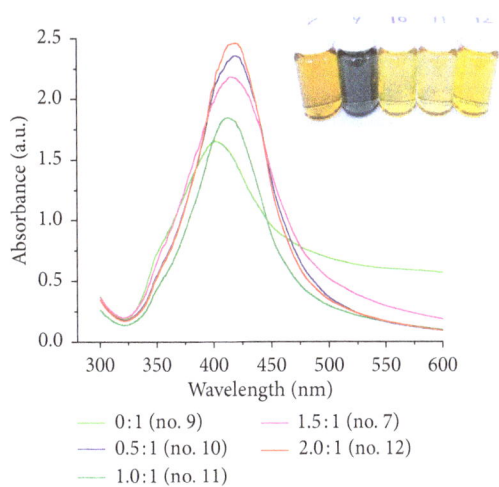

(a)

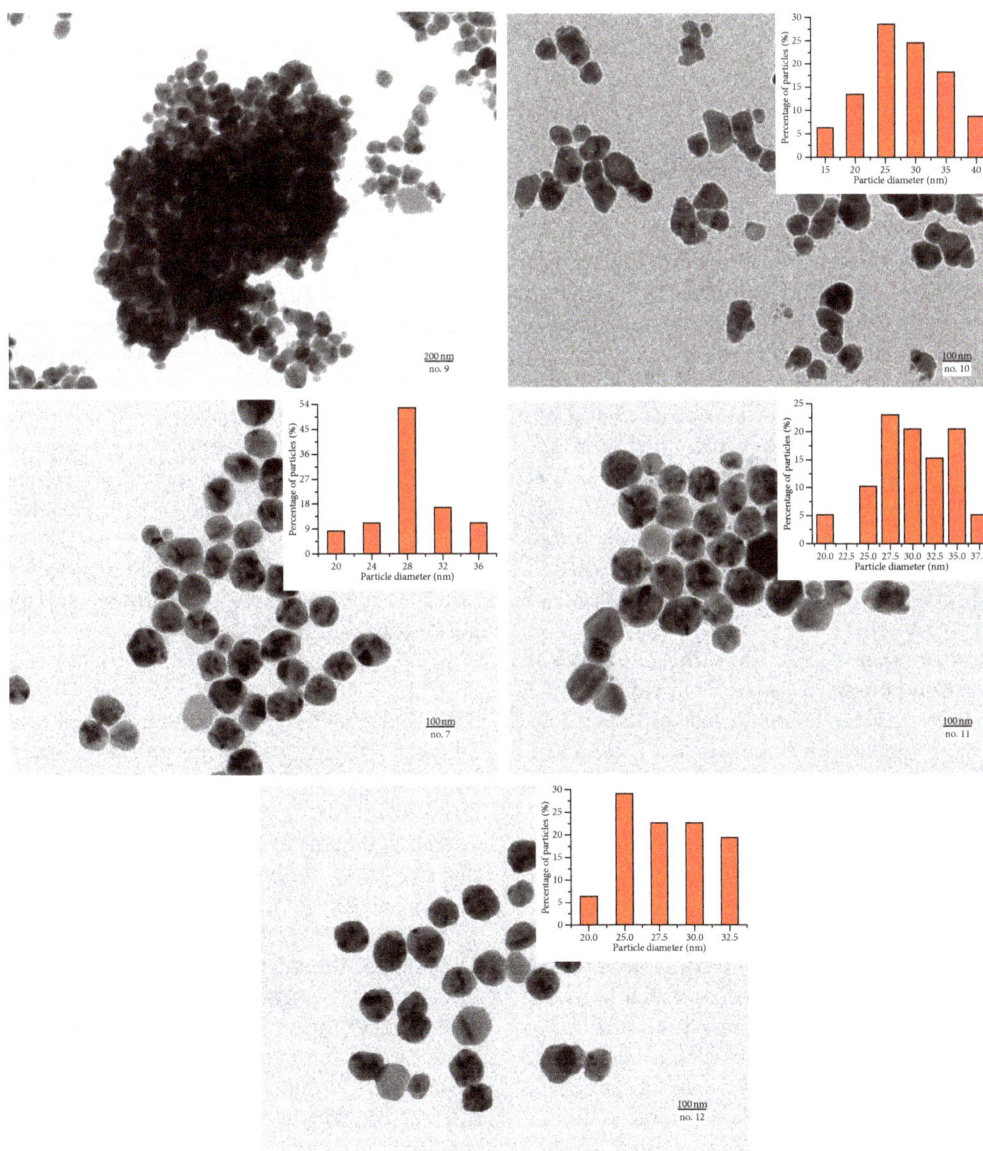

(b)

FIGURE 5: UV-Vis absorption spectra (a) and TEM images (b) of five groups of AgNPs synthesized under different PVP dosages.

too large, the rate of reduction of silver atoms is too fast, which will give rise to the increase of internuclear collision probability of ArgentCrystal and the formation of island agglomerated particles. And it is due to the serious aggregation of AgNPs leading to the decrease in the concentration and the maximum absorbance of AgNPs.

In summary, the glucose concentration affects the synthesis of AgNPs a lot, especially has a great influence on the yield, morphology, and monodispersity of AgNPs. Among the five groups of synthesis experiments launched in lab, samples of no. 3 and no. 7 all have a better synthesis effect. But in view of higher yield of AgNPs in no. 7, the synthesis conditions of sample no. 7 were finally selected as the best synthesis conditions, namely, the optimal concentration of $C_6H_{12}O_6$ solution in the experiment is 4 mM ($C_{C_6H_{12}O_6} : C_{AgNO_3} = 4 : 1$).

3.2.3. Effect of PVP Dosage on the Synthesis of AgNPs.

In order to study the effect of the PVP dosage on the synthesis of AgNPs, the related synthesis experiments were carried out. The basic experiment conditions are the same as that of above Section 3.2.1 (among them, $C_{NaOH} : C_{AgNO_3} = 2 : 1$ and $C_{C_6H_{12}O_6} : C_{AgNO_3} = 4 : 1$). According to the different amount of PVP used in the experiment, the synthesis experiments were designed and divided into five groups. Through relevant experiment testing, the UV-Vis absorption spectra and TEM images of five groups of AgNPs synthesized based on different PVP dosages as shown in Figure 5 were obtained. The detailed test results are listed in Table 3.

Analyzing the changes of the maximum absorption wavelengths λ_m, we can observe that samples no. 9 and no. 10 show an obvious blue shift, while the maximum absorption wavelengths of other samples vary little or almost no change. It shows that with the increase of PVP dosage, the particle size of AgNPs shows a rising trend at the initial stage. But, when the amount of PVP increases to a certain value, it has little effect on the particle size of synthesized nanoparticles.

The changes of test data α_m also show that the effect of PVP addition on the synthesis of AgNPs is obvious. As the amount of PVP increases, the maximum absorption peaks α_m from the initial value 1.659 increases to 2.165. It can be seen that whether PVP is added or not has a great influence on the yield of AgNPs. However, when the mass ratio of PVP to Ag is more than 1, namely, $m_{PVP} : m_{Ag} \geq 1$, the amount of PVP has little effect on the yield of AgNPs. Similarly, the value of PWHM also shows a similar variation trend.

Through further observation of the sample photos and TEM images in Figures 5(a) and 5(b), it can be found that sample no. 9 becomes turbid, the synthesized nanoparticles produce serious aggregation, at the same time, the morphology and monodispersity of nanoparticles are poor, too. After a small amount of PVP ($m_{PVP} : m_{Ag} = 0.5 : 1$) is added, no obvious precipitation occurs in the sample solution. Although the morphology and monodispersity of synthesized nanoparticles have been improved, the overall situation is still relatively poor, and some nanoparticles are still agglomerated. Only when the amount of PVP in solution

TABLE 3: Test results of AgNPs synthesized under different PVP dosages.

Number	$m_{PVP} : m_{Ag}$	λ_m (nm)	α_m	PWHM (nm)
9	0 : 1 (no PVP)	402	1.659	114.47
10	0.5 : 1	413	1.848	87.11
11	1 : 1	419	2.181	90.89
7	1.5 : 1	421	2.194	91.84
12	2 : 1	419	2.165	91.05

exceeds a certain amount (e.g., $m_{PVP} : m_{Ag} = 1 : 1$, $m_{PVP} : m_{Ag} = 1.5 : 1$, and $m_{PVP} : m_{Ag} = 2 : 1$), the aggregation phenomenon of synthesized AgNPs no longer occurs, and the morphology and monodispersity of nanoparticles become better.

Comprehensively analyzing the test results, we find that the amount of PVP has a great influence on the synthesis of AgNPs, especially on the morphology and monodisperse properties of the synthesized nanoparticles. As for the action mechanism, we think that, as a stabilizer, PVP plays a role of surface protective agent in the synthesis of AgNPs. When no PVP is added, or the amount of PVP is less, the silver crystal nuclei formed in the solution cannot be completely encapsulated, and which directly leads to the results that most of synthesized nanoparticles cannot grow into a normal spherical or spherical-like shape. Furthermore, since the nanoparticles are incompletely wrapped, the smaller electrostatic repulsion forces between nanoparticles give rise to the partial agglomeration. While the amount of PVP is sufficient, the silver crystal nuclei fully coated by PVP are more likely to grow into spherical or spherical-like nanoparticles. At the same time, as the amide strong polar groups of PVP are negatively charged, this improves the electrostatic repulsion between nanoparticles. Hence, it is beneficial to synthesize the spherical or spherical-like AgNPs with good dispersibility.

Therefore, it is essential to add proper amount of PVP in order to synthesize AgNPs with regular morphology and good dispersibility.

3.2.4. Effect of Inlet Flow Rate of Microfluidic Chip on the Synthesis of AgNPs.

For the pulsating mixing microfluidic chip actuated by PZT, the working frequency and the inlet flow rate of mixing channels are very important and critical operating parameters that affect the working performance of the chip. No matter for two-phase or three-phase flow mixing, once the working frequency is determined, there will be an optimal flow rate to match it. In order to investigate the influence of the inlet flow rate on the synthesis of AgNPs, the working frequency was selected as 200 Hz [17], and the related experiments based on four groups of inlet flow rates were carried out. The test of the basic experimental conditions was as follows: the initial concentration of silver nitrate solution is 1 mM, the experimental temperature is 80°C, $C_{NaOH} : C_{AgNO_3} = 2 : 1$, $C_{C_6H_{12}O_6} : C_{AgNO_3} = 4 : 1$, and $m_{PVP} : m_{Ag} = 1.5 : 1$. Through series of experimental tests, the UV-Vis absorption spectra and TEM images of four groups of AgNPs synthesized based on different inlet flow rates as

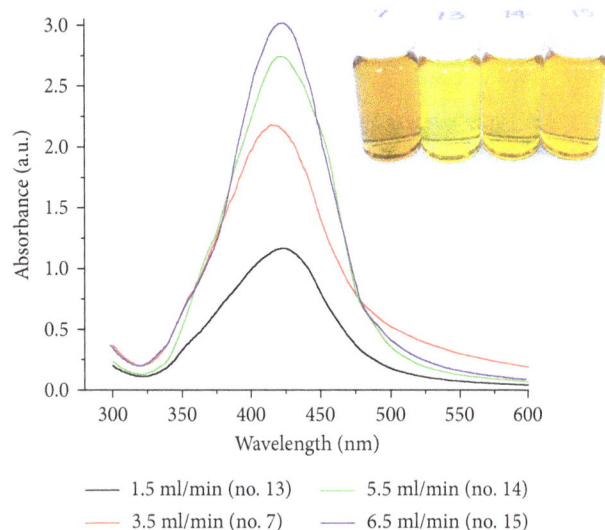

(a)

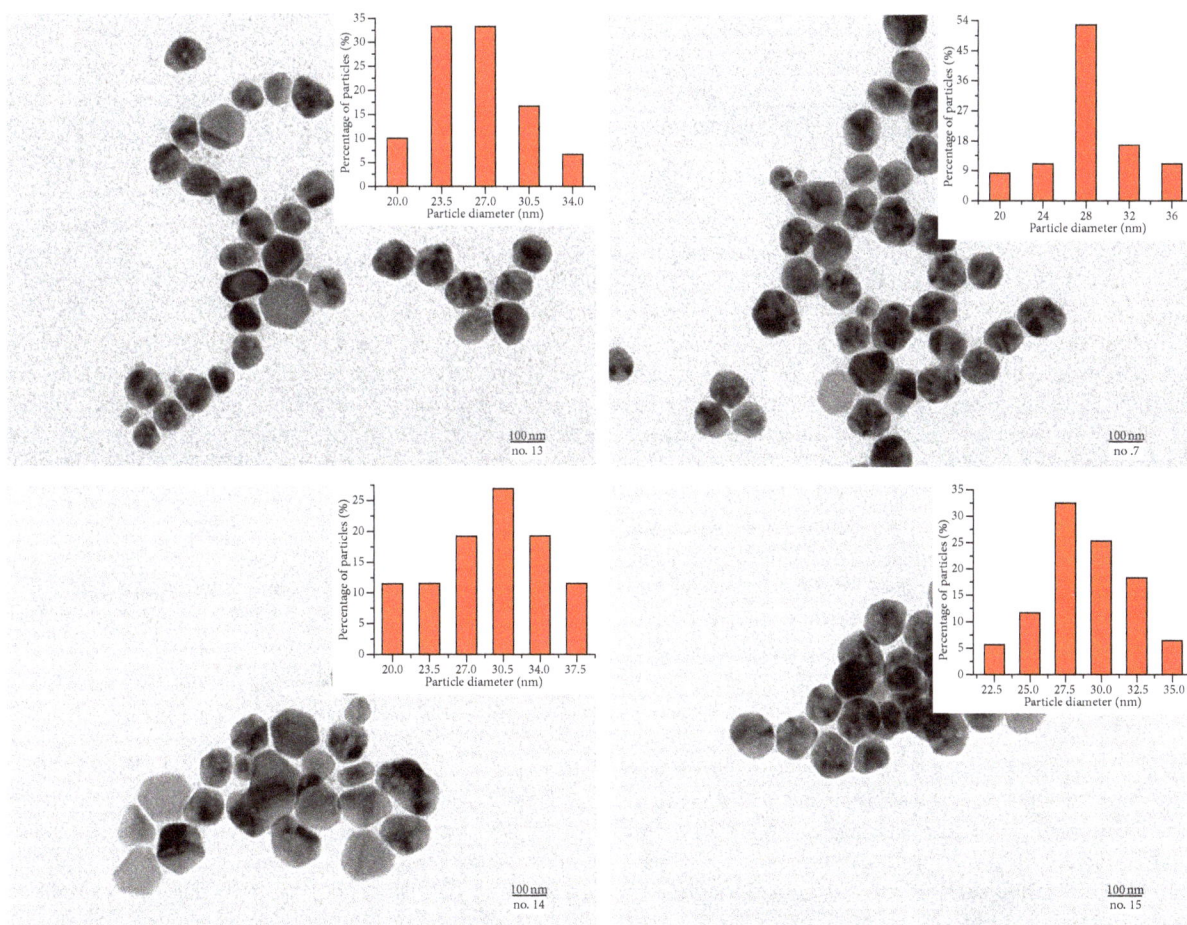

(b)

FIGURE 6: UV-Vis absorption spectra (a) and TEM images (b) of four groups of AgNPs synthesized under different inlet flow rates.

shown in Figure 6 were obtained, and the detailed test results are listed in Table 4.

Analyzing the UV-Vis absorption spectra and the related data, we find that when the inlet flow rate is 3.5 ml·min^{-1}, the

synthesized nanoparticles had a smaller mean particle size. While the flow rate is 6.5 ml·min^{-1}, the size of the synthesized nanoparticles is a little large. Observing the data of α_m, we can conclude that the yield of AgNPs increases with the

TABLE 4: Test results of AgNPs synthesized under different inlet flow rates.

Number	Q (ml·min^{-1})	λ_m (nm)	α_m	PWHM (nm)
13	1.5	424	1.165	88.58
7	3.5	421	2.194	91.84
14	5.5	430	2.742	93.16
15	6.5	426	3.020	91.39

increase of the inlet flow rate. From the PWHM point of view, we can see that the values of half peak width of four samples first increase and then decrease. But, in view of the overall values of PWHM, the size uniformity of the synthesized AgNPs is better relatively. Comparing the TEM photographs of the synthesized AgNPs, we find that when the flow rate is 5.5 ml·min^{-1}, the morphology of synthesized nanoparticles is relatively poor; while the flow rate reaches 6.5 ml·min^{-1}, part of nanoparticles produce agglomeration. For the occurrence of agglomeration, we think that this may be due to the excessive flow speed. Although the flow rate is large, the comprehensive mixing efficiency is not the best. Furthermore, the excessive flow speed directly leads to the increase of the collision probability between the nuclei of ArgentCrystal, which in turn makes it easy to form irregular nanostructures, even agglomerate.

In summary, the inlet flow rate of microfluidic chip is an important factor for the controllable synthesis of AgNPs, which affects the size, yield, morphology, and mono-dispersity of the synthesized nanoparticles. With the increase of the inlet flow rate, the AgNP yield gradually increases, but the high flow rate will lead to poor morphology of synthesized nanoparticles and even aggregation of some nanoparticles. Moreover, for the micromixing based on the pulsating mixing mechanism, once the working frequency is determined, there will be an optimal inlet flow rate to match it. Comprehensively analyzing and comparing the above experimental results, we conclude that when the flow rate is 3.5 ml·min^{-1}, a spherical AgNP with relatively high concentration, good morphology, and good mono-dispersity has been synthesized. Therefore, the optimum inlet flow rate is 3.5 ml·min^{-1}, when the working frequency is 200 Hz.

3.2.5. Effect of Temperature on the Synthesis of AgNPs. In order to study the effect of temperature on the synthesis of AgNPs, five groups of synthesis experiments at different temperatures were designed and carried out. The basic experiment conditions, such as the working frequency of PZT micropump, the initial concentration of AgNO$_3$ solution, the inlet flow rate, and the PVP dosage, are the same as those in Section 3.2.1. In addition, the concentration parameters of C$_6$H$_{12}$O$_6$ and NaOH were C$_{C_6H_{12}O_6}$: C$_{AgNO_3}$ = 4 : 1, and C$_{NaOH}$: C$_{AgNO_3}$ = 2 : 1, respectively. The UV-Vis absorption spectra and TEM images of five groups of AgNPs synthesized based on different temperatures as shown in Figure 7 were obtained. The detailed test results are listed in Table 5.

Analyzing the UV-Vis absorption spectra and the related data, we will find that although the test data are quite different, the maximum absorption wavelengths all change in the range of 400~430 nm, which well accord with specific optical characteristics of spherical AgNP sol. The positions of data λ_m show that with the increase of temperature, the maximum absorption wavelength of AgNPs appears to be red shifted first and then blue shifted. The results show that the size of AgNPs increases first and then decreases with the increase of temperature. While the test data α_m show that the maximum absorbance increases first and then decreases sharply with the increase of temperature, and the maximum absorbance reaches maximum when the temperature is 60°C and 80°C. When the temperature is set at 100°C and 120°C, the maximum absorbance α_m is lower, and the synthesis effect is relatively poorer, which indicates that the synthesis temperature of AgNPs should not be too high when synthesized using the microfluidic chip. Moreover, the TEM photos of Figure 7(b) also show that the size uniformity and monodispersity of the synthesized AgNPs are relatively poorer, when the temperature is 100°C and 120°C. The most likely cause is that the reaction rate of synthesis process of AgNPs increases too fast, with the increase of temperature. Meanwhile, as the collision probability of silver atom increases, the growth rate of crystal nuclei will exceed that of silver atoms. Therefore, the size of synthesized AgNPs is not uniform, the size deviation of AgNPs is large, and some nanoparticles are even aggregated.

In summary, too high or too low temperature is not conducive to the synthesis of AgNPs. In our synthesis experiments, when the temperature is set at 60°C and 80°C, the particle size, yield, morphology, monodispersity, and particle size uniformity of synthesized AgNPs are relatively good. After comprehensive comparison and analysis, we select 80°C as the optimum temperature of AgNPs' synthesis.

3.3. Size and Deviation of AgNPs Synthesized under Optimum Conditions. The AgNPs' synthesis using the liquid-phase reduction method has many influencing factors. In this paper, a self-made three-phase flow pulsating mixing microfluidic chip was presented. Using the microfluidic chip, a series of controllable synthesis experiments of AgNPs had been designed and carried out under various factors. By comparing and analyzing the test results, we obtained the optimal synthesis conditions under limited experimental conditions. The specific conditions are as follows: the working frequency of PZT micropump is 200 Hz, the initial concentration of AgNO$_3$ solution is 1 mM, the temperature is 80°C, the inlet flow rate Q is 3.5 ml·min^{-1}, C$_{NaOH}$: C$_{AgNO_3}$ = 2 : 1, C$_{C_6H_{12}O_6}$: C$_{AgNO_3}$ = 4 : 1, and m_{PVP} : m_{Ag} = 1.5 : 1. Under the above conditions, AgNPs with uniform size, good morphology, and monodispersity have been synthesized. Moreover, the size and deviation of synthesized AgNPs are about 29.11 ± 3.98 nm. The TEM image and histogram of particle size distribution of synthesized AgNPs are shown in Figure 8.

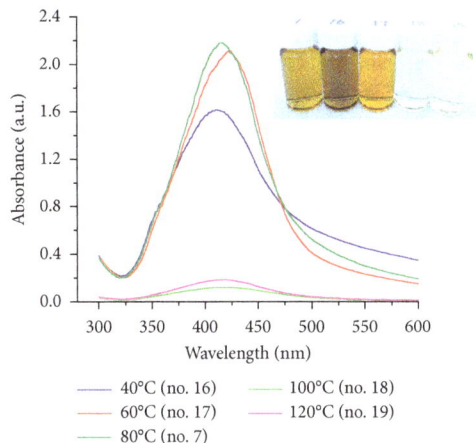

(a)

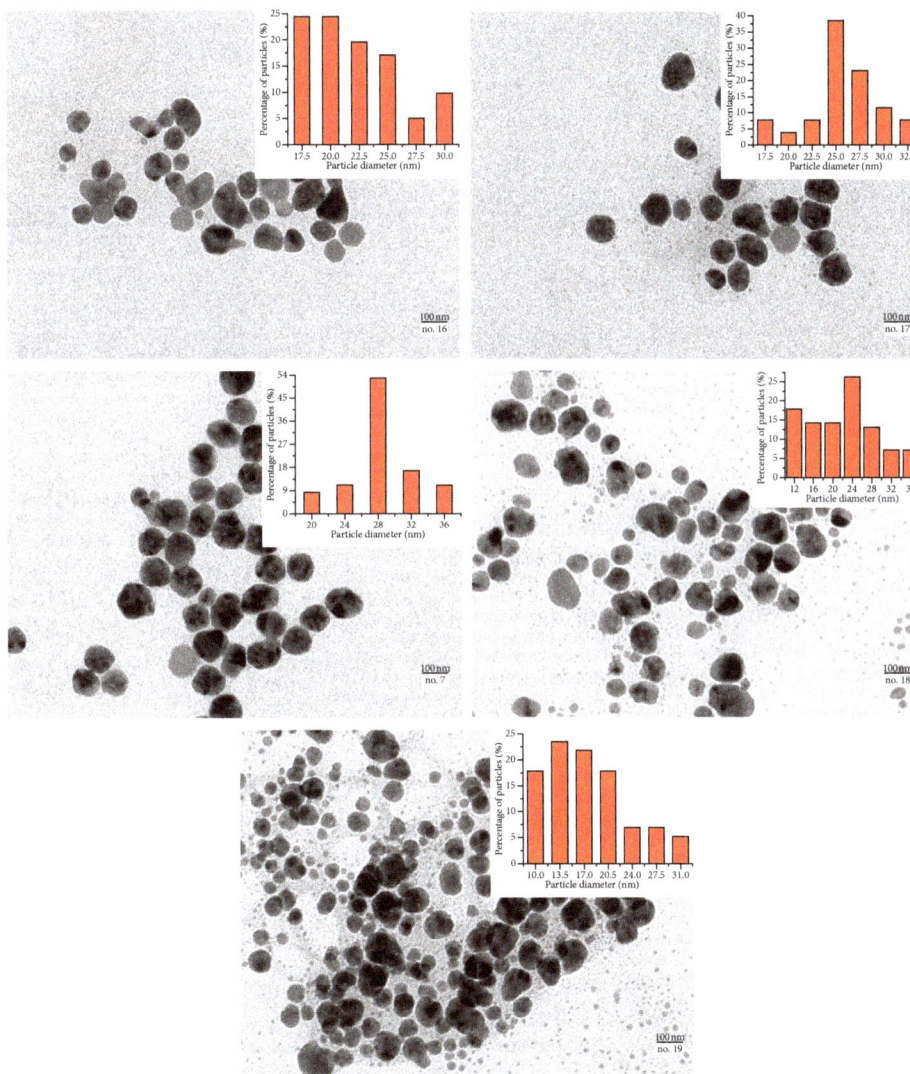

(b)

FIGURE 7: UV-Vis absorption spectra (a) and TEM images (b) of five groups of AgNPs synthesized under different temperatures.

TABLE 5: Test results of AgNPs synthesized under different temperatures.

Number	T (°C)	λ_m (nm)	α_m	PWHM (nm)
16	40	411	1.167	101.84
17	60	422	2.123	93.95
7	80	421	2.194	91.84
18	100	416	0.122	93.68
19	120	417	0.185	82.69

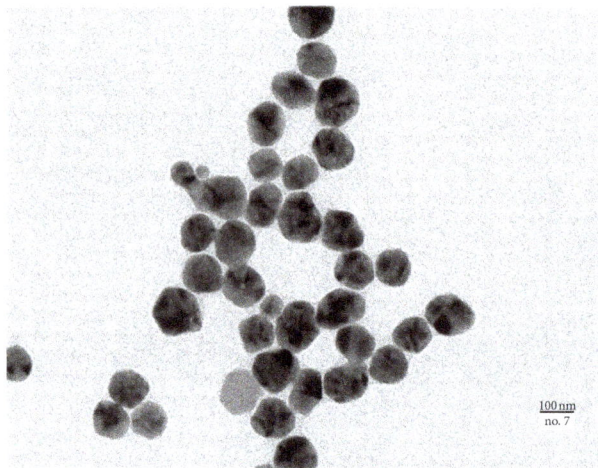

(a)

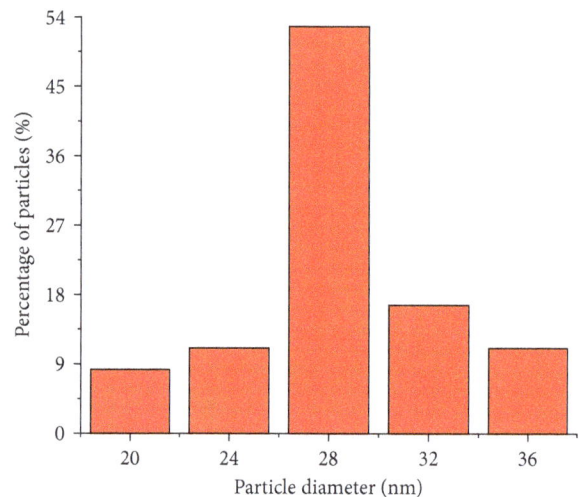

(b)

FIGURE 8: The TEM image (a) and histogram of particle size distribution (b) of synthesized AgNPs.

4. Conclusion

Based on the reduction of silver nitrate with glucose, the controllable synthesis experiments of AgNPs were carried out by using a three-phase flow pulsating mixing microfluidic chip. Through investigating and exploring the influence of different factors on the synthesis of AgNPs, the optimum synthesis conditions of AgNPs were obtained, and the following conclusions were drawn:

(1) It is an efficient and feasible experimental method to develop controllable synthesis experiments of AgNPs under multifactors by using the three-phase pulsating mixing microfluidic chip. This method can be extended to many research fields, such as the controllable synthesis of other nanoparticles and the control of complex chemical reactions.

(2) When AgNPs are synthesized by the microfluidic chip, the influence of different synthesis conditions on AgNPs is quite different. NaOH concentration, PVP dosage, microfluidic chip entrance flow, $C_6H_{12}O_6$ concentration, and reaction temperature have great influence on the aspects of the particle size, yield, morphology, dispersibility, and uniformity of the particle size. Under the limited experimental conditions, the above experimental parameters all have their own optimal value, and

too high or too low is not conducive to the controllable synthesis of AgNPs.

(3) When the working frequency of the PZT micropump is set as 200 Hz, the initial concentration of AgNO$_3$ solution is 1 mM, the temperature is 80°C, the inlet flow rate Q is set as 3.5 ml·min^{-1}, $C_{NaOH} : C_{AgNO_3} = 2 : 1, C_{C_6H_{12}O_6} : C_{AgNO_3} = 4 : 1, m_{PVP} : m_{Ag} = 1.5 : 1$, the AgNPs with uniform size, and good morphology and monodispersity have been synthesized. The size and deviation of synthesized AgNPs are about 29.11 ± 3.98 nm.

Conflicts of Interest

The authors declare that they have no conflicts of interest.

Acknowledgments

This research is supported and funded by the National Natural Science Foundation Projects (no. 51375207) from Ministry of Science and Technology of the People's Republic of China and the Jilin Province Natural Science Foundation Projects (no. 20170101136JC). The authors thank the Chain Drive Research Institution of Jilin University for the energetic support and help in the aspect of microdevice fabrication and electronic control unit (R&D).

References

[1] H.-J. Li, A.-Q. Zhang, L. Sui, D.-J. Qian, and M. Chen, "Hyaluronan/Tween 80-assisted synthesis of silver nano-particles for biological application," *Journal of Nanoparticle Research*, vol. 17, no. 2, p. 111, 2015.

[2] Y. Lu, Y. Mei, M. Drechsler, and M. Ballauff, "Thermosensitive core–shell particles as carriers for Ag nanoparticles: modulating the catalytic activity by a phase transition in networks," *Angewandte Chemie*, vol. 45, no. 5, pp. 813–816, 2006.

[3] A. C. Patel, S. Li, C. Wang, W. J. Zhang, and Y. Wei, "Electrospinning of porous silica nanofibers containing silver nanoparticles for catalytic applications," *Chemistry of Materials*, vol. 19, no. 6, pp. 1231–1238, 2007.

[4] S. Mishra and H. B. Singh, "Biosynthesized silver nano-particles as a nanoweapon against phytopathogens: exploring their scope and potential in agriculture," *Applied Microbiology and Biotechnology*, vol. 99, no. 3, pp. 1097–1107, 2015.

[5] Y.-C. Tsai, T.-M. Wu, P.-C. Hsu, and Y.-W. Lin, "Silver nanoparticles in multiwalled carbon nanotube–Nafion for surface-enhanced Raman scattering chemical sensor," *Sensors and Actuators B: Chemical*, vol. 138, no. 1, pp. 5–8, 2009.

[6] S. K. Laliwala, V. N. Mehta, J. V. Rohit, and S. K. Kailasa, "Citrate-modified silver nanoparticles as a colorimetric probe for simultaneous detection of four triptan-family drugs," *Sensors and Actuators B: Chemical*, vol. 197, pp. 254–263, 2014.

[7] J. R. Morones, J. L. Elechiguerra, A. Camacho et al., "The bactericidal effect of silver nanoparticles," *Nanotechnology*, vol. 16, no. 10, pp. 2346–2353, 2005.

[8] E. Navarro, F. Piccapietra, B. Wagner et al., "Toxicity of silver nanoparticles to *Chlamydomonas reinhardtii*," *Environmental Science and Technology*, vol. 42, no. 23, pp. 8959–8964, 2008.

[9] C. N. R. Rao, G. U. Kulkarni, P. J. Thomas, and P. P. Edwards, "Metal nanoparticles and their assemblies," *Chemical Society Reviews*, vol. 29, no. 1, pp. 27–35, 2001.

[10] N. K. Agrawal, R. Agarwal, D. Bhatia et al., "Synthesis of Al and Ag nanoparticles through ultra-sonic dissociation of thermal evaporation deposited thin films for promising clinical applications as polymer nanocomposite," *Advanced Materials Letters*, vol. 6, no. 4, pp. 301–308, 2015.

[11] S. Küünal, S. Kutti, P. Rauwel, M. Guha, D. Wragg, and E. Rauwel, "Biocidal properties study of silver nanoparticles used for application in green housing," *International Nano Letters*, vol. 6, no. 3, pp. 191–197, 2016.

[12] S. M. Roopan, Rohit, G. Madhumitha et al., "Low-cost and eco-friendly phyto-synthesis of silver nanoparticles using *Cocos nucifera* coir extract and its larvicidal activity," *Industrial Crops and Products*, vol. 43, no. 5, pp. 631–635, 2013.

[13] S. Ahmed, M. Ahmad, B. L. Swami, and S. Ikram, "A review on plants extract mediated synthesis of silver nanoparticles for antimicrobial applications: a green expertise," *Journal of Advanced Research*, vol. 7, no. 1, pp. 17–28, 2015.

[14] P. Rauwel, S. Küünal, S. Ferdov, and E. Rauwel, "A review on the green synthesis of silver nanoparticles and their mor-phologies studied via TEM," *Advances in Materials Science and Engineering*, vol. 2015, Article ID 682749, 9 pages, 2015.

[15] Q. Zhang, J.-J. Xu, Y. Liu, and H.-Y. Chen, "In-situ synthesis of poly(dimethylsiloxane)-gold nanoparticles composite films and its application in microfluidic systems," *Lab on a Chip*, vol. 8, no. 2, pp. 352–357, 2008.

[16] R. Lin, R. G. Freemantle, N. M. Kelly, T. R. Fielitz, S. O. Obare, and R. Y. Ofoli, "In situ immobilization of palladium nanoparticles in microfluidic reactors and assessment of their catalytic activity," *Nanotechnology*, vol. 21, no. 32, p. 32605, 2010.

[17] G. J. Liu, X. H. Yang, Y. Li, Z. G. Yang, W. Hong, and J. F. Liu, "Continuous flow controlled synthesis of gold nanoparticles using pulsed mixing microfluidic system," *Advances in Materials Science and Engineering*, vol. 2015, Article ID 160819, 11 pages, 2015.

[18] J. Ma, S. M. Lee, C. Yi, and C. W. Li, "Controllable synthesis of functional nanoparticles by microfluidic platforms for bio-medical applications-a review," *Lab on a Chip*, vol. 17, no. 2, pp. 209–226, 2017.

[19] R. Baber, L. Mazzei, N. T. K. Thanh, and A. Gavriilidis, "Synthesis of silver nanoparticles in a microfluidic coaxial flow reactor," *RSC Advances*, vol. 5, no. 116, pp. 95585–95591, 2015.

[20] S. T. He, Y. L. Liu, and H. Maeda, "Controlled synthesis of colloidal silver nanoparticles in capillary micro-flow reactor," *Journal of Nanoparticle Research*, vol. 10, no. S1, pp. 209–215, 2008.

[21] L. L. Lazarus, C. T. Riche, B. C. Marin, M. Gupta, N. Malmstadt, and R. L. Brutchey, "Two-phase microfluidic droplet flows of ionic liquids for the synthesis of gold and silver nanoparticles," *ACS Applied Materials & Interfaces*, vol. 4, no. 6, pp. 3077–3083, 2012.

[22] C.-H. Yang, L.-S. Wang, S.-Y. Chen et al., "Microfluidic assisted synthesis of silver nanoparticle-chitosan composite microparticles for antibacterial applications," *International Journal of Pharmaceutics*, vol. 510, no. 2, pp. 493–500, 2016.

[23] L. Xu, J. Peng, M. Yan, D. Zhang, and A.-Q. Shen, "Droplet synthesis of silver nanoparticles by a microfluidic device," *Chemical Engineering and Processing: Process Intensification*, vol. 102, pp. 186–193, 2016.

[24] K. Sugano, Y. Uchida, O. Ichihashi, H. Yamada, T. Tsuchiya, and O. Tabata, "Mixing speed-controlled gold nanoparticle synthesis with pulsed mixing microfluidic system," *Micro-fluidics and Nanofluidics*, vol. 9, no. 6, pp. 1165–1174, 2010.

[25] G. J. Liu, X. H. Yang, J. F. Liu, Z. G. Yang, X. B. Li, and T. Zhao, "Controlled synthesis of gold nanoparticles using pulsed mixing based on piezoelectric actuation," *Rare Metal Materials and Engineering*, vol. 45, no. 6, pp. 1625–1630, 2016.

[26] P. V. Kamat, A. M. Flumiani, and G. V. Hartland, "Picosecond dynamics of silver nanoclusters: photoejection of electrons and fragmentation," *Journal of Physical Chemistry B*, vol. 102, no. 17, pp. 3123–3128, 1998.

[27] L. M. Liz-Marzán and I. Ladotouriño, "Reduction and sta-bilization of silver nanoparticles in ethanol by nonionic surfactants," *Langmuir*, vol. 12, no. 15, pp. 3585–3589, 1996.

[28] M. O. Mennessier, G. Burki, and J. P. Cordoni, "Synthesis and study of silver nanoparticles," *Journal of Chemical Education*, vol. 84, no. 2, pp. 322–325, 2007.

[29] J. P. Yang, H. J. Yin, J. J. Jia, and Y. Wei, "Facile synthesis of high-concentration, stable aqueous dispersions of uniform silver nanoparticles using aniline as a reductant," *Langmuir*, vol. 27, no. 8, pp. 5047–5053, 2011.

[30] B. J. Wiley, T. Herricks, A. Y. Sun, and Y. Xia, "Polyol synthesis of silver nanoparticles: use of chloride and oxygen to promote the formation of single-crystal, truncated cubes and tetrahedrons," *Nano Letters*, vol. 4, no. 10, pp. 1733–1739, 2004.

11

Novel Preparation of Fe Doped TiO$_2$ Nanoparticles and Their Application for Gas Sensor and Photocatalytic Degradation

Sunil Babu Eadi,[1] **Sungjin Kim,**[1] **Soon Wook Jeong,**[1] **and Heung Woo Jeon**[2]

[1]*School of Advanced Materials & Engineering, Kumoh National Institute of Technology, 61 Daehak-Ro, Gumi 39177, Republic of Korea*
[2]*School of Electronic Engineering, Kumoh National Institute of Technology, 61 Daehak-Ro, Gumi 39177, Republic of Korea*

Correspondence should be addressed to Heung Woo Jeon; hwjeon@kumoh.ac.kr

Academic Editor: Fabrizio Pirri

The preparation of Fe doped TiO$_2$ is demonstrated as well as application for gas sensing and photocatalytic degradation. Fe doped TiO$_2$ nanopowder was prepared by the mechanochemical ball milling method. The results show the uniform doping of Fe in TiO$_2$ powder. The average size of the particle is observed to be ~28 nm. The H$_2$ gas sensor was fabricated using the Fe doped TiO$_2$ nanopowder and the effect of Fe doping on the sensing properties is investigated with various temperature ranges. The prepared Fe doped TiO$_2$ nanoparticles are also used to explore degradation properties of Rhodamine B dye under LED light.

1. Introduction

Titanium dioxide (TiO$_2$) is a well-known multifunctional metal oxide material because of its wide range of applications and benign properties such as stability and nontoxicity [1]. The wide range of uses of TiO$_2$ is due to its unique electronic and structural properties. TiO$_2$ exists in three main crystallographic forms: anatase, rutile, and brookite. However due to high band gap of TiO$_2$ (>3 eV) its optical application is limited to UV region of the electromagnetic spectrum [2]. Doping opens up the possibility of changing the electronic structure of TiO$_2$ nanoparticles, altering their chemical composition and optical properties. Much effort has been made by incorporating doping with metal ions, such as nickel, chromium, iron, vanadium, and zinc [3–7].

Iron metal ions have been considered as a suitable candidate, for doping owing to the fact that the radius of both Fe^{3+} (0.64 Å) and Ti^{4+} (0.68 Å) is similar in size. Therefore, it can be easily incorporated with the crystal lattice of TiO$_2$ [8]. One of the advantages of inclusion of Fe into TiO$_2$ lattice is its potential application in photocatalysis and hydrogen production due to reducing in the energy gap of TiO$_2$ and increasing the efficiency of absorbing visible light [9]. However, the application of Fe doped TiO$_2$ for gas sensing, especially hydrogen sensing, has still not been properly explored. It is also well recognized that doping Fe can improve the grain size, carrier concentration, and thermal stability of the TiO$_2$ sensor performance significantly. Furthermore, several researchers reported about the fabrication of Fe doped TiO$_2$ using the spin coating method Effendi and Bilalodin, reported the Fe doped TiO$_2$ powder using Titania and iron oxide as chemical precursors [10]. Kumar et al. reported the fabrication of Fe doped TiO$_2$ film by sol-gel technique for the application of gas sensor [11]. Othman et al., reported the Fe doped TiO$_2$ nanoparticles produced by MOCVD method for photocatalytic activity [12]. However, there are very few reports of preparing Fe doped TiO$_2$ nanoparticles by the mechanochemical ball milling method. Carneiro et al. studied the effect of Fe on TiO$_2$ powder on optical and photocatalytic properties [13].

In this work, we have developed Fe doped TiO$_2$ nanopowder using mechanochemical ball milling method. The effect of the ball milling procedure on Fe doping in TiO$_2$ is systematically studied. H$_2$ gas sensor was fabricated using the prepared powder and sensing properties are studied.

2. Experimental Procedure

Iron doped TiO$_2$ nanoparticles were prepared by ball milling of TiO$_2$ powders (TiO$_2$ P25) in a high-energy ball mill in the

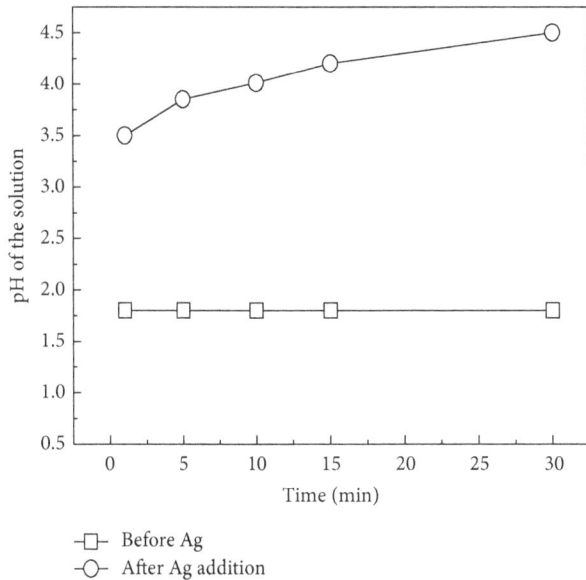

FIGURE 1: The pH variation during ball milling process.

FIGURE 2: Flow chart for preparing Fe doped TiO_2.

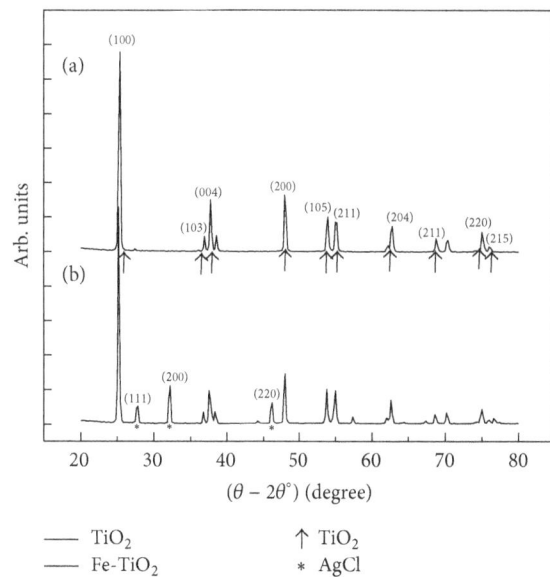

presence of fine $FeCl_3$ powder using Al_2O_3 balls. The slurry was prepared by mixing TiO_2 powder and iron chloride in DI water in 1:5 ratio for 120 min at 1200 rpm. After 30 min the Ag powder was added in the slurry for the removal of unreacted $FeCl_3$. The slurry was monitored by pH meter. After reaction the slurry was washed with DI water and ethanol for removing the impurities. Finally, the powder was dried at 80°C in an open atmosphere.

3. Results and Discussion

Initially, the TiO_2 and $FeCl_3$ powders were mixed in DI water. The pH of the slurry changed to acidic medium with value 1.8. This indicates iron chloride dissolution in water and formation of iron hydroxide, as shown in (1) in the following:

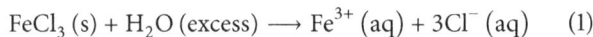

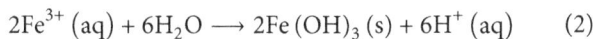

$$FeCl_3 (s) + H_2O (excess) \longrightarrow Fe^{3+} (aq) + 3Cl^- (aq) \qquad (1)$$

$$2Fe^{3+} (aq) + 6H_2O \longrightarrow 2Fe(OH)_3 (s) + 6H^+ (aq) \qquad (2)$$

The pH of the slurry remained steady for 30 min ball milling. To remove the Cl^- ions from the slurry Ag metal powder was added and the slurry was ball milled for another 30 min. During this time pH of slurry was increased gradually from 1.8 to 4.5 as shown in Figure 1. The experimental flow chart of fabrication of iron doped TiO_2 nanoparticles is shown in Figure 2. It is also noticed that amount of Fe dopant concentration in the TiO_2 slurry is very important for the controlling the pH and size of the particle. Increase in the concentration ratio of $FeCl_3$ to TiO_2 will lead to more lattice distortion in the crystal. It causes a decrease in the crystalline density, yielding an increase in the size of the nanoparticles. To characterize the surface area of the Fe doped TiO_2 nanoparticles BET measurement was used. The BET surface area obtained was $51.856 \, m^2/g$ and the total pore volume and average pore size were $0.462 \, cm^3/g$ and $357.246 \, \text{Å}$, respectively. It was noted that the surface area

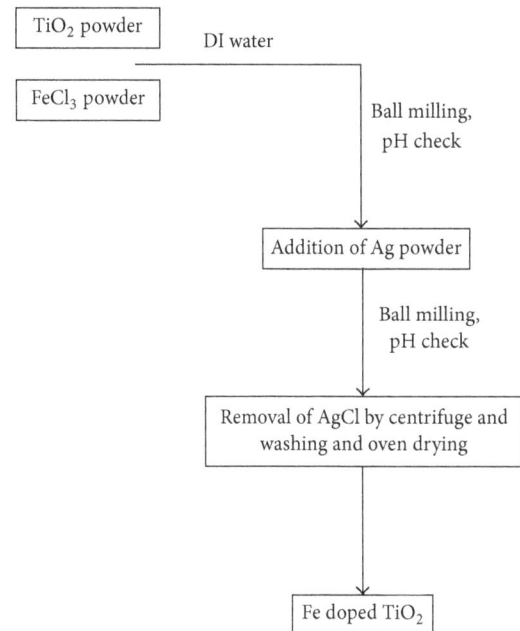

FIGURE 3: X-ray diffraction pattern of (a) pure TiO_2 and (b) iron doped TiO_2.

decreased with increasing in the annealing temperature of the powder. Table 1 shows the surface area, pore volume, and pore size of Fe doped TiO_2 at different annealing temperature.

Figure 3 shows the X-ray analysis of pure TiO_2 and synthesized Fe doped TiO_2 particles. The XRD profiles relieves the anatase phase (JCPDS number 84-1157) for pure TiO_2 and additional peaks of AgCl (JCPDS number 71-5209) for the doped sample as shown in the figure. All the peaks correspond to the (1 0 0), (0 0 4), (2 0 0), (1 0 5), (2 1 1), and (2 0 4) planes of anatase TiO_2. It is observed that the intensity of anatase phase in iron doped TiO_2 particles decreased with

TABLE 1: The surface area, pore volume, and pore size of Fe doped TiO$_2$ at different annealing temperature.

Sample name	BET surface area	Total pore volume	Average pore size
RT	51.856 m^2/g	0.462 cm^3/g	357.246 Å
400	50.514 m^2/g	0.455 cm^3/g	367.671 Å
500	43.288 m^2/g	0.338 cm^3/g	334.475 Å
600	23.132 m^2/g	0.071 cm^3/g	136.759 Å

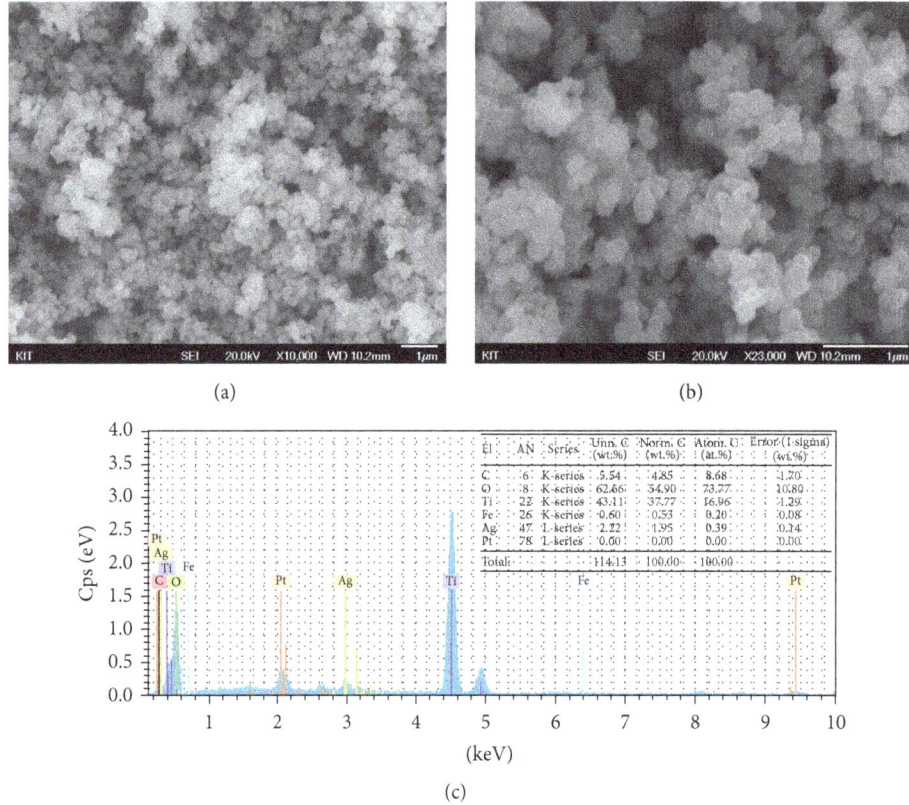

(a)

(b)

El	AN	Series	Unn. C (wt.%)	Norm. C (wt.%)	Atom. C (at.%)	Error (1 sigma) (wt.%)
C	6	K-series	5.54	4.85	8.68	1.20
O	8	K-series	62.66	54.90	73.77	10.80
Ti	22	K-series	43.11	37.77	16.96	1.29
Fe	26	K-series	0.60	0.53	0.20	0.08
Ag	47	L-series	2.22	1.95	0.39	0.14
Pt	78	L-series	0.00	0.00	0.00	0.00
Total:			114.13	100.00	100.00	

(c)

FIGURE 4: FESEM images of Fe doped TiO$_2$ (a) low magnification and (b) high magnification and (c) EDS spectrum table of the Fe doped TiO$_2$ particle.

insertion of Fe ions. However, in addition, it is worth noting that no crystalline phase attributed to iron oxide can be found in the XRD patterns.

One possible reason can be that the Fe^{3+} content in the Fe-TiO$_2$ is below the detection limit. The ionic radii of Ti^{+4} (0.68 Å) and Fe^{+3} (0.64 Å) are almost the same; hence there is a possibility of Fe ions occupying some of the lattice sites of TiO$_2$ [14]. The crystallite size of the Fe doped TiO$_2$ monodispersed nanoparticles is ~28 nm. Determined from the full-width at half maximum of the anatase peak by Scherrer's formula

$$Dp = \frac{K\lambda}{\beta \cos\theta}, \qquad (3)$$

where Dp is average crystallite size, β is line broadening in radians, θ is Bragg angle, and λ is X-ray wavelength, respectively.

FESEM images of Fe doped TiO$_2$ powder obtained after annealing at 450°C for 1 hour are shown in Figures 4(a) and 4(b). The avg. diameters of the nanoparticles are ~28~30 nm. The nanoparticles are of uniform size and spherical shape. It is noticed that agglomeration of particles increases with the annealing temperature. Figure 4(c) shows the EDS spectrum of Fe doped TiO$_2$ nanoparticles. The peaks of titanium and iron can be clearly seen in the spectrum with a dotted spot in the FESEM image. From the spectrum table, we can see the Fe (at. % 0.53) in the TiO$_2$.

Gas sensing measurement was performed with sensor fabricated by Fe doped TiO$_2$ for the detection of H$_2$ gas. Figures 5(a) and 5(b) show schematic diagram of sensor fabrication and illustration of hydrogen sensor measurement instrument. The sensor was prepared by coating the paste of nanoparticles powder on precoated Pt electrode alumina substrate. Initially 500 ppm of H$_2$ reducing gas diluted with nitrogen was used as the source gas. The typical dynamic

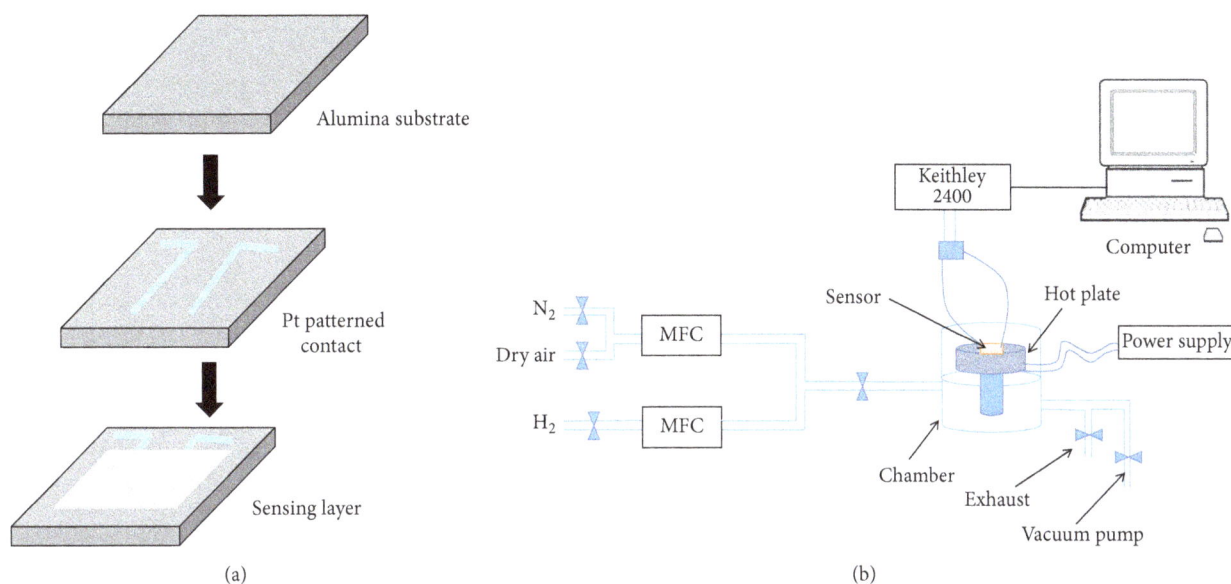

(a)

(b)

FIGURE 5: (a) Schematic diagram of sensor fabrication and (b) shows the schematic illustration of hydrogen sensor measurement instrument.

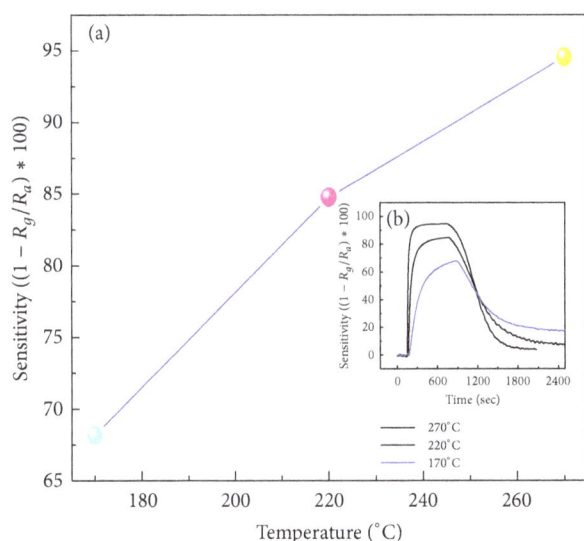

FIGURE 6: (a) Sensitivity plot at different operation temperatures 170, 220, and 270°C. (b) Dynamic responses from the sensor fabricated using Fe doped TiO$_2$ sample under the exposure of 500 ppm H$_2$.

responses of sensor upon exposure to 500 ppm H$_2$ (Figure 6(a)) diluted in nitrogen were measured at different operating temperature of 170, 220, and 270°C. TiO$_2$ shows typical gas sensing behavior of n-type semiconductor where first the resistance of the sensor increases the operating temperature and decreases upon exposure to the H$_2$ gas [15–17].

A plot of the sensitivity to 500 ppm H$_2$ concentration measured as a function of operating temperature is shown in Figure 6(b). The effect of temperature on sensitivity of sensor on the H$_2$ response was higher with an increase in

temperature. A significant increase of the H$_2$ response was observed at 270°C. The sensitivities of 68.1, 84.7, and 94.5 were observed at temperatures of 170, 220, and 270°C, respectively. It is also noticed that at lower temperatures 170 and 220°C sensor took longer time to stabilize to normal state after H$_2$ gas is off. At 270°C sensor recovers fast due to rapid desorption from the surface of nanoparticles.

The H$_2$ gas sensing mechanism of the resistive sensor base on metal oxide semiconductors was proposed in previous reports [18–20]. In current measurement (temperature range 160°C < $T_{substrate}$ < 270°C), the O$^-$ and O^{2-} oxygen ions are the dominant species on the surface of metal oxide semiconductors. The oxygen ions species extract electrons from the conduction of TiO$_2$. Thereby resistivity of the sensing layer increases and reaches steady state and when the H$_2$ gas is purged in; electrons are injected back to the conduction band thereby decreasing in resistance. In our case, sensing properties depend on the large surface area of the separated Fe doped TiO$_2$ particles, which provide more adsorption-desorption sites of gas molecules leading to improve the interaction between H$_2$ gas and adsorbed oxygen ions and thereby higher sensitivity was achieved.

The photodegradability of Rhodamine B dye solution was done using Fe doped TiO$_2$ powder. The solution was prepared by dissolving 0.2 g of Fe doped TiO$_2$ powder into 50 ml of 1 ppm concentration Rhodamine dye. Initially the solution was stirred for 30 min under dark to ensure complete surface adsorption of Rhodamine B onto the nanoparticles. After that LED white light was illuminated on the solution, for every 10 min samples were taken for checking the degradation property. Figure 7(a) shows the UV-Vis spectra of the dye solution with different interval time interval. The spectra data clearly demonstrates the photocatalytic efficiency of Fe doped TiO$_2$ powder in decolorization of Rhodamine B from pink to white under 60 min time interval. The gradual decrease

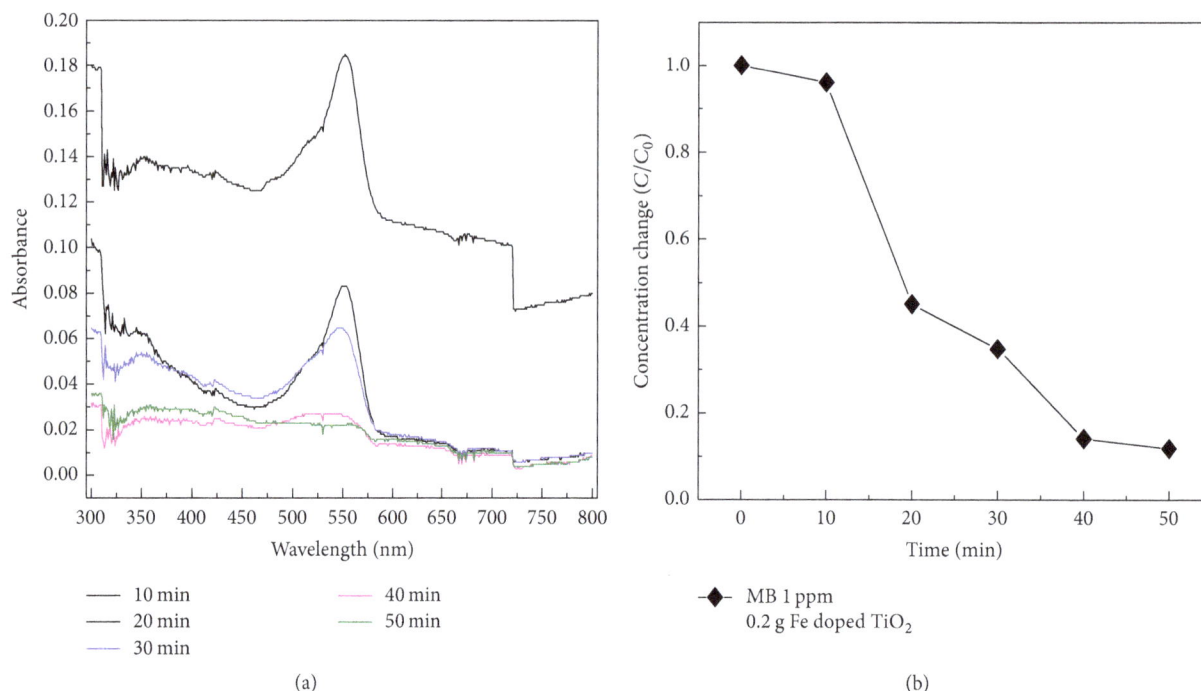

FIGURE 7: (a) UV-Vis spectra of the dye solution with different interval time and (b) shows the plot of concentration decreasing with time.

in the concentration ratio from initial to final decoloration is plotted in Figure 7(b). The photocatalytic mechanism of TiO_2 under UV light is well documented [21, 22]. Under the illumination of LED light, the pair of electron and hole released react with water and the electron-hole pairs migrate to the catalyst surface where they participate in redox reactions with adsorbed dye species. The holes (h^+) in valance band react with surface-bound H_2O to produce the hydroxyl radical and (e^-) in conduction band react with oxygen to generate superoxide radical anions. The formed hydroxyl radical ($^\bullet OH$) and superoxide radical anions ($O^{2\bullet -}$) degrade the pollutants in the photocatalytic oxidation processes. In our case, doping of Fe into TiO_2 enhances the electron-hole pair density thereby increasing the concentration of hydroxyl radical ($^\bullet OH$) and superoxide radical anions thereby showing efficient catalytical reaction.

4. Conclusions

In this study, Fe doped TiO_2 nanopowder is prepared by the mechanochemical ball milling method. The prepared powder was characterized by using field emission scanning electron microscope (FESEM) and X-ray diffraction studies. The results show the uniform doping of Fe in TiO_2 powder. The avg. size of the particle is observed to be 28 nm. The H_2 gas sensor fabricated using the Fe doped TiO_2 nanopowder shows high sensitivity of 86% and it was noticed that sensitivity depends on the temperature. Also the effect of Fe doping on photodegradation of Rhodamine B under LED light was investigated. The results show that doping helps in decreasing degradation time lower than 60 min.

Conflicts of Interest

The authors declare that they have no conflicts of interest.

Acknowledgments

This study was fully funded by Kumoh National Institute of Technology, Korea. The authors would like to acknowledge Ministry of Trade, Industry and Energy (MOTIE) Project 10063553 in 2016. They would also like to acknowledge WC 300 Project by Business for Global Cooperative R&D by Korea Small and Medium Business Administration in 2015.

References

[1] O. Ola and M. Mercedes Maroto-Valer, "Review of material design and reactor engineering on TiO_2 photocatalysis for CO_2 reduction," *Journal of Photochemistry and Photobiology C: Photochemistry Reviews*, vol. 24, pp. 16–42, 2015.

[2] Landmannm M., E. Rauls, and W. G. Schmidt, "The electronic structure and optical response of rutile, anatase and brookite TiO2," *Journal of Physics: Condensed Matter*, vol. 24, no. 19, 2012.

[3] A. Zaleska, "Doped-TiO_2: a review," *Recent Patents on Engineering*, vol. 2, no. 3, pp. 157–164, 2008.

[4] Y. Wang, H. Cheng, Y. Hao et al., "The photoelectrochemistry of transition metal-ion-doped TiO2 nanocrystalline electrodes and higher solar cell conversion efficiency based on Zn2+-doped TiO2 electrode," *Journal of Materials Science*, vol. 34, 1999.

[5] M. S. Jeon, W. S. Yoon, H. K. Joo, and T. K. Lee, "Preparation and characterization of a nano-sized Mo/Ti mixed photocatalyst," *Applied Surface Science*, vol. 165, no. 2-3, pp. 209–216, 2000.

[6] E. Traversa, M. L. Di Vona, P. Nunziante, S. Licoccia, T. Sasaki, and N. Koshizaki, "Sol-gel preparation and characterization of Ag-TiO2 nanocomposite thin films," *Journal of Sol-Gel Science and Technology*, vol. 19, no. 1-3, pp. 733–736, 2000.

[7] J. Moon, H. Takagi, Y. Fujishiro, and M. Awano, "Preparation and characterization of the Sb-doped TiO2 photocatalysts," *Journal of Materials Science*, vol. 36, no. 4, pp. 949–955, 2001.

[8] P. B. Nair, V. B. Justinvictor, G. P. Daniel et al., "Structural, optical, photoluminescence and photocatalytic investigations on Fe doped TiO$_2$ thin films," *Thin Solid Films*, vol. 550, pp. 121–127, 2014.

[9] R. D. S. Santos, G. A. Faria, C. Giles et al., "Iron insertion and hematite segregation on Fe-doped TiO 2 nanoparticles obtained from sol-gel and hydrothermal methods," *ACS Applied Materials & Interfaces*, vol. 4, no. 10, pp. 5555–5561, 2012.

[10] M. Effendi and Bilalodin, "Effect of doping Fe on TiO2 thin films prepared by spin coating method," *International Journal of Basic & Applied Sciences IJBAS-IJENS*, vol. 12, no. 2, 2012.

[11] M. Kumar, D. Kumar, and A. K. Gupta, "Undoped and Fe doped TiO2 thin films fabricated by sol-gel technique: an approach to gas sensing applications," *International Journal of Enhanced Research in Science, Technology & Engineering*, vol. 5, pp. 2319–7463, 2016.

[12] S. H. Othman, S. Abdul Rashid, T. I. Mohd Ghazi, and N. Abdullah, "3D CFD simulations of MOCVD synthesis system of titanium dioxide nanoparticles," *Journal of Nanomaterials*, vol. 2013, Article ID 123256, 11 pages, 2013.

[13] J. O. Carneiro, S. Azevedo, F. Fernandes et al., "Synthesis of iron-doped TiO$_2$ nanoparticles by ball-milling process: the influence of process parameters on the structural, optical, magnetic, and photocatalytic properties," *Journal of Materials Science*, vol. 49, no. 21, pp. 7476–7488, 2014.

[14] N. Nasrallaa, M. Yeganehb, Y. Astuti et al., "Structural and spectroscopic study of Fe-doped TiO2 nanoparticles prepared by sol-gel method," *Scientia Iranica*, vol. 20, pp. 1018–1022, 2013.

[15] G. F. Fine, L. M. Cavanagh, A. Afonja, and R. Binions, "Metal oxide semi-conductor gas sensors in environmental monitoring," *Sensors*, vol. 10, no. 6, pp. 5469–5502, 2010.

[16] G. Jiménez-Cadena, J. Riu, and F. X. Rius, "Gas sensors based on nanostructured materials," *Analyst*, vol. 132, no. 11, pp. 1083–1099, 2007.

[17] W. Guo, T. Liu, H. Zhang et al., "Gas-sensing performance enhancement in ZnO nanostructures by hierarchical morphology," *Sensors and Actuators B: Chemical*, vol. 166-167, pp. 492–499, 2012.

[18] N. M. Vuong, H. Jung, D. Kim, H. Kim, and S.-K. Hong, "Realization of an open space ensemble for nanowires: A strategy for the maximum response in resistive sensors," *Journal of Materials Chemistry*, vol. 22, no. 14, pp. 6716–6725, 2012.

[19] H. Gu, Z. Wang, and Y. Hu, "Hydrogen gas sensors based on semiconductor oxide nanostructures," *Sensors*, vol. 12, no. 5, pp. 5517–5550, 2012.

[20] C.-Y. Liu, C.-F. Chen, and J.-P. Leu, "Fabrication and CO sensing properties of mesostructured ZnO gas sensors," *Journal of The Electrochemical Society*, vol. 156, no. 1, pp. J16–J19, 2009.

[21] S. N. Frank and A. J. Bard, "Heterogeneous photocatalytic oxidation of cyanide and sulfite in aqueous solutions at semiconductor powders," *The Journal of Physical Chemistry*, vol. 81, 1484 pages, 1977.

[22] A. Fujishima and K. Honda, "Electrochemical photolysis of water at a semiconductor electrode," *Nature*, vol. 238, no. 5358, pp. 37-38, 1972.

12

Preparation of Metal-Containing Diamond-Like Carbon Films by Magnetron Sputtering and Plasma Source Ion Implantation and Their Properties

Stefan Flege,[1] Ruriko Hatada,[1] Andreas Hanauer,[1] Wolfgang Ensinger,[1] Takao Morimura,[2] and Koumei Baba[2,3]

[1]*Department of Materials Science, Technische Universität Darmstadt, Alarich-Weiss-Str. 2, 64287 Darmstadt, Germany*
[2]*Nagasaki University, Graduate School of Engineering, 1-14 Bunkyo, Nagasaki 852-8521, Japan*
[3]*Industrial Technology Center of Nagasaki, Omura, Nagasaki 856-0026, Japan*

Correspondence should be addressed to Stefan Flege; flege@ma.tu-darmstadt.de

Academic Editor: Hossein Moayedi

Metal-containing diamond-like carbon (Me-DLC) films were prepared by a combination of plasma source ion implantation (PSII) and reactive magnetron sputtering. Two metals were used that differ in their tendency to form carbide and possess a different sputter yield, that is, Cu with a relatively high sputter yield and Ti with a comparatively low one. The DLC film preparation was based on the hydrocarbon gas ethylene (C_2H_4). The preparation technique is described and the parameters influencing the metal content within the film are discussed. Film properties that are changed by the metal addition, such as structure, electrical resistivity, and friction coefficient, were evaluated and compared with those of pure DLC films as well as with literature values for Me-DLC films prepared with a different hydrocarbon gas or containing other metals.

1. Introduction

Diamond-like carbon films offer many advantages such as high hardness and wear resistance, low friction coefficient, low surface roughness, chemical inertness, and optical transparency. Accordingly, DLC is used in a variety of applications such as protection layer on magnetic media, lubricating layer on sliding parts, and coating on biomedical implants, as antireflective coating [1] or diffusion barrier [2]. Disadvantages of DLC are its poor thermal stability [3], high internal stresses which might lead to poor adhesion on some substrates such as steel and copper [4], and an increase of friction coefficient with humidity [5]. Most of those limitations can be circumvented by a modification of the DLC film, that is, by the addition of one or more elements to the film [6]. Apart from a few light elements, for example, B [7], N [8], or F [9], mainly transition metals are added for this purpose [6, 10].

DLC films can be produced by a variety of methods [11], which include plasma source ion implantation (PSII) [12, 13]. This technique makes it possible to coat complex shaped three-dimensional samples homogeneously without the use of sample manipulation [14]. The sample is placed within a plasma and the ions of this plasma are accelerated towards the sample by a negative voltage applied to the sample or to the sample holder. By using a hydrocarbon gas, DLC films can be produced [15]. Usually, a pulsed high voltage is used to avoid excessive heating of the substrate.

The choice of hydrocarbon gas whose molecules (or fragments thereof) will form the DLC layer depends on the desired properties, and thus the application, of the DLC film [16]. Considerations are the implantation depth, the deposition rate, the hydrogen content, and the structure of the DLC film. While small molecules such as methane (CH_4) are the preferred choice for implantation, they tend to provide only low deposition rates. The use of bigger molecules is equivalent to higher deposition rates. To keep the hydrogen content in the DLC film low, usually molecules with a parity or near-parity of the number of C and H atoms are chosen; that is, acetylene (C_2H_2) and toluene (C_7H_8) are commonly used. A higher hydrogen content will decrease the hardness of the films [17]. Then again, the friction coefficient is generally lower when the hydrogen content is higher [18]. Here, we will show that the use of ethylene (C_2H_4) is a suitable choice for the preparation of DLC films, too.

A metal can be added to the DLC film by sputtering. Magnetron sputter sources are a mature and affordable technology. However, most of the sputtered material is provided in a neutral state [19, 20], provided that no high power impulse sputtering is used [21]. Below, evidence will be provided that in combination with the PSII process some of the sputtered metal atoms are ionized by the plasma and can thus be implanted into the substrate. The combination of a magnetron sputter source with plasma source ion implantation is sometimes called magnetron plasma source ion implantation [22].

When metal sputtering is combined with DLC deposition by PSII, the setup transforms into a reactive sputtering system. Even though argon is commonly used for the sputtering of the metal, the precursor for the DLC film is also in the vacuum chamber at the same time. Hence, the hydrocarbon will deposit onto the sputtering target, too. This leads to target poisoning which reduces the deposition rate of the metal but in turn provides more carbon to the sample. One way to change the amount of metal within the deposited film is to vary the ratio of the flow rates of argon and hydrocarbon gas [23]. As this process also depends on the sputter yield of the metal, the incorporation of different metals into the DLC film requires different preparation parameters. This will be exemplified by the use of copper and titanium.

2. Materials and Methods

In this study, reactive magnetron sputtering with a mixture of Ar and a hydrocarbon gas (C_2H_4) in combination with PSII was used to produce Me-DLC films on silicon wafer and glass substrates. Metal discs were used as targets for the sputter source, which was operated in RF power mode. A negative high voltage pulse was applied to the substrate holder. This voltage was also used to generate the plasma; that is, there was no additional plasma source (apart from the sputter source). The distance between the sputter source and the sample holder was in the range of 10 cm. For a schematic of the setup, see Figure 1. The content of metal was varied by changing the flow rate ratio of Ar and the hydrocarbon gas, that is, retaining an Ar flow rate of 20 sccm and varying the C_2H_4 flow rate, while keeping the same pressure of 0.7 Pa by changing the pumping rate.

The metal content and the distribution with depth were measured by Auger electron spectroscopy (AES) and X-ray photoelectron spectroscopy (XPS). Additional depth profiles were recorded by secondary ion mass spectrometry (SIMS). The structure of the films was investigated by glancing incidence X-ray diffraction (XRD), Raman spectroscopy (514 nm wavelength), and XPS. The electrical resistivity was measured with a four-point probe and the tribological properties with a ball-on-disk setup using a tungsten carbide ball (6 mm diameter, force 1 N, linear speed 10 cm/s, test at room temperature and ~20% relative humidity). The film adhesion was

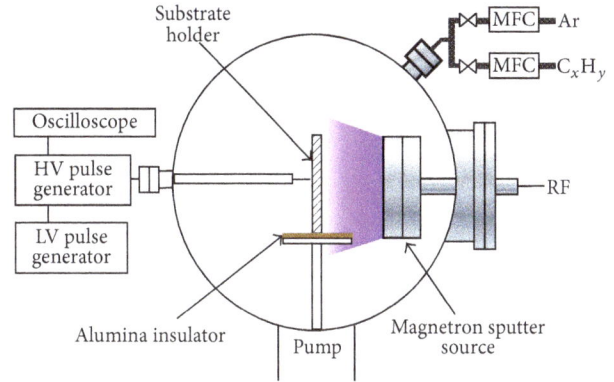

FIGURE 1: Schematic diagram of the magnetron PSII system. MFC = mass flow controller.

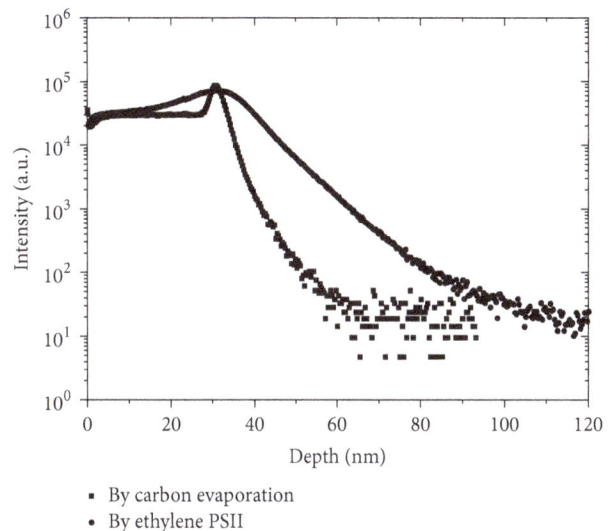

- By carbon evaporation
- By ethylene PSII

FIGURE 2: SIMS depth profile of a carbon film, deposited by evaporation (black) and by ethylene PSII at −15 kV (red).

measured with a pull-off test of an Al pin that was glued to the surface by an epoxy resin.

3. Results and Discussion

3.1. Preparation Process and Metal Content. With a combination of sputtering and PSII both parts will influence the film properties.

In accordance with the concept of PSII, ions generated from the surrounding gas are implanted into the substrate. Using exclusively a hydrocarbon gas, some of its molecules will be ionized when the high voltage pulse is applied to the substrate. As a consequence, a DLC film will be deposited on the substrate. With C_2H_4 DLC films can be grown. Figure 2 shows that the carbon from the C_2H_4 is not only deposited onto the substrate, but also implanted into it (process conditions: −15 kV, pulse length 10 μs, and 1 kHz repetition rate). The SIMS depth profile of the carbon shows a slowly declining signal that reaches several ten nm into the silicon substrate. In the case of a carbon film of similar

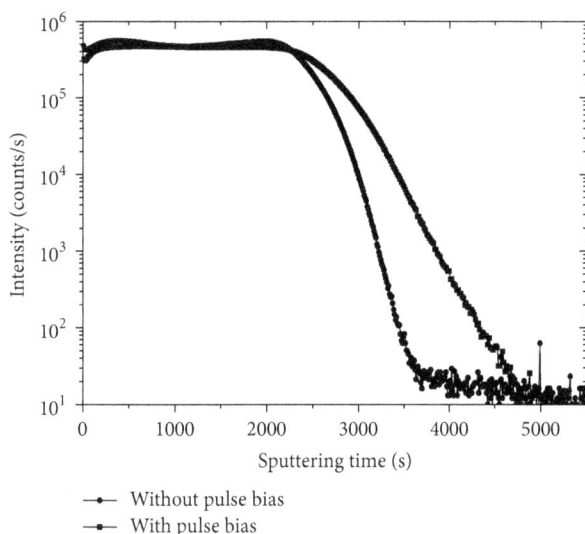

FIGURE 3: SIMS depth profile of the titanium signal of a sputter deposited Ti film with and without pulse bias.

thickness, which was deposited by evaporation of a carbon thread in a carbon coater, the slope of the signal is much higher.

Adding argon to the plasma gas changes the film structure. Now, not only the hydrocarbon molecules are ionized but also some of the argon atoms. When the argon ions hit the substrate, they will introduce more energy into the film. On the one hand, this leads to a higher disorder within the film caused, for example, by broken bonds. On the other hand, the argon will remove some of the already deposited film by sputtering, thereby decreasing the deposition rate.

Sputtered particles from a magnetron sputter source in proximity of the substrate can be considered to contribute to the gas phase. Hence, at least some of the sputtered particles should be ionized by the plasma from the PSII part of the setup. An easy way to demonstrate this is the preparation of a metal layer on a flat substrate by using only argon as process gas, one time without and one time with pulse bias applied to the sample holder. Depth profiling through the interface of the samples reveals that the interface is wider in the case when a pulse bias is used. In Figure 3 SIMS depth profiles are shown with only the signal of ^{48}Ti of both samples. The wider interface is apparent for the case of the pulse bias (-10 kV with 1 kHz repetition rate and 10 μs pulse length). The metal was implanted into the substrate, even if some part of this wider profile might be attributed to knock-on effects of the argon ions that impinge onto the sample during the high voltage pulse. Because of the metal gradient within the interface the adhesion of the metal layer is higher than in the case of no pulse bias. The adhesion of a sputtered copper film on silicon increases by about 15% when a pulse bias is used, from 29 to 34 MPa.

The second important process is occurring at the sputter source. The main parameter that influences not only the deposition rate of the Me-DLC film but also the metal content within the film is the ratio of the flow rates of the process gases. Usually, the flow rate of argon, which is used for sputtering of the metal target, is much higher than that of the hydrocarbon. Starting with a flow rate of zero for the hydrocarbon, there is obviously only the deposition of metal onto the substrate. Increasing the relative flow of the hydrocarbon to a few percent, there is a sharp decrease in metal content [24]. Because the hydrocarbon in front of the sputter source is ionized in the same way as the argon is, it is also accelerated onto the sputter target. In contrast to argon, the hydrocarbon tends to form a film on the target. Depending on the relative amount of incoming and deposited carbon ions and of the impinging argon ions, the thickness of the carbon film on the target is changed. When the target is completely covered with carbon, no metal will be deposited onto the substrate any more but only carbon which is sputtered from the target. To avoid this, the relative amount of argon is kept high. In Figure 4 the resulting metal contents for Cu and Ti as measured by XPS are presented for different flow rates of C_2H_4 at a fixed flow rate of 20 sccm for Ar. While at 2 sccm of C_2H_4 there is nearly no titanium in the samples because of the target poisoning, there is still about one-third of metal content in the case of copper. These values refer to the surface of the Me-DLC films. The sputter yield of copper is much higher than that of titanium. For argon ions at an energy of 1 keV, the yield is 3.8 for copper but only 0.9 for titanium, according to SRIM [25] simulations. No metal was found in the surface of different metal DLC films when the relative amount of methane in the process gas constituted 25% or more [24]. Here, a similar observation can be made. At above 5 sccm C_2H_4 flow, the copper content in the film is not more than 1 at.%.

Depending on the time scale of the target poisoning, a gradient of the metal can develop during the deposition. At first, that is, near the substrate, the metal content will be high. With time the hydrocarbon deposits onto the target, diminishing the amount of sputtered metal. In Figure 5 SIMS depth profiles of two Cu-DLC films prepared with different flow rates of C_2H_4 and a constant flow rate of 20 sccm of argon are shown. The deposition time was 20 minutes. For the higher metal content, a more or less constant concentration throughout the film was found. Higher flow rates of the hydrocarbon lead to a decrease of the metal content towards the surface. In the case of 6 sccm C_2H_4 flow it can be seen that it takes about 15 minutes until the process of target poisoning is completed. For certain applications this metal gradient might be desirable and can be produced in this way. If a constant metal concentration throughout the film is required, the process should initially be run with a closed shutter in front of the magnetron source and with no high voltage applied to the sample holder until a steady state is reached. The state of the target poisoning can be controlled by means of optical spectrometry [26], for example, provided that there is a line of sight to the plasma in front of the magnetron sputter source.

3.2. Structure of Me-DLC Films. When considering the structural changes within the DLC film, it can be found that the results depend on the type of metal. When a carbide forming metal such as Ti, Ta, W, Mo, or Nb is added in sufficient concentration, small crystalline metal carbide clusters are

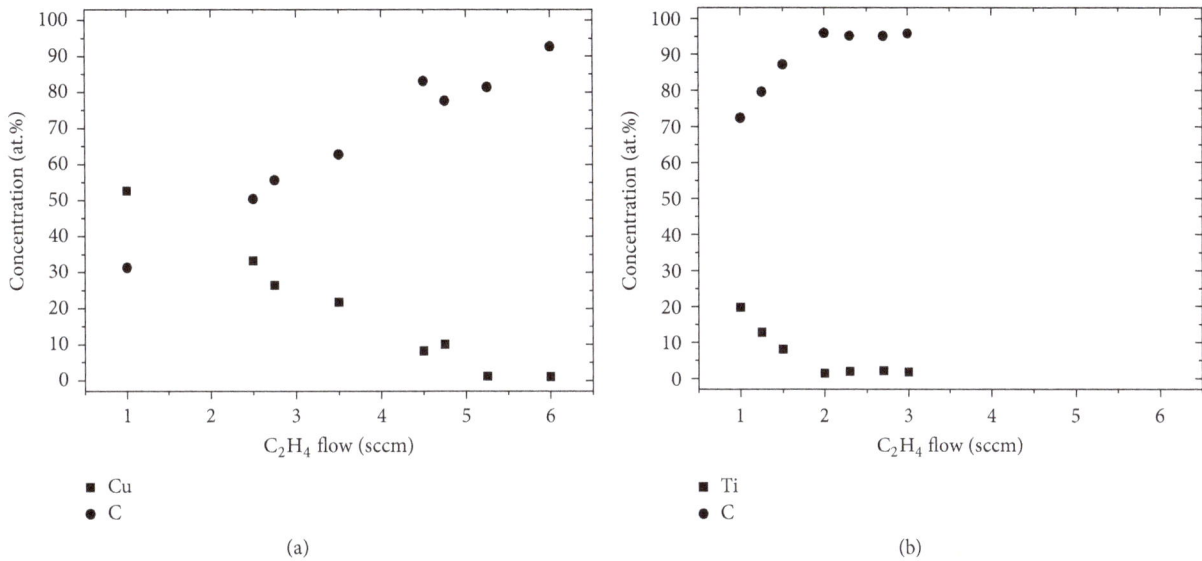

FIGURE 4: Composition of (a) Cu-DLC and (b) Ti-DLC films as a function of C_2H_4 flow rate, with a fixed Ar flow rate of 20 sccm.

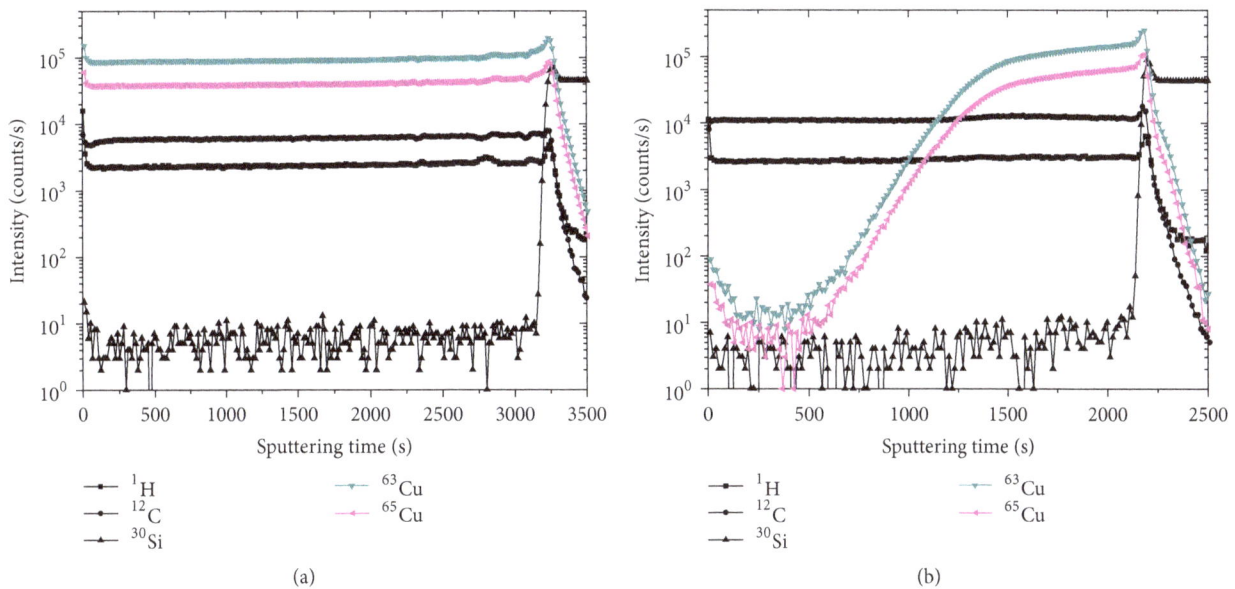

FIGURE 5: SIMS depth profiles of Cu-DLC films prepared by a flow of 20 sccm argon and different C_2H_4 flow rates of (a) 2 sccm and (b) 6 sccm.

formed. While no signal for TiC could be found in XRD diffractograms of our Ti-DLC samples (not shown here), with a maximum Ti content of about 20 at.%, a higher metal concentration leads to signals stemming from the carbide phase. In the case of tungsten, there are two phases involved, that is, the WC and the β-WC_{1-x} phase [27]. In the case of Mo, the carbide phase Mo_3C_2 was detected for Mo concentrations of 11 at.% and above [28]. There is a transition from the metal-containing amorphous phase for low concentrations to the metal carbide phase for higher concentrations.

In XPS spectra of the metal signal, here Ti2p, its position is at the one of the carbide, TiC; see Figure 6(a). The presence of the carbide phase can also be seen for higher metal contents in the C1s signal. For a metal content of about 20 at.% a

shoulder in the signal towards lower binding energies can be seen. This feature is absent in the case of smaller metal concentrations, for example, 2 at.% in Figure 6(b). Similar results were found based on Ti-DLC films prepared by C_2H_2 [29]. At around 19 at.% Ti a shoulder in the C1s signal is apparent which indicates the formation of TiC. At higher contents, such as 42 at.%, most of the C1s signal represents TiC, while only a small peak for pure C is left. The binding energy of the Ti2p signal shifts accordingly from the Ti to the TiC position. It was concluded that the amorphous metal-containing DLC phase and carbide nanoparticles coexist in a range of roughly 10–30% metal [22, 28], depending on the type of metal. With Ni, Fe and Co carbides were only observed for metal contents of 60% or more [28]. In the case

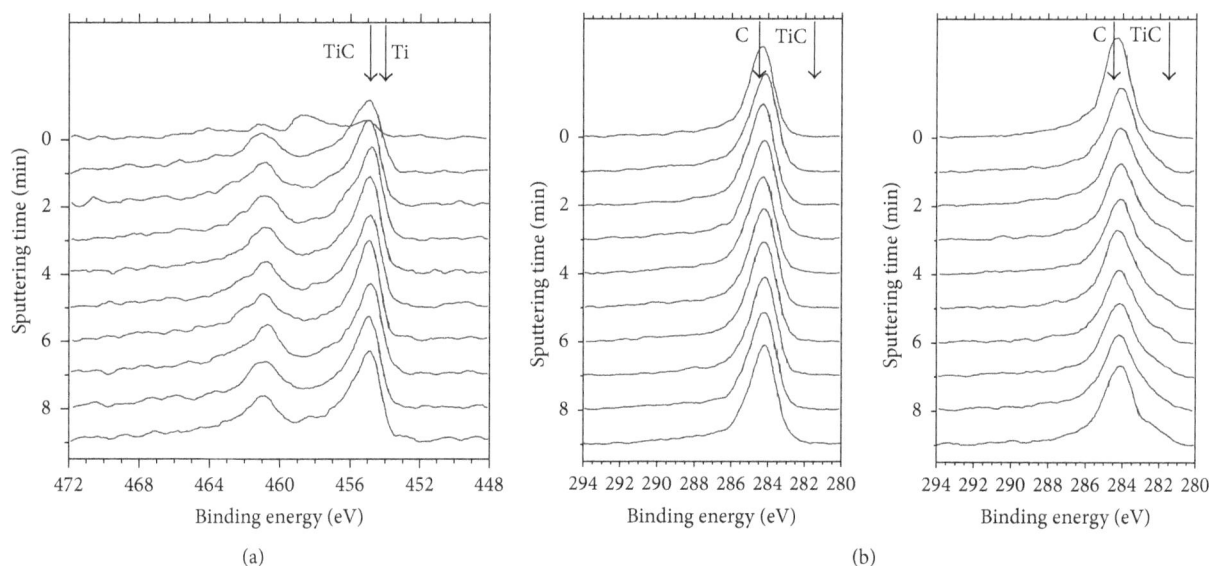

FIGURE 6: XPS spectra of (a) Ti2p of a Ti-DLC film with about 20 at.% Ti and (b) C1s of Ti-DLC films with 2 at.% (left) and 20 at.% (right) titanium content, respectively.

of copper, the XPS signal positions of C and Cu do not change with metal content (not shown here).

Generally, smaller metal particles within a DLC film are amorphous, whereas larger nanoparticles are of a (poly)crystalline nature [30]. Particle radii increase monotonically with increasing metal content and so do the particle distances [31]. For Nb-DLC and Mo-DLC a crystallite size of 4–6 nm was reported [24]. For Ti-DLC films crystallites of TiC of a similar size (about 10 nm) were found for an amount of 30% of Ti within the films [32]. For Ni the crystalline size was estimated as about 4 nm for a 13% Ni-DLC film [28].

Metals that do not form any carbide, or only weakly, are present as nanocrystalline metal particles within the DLC film. The Ag particles in a film with 3.8 at.% were of several nm size [33]. For copper, nanocrystals with sizes in the range of 15–30 nm were found in samples containing 11–23% Cu [34].

In addition to the formation of carbide, the structure of the carbon film is also affected by the metal incorporation. The films exhibit the two typical features of amorphous carbon films, the D peak around $1350\,cm^{-1}$, and the G peak in the vicinity of $1550\,cm^{-1}$ [35]. Pure DLC films prepared with C_2H_4 and a pulse voltage of −10 kV possess an intensity ratio of the D peak and the G peak (I_D/I_G) of around 1. This value is a bit higher than the comparison value of 0.86 for films prepared at −10 kV with a pulse of $5\,\mu s$ length and 1 kHz repetition rate at a pressure of 2 Pa C_2H_2 [36]. Evaluating the hydrogen content of the films by the slope of the luminescence background from the Raman spectra and an empirical formula which links it to the hydrogen content [37], a value of around 30 at.% is derived. This is more than the 26% found for samples made by C_2H_2 [38]; this is expected, however, since the precursor C_2H_4 contains double the amount of hydrogen.

When adding the metal, the I_D/I_G ratio increases, to values of about 1.7 in the case of copper and to values

between 2 and 2.5 in the case of titanium. There is no clear dependence on the metal content. The position of the G peak is in the range of $1570\,cm^{-1}$ which indicates, together with the increased intensity ratio and a narrower G peak, an increase of the average crystallite size of sp^2-bonded clusters [11, 37, 39]. Similar observations were made before for carbide formers like Ti [22] and W [27], but also for noncarbide formers like Ag [33].

The ratio of sp^2- and sp^3-bonded carbon is also affected by the incorporation of metal into the film. For vanadium it was found that the ratio shifts from 75 : 25 for a pure DLC film to 38 : 62 for a sample with 48% vanadium carbide and to 20 : 80 for a film with 60% vanadium carbide [40]. In some cases, for example, for copper, an initial reduction of sp^2 content was followed by an increase again with higher copper concentrations [34].

The addition of clusters of metal or metal carbide particles leads to an increase of the surface roughness of the Me-DLC films as compared to pure DLC films. While the latter are generally very smooth with R_a values in the range of 0.1 [41] to 0.2 nm [33], Me-DLC films can exhibit a roughness up to several nm. Ti- and Mo-DLC films seem to be not so rough with values of 0.4 [24] and 0.3–1.1 nm [24, 28], respectively. Nb- and W-DLC films [24] as well as Ni-DLC films [28] can reach several nm of roughness. Especially for metals that tend to form islands when a film is grown, such as silver, the roughness increases with silver content and globular domains can be seen on the film surface [41, 42].

3.3. Electrical Resistivity. Connected with the addition of metal to the DLC film is a drop in electrical resistivity. In Figure 7 this is displayed for the Ti and Cu containing DLC films as a function of the carbon concentration. The starting values for a pure DLC film are around 10 Ωm, which is in the same range as for films prepared by C_2H_2 [22, 29]. With a

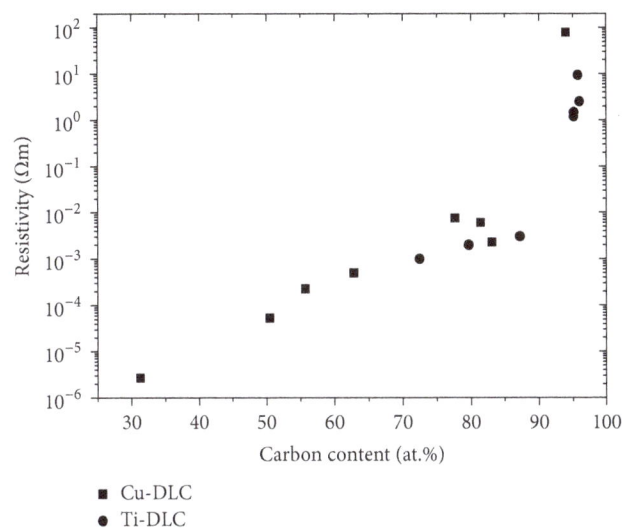

FIGURE 7: Electrical resistivity of Ti-DLC and Cu-DLC films as a function of the carbon content.

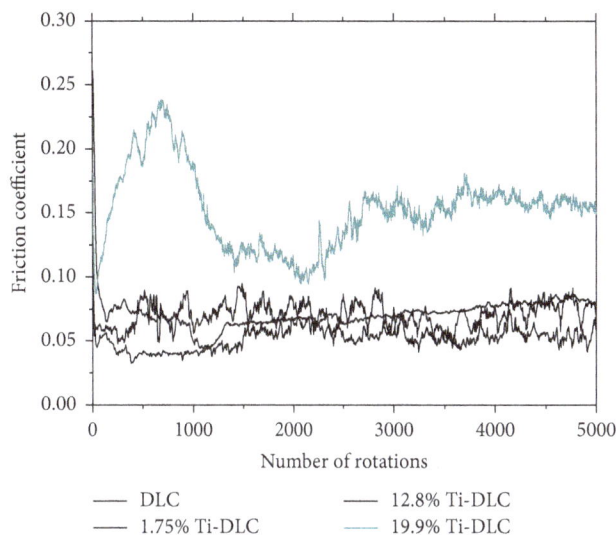

FIGURE 8: Friction coefficients as a function of number of rotations of Ti-DLC films.

few percent of metal added the resistivity drops sharply. With metal concentrations around 20–25% there is already a drop of about two or three orders of magnitude, which has also been noted for Co and Au additions [43]. When about half of the film consists of metal, the resistivity is merely around 10^{-5} Ωm. From the values in Figure 7 no distinct differences can be seen when comparing Ti- and Cu-DLC films. The decrease of the resistivity with metal content is in most cases nonlinear. This is connected with the structural changes as outlined above and with the changes in the conduction process. With the addition of metal there is a transition from thermally activated conduction along linkages or chains of sp^2 carbon atoms [44] to a process following a percolation law [45]. With metal clusters surrounded by a carbon matrix, there is a combination of percolation and hopping going on. Especially for nickel and iron the different conduction processes lead to a change of electrical resistivity with strain, which can be exploited to manufacture strain sensors [46].

3.4. *Tribological Performance.* Figure 8 shows the friction coefficient of a pure DLC film and three Ti-DLC films with different Ti contents during a tribology test with a ball-on-disk setup. With a value of 0.05–0.08 the friction coefficient of the pure DLC film is in the same range as the one of the DLC film prepared by C_2H_2 [47]. Even though the film here contains more hydrogen, the difference is apparently not large enough to influence the friction coefficient markedly. With the addition of titanium, the friction coefficient can be decreased. The addition of less than 2 at.% Ti reduces the friction coefficient slightly, to values of around 0.05. A similar effect was reported for Ag-containing films, where 1.8 at.% Ag reduced the friction coefficient to about 0.06 [42]. In case of the Cu-DLC films the friction coefficient of a film with about 1 at.% Cu is similar to the one of the pure DLC film with about 0.06. For a higher Cu content of 9 at.% the friction coefficient is higher, about 0.22.

Values below 0.1 have been reported for other metal-containing DLC films, for example, for 11 at.% Mo [28] and 27 at.% Ta, although there is a large range of Ta content with similar effects [22], and with 9 at.% W [29]. For Ni, Fe, and Co [28] no improvements were found. For Ti-DLC films prepared with C_2H_2 no reduction of the friction coefficient with the addition of Ti was found [29], in contrast to the results reported here.

Factors for the higher friction coefficient with larger amounts of metal are the hardness and the surface roughness. The hardness increases for additions of carbide forming metals, although it takes a volume fraction of about 25% of the metal carbide for distinct changes to be observed [43]. With the addition of noncarbide forming metals, the hardness decreases with metal content, for example, for Ag-DLC films [48, 49]. The surface roughness usually increases with metal content. At low metal contents, small metal precipitates are present, whereas the surface roughness and the hardness are only slightly affected. As a consequence of the nanoscale precipitates [50], the friction coefficient can decrease, as is shown in Figure 8.

The wear of the metal-containing films is slightly higher than that of the pure DLC film. The wear track of the Me-DLC samples after 5000 rotations is about 20–30 nm deep, as measured with a profilometer, with no distinct dependence on metal content. The pure DLC film features a wear track with about 15 nm depth after the test. The higher wear of W containing DLC films was noted before [51].

4. Conclusions

By the use of ethylene gas in combination with reactive magnetron sputtering, metal-containing DLC films can be prepared. The films prepared by C_2H_4 possess a different film structure as compared to ones made by C_2H_2 as evidenced by a higher I_D/I_G ratio and a higher hydrogen content.

The different sputter yields of the metals require different flow ratios of the argon and the hydrocarbon gas to prepare Me-DLC films with similar metal contents. The role of target poisoning of the metal target by the hydrocarbon was exemplarily shown via depth profiles of the Cu-DLC films. The Ti forms carbide within the Ti-DLC films and leads to more distinct changes in the Raman spectra, in contrast to the Cu-DLC films. The effect on the electrical resistivity of the films was similar for both metals, whereas a decrease in friction coefficient could be observed for low content Ti-DLC films and an unchanged friction coefficient for low content Cu-DLC films. Higher metal concentrations lead to a considerably higher friction coefficient, with the value of the Cu-DLC films being higher than the ones of the Ti-DLC films.

Competing Interests

The authors declare that there is no conflict of interests regarding the publication of this paper.

References

[1] A. Grill, "Diamond-like carbon: state of the art," *Diamond and Related Materials*, vol. 8, no. 2-5, pp. 428–434, 1999.

[2] Z. Zhang, R. Song, G. Li, G. Hu, and Y. Sun, "Improving barrier properties of PET by depositing a layer of DLC films on surface," *Advances in Materials Science and Engineering*, vol. 2013, Article ID 861804, 6 pages, 2013.

[3] D. R. Tallant, J. E. Parmeter, M. P. Siegal, and R. L. Simpson, "The thermal stability of diamond-like carbon," *Diamond and Related Materials*, vol. 4, no. 3, pp. 191–199, 1995.

[4] K. C. Walter, M. Nastasi, and C. Munson, "Adherent diamond-like carbon coatings on metals via plasma source ion implantation," *Surface and Coatings Technology*, vol. 93, no. 2-3, pp. 287–291, 1997.

[5] D. Neerinck, P. Persoone, M. Sercu et al., "Diamond-like nanocomposite coatings for low-wear and low-friction applications in humid environments," *Thin Solid Films*, vol. 317, no. 1-2, pp. 402–404, 1998.

[6] J. C. Sánchez-López and A. Fernández, "Doping and alloying effects on DLC coatings," in *Tribology of Diamond-Like Carbon Films*, pp. 311–338, Springer, New York, NY, USA, 2008.

[7] X. Wang and Y. Zhao, "Study of electrical conductivity and microcosmic structure of tetrahedral amorphous carbon films doped by boron," *Advances in Materials Science and Engineering*, vol. 2015, Article ID 727285, 6 pages, 2015.

[8] S. Flege, R. Hatada, M. Hoefling et al., "Modification of diamond-like carbon films by nitrogen incorporation via plasma immersion ion implantation," *Nuclear Instruments and Methods in Physics Research, Section B: Beam Interactions with Materials and Atoms*, vol. 365, pp. 357–361, 2015.

[9] A. Jiang, J. Xiao, X. Li, and Z. Wang, "Effect of structure, composition, and micromorphology on the hydrophobic property of F-DLC film," *Journal of Nanomaterials*, vol. 2013, Article ID 690180, 7 pages, 2013.

[10] A. Markwitz, J. Leveneur, P. Gupta, K. Suschke, J. Futter, and M. Rondeau, "Transition metal ion implantation into diamond-like carbon coatings: development of a base material for gas sensing applications," *Journal of Nanomaterials*, vol. 2015, Article ID 706417, 7 pages, 2015.

[11] J. Robertson, "Diamond-like amorphous carbon," *Materials Science and Engineering: R: Reports*, vol. 37, no. 4–6, pp. 129–281, 2002.

[12] A. Anders, Ed., *Handbook of Plasma Immersion Ion Implantation and Deposition*, John Wiley & Sons, New York, NY, USA, 2000.

[13] J. Pelletier and A. Anders, "Plasma-based ion implantation and deposition: a review of physics, technology, and applications," *IEEE Transactions on Plasma Science*, vol. 33, no. 6, pp. 1944–1959, 2005.

[14] J. R. Conrad, R. A. Dodd, F. J. Worzala, and X. Qiu, "Plasma source ion implantation: a new, cost-effective, non-line-of-sight technique for ion implantation of materials," *Surface and Coatings Technology*, vol. 36, no. 3-4, pp. 927–937, 1988.

[15] J. Chen, J. Blanchard, J. R. Conrad, and R. A. Dodd, "Structure and wear properties of carbon implanted 304 stainless steel using plasma source ion implantation," *Surface and Coatings Technology*, vol. 53, no. 3, pp. 267–274, 1992.

[16] W. Ensinger, "Formation of carbides and diamond-like carbon films by hydrocarbon plasma immersion ion implantation," in *Plasma Surface Engineering Research and Its Practical Applications*, R. Wei, Ed., pp. 135–178, Research Signpost, Trivandrum, India, 2008.

[17] M. Kamiya, H. Tanoue, H. Takikawa, M. Taki, Y. Hasegawa, and M. Kumagai, "Preparation of various DLC films by T-shaped filtered arc deposition and the effect of heat treatment on film properties," *Vacuum*, vol. 83, no. 3, pp. 510–514, 2008.

[18] J. Fontaine, C. Donnet, and A. Erdemir, "Fundamentals of the tribology of DLC coatings," in *Tribology of Diamond-Like Carbon Films*, C. Donnet and A. Erdemir, Eds., pp. 139–154, Springer, Berlin, Germany, 2008.

[19] P. J. Kelly and R. D. Arnell, "Magnetron sputtering: a review of recent developments and applications," *Vacuum*, vol. 56, no. 3, pp. 159–172, 2000.

[20] G. Bräuer, B. Szyszka, M. Vergöhl, and R. Bandorf, "Magnetron sputtering—milestones of 30 years," *Vacuum*, vol. 84, no. 12, pp. 1354–1359, 2010.

[21] K. Sarakinos, J. Alami, and S. Konstantinidis, "High power pulsed magnetron sputtering: a review on scientific and engineering state of the art," *Surface and Coatings Technology*, vol. 204, no. 11, pp. 1661–1684, 2010.

[22] K. Baba and R. Hatada, "Preparation and properties of metal containing diamond-like carbon films by magnetron plasma source ion implantation," *Surface and Coatings Technology*, vol. 158-159, pp. 373–376, 2002.

[23] S. Berg and T. Nyberg, "Fundamental understanding and modeling of reactive sputtering processes," *Thin Solid Films*, vol. 476, no. 2, pp. 215–230, 2005.

[24] C. Corbella, M. Vives, A. Pinyol et al., "Preparation of metal (W, Mo, Nb, Ti) containing a-C:H films by reactive magnetron sputtering," *Surface and Coatings Technology*, vol. 177-178, pp. 409–414, 2004.

[25] J. F. Ziegler, SRIM—The stopping and range of ions in matter, 2013, http://srim.org/.

[26] S. Konstantinidis, F. Gaboriau, M. Gaillard, M. Hecq, and A. Ricard, "Optical plasma diagnostics during reactive magnetron sputtering," in *Reactive Sputter Deposition*, D. Depla and S. Mahieu, Eds., pp. 301–336, Springer, Berlin, Germany, 2008.

[27] K. Baba, R. Hatada, and Y. Tanaka, "Preparation and properties of W-containing diamond-like carbon films by magnetron plasma source ion implantation," *Surface and Coatings Technology*, vol. 201, no. 19-20, pp. 8362–8365, 2007.

[28] K. Baba and R. Hatada, "Preparation and properties of metal-containing diamond-like carbon films by magnetron plasma source ion implantation," *Surface and Coatings Technology*, vol. 196, no. 1-3, pp. 207–210, 2005.

[29] K. Baba and R. Hatada, "Deposition and characterization of Ti- and W-containing diamond-like carbon films by plasma source ion implantation," *Surface and Coatings Technology*, vol. 169-170, pp. 287–290, 2003.

[30] W.-Y. Wu and J.-M. Ting, "Growth and characteristics of carbon films with nano-sized metal particles," *Thin Solid Films*, vol. 420-421, pp. 166–171, 2002.

[31] K. I. Schiffmann, M. Fryda, G. Goerigk, R. Lauer, P. Hinze, and A. Bulack, "Sizes and distances of metal clusters in Au-, Pt-, W- and Fe-containing diamond-like carbon hard coatings: a comparative study by small angle X-ray scattering, wide angle X-ray diffraction, transmission electron microscopy and scanning tunneling microscopy," *Thin Solid Films*, vol. 347, no. 1-2, pp. 60–71, 1999.

[32] A. A. Voevodin, S. V. Prasad, and J. S. Zabinski, "Nanocrystalline carbide/amorphous carbon composites," *Journal of Applied Physics*, vol. 82, no. 2, pp. 855–858, 1997.

[33] K. Baba, R. Hatada, S. Flege et al., "Preparation and antibacterial properties of Ag-containing diamond-like carbon films prepared by a combination of magnetron sputtering and plasma source ion implantation," *Vacuum*, vol. 89, no. 1, pp. 179–184, 2013.

[34] C.-C. Chen and F. C.-N. Hong, "Structure and properties of diamond-like carbon nanocomposite films containing copper nanoparticles," *Applied Surface Science*, vol. 242, no. 3-4, pp. 261–269, 2005.

[35] A. C. Ferrari and J. Robertson, "Interpretation of Raman spectra of disordered and amorphous carbon," *Physical Review B*, vol. 61, no. 20, pp. 14095–14107, 2000.

[36] K. Baba and R. Hatada, "Deposition of diamond-like carbon films on polymers by plasma source ion implantation," *Thin Solid Films*, vol. 506-507, pp. 55–58, 2006.

[37] J. Choi and T. Hatta, "Structural changes of hydrogenated amorphous carbon films deposited on steel rods," *Applied Surface Science*, vol. 357, pp. 814–818, 2015.

[38] R. Hatada, K. Baba, S. Flege, and W. Ensinger, "Long-term thermal stability of Si-containing diamond-like carbon films prepared by plasma source ion implantation," *Surface and Coatings Technology*, vol. 305, pp. 93–98, 2016.

[39] Y. Hirata and J. Choi, "Microstructure of a-C:H films prepared on a microtrench and analysis of ions and radicals behavior," *Journal of Applied Physics*, vol. 118, no. 8, Article ID 085305, 2015.

[40] A. Shigemoto, T. Amano, and R. Yamamoto, "Work function measurements of vanadium doped diamond-like carbon films by ultraviolet photoelectron spectroscopy," https://arxiv.org/abs/1402.1911.

[41] L. Kolodziejczyk, W. Szymanski, D. Batory, and A. Jedrzejczak, "Nanotribology of silver and silicon doped carbon coatings," *Diamond and Related Materials*, vol. 67, pp. 8–15, 2016.

[42] K. Baba, R. Hatada, S. Flege, and W. Ensinger, "Preparation and properties of Ag-containing diamond-like carbon films by magnetron plasma source ion implantation," *Advances in Materials Science and Engineering*, vol. 2012, Article ID 536853, 5 pages, 2012.

[43] C. P. Klages and R. Memming, "Microstructure and physical properties of metal-containing hydrogenated carbon films," *Materials Science Forum*, vol. 52-53, pp. 609–644, 1990.

[44] A. Grill, "Electrical and optical properties of diamond-like carbon," *Thin Solid Films*, vol. 355-356, pp. 189–193, 1999.

[45] R. Sanjinés, M. D. Abad, C. Vâju, R. Smajda, M. Mionić, and A. Magrez, "Electrical properties and applications of carbon based nanocomposite materials: an overview," *Surface and Coatings Technology*, vol. 206, no. 4, pp. 727–733, 2011.

[46] U. Heckmann, R. Bandorf, H. Gerdes, M. Lübke, S. Schnabel, and G. Bräuer, "New materials for sputtered strain gauges," *Procedia Chemistry*, vol. 1, no. 1, pp. 64–67, 2009.

[47] K. Baba, R. Hatada, S. Flege, and W. Ensinger, "Deposition of silicon-containing diamond-like carbon films by plasma-enhanced chemical vapour deposition," *Surface and Coatings Technology*, vol. 203, no. 17-18, pp. 2747–2750, 2009.

[48] H. W. Choi, J.-H. Choi, K.-R. Lee, J.-P. Ahn, and K. H. Oh, "Structure and mechanical properties of Ag-incorporated DLC films prepared by a hybrid ion beam deposition system," *Thin Solid Films*, vol. 516, no. 2-4, pp. 248–251, 2007.

[49] F. R. Marciano, L. F. Bonetti, L. V. Santos, N. S. Da-Silva, E. J. Corat, and V. J. Trava-Airoldi, "Antibacterial activity of DLC and Ag-DLC films produced by PECVD technique," *Diamond and Related Materials*, vol. 18, no. 5–8, pp. 1010–1014, 2009.

[50] C. P. Lungu, "Nanostructure influence on DLC-Ag tribological coatings," *Surface and Coatings Technology*, vol. 200, no. 1–4, pp. 198–202, 2005.

[51] K. Bewilogua, R. Wittorf, H. Thomsen, and M. Weber, "DLC based coatings prepared by reactive d.c. magnetron sputtering," *Thin Solid Films*, vol. 447-448, pp. 142–147, 2004.

13

Inverse Strategies for Identifying the Parameters of Constitutive Laws of Metal Sheets

P. A. Prates, A. F. G. Pereira, N. A. Sakharova, M. C. Oliveira, and J. V. Fernandes

CEMUC, Department of Mechanical Engineering, University of Coimbra, Rua Luís Reis Santos, Pinhal de Marrocos, 3030-788 Coimbra, Portugal

Correspondence should be addressed to P. A. Prates; pedro.prates@dem.uc.pt

Academic Editor: Sutasn Thipprakmas

This article is a review regarding recently developed inverse strategies coupled with finite element simulations for the identification of the parameters of constitutive laws that describe the plastic behaviour of metal sheets. It highlights that the identification procedure is dictated by the loading conditions, the geometry of the sample, the type of experimental results selected for the analysis, the cost function, and optimization algorithm used. Also, the type of constitutive law (isotropic and/or kinematic hardening laws and/or anisotropic yield criterion), whose parameters are intended to be identified, affects the whole identification procedure.

1. Introduction

Finite Element Analysis (FEA) is now a well-established computational tool in industry for the optimization of sheet metal forming processes. The accurate modelling of these processes is a complex task due to the nonlinearities involved, such as those associated with (i) the kinematics of large deformations, (ii) the contact between the sheet and the tools, and (iii) the plastic behaviour of the metal sheet.

The description of the plastic behaviour of metal sheets is usually performed using phenomenological constitutive models. In this context, the emergence of new steels and aluminium, magnesium, and other alloys, as well as their increasingly widespread use in the automotive and aeronautical industries, has encouraged the development of more reliable models, with increasing flexibility associated with a larger number of parameters to identify [1–14]. In fact, the accuracy of the numerical simulation results of sheet metal forming processes depends on the flexibility of a constitutive material model but also on the procedure adopted to identify its parameters. The complex nature of the plastic behaviour of metal sheets makes their characterization dependent upon factors such as (i) the constitutive model; (ii) the experimental tests performed, comprising the sample geometry, the testing conditions, and the analysis methodologies; (iii) the strategy for identifying the constitutive parameters.

The strategy for identifying the model parameters is generally seen as an optimization problem, where the purpose is to minimise the difference between computed and experimental results of one or more experiments. Two main types of strategies for the identification of the constitutive parameters can be recognised in literature: classical and inverse strategies. The classical identification strategies for the constitutive parameters make use of a large number of standardised mechanical tests, with well-defined geometry and loading conditions, such that homogeneous stress and strain distribution develop in the region of interest (e.g., [15, 16]); nonstandard mechanical tests can also be performed to properly describe other biaxial stress states in the sheet plane (e.g., [17, 18]). However, sheet metal forming processes are carried out under strongly nonhomogeneous stresses and strains fields. Therefore, limiting the characterization of the mechanical behaviour of metal sheets to a restricted number of tests with linear strain paths and homogeneous deformation can lead to a somewhat incomplete characterization of the overall plastic behaviour of the material [19].

Recent developments and accessibility of optical full-field measurement techniques, such as digital image correlation (DIC) technique coupled with FEA, make the inverse identification strategies a common current place. The full-field measurements allow the acquisition of enriched information from mechanical tests, such as displacement and strain fields;

an overview on this topic can be found in [20]. This allows attenuating the constraints on the geometry and loading conditions of the mechanical tests used for the identification of materials parameters, so that nonhomogeneous stress and strain distributions can be developed in the region of interest (e.g., [21–29]). In this sense, the identification of constitutive parameters from nonhomogeneous strain fields and complex loading conditions provides a more reliable description of the material behaviour during real sheet metal forming processes [21]. In such complex mechanical tests, it is no longer possible to identify the constitutive parameters based on simple assumptions on the stress and/or strain states, as in the classical identification strategies. Instead, a finite element model of the mechanical test is established and cost functions are defined to minimise the gap between numerical and experimental results of the mechanical test, which demands efficient optimization algorithms. However, the efficiency of any inverse identification strategy directly depends on the information contained by the objective function. In the context of constitutive parameters identification, this is related to the type of experimental results included (e.g., loads, displacements, and strains) but also to the strain paths and levels of deformation attained by the experimental test. It turns out that there is no consensus about the experimental tests (sample geometry and loading conditions), the cost functions, and the optimization procedure that will lead to accurate constitutive parameters identification. Also, a major obstacle to the widespread use of advanced constitutive models in industrial simulations seems to result from the lack of an efficient strategy for parameters identification. In this sense, the developed strategy must be simple, from an experimental point of view, and allow evaluating to what extent the selected constitutive model allows perfectly describing the behaviour of a given material.

The present paper describes recent inverse strategies coupled with FE simulations for the identification of the parameters of constitutive laws that describe the plastic behaviour of metal sheets, resorting to mechanical tests leading to nonuniform strain and stress states. Following this introduction, the paper addresses general concepts for the constitutive modelling and the optimization problem. Afterwards, an overview of inverse identification strategies for the constitutive parameters is presented, with emphasis on inverse identification strategies resorting to FE simulations.

2. Constitutive Modelling

Constitutive models have been developed to predict the onset and evolution of the plastic deformation of a deformable body undergoing a general state of stress. A phenomenological constitutive model is typically a combination of the following components:

(i) Yield criterion that describes the yield surface of the material in a multidimensional stress space: The metal sheets are usually assumed to be orthotropic, with invariant anisotropy during plastic deformation. With high incidence in the last decades, the emergence of anisotropic yield criteria with an increasing number of material parameters has been witnessed. They provide the flexibility required for accurately modelling the plastic behaviour of advanced metallic alloys, which are frequently used in automotive and aeronautical industries. Several approaches have been used for deriving yield criteria, based on

(1) high-order polynomial functions (e.g., [1, 2]);

(2) the generalization to anisotropy of the second and third invariants of the deviatoric stress tensor, J_2 and J_3, respectively (e.g., [3]);

(3) one or more isotropic yield functions, using the linear Isotropic Plasticity Equivalent (IPE) stress space concept (e.g., [3–9]);

(4) the construction of weighted sums of anisotropic yield criteria (e.g., [7]);

(5) the capability to model the tension-compression asymmetry, particularly devoted to specific magnesium and titanium alloys (e.g., [3, 6, 10]);

(6) the capability to model kinematic hardening [11];

(7) the interpolation of second-order Bézier curves [12].

(ii) Hardening laws that express the evolution of the yield surface during plastic deformation, as schematized in Figure 1: The isotropic hardening law refers to the homothetic expansion of the yield surface (see Figure 1(a)) while the kinematic hardening law describes its translation in the stress space (see Figure 1(b)). Kinematic hardening laws are recommended for describing plastic deformation under strain path changes, mainly strain path reversal, in materials that exhibit Bauschinger effect (e.g., [14]). The combination of isotropic and kinematic hardening laws provides a flexible model, for simultaneously describing the change in size and the position of the centre of the yield surface, during plastic deformation. Isotropic hardening laws described by power laws (e.g., [32–37]), saturation laws (e.g., [38, 39]), and weighted combinations of isotropic hardening laws (e.g., [40, 41]) have been proposed. Linear (e.g., [42, 43]) and nonlinear (e.g., [13, 14, 44–47]) kinematic hardening laws were proposed, with the latter being more appropriate to describe the Bauschinger effect.

(iii) Flow rule, to establish a relationship between the stress state and the plastic strain increment: Typically, an associated flow rule is adopted, that is, using the yield function as plastic potential, although some exceptions can be found in literature (see, e.g., [48]).

The general representation of a constitutive model can be described through a function $\mathscr{F} = \mathscr{F}(\boldsymbol{\sigma}' - \mathbf{X}', \bar{\varepsilon}^{\mathrm{p}}, \alpha, \beta)$:

$$\mathscr{F}\left(\boldsymbol{\sigma}' - \mathbf{X}', \bar{\varepsilon}^{\mathrm{p}}, \alpha, \beta\right) = \bar{\sigma}\left(\boldsymbol{\sigma}' - \mathbf{X}', \alpha\right) - Y\left(\bar{\varepsilon}^{\mathrm{p}}, \beta\right), \quad (1)$$

where $\bar{\sigma}(\boldsymbol{\sigma}' - \mathbf{X}', \alpha)$ is the equivalent stress defined by a given yield criterion and $Y(\bar{\varepsilon}^{\mathrm{p}}, \beta)$ is the hardening law

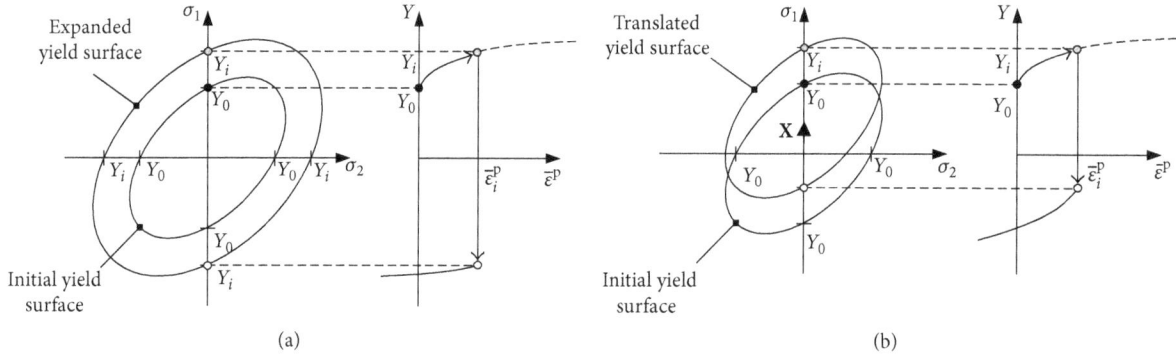

FIGURE 1: Representation of the plastic behaviour of materials in tension-compression. The left side shows generic yield surfaces in the plane $(\sigma_1; \sigma_2)$ and the right side shows the corresponding hardening law: (a) isotropic hardening and (b) kinematic hardening. See text for details. Adapted from [30].

that represents the evolution of the yield stress during the deformation. The equivalent stress, $\overline{\sigma}(\boldsymbol{\sigma}' - \mathbf{X}', \alpha) = \overline{\sigma}$, is a function of the effective stress tensor, $(\boldsymbol{\sigma}' - \mathbf{X}')$, that includes the parameters of the yield criterion, α, for describing the anisotropy ($\boldsymbol{\sigma}'$ and $\mathbf{X}'$ are the deviatoric Cauchy stress and the deviatoric backstress tensors, resp.) and $Y(\overline{\varepsilon}^P, \beta) = Y$ is a function of the equivalent plastic strain, $\overline{\varepsilon}^P$, in which the parameters are represented by β. The yielding is defined based on the function $\mathcal{F}$ of (1) and can be written as follows:

$$\mathcal{F} = \overline{\sigma} - Y = 0 \implies$$
$$\overline{\sigma} = Y. \tag{2}$$

If $\overline{\sigma} < Y$, the stress state of the material remains inside the yield surface and only elastic deformation occurs. When plastic deformation occurs, the associated flow rule states that the increment of the plastic strain tensor is normal to the yield surface, for a stress state such that $\overline{\sigma} = Y$. The normality condition, defined by the associated flow rule, assumes that the increment of the plastic strain tensor is normal to the yield surface and is expressed by

$$d\boldsymbol{\varepsilon}^P = d\lambda \frac{\partial \overline{\sigma}(\boldsymbol{\sigma}' - \mathbf{X}')}{\partial(\boldsymbol{\sigma}' - \mathbf{X}')}, \tag{3}$$

where $d\boldsymbol{\varepsilon}^P$ is the increment of the plastic strain tensor, $d\lambda$ is a scalar multiplier, and $\overline{\sigma}(\boldsymbol{\sigma}' - \mathbf{X}') = \overline{\sigma}$ is the equivalent stress function, representing the plastic potential.

Even though a number of advanced constitutive models are available in literature, sheet metal forming simulations are still mostly performed in industry not taking into account kinematic hardening and with the well-known Hill'48 yield criterion [49], whose parameters identification can be easily assessed by uniaxial tensile tests. Mattiasson and Sigvant [50] mentioned some plausible explanations, still valid today, for this reality:

(i) The relative simplicity of the Hill'48 model that makes it attractive to use

(ii) The unavailability of industry analysts for understanding to what extent the modelling of the material influences the simulation results

(iii) The lack of knowledge, time, and money for performing the multiaxial tests required to identify reliable hardening curves and parameters of advanced yield criteria

(iv) The additional cost in terms of CPU time for using more advanced constitutive models which is considered to be an effort that is not worth it.

Nevertheless, in our view, the major obstacle to the widespread use of advanced constitutive models in industrial simulations comes from the large number of linear strain path tests, including multiaxial tests, required for the parameters identification. To overcome this barrier, a potential approach is to look for new constitutive parameters identification strategies that are alternative to the classical ones. In this sense, an accurate description of the material plastic behaviour could be attained from (i) a minimum number of mechanical tests and experimental data; (ii) flexible and user-friendly constitutive models; and (iii) an accessible identification procedure for the constitutive parameters, coupled with robust optimization algorithms. Therefore, Section 4 will discuss some identification procedures for the constitutive parameters based on inverse analysis, as an alternative to the classical approaches.

3. The Optimization Problem

The inverse identification of constitutive model parameters is generally seen as an optimization problem. The purpose is to minimise the difference between computed and experimental results of one or more experiments. This difference is expressed by a cost function and its minimisation is performed using optimization algorithms, which automatically operate on the values of the constitutive parameters.

3.1. Cost Function. A wide number of cost function formulations for the identification of constitutive parameters have been proposed in literature (e.g., [57, 58]). According to Cao and Lin [57], the cost function should operate as an "efficient guide" of the optimization procedure, in order to search for the best fit to the experimental results; therefore, the ideal cost function should comprise the following conditions:

(i) All measured points of a given experiment should be part of the optimization procedure and have equal opportunity to be optimized, provided that experimental errors are eliminated.

(ii) All experiments should have equal opportunity to be equally optimized, and so the optimization should not depend on the number of points considered in each experiment.

(iii) Different units of measure in the cost function should not affect the performance of the optimization.

(iv) The identification procedure should not be dependent of the user, and so the values of the weighting factors should be optimized to achieve the abovementioned conditions.

Cost functions are typically formulated under the concept of weighted least-squares, as follows:

$$F(\mathbf{A}) = \frac{1}{m} \frac{1}{n} \sum_{i=1}^{m} w_i \sum_{j=1}^{n} w_j \left[r_{ij}(\mathbf{A}) \right]^2, \qquad (4)$$

where $F(\mathbf{A})$ is the cost function to minimise; $\mathbf{A}$ is the vector of constitutive parameters to optimize; m is the total number of experiments and n is the total number of points, considered in each experiment i; $r_{ij}(\mathbf{A})$ is the residual between the numerically predicted results and those of the experiment i at point j; w_i and w_j are the weighting factors for each experiment i and for each point j, respectively. Within the context of inverse parameter identification, $r_{ij}(\mathbf{A})$ can contain variables such as loads, pressures, angular moments, or those arising from full-field measurements (displacements or strains), as will be seen in detail later.

The residuals can be expressed in terms of relative differences,

$$r_{ij}(\mathbf{A}) = \frac{u_{ij}^{\text{Num}}(\mathbf{A}) - u_{ij}^{\text{Exp}}}{u_{ij}^{\text{Exp}}}, \qquad (5)$$

$$\text{with } i = 1, \ldots, m, \ j = 1, \ldots, n,$$

or in terms of absolute differences,

$$r_{ij}(\mathbf{A}) = u_{ij}^{\text{Num}}(\mathbf{A}) - u_{ij}^{\text{Exp}}, \qquad (6)$$

where u_{ij}^{Num} and u_{ij}^{Exp} are, respectively, the numerically predicted and the experimental results at point j of experiment i. Residuals are often expressed using relative differences, which allows the use, in the same cost function, of several kinds of quantities exhibiting various orders of magnitude and units of measure [59]. When u_{ij}^{Exp} admits values close to or equal to zero, the residuals should be expressed using absolute differences.

3.2. Optimization Algorithms.
The minimization of the least-squares cost function, presented in (4), requires efficient and robust optimization algorithms, due to the strongly nonlinear nature of the least-squares cost function [60]. For this purpose, several optimization algorithms are described in the literature, which are commonly divided into two categories: gradient-free algorithms and gradient-based algorithms. Hybrid optimization strategies using both gradient-free and gradient-based algorithms are also proposed (e.g., [15, 60]).

Gradient-free algorithms, such as evolutionary and SIMPLEX algorithms, have a great probability of achieving a global minimum due to their random search capability. They require a large number of cost function evaluations (i.e., iterations) and therefore the convergence can be very time-consuming. Because of this, gradient-free algorithms are not recommended within the context of inverse identification strategies, since they require a large number of finite element simulations and analyses [61].

Gradient-based algorithms are most popular within inverse identification strategies, as they require far less cost function evaluations than gradient-free algorithms. As local optimizers, these algorithms use information of the gradient to update the vector of constitutive parameters in an adequate search direction [62]. Therefore, there is no guarantee that these algorithms converge to the global minimum, with the possibility of converging to undesirable local minima. This makes the optimization procedure dependent on the initial estimate for the parameters, and therefore the choice of convenient initial estimates for the constitutive parameters can be essential.

Examples of gradient-based algorithms commonly used within the context of inverse identification strategies are the Gauss-Newton and Levenberg-Marquardt algorithms. The Gauss-Newton algorithm is described as follows:

$$\mathbf{A}^{s+1} = \mathbf{A}^s - \left(\mathbf{J}^{\mathrm{T}} \mathbf{W} \mathbf{J} \right)^{-1} \mathbf{J}^{\mathrm{T}} \mathbf{W} \mathbf{r}(\mathbf{A}^s), \qquad (7)$$

where s is the iteration step, $\mathbf{A}$ is the vector of constitutive parameters, $\mathbf{W}$ is the vector of weighting factors, $\mathbf{J}$ is the Jacobian matrix that expresses the sensitivity of the computed results to the constitutive parameters, and $\mathbf{r}(\mathbf{A}^s)$ is the vector of residuals, which can be expressed in terms of relative or absolute differences (see (5) and (6), resp.). The dimension of the vector of residuals depends on the total number of experiments m and the total number of points n, in each experiment, that is, the dimension $n_r = m \times n$. Considering that the total number of constitutive parameters to be identified is n_p, with $n_r \geq n_p$, the Jacobian containing the partial derivatives of the residuals with respect to the constitutive parameters is defined:

$$J_{l,p} = \frac{\partial r_l(A_p)}{\partial A_p}, \quad \text{with } l = 1, \ldots, n_r, \ p = 1, \ldots, n_p. \qquad (8)$$

An efficient method to compute the Jacobian matrix is finite differentiation. In order to improve convergence, the Jacobian matrix must be updated at each iteration step s. However, the calculation in each step requires high computational cost (at least one numerical simulation per constitutive parameter). To overcome this inconvenient, Endelt et al. [63, 64] and Cooreman [62] highlighted the possibility of computing the sensitivity matrix analytically, with the latter author having concluded the inability of this approach for computing the sensitivities of strain fields to the material parameters, in mechanical tests involving complex and/or heterogeneous deformation.

In some cases, the Gauss-Newton algorithm can become unstable in the neighbourhood of the minimum, and so a stabilisation procedure is required. The Levenberg-Marquardt algorithm is similar to Gauss-Newton one, but includes a stabilising term, as follows [65]:

$$\mathbf{A}^{s+1} = \mathbf{A}^s - \left(\mathbf{J}^\mathrm{T}\mathbf{W}\mathbf{J} + \lambda \operatorname{diag}\left(\mathbf{J}^\mathrm{T}\mathbf{W}\mathbf{J}\right)\right)^{-1}\mathbf{J}^\mathrm{T}\mathbf{W}\mathbf{r}\left(\mathbf{A}^s\right), \quad (9)$$

where λ is the stabilising parameter that is updated in each iteration according to the convergence rate [65]. When the Levenberg-Marquardt method shows stability, small values for λ are recommended for fast convergence; otherwise, large values of λ usually allow stable convergence, although slower, towards the minimum. Note that for $\lambda = 0$ the Levenberg-Marquardt algorithm is equal to the Gauss-Newton one.

A different type of optimization technique that has been recently used in the identification of constitutive parameters is the Response Surface Methodology (RSM) (e.g., [54, 66]). RSM is an optimization technique for generating smooth approximations of complex functions in a multidimensional design space. In the context of parameter identification, the design space contains all possible combinations for the constitutive parameters and related values of the cost function. The prohibitive size of the full design space requires a Design of Experiments (DoE), to efficiently construct an approximated design space from a few number of representative points (i.e., sets of constitutive parameters). The responses of the representative points (i.e., the values of the cost function) are used to fit a response surface, which is typically obtained from second-order polynomial regression, for the sake of simplicity. Finally, the minimum of the response surface is calculated using a gradient-based optimization procedure, which leads to an estimate of the optimal set of constitutive parameters. In brief, the RSM technique can be summarised as follows:

(1) An initial guess for the design space for the material parameters is selected.

(2) Numerical simulations are performed with the different sets of parameters, representing the experimental design points needed for filling the design space.

(3) For each simulation, the predicted results are compared with the experimental ones, and the cost function values are calculated according to (4).

(4) A response surface is constructed to approximate the values of the cost function; typically, least-squares approximations are used to determine second-order polynomials.

(5) An optimization algorithm is applied to determine the minimum point of the response surface (i.e.,

where $F(\mathbf{A})$ is minimum), providing the optimal set of material parameters.

If a converged solution is not found, the process starts all over again, adding a new region of interest to the design space.

4. Inverse Identification Strategies

While classical strategies make use of global measurements from experiments to infer the values of the constitutive parameters, using simple analytical relations to estimate the material response under the assumption of homogeneous stress and strain fields in the region of interest, the inverse identification strategies are much more flexible [20]. They make use of experiments allowing heterogeneous deformation and/or strain path changes, as close as possible of the conditions usually found during real sheet metal forming processes. In this perspective, some authors even proposed tests involving contact with friction, such as the punch stretch test (e.g., [67]) and the cylindrical cup test (e.g., [68]), for performing the inverse parameter identification. In these latter cases, the adequate description of the local contact with friction is of paramount importance because it can affect the final results of the parameter identification (e.g., see [69]).

The inverse identification strategies make use of global measurements, such as tool loads and tool displacements, which are usually coupled with local measurements, represented as full-field states of displacements and/or strains on the surface of the sample. Then, a numerical analysis of the mechanical test is performed, assuming a constitutive model chosen *a priori* and an initial estimate for its parameters. Finally, the experimental results of the mechanical test are iteratively compared with numerical results by acting on the values of the constitutive parameters until there is an adequate agreement between experimental and numerical results. Figure 2 shows a schematic representation of inverse identification strategies based on the comparison between experimental and numerical measurements using FE simulations.

The advantages of this identification approaches include substantial amount of reliable data extracted from a single mechanical test, using full-field measurements, which enables the accurate identification of large sets of constitutive parameters taking into account a wide range of strain levels and strain paths; therefore, it does not require uniform stress and strain distributions, in the region of interest, and no particular restrictions to the sample geometry and/or loading conditions are imposed.

Nevertheless, due to the design of the sample geometry, loading conditions, and induced strain paths, the inverse identification requires proper computational strategies [20]. The most common strategy uses Finite Element Model Updating (FEMU) and consists of performing successive finite element (FE) simulations of the physical experiment; the set of parameters are obtained by minimising the difference between the experimental and the numerical measurements. This difference is expressed by a cost function and its minimisation is performed using optimization algorithms, which automatically operate on the values of the constitutive

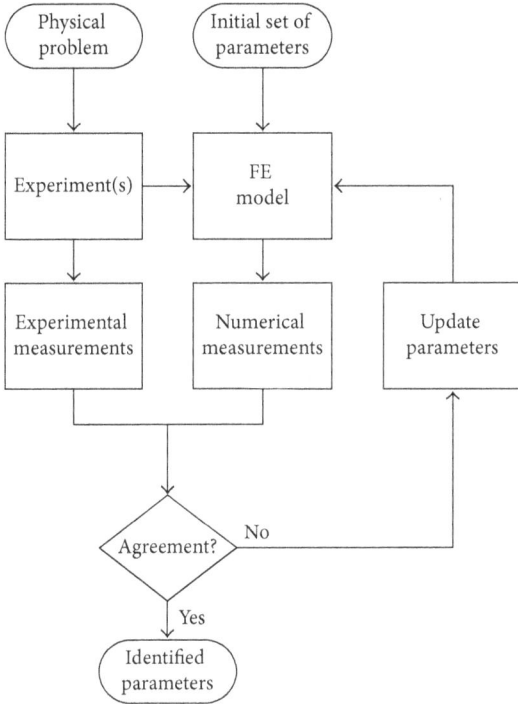

FIGURE 2: Schematic representation of inverse identification strategies. Adapted from [31].

parameters. Usually, cost functions compare the experimental and simulated loads and full-field measurements (e.g., [23, 59, 70–72]); less frequently, some authors propose to use only the load (e.g., [54, 66]), or full-field displacements (e.g., [24]) or strains (e.g., [21, 27–29]), at a given moment of loading.

A promising alternative to the use of FEMU is the Virtual Fields Method (VFM), which is based on the principle of virtual work. This approach does not require using time-consuming FE analysis and therefore avoids potential drawbacks related to the accuracy of FE models, namely, the representation of the geometry and boundary conditions [73]. The VFM was successfully used in the identification of parameters of constitutive laws describing the plastic behaviour of metal sheets [73–76]. However, the accuracy of the parameter identification using VFM depends on the adequate choice of the virtual field, which is currently a challenge for problems involving large heterogeneity of deformation of anisotropic materials, as well as large plastic deformations. In fact, in this type of problems, the optimal virtual field has to be evaluated for each time increment, which makes it less attractive than that for linear problems.

4.1. Overview of FEMU Strategies. This subsection provides an overview of the literature on inverse methodologies for the identification of parameters of constitutive laws based on inverse strategies coupled with FE simulations. These cases highlight that the identification procedure is dictated by the loading conditions, the geometry of the sample, the type of experimental results selected for the analysis, the cost functions, and the optimization algorithm used.

In this context, inverse strategies were developed for the identification of the anisotropy and hardening behaviour of metal sheets, simultaneously or separately. Table 1 shows a comparative outline of these strategies, whose key details are highlighted in the following subsections.

4.1.1. Identification of Isotropic Hardening and/or Yield Criterion Parameters

(1) Strategies Using Cruciform Specimens. There has been a steadily growing interest in developing inverse identification strategies supported by the use of the biaxial tensile test of cruciform specimens, coupled with full-field displacement or strain measurements (e.g., [21, 24–29]). In general, this test allows (i) strain paths ranging from uniaxial tension (in the arms region of the specimen) to balanced biaxial tension (in the centre section of the specimen), (ii) high strain gradients from the centre region of the specimen to the end of the arms region, and (iii) no contact between surfaces and therefore no friction. Also, by changing the load and/or displacement ratio over the two normal loading axes, it is possible to achieve several biaxial stress states in the central region of the specimen. However, this kind of test only permits attaining low values of equivalent plastic strain (close to those obtained in uniaxial tension) before instability occurs and no occurrence of out-of-plane shear stresses is observed (i.e., the test is insensitive to the material parameters associated with these stresses). Figure 3 shows a set of cruciform geometries proposed in the literature for identifying material parameters using FEMU strategies.

In this context, Cooreman et al. [21] proposed the use of the biaxial tensile test on a perforated cruciform specimen (see Figure 3(b)), to simultaneously identify the material parameters of Swift hardening law [34] and Hill'48 yield criterion that describe the plastic behaviour of a 0.8 mm thick DC06 sheet steel. The identification makes use of strain field data measured at the central region of the specimen (see dashed area in Figure 3(b)) taken at 7 distinct load steps and iteratively compared with their numerical counterparts using Gauss-Newton algorithm, using the following cost function:

$$F(\mathbf{A}) = \frac{1}{3n} \left\{ \sum_{i=1}^{n} \left[\left(\left(\varepsilon_{xx}^{\text{Exp}} \right)_i - \left(\varepsilon_{xx}^{\text{Num}}(\mathbf{A}) \right)_i \right)^2 \right. \right.$$
$$+ \left(\left(\varepsilon_{yy}^{\text{Exp}} \right)_i - \left(\varepsilon_{yy}^{\text{Num}}(\mathbf{A}) \right)_i \right)^2 \tag{10}$$
$$\left. \left. + \left(\left(\varepsilon_{xy}^{\text{Exp}} \right)_i - \left(\varepsilon_{xy}^{\text{Num}}(\mathbf{A}) \right)_i \right)^2 \right]^{1/2} \right\},$$

where $(\varepsilon^{\text{Exp}})_i$ and $(\varepsilon^{\text{Num}}(\mathbf{A}))_i$ are the experimentally determined and numerical values of the strain components ε_{xx}, ε_{yy}, and ε_{xy} at point i, respectively, and n is the total number of measuring points. According to the authors, the results from the inverse strategy are similar to those from classical strategies, except for the parameter ε_0 of the Swift law, which leads to clearly different yield stress values. The authors attribute the discrepancy of ε_0 results to the use of strain field results from loading steps that neglect the

TABLE 1: Comparative overview of inverse identification strategies coupled with FE simulations.

Author	Type of test	Strain paths	Strain path reversal	Constitutive model	Optimization steps	Number of cost functions	Analysed results	Measuring instants
Cooreman et al. [21]	Cruciform biaxial tensile test	Uniaxial tension to biaxial stretching	No	Hill'48 criterion + Swift isotropic law	One	One	Strain field (2D)	Various (7)
Schmaltz and Willner [24]	Cruciform biaxial tensile test	Simple shear to biaxial stretching (but depending on the test geometry)	No	Hill'48 criterion + Hockett-Sherby isotropic law (only 2 of 4 parameters identified)	Two steps using the same cost function	One	Displacement field (2D)	One
Prates et al. [25]	Cruciform biaxial tensile test	Uniaxial tension to biaxial stretching	No	Hill'48 criterion + Swift isotropic law	Five steps using four cost functions	Four	Load + equivalent strain field (1D)	During the test / One
Prates et al. [26]	Cruciform biaxial tensile test	Uniaxial tension to biaxial stretching	No	Various criteria (4) + Swift and Voce isotropic laws	Two (Hill'48 criterion) or three (other criteria)	Two or three	Load + equivalent strain + strain path fields (1D)	During the test / One
Zhang et al. [28]	Cruciform biaxial tensile test	Uniaxial tension to biaxial stretching	No	Bron and Besson criterion	One	One	Strain field (1D)	One
Liu et al. [29]	Cruciform biaxial tensile test / Uniaxial tensile	Uniaxial tension to biaxial stretching / Uniaxial tension	No	Voce-type isotropic law	One	One	Strain field (2D)	During the test
Güner et al. [22]	Uniaxial tensile test on a notched sample	Near uniaxial tension	No	Yld2000 (2D) criterion	One	One	Strain field (2D) + load	During the test
Pottier et al. [23]	Out-of-plane test	Simple shear to biaxial stretching	No	Hill'48 Criterion + Ludwick isotropic law	One	One	Load + displacement field (3D)	Various (6)
Chamekh et al. [51]	Circular and elliptical bulge tests	Biaxial stretching	No	Ludwick isotropic law + r_0, r_{45}, and r_{90}	Two	One	Pressure	During the test
Bambach [52]	Circular bulge test	Biaxial stretching	No	Voce isotropic law	Two	Four	Pressure, pole strain, and pole thickness	During the test
Reis et al. [53]	Circular bulge test	Biaxial stretching	No	Swift isotropic law	Four	One	Pressure	During the test
Eggertsen and Mattiasson [54]	Three-point bending test	Tension-compression	Yes	Various kinematic laws (4)	One	One	Load	During the test

TABLE 1: Continued.

Author	Type of test	Strain paths	Strain path reversal	Constitutive model	Optimization steps	Number of cost functions	Analysed results	Measuring instants
Yin et al. [55]	Twin bridge shear test	Pure shear to uniaxial tension	Yes	Armstrong-Frederick kinematic law + Voce isotropic law (only 2 of 3 parameters identified)	One	One	Angular moment	During the test
Pereira et al. [56]	Shear test	Shear with reversal	Yes	Lemaitre-Chaboche kinematic law + Swift isotropic law	One	One	Load	During the test

Note: 1D: in one direction in the sheet plane; 2D: on the surface of the sheet plane; 3D: three-dimensional.

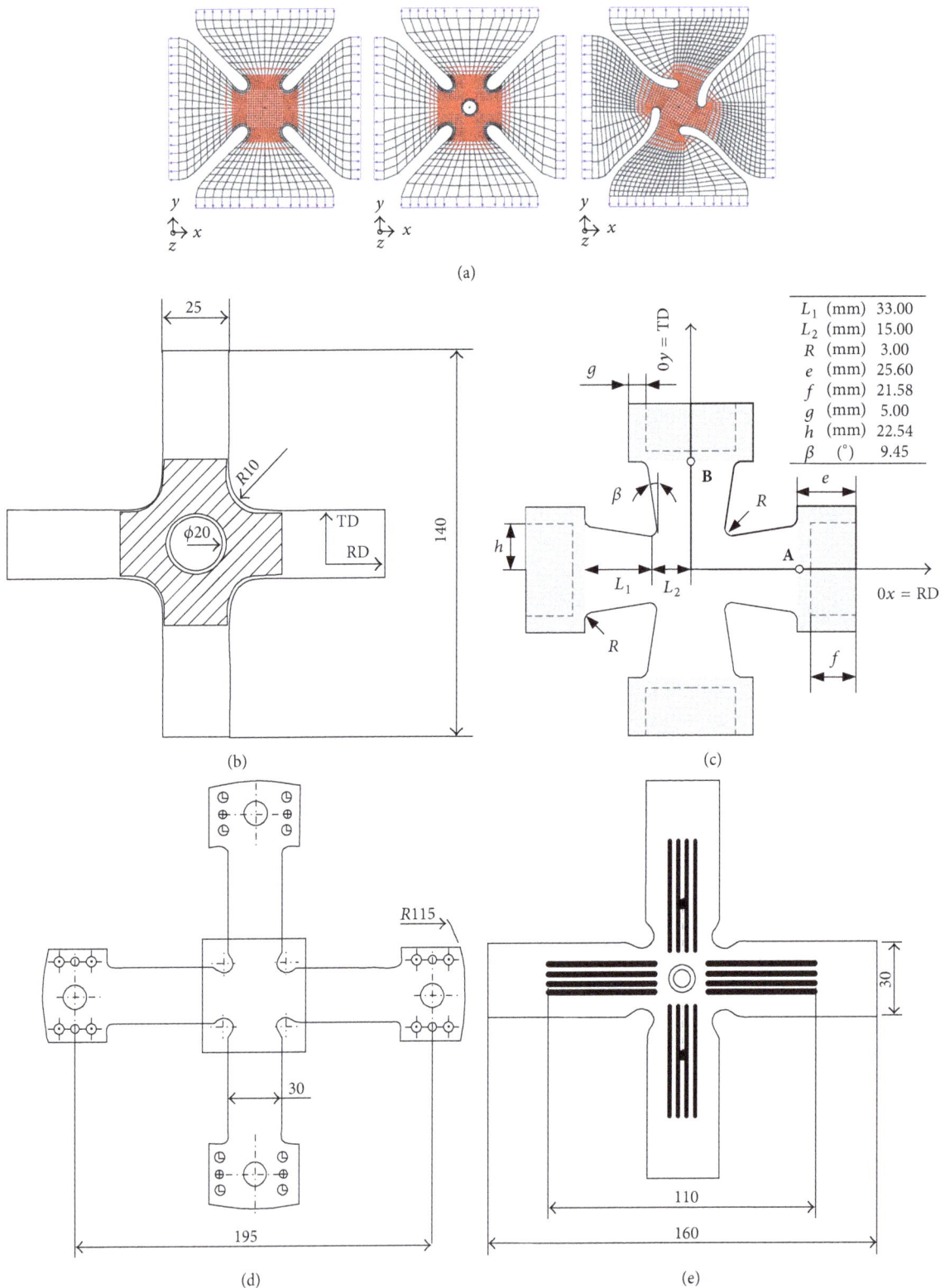

FIGURE 3: Cruciform specimen geometries proposed in the literature for FEMU strategies: (a) Schmaltz and Willner [24]; (b) Cooreman et al. [21]; (c) Prates et al. [25, 26]; (d) Zhang et al. [27, 28]; and (e) Liu et al. [29].

beginning of the test. They suggest performing the inverse identification using additional strain fields from loading steps located near the onset of plastic deformation, which in our opinion can lead to high relative errors. A simpler alternative, from the experimental point of view, would be to include in the identification a cost function considering the load-displacement curves for both axes of the cruciform sample, as adopted in other identification strategies later described.

Schmaltz and Willner [24] explored the usability of the biaxial tensile test on three cruciform specimen geometries, promoting different types of heterogeneous strain fields (see Figure 3(a)), in order to identify the plastic behaviour of a 2.0 mm thick DC04 sheet steel modelled via Hill'48 yield criterion and Hockett-Sherby hardening law [39]. The experimental and numerical displacement fields, in the regions in red (see Figure 3(a)) were compared and their difference was minimised using the Levenberg-Marquardt algorithm, with the following cost function:

$$F(\mathbf{A}) = \sum_{l=1}^{2} \sum_{i=1}^{n} \left[\left(u_l^{\text{Exp}} \right)_i - \left(u_l^{\text{Num}} (\mathbf{A}) \right)_i \right]^2, \qquad (11)$$

where $(u_l^{\text{Exp}})_i$ and $(u_l^{\text{Num}}(\mathbf{A}))_i$ are the experimentally determined and the numerically predicted values of the displacements at point i, respectively, in the $0x$ and $0y$ directions of the sheet plane ($l = 1, 2$), at a given moment of the test.

The optimization procedure starts with three different initial sets of parameters that lead to quite similar optimized values, for each test geometry, suggesting that the global optimum is reached. Moreover, the identification is split in two sequential steps (using the same cost function): (i) the first step identifies two of the Hockett-Sherby hardening law parameters (the yield stress and the saturation stress), under the assumption of isotropic (von Mises) material; the two remaining parameters of this law (contained in the exponent) are obtained by fitting the experimental results determined from the biaxial tensile test and are kept fixed during the identification procedure; (ii) in the second step, the Hill'48 parameters are identified. The cruciform geometry with the central hole (in the middle of Figure 3(a)) allows achieving the best results in terms of convergence of the iterative procedure and accuracy of the identified material parameters, which was assigned to the strong heterogeneity of the kinematic field.

Prates et al. [25, 26] designed a cruciform sample and developed two strategies for simultaneously identifying the parameters of the anisotropic yield criteria and isotropic hardening law of sheet metals. Both strategies use the results of the load evolution during the test and of the major and minor principal strains distributions, along the axes of the sample, at a given moment of the test, preceding the maximum load. Both strategies were numerically tested. The optimization of the design of the cruciform sample (see Figure 3(c)) was performed by means of a numerical study with the purpose of maximising the sensitivity of the test results to the values of the constitutive parameters and for allowing a wide range of strain paths, from uniaxial tension, at the arms of the sample, to near equibiaxial tension, at the centre of the sample.

The work was initially addressed for the identification of the parameters of the Hill'48 yield criterion and the isotropic Swift hardening law [25]. An inverse identification was performed without resorting to the traditional optimization algorithms (e.g., gradient-based algorithms, or others); that is, a specific algorithm was built for this purpose. The inverse analysis algorithm consists of a sequence of five optimization

steps, using the results of load evolution with the sample boundaries displacement during the test, for the axes $0x$ and $0y$ (see Figure 3(c)) and the distributions of von Mises equivalent strain along both axes of the sample, for an instant preceding and close to the maximum load. At each step, one or more parameters or a relationship between them is identified. The strategy was tested using numerically generated results of fictitious materials, which proved to be competitive, when compared with classical strategies. This allowed understanding that a sequential optimization, since properly elaborated, is clearly advantageous when compared to most commonly inverse identifications, consisting of using a unique cost function including different types of results.

The inverse analysis strategy mentioned above [25] enabled a good understanding of the issues involved, namely, concerning the delineation of the sequential algorithm leading to upper accuracy. This allowed extending the strategy to more complex constitutive models (yield criteria and isotropic hardening laws). Therefore, a general inverse identification strategy that sequentially uses three distinct cost functions was developed [26]. It resorts to the Levenberg-Marquardt algorithm for sequential optimization of the parameters of the yield criteria and isotropic hardening laws. More importantly, this strategy allows the identification of parameters of several yield criteria and hardening laws. It can be used directly for a given criterion or, sequentially, starting from the Hill'48 yield criterion and then using the Hill'48 criterion solution as an initial estimate for identifying the parameters of other criteria, on the condition that can be converted into the Hill'48 yield criterion for particular values of the parameters. In the last case, this strategy is detached in two stages and has the advantage of enabling the assessment of the adequacy of a number of constitutive models to describe the experimental results, starting from a simple anisotropic criterion, the Hill'48. The first stage consists of the simultaneous identification of the hardening law, Swift and/or Voce [38], in the case, and Hill'48 yield parameters, using the results of the load evolution in function of the displacement of the grips and the equivalent strain distribution at a given moment of the test, along the axes of the sample. The hardening parameters must be separately identified for the Swift and Voce laws and the one (Swift or Voce) that better describes the results of the cruciform test (if it is possible to distinguish) is selected for further optimization. The first stage involves the sequential minimisation of the following cost functions:

$$F_1(\mathbf{A}) = \frac{1}{Q_1} \sum_{i=1}^{Q_1} \left(\frac{P_i^{\text{Num}}(\mathbf{A}) - P_i^{\text{Exp}}}{P_i^{\text{Exp}}} \right)^2_{0x}$$

$$+ \frac{1}{Q_2} \sum_{i=1}^{Q_2} \left(\frac{P_i^{\text{Num}}(\mathbf{A}) - P_i^{\text{Exp}}}{P_i^{\text{Exp}}} \right)^2_{0y},$$

$$F_2(\mathbf{B}) = \frac{1}{R_1} \sum_{i=1}^{R_1} \left(\frac{\bar{\varepsilon}_i^{\text{Num}}(\mathbf{B}) - \bar{\varepsilon}_i^{\text{Exp}}}{\bar{\varepsilon}_i^{\text{Exp}}} \right)^2_{0x}$$

$$+ \frac{1}{R_2} \sum_{i=1}^{R_2} \left(\frac{\overline{\varepsilon}_i^{\text{Num}}(\mathbf{B}) - \overline{\varepsilon}_i^{\text{Exp}}}{\overline{\varepsilon}_i^{\text{Exp}}} \right)_{0y}^2 ,$$

$$(12)$$

where $P_i^{\text{Num}}(\mathbf{A})$ and P_i^{Exp} are the experimentally determined and numerical values of the load, respectively, and $\overline{\varepsilon}_i^{\text{Num}}(\mathbf{B})$ and $\overline{\varepsilon}_i^{\text{Exp}}$ are the experimentally determined and numerical values of the equivalent strain, respectively; Q_1, Q_2, R_1, and R_2 are the total number of measuring points for axes $0x$ and $0y$ of the sample; $\mathbf{A}$ and $\mathbf{B}$ are vectors of isotropic hardening law and yield criteria parameters, respectively.

The second stage allows extending the parameters identification procedure to more complex yield functions, such as Barlat'91 [8], Karafillis and Boyce [9], and Drucker+L [3], the cases studied in the work. This second stage should be performed whenever the identification carried out during the first stage is found to be not enough satisfactory to capture the experimental strain paths results, along the axes of the sample, which are not considered in minimisation during the first stage. The second stage of this inverse strategy involves the minimisation of the following cost function:

$$F_3(\mathbf{C}) = \frac{1}{S_1} \sum_{i=1}^{S_1} \left(\rho_i^{\text{Num}}(\mathbf{C}) - \rho_i^{\text{Exp}} \right)_{0x}^2$$

$$(13)$$

$$+ \frac{1}{S_2} \sum_{i=1}^{S_2} \left(\rho_i^{\text{Num}}(\mathbf{C}) - \rho_i^{\text{Exp}} \right)_{0y}^2 ,$$

where $\rho_i^{\text{Num}}(\mathbf{C})$ and ρ_i^{Exp} are the experimentally determined and numerical values of the strain path, respectively, S_1 and S_2 are the total number of measuring points for axes $0x$ and $0y$ of the sample, and $\mathbf{C}$ is the vector of yield criteria parameters. This sequential optimization procedure is a successful alternative to the parameter identification by minimising a single cost function comprising all material parameters and results of different types, as commonly found in the literature. Namely, it is concluded that this last approach can deteriorate the description of the material behaviour, concerning the load versus displacement results, and therefore the parameters and the choice of the hardening law, without apparent improvement of the description of the results of equivalent strain and strain path distributions.

Zhang et al. [28] identified the parameters of Bron and Besson yield criterion [4] for both AA5086 aluminium alloy and DP980 dual-phase steel sheets, using a cruciform sample previously designed by the authors [27] and shown in Figure 3(d). The inverse identification strategy consists of minimising the gap between the experimental and numerical distributions of the major and minor strains along the diagonal direction of the sample central area, at an instant immediately before rupture, using a SIMPLEX optimization algorithm. The cost function used is defined as follows:

$$F(\mathbf{A}) = \sum_{l=1}^{2} \frac{\sum_{i=1}^{n} \left(\left(\varepsilon_l^{\text{Exp}} \right)_i - \left(\varepsilon_l^{\text{Num}}(\mathbf{A}) \right)_i \right)^2}{\sum_{i=1}^{n} \left(\left(\varepsilon_l^{\text{Exp}} \right)_i \right)^2} , \qquad (14)$$

where $(\varepsilon_l^{\text{Exp}})_i$ and $(\varepsilon_l^{\text{Num}}(\mathbf{A}))_i$ are the experimentally determined and numerical values of the principal strain components ($l = 1, 2$) at point i, respectively, n is the total number of measuring points along the diagonal path, and $\mathbf{A}$ is the vector of 13 parameters to be identified: four isotropy and eight anisotropy parameters of the Bron and Besson yield model and the yield stress value. The hardening of the material is modelled with isotropic hardening described by the Voce law, in case of AA5086 aluminium, and an equation based on Swift and Voce laws, for DP980 steel. In both cases, the hardening parameters are directly identified from results of the tensile test in the rolling direction and were kept fixed during the inverse identification procedure (except the yield stress).

Liu et al. [29] designed a cruciform sample with a thickness-reduced central zone and four slots at each arm (see Figure 3(e)), to perform the parameters identification of a modified Voce law, describing the hardening behaviour of AA5086 aluminium sheet. The inverse analysis procedure makes use of the SIMPLEX algorithm, to minimise the difference between experimental and numerical strains measured at the centre region of the sample during the test, expressed by the following cost function:

$$F(\mathbf{A}) = \sum_{l=1}^{2} \left[\frac{\sum_{i=1}^{n} \left(\left(\varepsilon_l^{\text{Exp}} \right)_i - \left(\varepsilon_l^{\text{Num}}(\mathbf{A}) \right)_i \right)^2}{\sum_{i=1}^{n} \left(\left(\varepsilon_l^{\text{Exp}} \right)_i \right)^2} \right]^{1/2} , \quad (15)$$

where $(\varepsilon_l^{\text{Exp}})_i$ and $(\varepsilon_l^{\text{Num}}(\mathbf{A}))_i$ are the experimentally determined and numerical values of the principal strain components ($l = 1, 2$) at point i, respectively, at the central point of the sample, n is the total number of the time points of simulation, and $\mathbf{A}$ is the vector of three hardening parameters of the modified Voce law. The experimental force evolution along the two arms of the cruciform specimen was applied to the FE model for the numerical simulations, taking into account the lack of synchronization observed in the two tensile forces on each axis of the sample. In this inverse identification of the modified Voce law parameters, three yield functions were considered: von Mises, Hill'48, and the more advanced Bron and Besson criterion, whose parameters were previously identified [27]. The identified biaxial flow stress curves were then compared with an experimental curve obtained from a uniaxial tensile test, showing that a good agreement can be achieved if an adequate yield function is used in the FE model.

(2) Strategies Using Bulge Test. Most of the strain paths observed in deep-drawing components are in the range between simple tension and balanced biaxial tension, which justifies the widespread use of the cruciform specimen in the framework of the methodologies of inverse analysis. Nevertheless, this test has a strong drawback related to low deformation levels achieved, particularly in the central region of the sample, in which the strain and stress paths can be close to equibiaxial, although being dependent on the applied load or displacement ratios along the two axes of the specimen. In contrast, the bulge test allows obtaining relatively high strain values before necking, and so the flow stress curves can be assessed up to large strain values, for several biaxial strain (or

stress) paths depending on the geometry of the die (circular or elliptical). In this context, a few cases of inverse analysis methodologies were developed, with the aim of identifying the parameters of work-hardening laws.

Chamekh et al. [51] describe an inverse approach, based on Artificial Neural Networks (ANN), to identify the material parameters of a stainless steel (AISI 304). They use the results of the evolutions of pressure with the pole height, which are transferred to the neural network. The ANN is trained by means of curves of pressure versus displacement of the central point of the cap, generated by finite element simulations of the circular bulge test. During the training process, the network computes the weight connections, minimising the total mean squared error between the actual output and the desired output. The neural network generates an approximated function for the material parameters depending on the profile of the evolution of pressure with the pole height curve. Then, it was exploited for the identification of material parameters from experimental results. The Ludwick hardening law [33] and the Hill'48 yield criterion were selected. Therefore, the set of parameters to be identified also comprise the Lankford coefficients, r_0, r_{45}, and r_{90}. The material parameters are identified according to the following two steps: (i) the first step, using the circular bulge test, is to find the Ludwick hardening law parameters (assuming the knowledge of Lankford's coefficients determined from the tensile tests); (ii) the second step, using the elliptical bulge test for an off axis angle of $0°$, is to recalculate the Lankford's coefficients. An elliptical die for an off axis angle of $45°$ is used for the validation of the parameters identification. The authors conclude the following: (i) the ANN methodology can predict acceptable combination of the values of the material parameters; (ii) once the ANN was trained, output results for a given set of input data are available almost instantaneously. Despite these conclusions, it should be noted that the values of the experimental (from the tensile test) and identified hardening coefficients are far away (the experimental and identified values are 0.67 and 0.4, resp.). The remaining identified values of the parameters differ between 20 and 30% when compared with the tensile test except for the yield stress, whose values are approximately equal.

Bambach [52] explored the usability of the circular bulge test to identify the parameters of the Voce law of a fictitious material, which is considered isotropic. Initially, the membrane theory is applied to the results as in experimental cases, in order to obtain a set of parameters of the Voce, by fitting the stress versus strain results. The inverse analysis strategy proposed resorts to a gradient-based optimization algorithm, which is known for being sensitive to the initial solution. Thus, by using an initial solution, the one previously obtained with the membrane theory, it is expected to avoid convergence problems. The work gives special focus on the choice of the cost function to be minimised, making use, separately or simultaneously, of results of pressure versus pole height, pole strain versus pole height, and pole thickness versus pole height, and formulated as follows:

$$F_p(\mathbf{A}) = \int_0^{h_{\max}} \left[p^{\mathrm{Exp}}(h) - p^{\mathrm{Num}}(h, \mathbf{A}) \right]^2 dh,$$

$$F_\varepsilon(\mathbf{A}) = \int_0^{h_{\max}} \left[\varepsilon^{\mathrm{Exp}}(h) - \varepsilon^{\mathrm{Num}}(h, \mathbf{A}) \right]^2 dh,$$

$$F_t(\mathbf{A}) = \int_0^{h_{\max}} \left[t^{\mathrm{Exp}}(h) - t^{\mathrm{Num}}(h, \mathbf{A}) \right]^2 dh,$$

(16)

where $F_p(\mathbf{A})$, $F_\varepsilon(\mathbf{A})$, and $F_t(\mathbf{A})$ are cost functions defined by the pressure, p, strain, ε, and thickness, t, with pole height, h, respectively. The author concluded that the reidentification procedure (so called by the author) is significantly improved when combining the first two types of results. This significantly improves the reidentification, since it will contribute to reducing the search area where the minimum value of the objective function is located. It should be noted that this proposal is a reidentification, which has its starting point in the parameters previously obtained from the use of the membrane theory, such as in the traditional procedure recommended by the ISO standard [77]. Furthermore, it needs to resort to strain results in the pole of the cap during the test, which does not simplify the experimental procedure.

Reis et al. [53] proposed an inverse methodology for determining the hardening law of metal sheets, from the results of pressure versus pole height obtained in the bulge test, involving the identification of the parameters of the Swift law. The starting point of this analysis was to realize that it is possible to achieve a unified description (i.e., overlapping) of the evolution of the pressure with the pole height, for a given value of the hardening parameter of the Swift law, regardless of the yield stress and anisotropy of the material and sheet thickness. To achieve the overlapping of such curves, appropriate multiplying factors must be used for the values of pressure and pole height, depending on the yield stresses and thicknesses ratios of the sheets and also on their anisotropy. Thereafter, an inverse analysis methodology was developed, which consists in the search for the best coincidence between pressure versus pole height experimental and reference curves, with the latter being obtained by numerical simulation assuming isotropic material behaviour with various values of the hardening parameter in the range of the material under study. This methodology, when compared with the classical strategy, proves to be an efficient alternative avoiding the use of complex devices for measuring the radius of curvature and strain at the pole of the cap, during the bulge test. Moreover, the authors claim that it is easy to implement and it is more efficient than classical approach, since (i) a unique set of numerical curves can be used within a relatively wide range of hardening coefficients, that is, covering the values usually found within one or several class of materials, without having to remake the simulations every time an identification is performed; (ii) it is not exposed to experimental errors related to the evaluation of the strain at the pole of the bulge and the use of membrane theory approach for assessment of the stress from the radius of curvature, which is usually the major source of errors.

(3) Other Specimens. Güner et al. [22] proposed an inverse analysis procedure for the identification of the Yld2000-2D

yield criterion parameters [7]. This study uses a notched specimen submitted to a tensile test, in order to obtain an inhomogeneous deformation field. Moreover, a layer compression test is used in order to supply additional information, that is, the equibiaxial yield stress. The required data for the inverse identification are the major and minor principal strains in the sheet plane and the load and the equibiaxial yield stress. The cost function is minimised using the Levenberg-Marquardt algorithm and is written as follows:

$$
\begin{aligned}
F(\mathbf{A}) \\
= \sum_{i=1}^{n_{\text{inc}}} \sum_{j=1}^{n_{\text{elem}}} \left[\left(\varepsilon_{1,ij}^{\text{Num}}(\mathbf{A}) - \varepsilon_{1,ij}^{\text{Exp}} \right)^2 + \left(\varepsilon_{2,ij}^{\text{Num}}(\mathbf{A}) - \varepsilon_{2,ij}^{\text{Exp}} \right)^2 \right] \\
+ C_1 \sum_{i=1}^{n_{\text{inc}}} \left[\left(P_i^{\text{Num}}(\mathbf{A}) - P_i^{\text{Exp}} \right)^2 \right] \\
+ C_2 \left(\sigma_b(\mathbf{A}) - \sigma_b^{\text{Exp}} \right)^2,
\end{aligned}
\tag{17}
$$

where ε_1 and ε_2 are the principal strains measured at each tool displacement increment i, at each element of the optical measurements, j; P represents the load and σ_b the equibiaxial yield stress; C_1 and C_2 are scale factors. The value of $\sigma_b(\mathbf{A})$ is analytically calculated at each iteration with the values of $\mathbf{A}$ predicted by the optimization algorithm. Different alternative orientations of the specimen with the rolling direction (0°, 45°, and 90°) and configurations of the cost function (setting one of the three terms of the cost function equal to zero) were considered to test the inverse procedure. The authors highlight the importance of including strain information and the equibiaxial stress on the cost function, in order to improve the characterization of the anisotropy coefficients.

Pottier et al. [23] developed an out-of-plane testing procedure for the simultaneous identification of Hill'48 yield criterion and Ludwick hardening law parameters of a rolled titanium sheet. Figure 4 illustrates the experimental setup and the geometry of the sample developed by the authors. A hemispherical punch applies a prescribed displacement normal to the sheet plane, at the centre of the surface of the sample, using a simple uniaxial tensile test machine. Two cameras are located on the opposite side of the sample and the components of the displacement fields along the $0x$, $0y$, and $0z$ axes are captured during the test using stereo digital image correlation. The sample was designed in order to exhibit multiaxial stress states, including shear, tension, and biaxial stretching. The numerical displacements fields along the $0x$, $0y$, and $0z$ axes and the global load are obtained from a numerical model of the test and compared to the experimental ones using a single cost function, formulated as follows:

$$
F(\mathbf{A}) = \left[\sum_{i,j=1}^{N_t,n} \left(\frac{u_{x,ij}^{\text{Num}}(\mathbf{A}) - u_{x,ij}^{\text{Exp}}}{\max_{i,j} \left(u_x^{\text{Exp}} \right)} \right)^2 \right.
$$

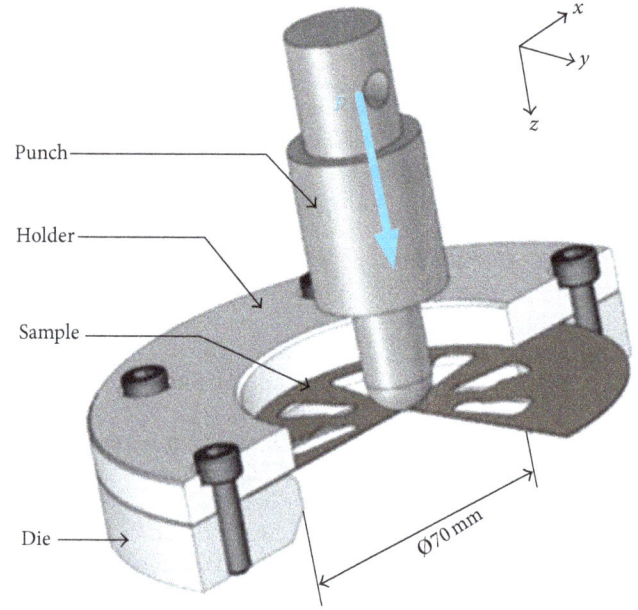

FIGURE 4: Schematic representation of the out-of-plane test developed by Pottier et al. [23].

$$
\begin{aligned}
&+ \sum_{i,j=1}^{N_t,n} \left(\frac{u_{y,ij}^{\text{Num}}(\mathbf{A}) - u_{y,ij}^{\text{Exp}}}{\max_{i,j} \left(u_y^{\text{Exp}} \right)} \right)^2 \\
&+ \sum_{i,j=1}^{N_t,n} \left(\frac{u_{z,ij}^{\text{Num}}(\mathbf{A}) - u_{z,ij}^{\text{Exp}}}{\max_{i,j} \left(u_z^{\text{Exp}} \right)} \right)^2 \\
&\left. + n \left(\frac{P_i^{\text{Num}}(\mathbf{A}) - P_i^{\text{Exp}}}{\max_i \left(P^{\text{Exp}} \right)} \right)^2 \right]^{1/2},
\end{aligned}
\tag{18}
$$

where N_t is the number of time steps considered and n is the number of measured points; u_x, u_y, and u_z are the displacements along the $0x$, $0y$, and $0z$ axes, respectively, P is the global load and $\mathbf{A}$ is the vector of parameters to identify; the subscripts Num and Exp refer to numerical and experimental results and i and j refer to the number of time steps and measured points, respectively. The minimisation of the cost function is performed with the Levenberg-Marquardt algorithm. To assess the quality of the identified set of constitutive parameters, the authors performed deep-drawing tests of a circular cup. Moreover, additional identifications of the constitutive parameters of the material were performed, following two different strategies: a classic strategy, based on three tensile tests cut along three different directions in the sheet plane, and an inverse identification strategy using heterogeneous planar shear-like tests, previously proposed [59]. The experimental results of the earing profile of the circular cup were then compared with the numerically predicted results from the different parameter identification strategies. The authors concluded that the use of the nonplanar sample allows a more accurate

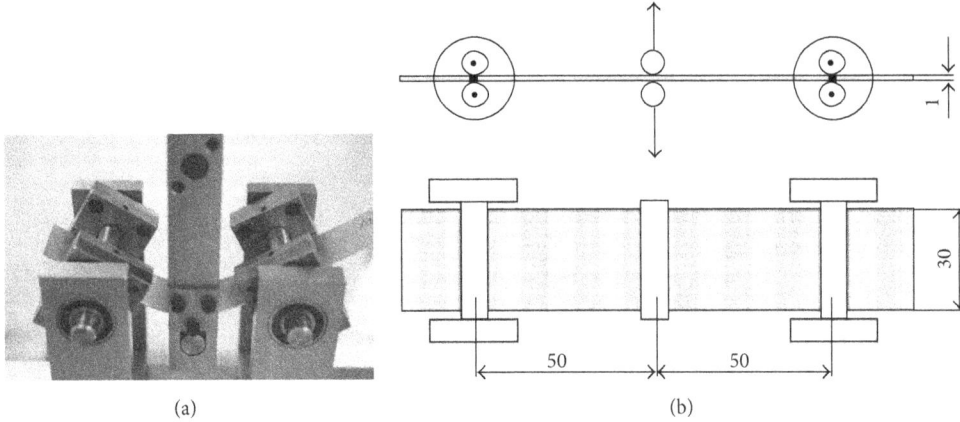

FIGURE 5: Device for three-point bending test [54]: (a) experimental setup and (b) schematic representation of horizontal and vertical views.

prediction of the earing profile than the planar shear-like tests and the three tensile tests, since the nonplanar sample test covers a wider range of strain paths. However, in our opinion, the chosen mechanical test shows a relative degree of complexity and has the inconvenience of presenting contact with friction between the punch and the metal sheet, which always raises questions concerning the impact of contact with friction on the identification of the constitutive parameters.

4.1.2. Identification of Kinematic Hardening. The identification of the kinematic hardening plays a significant role when phenomena such as the Bauschinger effect and permanent softening, due to reverse of strain path and other strain path changes, are relevant for the subsequent plastic deformation behaviour. Depending on the material and the strain path changes occurring in the deep-drawing process, these phenomena can be more or less noticeable and relevant. The Bauschinger effect is also associated with the springback, due to premature yielding after reversal of strain path [14]. Springback is the strain recovery when forming loads are removed, and its magnitude, which depends on the flow stress value [11] and Young's modulus, can also be influenced by the Bauschinger effect [78]. Therefore, the proper modelling and identification of the kinematic hardening are also important in order to efficiently predict the springback.

In this sense, Eggertsen and Mattiasson [54] were focused on the identification of the kinematic hardening law with the main concern of the accurate prediction of springback. The goal was to select the model able to accurately describe kinematic hardening features, such as the early reyielding, transient behaviour, work-hardening stagnation, and permanent softening, taking also into account the complexity on the evaluation of its parameters. Three-point bending tests (see Figure 5) on four typical materials from car manufacturing industry were performed: two dual-phase steels (TKS-DP600HF and SSAB-DP600), from different suppliers and with different thicknesses, a mild steel (Voest-DX56D), and an interstitial-free steel (TKS-220IF). Five different hardening models were considered: (i) a pure isotropic hardening law, used as comparative reference; (ii) a mixed isotropic-kinematic law [79, 80]; (iii) the Armstrong-Frederick model

[44]; (iv) the Geng-Wagoner hardening law [47]; and (v) the Yoshida-Uemori hardening law [14]. The hardening parameters of all models were determined by inverse analysis, where the difference between the experimentally and numerically generated load-displacement curves of the three-point bending test is minimised. The inverse identification of the material parameters was performed resorting to Response Surface Methodology, using the following cost function:

$$F(\mathbf{A}) = \frac{1}{n}\sum_{i=1}^{n} w_i \left(\frac{P_i^{\text{Num}}(\mathbf{A}) - P_i^{\text{Exp}}}{s_i} \right)^2, \qquad (19)$$

where $P_i^{\text{Num}}(\mathbf{A})$ and P_i^{Exp} represent the calculated and measured values of the punch load as a function of the vector of hardening parameters $\mathbf{A}$, respectively; s_i is the residual scale factor; and w_i is the weight applied to each component of the cost function. Both s_i and w_i were set equal to 1. The authors conclude that Yoshida-Uemori model provides the best result for all materials, while the isotropic hardening model gives the worst result. However, taking into account the accuracy and the complexity of the hardening model, the authors point out that Geng-Wagoner law corresponds to a better compromise. In fact, they state that about 30 simulations are needed to optimize the parameters of the mixed isotropic-kinematic hardening law, while up to 170 simulations are required to optimize the parameters of Yoshida-Uemori hardening law.

Pereira et al. [56] outline an inverse analysis methodology for simultaneously identifying the parameters of the isotropic Swift law and the Lemaître-Chaboche kinematic law [13] of metal sheets, using a reverse shear test. The outlined strategy uses a modified shear sample with a cylindrical notch along the axis of the sample, in order to confine the plastic deformation within the entire gauge section, which is not the case for the classical shear samples with constant thickness (see Figure 6). The geometry of the cylindrical notch was defined in order to ensure that all strain values, between the maximum (in the centre of the notch) and the minimum (zero, along the edge of the notch), are as much as possible represented at the moment of the strain path reversing. The geometry of the sample allows that the

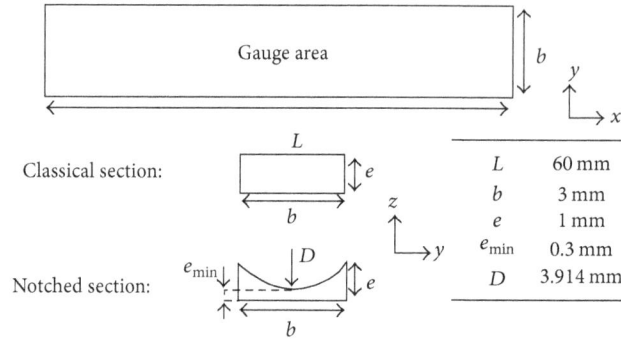

FIGURE 6: Geometry and dimensions of the classical and proposed notched shear specimens used in [56].

boundary conditions of experimental tests are accurately reproduced numerically and avoids the errors in the experimental determination of the stress versus strain curves, used in traditional methodologies, whose accuracy requires homogeneity of the stress and strain fields in the sample. The inverse analysis methodology consists of minimising the gap between experimental and numerical load versus displacement curves by making variations of the constitutive parameters, using the Levenberg-Marquardt algorithm. The following cost function is used:

$$F(\mathbf{A}) = \frac{1}{q}\sum_{i=1}^{q}\left(\frac{P_i^{\mathrm{Exp}} - P_i^{\mathrm{Num}}(\mathbf{A})}{P_i^{\mathrm{Exp}}}\right)^2$$
$$+ \frac{1}{p}\sum_{i=1}^{p}\left(\frac{P_i^{\mathrm{Exp}} - P_i^{\mathrm{Num}}(\mathbf{A})}{P_i^{\mathrm{Exp}}}\right)^2, \quad (20)$$

where P_i^{Exp} and $P_i^{\mathrm{Num}}(\mathbf{A})$ are, respectively, the experimental and numerical values of load for the same tool displacement; q and p are the total number of points in the forward and reverse paths, respectively; and $\mathbf{A}$ is the vector of the parameters to be identified. The parameters of the Hill'48 yield criterion that best describe the anisotropy of the fictitious materials used in this work (described by Drucker+L yield criterion) were identified following the methodology mentioned above, proposed by Prates et al. [25]. This methodology also allows identifying the parameters of the isotropic Swift law, which were used as first estimate in this inverse analysis. If no identification of the Swift law parameters was previously performed, the first estimate of isotropic Swift hardening parameters can be obtained adopting typical values for the material under study. It is appropriate to experimentally test this methodology.

Yin et al. [55] proposed the use of the twin bridge cyclic shear test proposed by Brosius et al. [81] to evaluate the Bauschinger effect on three classes of steel sheets, DC06, DP600, and TRIP700. The twin bridge shear specimen has two gauge areas that are simultaneously deformed when a moment is applied. Due to the moment application, instead of load, no unwanted reaction moment is created, in contrast with the ASTM shear specimen. The inverse identification strategy involves minimising the difference between experimental and numerical angular moment results, resorting

to a Trust Region Reflective Method [82], to identify the parameters of isotropic Voce hardening law and Armstrong-Frederick kinematic hardening law. The cost function is described as follows:

$$F(\mathbf{A}) = \sum_{i=1}^{q}\left(M_i^{\mathrm{Exp}} - M_i^{\mathrm{Num}}(\mathbf{A})\right)^2$$
$$+ \sum_{i=1}^{p}\left(M_i^{\mathrm{Exp}} - M_i^{\mathrm{Num}}(\mathbf{A})\right)^2, \quad (21)$$

where M_i^{Exp} and $M_i^{\mathrm{Num}}(\mathbf{A})$ are, respectively, the experimental and numerical values of moment at the same rotation angles; q and p are the total number of points in the forward and backward paths, respectively; and $\mathbf{A}$ is the vector of kinematic hardening parameters of the Lemaître-Chaboche law and two of the three parameters of the Voce law. The remaining Voce law parameter, the yield stress, is obtained from uniaxial tensile tests in the rolling direction and is kept fixed during the identification procedure.

5. Final Remarks

This review shows that a great investment has been made lately in the development of inverse strategies, namely, in FEMU strategies for the identification of parameters of constitutive laws describing the plastic behaviour of metal sheets. This includes the sample design, loading conditions, and optimization procedure and intends to make the identification of the material parameters easier and more reliable. Currently, some of these strategies allows determining simultaneously the parameters of isotropic hardening law and/or anisotropic yield criteria, using only a single test. Others only allow the evaluation of the parameters of kinematic hardening law. Also, most strategies are limited to specific types of work-hardening laws and plasticity criteria. In this context, the investment in such strategies should be directed towards the simultaneous identification of parameters of any constitutive law, including isotropic and kinematic hardening, and any anisotropic yield criterion. This must be accomplished using the results of the minimum number of mechanical tests and results, for example, from the biaxial test of the cruciform sample and the shear test, and building their own optimization procedure, eventually sequential. In fact, given that the optimization procedure for parameters identification can influence the solution, it is important to examine the possibility of resorting to sequential optimization procedures, especially when using different types of results.

Competing Interests

Acknowledgments

The authors gratefully acknowledge the financial support of the Portuguese Foundation for Science and Technology (FCT), Portugal, via Projects PTDC/EMS-TEC/0702/2014 (POCI-01-0145-FEDER-016779), PTDC/EMS-TEC/6400/ 2014 (POCI-01-0145-FEDER-016876), and UID/EMS/00285/ 2013, by UE/FEDER through Program COMPETE2020. P. A. Prates, A. F. G. Pereira, and N. A. Sakharova were supported by a grant for scientific research from the Portuguese Foundation for Science and Technology (refs. SFRH/BPD/ 101465/2014, SFRH/BD/102519/2014, and SFRH/BPD/107888/ 2015, resp.). All supports are gratefully acknowledged.

References

[1] W. Hu, "An orthotropic yield criterion in a 3-D general stress state," *International Journal of Plasticity*, vol. 21, no. 9, pp. 1771–1796, 2005.

[2] S. Soare, *On the use of homogeneous polynomials to develop anisotropic yield functions with applications to sheet metal forming [Ph.D. thesis]*, University of Florida, 2007.

[3] O. Cazacu and F. Barlat, "Generalization of Drucker's yield criterion to orthotropy," *Mathematics and Mechanics of Solids*, vol. 6, no. 6, pp. 613–630, 2001.

[4] F. Bron and J. Besson, "A yield function for anisotropic materials—application to aluminum alloys," *International Journal of Plasticity*, vol. 20, no. 4-5, pp. 937–963, 2004.

[5] F. Barlat, H. Aretz, J. W. Yoon, M. E. Karabin, J. C. Brem, and R. E. Dick, "Linear transfomation-based anisotropic yield functions," *International Journal of Plasticity*, vol. 21, no. 5, pp. 1009–1039, 2005.

[6] B. Plunkett, O. Cazacu, and F. Barlat, "Orthotropic yield criteria for description of the anisotropy in tension and compression of sheet metals," *International Journal of Plasticity*, vol. 24, no. 5, pp. 847–866, 2008.

[7] F. Yoshida, H. Hamasaki, and T. Uemori, "A user-friendly 3D yield function to describe anisotropy of steel sheets," *International Journal of Plasticity*, vol. 45, pp. 119–139, 2013.

[8] F. Barlat, D. J. Lege, and J. C. Brem, "A six-component yield function for anisotropic materials," *International Journal of Plasticity*, vol. 7, no. 7, pp. 693–712, 1991.

[9] A. P. Karafillis and M. C. Boyce, "A general anisotropic yield criterion using bounds and a transformation weighting tensor," *Journal of the Mechanics and Physics of Solids*, vol. 41, no. 12, pp. 1859–1886, 1993.

[10] O. Cazacu, B. Plunkett, and F. Barlat, "Orthotropic yield criterion for hexagonal closed packed metals," *International Journal of Plasticity*, vol. 22, no. 7, pp. 1171–1194, 2006.

[11] F. Barlat, J. J. Gracio, M.-G. Lee, E. F. Rauch, and G. Vincze, "An alternative to kinematic hardening in classical plasticity," *International Journal of Plasticity*, vol. 27, no. 9, pp. 1309–1327, 2011.

[12] H. Vegter and A. H. Van Den Boogaard, "A plane stress yield function for anisotropic sheet material by interpolation of biaxial stress states," *International Journal of Plasticity*, vol. 22, no. 3, pp. 557–580, 2006.

[13] J. L. Chaboche, "A review of some plasticity and viscoplasticity constitutive theories," *International Journal of Plasticity*, vol. 24, no. 10, pp. 1642–1693, 2008.

[14] F. Yoshida and T. Uemori, "A model of large-strain cyclic plasticity and its application to springback simulation," *International Journal of Mechanical Sciences*, vol. 45, no. 10, pp. 1687–1702, 2003.

[15] B. M. Chaparro, S. Thuillier, L. F. Menezes, P. Y. Manach, and J. V. Fernandes, "Material parameters identification: gradient-based, genetic and hybrid optimization algorithms," *Computational Materials Science*, vol. 44, no. 2, pp. 339–346, 2008.

[16] M. Rabahallah, T. Balan, S. Bouvier et al., "Parameter identification of advanced plastic strain rate potentials and impact on plastic anisotropy prediction," *International Journal of Plasticity*, vol. 25, no. 3, pp. 491–512, 2009.

[17] T. Kuwabara, "Advances in experiments on metal sheets and tubes in support of constitutive modeling and forming simulations," *International Journal of Plasticity*, vol. 23, no. 3, pp. 385–419, 2007.

[18] N. Deng, T. Kuwabara, and Y. P. Korkolis, "Cruciform specimen design and verification for constitutive identification of anisotropic sheets," *Experimental Mechanics*, vol. 55, no. 6, pp. 1005–1022, 2015.

[19] I. Zidane, D. Guines, L. Léotoing, and E. Ragneau, "Development of an in-plane biaxial test for forming limit curve (FLC) characterization of metallic sheets," *Measurement Science and Technology*, vol. 21, no. 5, Article ID 055701, pp. 1–11, 2010.

[20] S. Avril, M. Bonnet, A.-S. Bretelle et al., "Overview of identification methods of mechanical parameters based on full-field measurements," *Experimental Mechanics*, vol. 48, no. 4, pp. 381–402, 2008.

[21] S. Cooreman, D. Lecompte, H. Sol, J. Vantomme, and D. Debruyne, "Identification of mechanical material behavior through inverse modeling and DIC," *Experimental Mechanics*, vol. 48, no. 4, pp. 421–433, 2008.

[22] A. Güner, C. Soyarslan, A. Brosius, and A. E. Tekkaya, "Characterization of anisotropy of sheet metals employing inhomogeneous strain fields for Yld2000-2D yield function," *International Journal of Solids and Structures*, vol. 49, no. 25, pp. 3517–3527, 2012.

[23] T. Pottier, P. Vacher, F. Toussaint, H. Louche, and T. Coudert, "Out-of-plane testing procedure for inverse identification purpose: application in sheet metal plasticity," *Experimental Mechanics*, vol. 52, no. 7, pp. 951–963, 2012.

[24] S. Schmaltz and K. Willner, "Comparison of different biaxial tests for the inverse identification of sheet steel material parameters," *Strain*, vol. 50, no. 5, pp. 389–403, 2014.

[25] P. A. Prates, M. C. Oliveira, and J. V. Fernandes, "A new strategy for the simultaneous identification of constitutive laws parameters of metal sheets using a single test," *Computational Materials Science*, vol. 85, pp. 102–120, 2014.

[26] P. A. Prates, M. C. Oliveira, and J. V. Fernandes, "Identification of material parameters for thin sheets from single biaxial

tensile test using a sequential inverse identification strategy," *International Journal of Material Forming*, vol. 9, no. 4, pp. 547–571, 2016.

[27] S. Zhang, L. Leotoing, D. Guines, S. Thuillier, and S.-L. Zang, "Calibration of anisotropic yield criterion with conventional tests or biaxial test," *International Journal of Mechanical Sciences*, vol. 85, pp. 142–151, 2014.

[28] S. Zhang, L. Léotoing, D. Guines, and S. Thuillier, "Potential of the cross biaxial test for anisotropy characterization based on heterogeneous strain field," *Experimental Mechanics*, vol. 55, no. 5, pp. 817–835, 2015.

[29] W. Liu, D. Guines, L. Leotoing, and E. Ragneau, "Identification of sheet metal hardening for large strains with an in-plane biaxial tensile test and a dedicated cross specimen," *International Journal of Mechanical Sciences*, vol. 101-102, pp. 387–398, 2015.

[30] F. Dunne and N. Petrinic, *Introduction to Computational Plasticity*, Oxford University Press, 2005.

[31] P. Ienny, A.-S. Caro-Bretelle, and E. Pagnacco, "Identification from measurements of mechanical fields by finite element model updating strategies. A review," *European Journal of Computational Mechanics*, vol. 18, no. 3-4, pp. 353–376, 2009.

[32] J. H. Hollomon, "Tensile deformations," *Transactions of the Metallurgical Society of AIME*, vol. 162, pp. 268–290, 1945.

[33] P. Ludwick, *Elemente der Technologischen Mechanik*, Springer, Berlin, Germany, 1909.

[34] H. W. Swift, "Plastic instability under plane stress," *Journal of the Mechanics and Physics of Solids*, vol. 1, no. 1, pp. 1–18, 1952.

[35] D. C. Ludwigson, "Modified stress–strain relation for FCC metals and alloys," *Metallurgical Transactions*, vol. 2, no. 10, pp. 2825–2828, 1971.

[36] A. K. Ghosh, "Tensile instability and necking in materials with strain hardening and strain-rate hardening," *Acta Metallurgica*, vol. 25, no. 12, pp. 1413–1424, 1977.

[37] J. V. Fernandes, D. M. Rodrigues, L. F. Menezes, and M. F. Vieira, "A modified Swift law for prestrained materials," *International Journal of Plasticity*, vol. 14, no. 6, pp. 537–550, 1998.

[38] E. Voce, "The relationship between stress and strain for homogeneous deformation," *Journal of the Institute of Metals*, vol. 74, pp. 537–562, 1948.

[39] J. E. Hockett and O. D. Sherby, "Large strain deformation of polycrystalline metals at low homologous temperatures," *Journal of the Mechanics and Physics of Solids*, vol. 23, no. 2, pp. 87–98, 1975.

[40] D. Banabic, *Sheet Metal Forming Processes—Constitutive Modelling and Numerical Simulation*, Springer, 2010.

[41] P. Larour, *Strain rate sensitivity of automotive sheet steels: influence of plastic strain, strain rate, temperature, microstructure, bake hardening and pre-strain [Ph.D. thesis]*, RWTH Aachen University, 2010.

[42] W. Prager, "Recent developments in the mathematical theory of plasticity," *Journal of Applied Physics*, vol. 20, pp. 235–241, 1949.

[43] H. Ziegler, "A modification of Prager's hardening rule," *Quarterly of Applied Mathematics*, vol. 17, pp. 55–65, 1959.

[44] P. J. Armstrong and C. O. Frederick, "A mathematical representation of the multiaxial Bauschinger effect," GEGB Report RD/B/N 731, 1966.

[45] C. Teodosiu and Z. Hu, "Evolution of the intragranular microstructure at moderate and large strains: modelling and computational significance," in *Proceedings of the 5th International Conference on Numerical Methods in Industrial Forming Processes (NUMIFORM '95)*, pp. 173–182, Ithaca, NY, USA, 1995.

[46] C. Teodosiu and Z. Hu, "Microstructure in the continuum modelling of plastic anisotropy," in *Proceedings of the 19th Riso International Symposium on Materials Science: Modelling of Structure and Mechanics of Materials from Microscale to Products*, pp. 149–168, Riso National Laboratory, 1998.

[47] L. Geng, Y. Shen, and R. H. Wagoner, "Anisotropic hardening equations derived from reverse-bend testing," *International Journal of Plasticity*, vol. 18, no. 5-6, pp. 743–767, 2002.

[48] A. Khalfallah, H. B. H. Salah, and A. Dogui, "Anisotropic parameter identification using inhomogeneous tensile test," *European Journal of Mechanics, A/Solids*, vol. 21, no. 6, pp. 927–942, 2002.

[49] R. Hill, "A theory of the yielding and plastic flow of anisotropic metals," *Proceedings of the Royal Society of A: Mathematical, Physical and Engineering Sciences*, vol. 193, pp. 281–297, 1948.

[50] K. Mattiasson and M. Sigvant, "An evaluation of some recent yield criteria for industrial simulations of sheet forming processes," *International Journal of Mechanical Sciences*, vol. 50, no. 4, pp. 774–787, 2008.

[51] A. Chamekh, H. BelHadjSalah, R. Hambli, and A. Gahbiche, "Inverse identification using the bulge test and artificial neural networks," *Journal of Materials Processing Technology*, vol. 177, no. 1–3, pp. 307–310, 2006.

[52] M. Bambach, "Comparison of the identifiability of flow curves from the hydraulic bulge test by membrane theory and inverse analysis," *Key Engineering Materials*, vol. 473, pp. 360–367, 2011.

[53] L. C. Reis, P. A. Prates, M. C. Oliveira, A. D. Santos, and J. V. Fernandes, "Inverse identification of the Swift law parameters using the bulge test," *International Journal of Material Forming*, 2016.

[54] P.-A. Eggertsen and K. Mattiasson, "On the modelling of the bending-unbending behaviour for accurate springback predictions," *International Journal of Mechanical Sciences*, vol. 51, no. 7, pp. 547–563, 2009.

[55] Q. Yin, C. Soyarslan, A. Güner, A. Brosius, and A. E. Tekkaya, "A cyclic twin bridge shear test for the identification of kinematic hardening parameters," *International Journal of Mechanical Sciences*, vol. 59, no. 1, pp. 31–43, 2012.

[56] A. F. G. Pereira, P. A. Prates, N. A. Sakharova, M. C. Oliveira, and J. V. Fernandes, "On the identification of kinematic hardening with reverse shear test," *Engineering with Computers*, vol. 31, no. 4, pp. 681–690, 2015.

[57] J. Cao and J. Lin, "A study on formulation of objective functions for determining material models," *International Journal of Mechanical Sciences*, vol. 50, no. 2, pp. 193–204, 2008.

[58] A. Andrade-Campos, R. De-Carvalho, and R. A. F. Valente, "Novel criteria for determination of material model parameters," *International Journal of Mechanical Sciences*, vol. 54, no. 1, pp. 294–305, 2012.

[59] T. Pottier, F. Toussaint, and P. Vacher, "Contribution of heterogeneous strain field measurements and boundary conditions modelling in inverse identification of material parameters," *European Journal of Mechanics—A/Solids*, vol. 30, no. 3, pp. 373–382, 2011.

[60] J.-P. Ponthot and J.-P. Kleinermann, "A cascade optimization methodology for automatic parameter identification and shape/process optimization in metal forming simulation," *Computer Methods in Applied Mechanics and Engineering*, vol. 195, no. 41–43, pp. 5472–5508, 2006.

[61] M. Meuwissen, *An inverse method for the mechanical characterisation of metals [Ph.D. thesis]*, Technische Universiteit Eindhoven, 1998.

[62] S. Cooreman, *Identification of the plastic material behaviour through full-field displacement measurements and inverse methods [Ph.D. thesis]*, Vrije Universiteit Brussel, 2008.

[63] B. Endelt and K. B. Nielsen, "Inverse modeling based on an analytical definition of the Jacobian matrix associated with Hill's 48 yield criterion," in *Proceedings of the 7th Esaform Conference on Material Forming (ESAFORM '04)*, S. Stören, Ed., Trondheim, Norway, April 2004.

[64] B. Endelt, K. B. Nielsen, and J. Danckert, "Analytic differentiation of Barlat's 2D criteria for inverse modeling," in *Proceedings of the Analytic differentiation of Barlat's 2D criteria for inverse modeling*, vol. 778 of *AIP Conference Proceedings*, pp. 789–794, Detroit, Mich, USA, August 2005.

[65] D. W. Marquardt, "An algorithm for least-squares estimation of nonlinear parameters," *Journal of the Society for Industrial and Applied Mathematics*, vol. 11, pp. 431–441, 1963.

[66] P.-A. Eggertsen and K. Mattiasson, "An efficient inverse approach for material hardening parameter identification from a three-point bending test," *Engineering with Computers*, vol. 26, no. 2, pp. 159–170, 2010.

[67] X.-Q. Li and D.-H. He, "Identification of material parameters from punch stretch test," *Transactions of Nonferrous Metals Society of China*, vol. 23, no. 5, pp. 1435–1441, 2013.

[68] A. Chamekh, H. Bel Hadj Salah, and R. Hambli, "Inverse technique identification of material parameters using finite element and neural network computation," *International Journal of Advanced Manufacturing Technology*, vol. 44, no. 1-2, pp. 173–179, 2009.

[69] M. C. Oliveira, J. L. Alves, B. M. Chaparro, and L. F. Menezes, "Study on the influence of work-hardening modeling in springback prediction," *International Journal of Plasticity*, vol. 23, no. 3, pp. 516–543, 2007.

[70] J. Kajberg and G. Lindkvist, "Characterisation of materials subjected to large strains by inverse modelling based on in-plane displacement fields," *International Journal of Solids and Structures*, vol. 41, no. 13, pp. 3439–3459, 2004.

[71] L. Robert, V. Velay, N. Decultot, and S. Ramde, "Identification of hardening parameters using finite element models and fullfield measurements: some case studies," *Journal of Strain Analysis for Engineering Design*, vol. 47, no. 1, pp. 3–17, 2012.

[72] H. Haddadi and S. Belhabib, "Improving the characterization of a hardening law using digital image correlation over an enhanced heterogeneous tensile test," *International Journal of Mechanical Sciences*, vol. 62, no. 1, pp. 47–56, 2012.

[73] F. Pierron, S. Avril, and V. T. Tran, "Extension of the virtual fields method to elasto-plastic material identification with cyclic loads and kinematic hardening," *International Journal of Solids and Structures*, vol. 47, no. 22-23, pp. 2993–3010, 2010.

[74] Y. Pannier, S. Avril, R. Rotinat, and F. Pierron, "Identification of elasto-plastic constitutive parameters from statically undetermined tests using the virtual fields method," *Experimental Mechanics*, vol. 46, no. 6, pp. 735–755, 2006.

[75] J.-H. Kim, F. Barlat, F. Pierron, and M.-G. Lee, "Determination of anisotropic plastic constitutive parameters using the virtual fields method," *Experimental Mechanics*, vol. 54, no. 7, pp. 1189–1204, 2014.

[76] J. Fu, F. Barlat, and J.-H. Kim, "Parameter identification of the homogeneous anisotropic hardening model using the virtual fields method," *International Journal of Material Forming*, 2015.

[77] BSI, "Metallic materials—sheet and strip—determination of biaxial stress-strain curve by means of bulge test with optical measuring systems," DIN EN ISO 16808:2014-11 (E), BSI, 2014.

[78] S. Chatti and N. Hermi, "The effect of non-linear recovery on springback prediction," *Computers and Structures*, vol. 89, no. 13-14, pp. 1367–1377, 2011.

[79] P. G. Hodge, "A new method of analyzing stresses and strains in work hardening solids," *Journal of Applied Mechanics*, vol. 24, pp. 482–483, 1957.

[80] M. A. Crisfield, "More plasticity and other material non-linearity-II," in *Nonlinear Finite Element Analysis of Solids and Structures*, vol. 2, pp. 158–164, John Wiley & Sons, Chichester, UK, 2001.

[81] A. Brosius, Q. Yin, A. Güner, and A. E. Tekkaya, "A new shear test for sheet metal characterization," *Steel Research International*, vol. 82, no. 4, pp. 323–328, 2011.

[82] T. F. Coleman and Y. Li, "An interior trust region approach for nonlinear minimization subject to bounds," *SIAM Journal on Optimization*, vol. 6, no. 2, pp. 418–445, 1996.

14

Cold-Sprayed Metal Coatings with Nanostructure

Shuo Yin ⓘ,[1] Chaoyue Chen,[2] Xinkun Suo ⓘ,[3] and Rocco Lupoi ⓘ[1]

[1]*Department of Mechanical and Manufacturing Engineering, Trinity College Dublin, The University of Dublin, Parsons Building, Dublin 2, Ireland*
[2]*ICB UMR 6303, CNRS, Univ. Bourgogne Franche-Comté, UTBM, 90010 Belfort, France*
[3]*Key Laboratory of Marine Materials and Related Technologies, Ningbo Institute of Materials Technology and Engineering, Chinese Academy of Sciences, Ningbo 315201, China*

Correspondence should be addressed to Shuo Yin; yins@tcd.ie; Xinkun Suo; suoxinkun@nimte.ac.cn and Rocco Lupoi; lupoir@tcd.ie

Academic Editor: Federica Bondioli

Cold spray is a solid-state coating deposition technology developed in the 1980s. In comparison with conventional thermal spray processes, cold spray can retain the original properties of feedstock, prevent the adverse influence on the underlying substrate materials, and produce very thick coatings. Coatings with nanostructure offer the potential for significant improvements in physical and mechanical properties as compared with conventional non-nanostructured coatings. Cold spray has also demonstrated great capability to produce coatings with nanostructure. This paper is aimed at providing a comprehensive overview of cold-sprayed metal coatings with nanostructure. A brief introduction of the cold spray technology is provided first. The nanocrystallization phenomenon in the conventional cold-sprayed metal coatings is then addressed. Thereafter, focus is switched to the microstructure and properties of the cold-sprayed nanocrystalline metal coatings, and the cold-sprayed nanomaterial-reinforced metal matrix composite (MMC) coatings. At the end, summary and future perspectives of the cold spray technology in producing metal coatings with nanostructure are concluded.

1. Introduction

Cold spray as an emerging coating technique has been developed for decades since its discovery in the 1980s [1]. In this process, powders are accelerated by the supersonic driving gas passing through a convergent-divergent nozzle and impacting onto a substrate at a very high velocity as schematized in Figure 1. Intensive plastic deformation induced by the high-velocity impact occurs in a cold sprayed particle, substrate (or already deposited coating), or both, enabling a low-oxidized cold-sprayed coating to be formed. Metals, metal matrix composites (MMCs), and even pure ceramics are able to deposit onto similar or dissimilar substrates with cold spray [2–4]. Unlike in conventional thermal spray, the feedstock used for cold spray remains solid state during the deposition process without any melting because of the relatively low temperature of the driving gas. Therefore, the inevitable defects emerging in the thermal-sprayed coating, for example, oxidation, thermal

residual stress, and phase transformation, can be considerably avoided in the cold-sprayed coating [5]. Besides, the coating growth is almost unlimited for most metals and MMCs, which allows cold spray to act as an additive manufacturing technique for fabricating bulk materials [6, 7].

As a low-temperature deposition technology, cold spray is primarily applied for producing metal-based coatings. In general, particle velocity prior to the impact is an important factor for cold spray because the successful deposition of cold-sprayed particles relies only on the kinetic energy rather than the combined effect of both kinetic and thermal energies available in conventional thermal spraying. It has been widely accepted that there exists a unique critical velocity for a given condition (e.g., specific particle size, temperature, and material properties), above which successful bonding can be achieved [8–10]. Therefore, the feedstock powders for cold spray must have a proper size range (normally between 10 and 100 μm) to achieve a high particle impact velocity [11, 12]. Nanoparticles, due to their low weights, are difficult

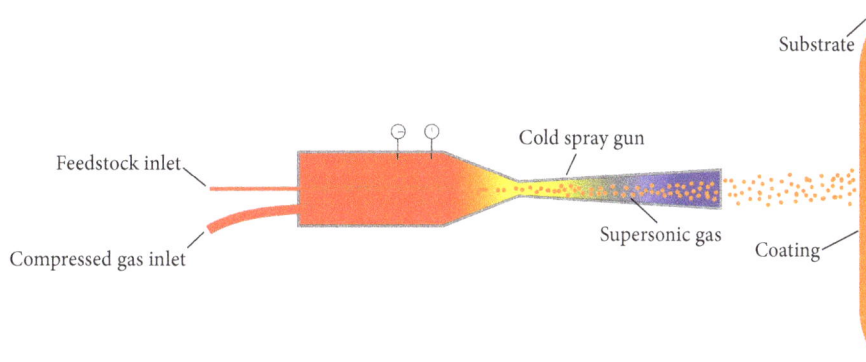

FIGURE 1: Schematic of the cold spray process.

to deposit on the substrate via cold spray. They are easy to be picked up by the driving gas and thereby suffer from dramatic deceleration when passing through the compressed bow-shock in front of the substrate [13]. Consequently, the impact velocity of nanoparticles is very low so that coating is hard to form on the substrate.

It is known that nanostructured materials generally have improved properties as compared with conventional materials [14]. Therefore, it is of importance for cold spray to gain the capability to produce coatings with nanostructure. Fortunately, although nanoparticles cannot be deposited directly by cold spray, cold-sprayed metal coatings can still exhibit nanostructure. Firstly, nanocrystallization in the form of grain refinement always occurs at the interparticle and coating-substrate interfacial regions during the deposition process due to the dynamic recrystallization, which can result in nanostructured grains within the cold-sprayed coating [10, 15–21]. Secondly, the starting feedstock for cold spray can be nanocrystalline powders [14, 22–31]; in this case, the coating retains the nanostructure of the starting powders. Thirdly, using nanomaterials to reinforce MMC coatings can also make the coating to present nanostructure [32–50]. Up till now, a large number of studies have been done to study the cold sprayed coatings with nanostructure. However, a systematic review of these studies still lacks. Therefore, this paper aims to provide an overview of the metal coatings with nanostructure produced by cold spray, particularly focusing on the coating microstructure and properties. The nanocrystallization phenomenon in the cold-sprayed metal coatings is addressed first. Thereafter, the microstructure and properties of the cold-sprayed nanocrystalline metal coatings are discussed. Then, focus is switched to review the cold-sprayed nanomaterial-reinforced MMC coatings. According to the dimensions of the reinforcements, the MMC coatings were classified as 1D material-reinforced, 2D material-reinforced, and 3D material-reinforced MMC coatings. The final part of this paper is a summary and further perspective of the cold spray technology in the fabrication of metal coatings with nanostructure.

2. Nanocrystallization in Cold-Sprayed Metal Coating

2.1. Nanocrystallization Phenomenon. During cold spray process, metal particles experience intensive plastic deformation

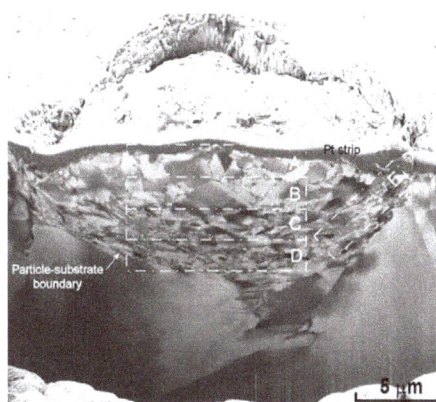

FIGURE 2: FIB-SEM imaging of the cross section of a Cu particle depositing onto a Cu substrate [17].

FIGURE 3: Euler angle EBSD patterns of the cross section of an Al 6061 particle in the cold-sprayed coating [18].

at the interparticle and coating-substrate interfacial regions due to the high-velocity impact. At highly deformed jetting areas, adiabatic shear instability takes place, which results in a significant temperature rise [51]. These rapid physical and

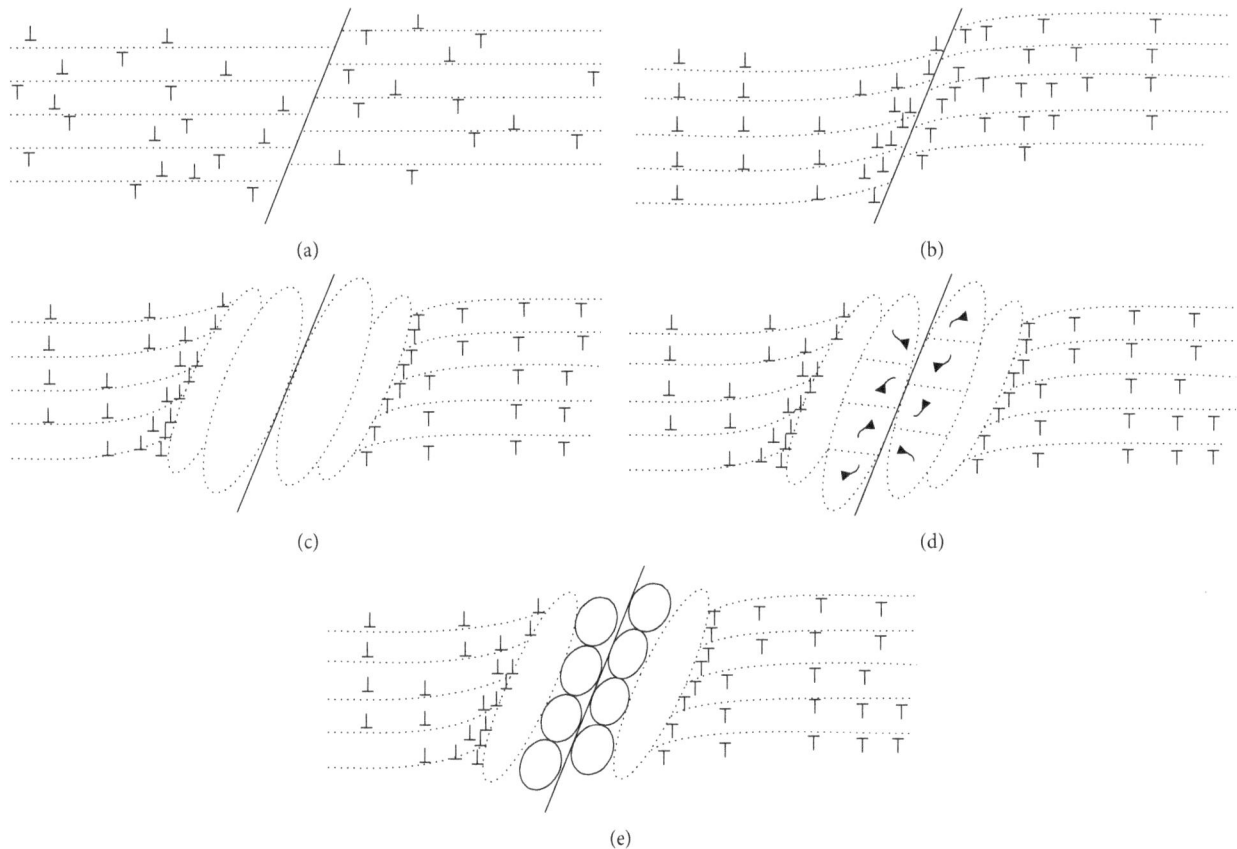

FIGURE 4: Schematic of the subgrains and ultrafine grain formation mechanism [20].

chemical changes work together, leading to the nano-crystallization of metal particles in the form of grain refinement [10, 15, 16]. Figure 2 shows the FIB-SEM imaging of the cross section of a Cu particle depositing onto a Cu substrate [17]. Clearly, particle grain structures showed an obvious change from the particle top surface towards the bottom, and such change was much dependent on the plastic deformation level. Based on the deformation level and grain size, the particle can be divided into three areas. The grains in the "A and B" areas had a large size because the top and inner parts of the particle did not experience too much plastic deformation. At the "C and D" areas, material underwent extensive high-strain/strain-rate plastic deformation; grains were highly deformed and elongated to subgrains. Particularly, at the peripheral "E" area where localized adiabatic shear instability took place, grains were significantly refined to ultrafine grains. Figure 3 shows the EBSD imaging of the cross section of an Al 6061 particle in the cold-sprayed coating, providing a clearer view of different regions [18]. Note that the cross section was perpendicular to the particle impact direction. As can be seen, in the center of the particle, grain size was much larger as compared with the surrounding area. Adjacent to the central zone, elongated subgrains can be clearly observed. At the far surrounding region, ultrafine grains marked by black dotted circle can be noticed. This microstructure is quite similar to the Cu particle grain structure as shown in Figure 2, which demonstrates the universality of the nanocrystallization of the cold-sprayed metal particles after deposition.

2.2. Nanocrystallization Formation Mechanism. In order to explain the substantial reason behind the nanocrystallization phenomenon, Figure 4 shows the schematic of the subgrains and equiaxed ultrafine grain formation mechanism [20]. At the beginning, the large grain of the original particle contains uniformly distributed low-density dislocations (Figure 4(a)). When the particle starts to plastically deform upon impact, dislocation multiplication takes place and the dislocation density begins to increase at highly deformed zone (Figure 4(b)). As the deformation continues, the accumulated dislocations produce a number of dislocation cells, forming the elongated subgrains (Figure 4(c)). The TEM imaging in Figure 5 shows some elongated subgrains in a cold-sprayed Al 7075 particle after deposition [52]. Following the formation of subgrains, adiabatic shear instability happens at extreme deformation areas, resulting in a rapid temperature rise to a value higher than recrystallization temperature. Plastic deformation and heating then work together to induce the dynamic recrystallization at these areas [15, 17, 18, 53, 54]. Basically, dynamic recrystallization is controlled by two mechanisms: rotational and migrational types [16, 19, 20]. In terms of the cold-sprayed particles, rotation has been found to be the dominant mechanism for the occurrence of dynamics recrystallization [20, 55]. Under the rotational dynamic recrystallization, elongated subgrains are further divided into equiaxed subgrains due to the increased dislocation density (Figure 4(d)). These equiaxed subgrains are rotated by additional shear forces to form the

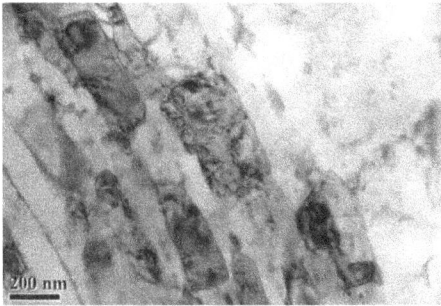

FIGURE 5: TEM imaging of elongated subgrains in a cold-sprayed Al 7075 particle [52].

FIGURE 6: TEM imaging of cold-sprayed nanocrystalline Al 2018 coating [23].

ultrafine grains (Figure 4(e)) [16, 20]. Due to the increased dislocations and grain boundaries, the nanohardness and strength at these refined areas are higher as compared to at the other areas [14, 21].

3. Nanocrystalline Metal Coatings via Cold Spray

Nanocrystalline metals are polycrystalline metals with a crystallite size of a few nanometers (normally smaller than 100 nm). They generally exhibit increased strength and hardness, improved toughness, reduced elastic modulus and ductility, enhanced diffusivity, higher specific heat, enhanced thermal expansion coefficient, and superior soft magnetic properties in comparison with conventional polycrystalline metals [56]. Cold spray, due to its low working temperature, has been found to be a robust tool to produce nanocrystalline metal coatings because nanocrystalline structure can be well retained in the coating after deposition. As an evidence, Figure 6 shows a TEM imaging of nanocrystalline Al 2018 coating produced via cold spray, where nanocrystalline grains can be clearly observed in the coating [23].

Preparation of nanocrystalline powders is one of the most important steps in the coating fabrication process. Mechanical ball milling is regarded as a simple, effective, and efficient method to produce nanocrystalline powders with a grain size as small as 20 nm or below [57]. It is a process where mixed powders are placed in a chamber and subjected to high-energy collision of balls to induce

mechanical alloying. So far, various nanocrystalline powders have been produced through mechanical ball milling for cold spray [14, 22–31]. Table 1 summarizes the nanocrystalline powders produced for cold spray and their milling conditions.

Due to the increase of grain boundaries, the nanocrystalline powder hardness was normally much higher than the conventional counterpart, which makes the nanocrystalline coating to be harder but more porous as compared to the conventional coating [23, 29, 30]. For the same reason, the work hardening effect in the nanocrystalline coatings was not as prominent as in the conventional coatings [23, 25]. Figure 7 shows a comparison of cross-sectional microstructure between conventional and nanocrystalline Al coatings produced under the same working parameters [23]. It is seen that nanocrystalline coating had higher porosity than conventional coating due to the lack of sufficient plastic deformation. However, the coating hardness showed an opposite trend; nanocrystalline coating had a hardness of 4.41 GPa which was higher than the hardness of conventional coating (3.75 GPa).

In terms of the mechanical properties, cold-sprayed nanocrystalline coatings have shown better wear-resistance performance than conventional coatings as a result of higher hardness [29]. However, due to the simultaneous higher porosity, fatigue strength was found to not improve significantly [30]. So far, the property investigation of cold-sprayed nanocrystalline coatings is still very limited; for example, investigations on coating cohesion strength are still lacking. Considering the unique advantages of hard nanocrystalline coatings, more mechanical property tests such as coating tensile stress and elongation are encouraged in the future work.

4. 1D Material-Reinforced MMC Coatings via Cold Spray

Carbon nanotube (CNT) is an allotrope of carbon with a cylindrical nanostructure. The diameter of CNT can be as small as 1 nm but the length can be up to several centimeters; thus, CNT is also recognized as 1D nanomaterial. As a member of carbon family, CNT has extraordinary thermal conductivity, electrical, and mechanical properties. These novel properties make CNT potentially valuable and useful in a wide variety of applications in nanotechnology, electronics, optics, thermal engineering, and other fields of material science [58]. Due to the 1D nanostructure, CNT cannot exist in the form of bulk state. Thereby, it is widely employed as reinforcements for improving the properties of pure metals. Currently, CNT-reinforced MMCs are mainly produced by powder metallurgy [59, 60] and thermal spray technologies [61, 62]. These processes generally require high temperature to melt the binder phase, resulting in damage and phase transformation of CNT during fabrication [63, 64]. Cold spray, due to its low working temperature, has been applied to produce CNT-reinforced MMC coatings in recent years.

A number of investigations have proved that dense and thick CNT-reinforced MMC coatings can be fabricated via cold spray [41–44]. Among all these works, mechanical ball

TABLE 1: Nanocrystalline powders for cold spray and the milling conditions [14, 22–31].

Materials	Milling time (h)	Speed (rpm)	Ball diameter (mm)	BPR	Control agent
Al alloy 2009	10	200	6.4	20 : 1	Liquid nitrogen
Al alloy 2618	8	180	6.4	32 : 1	Liquid nitrogen
Al alloy 5083	8	180	6.4	32 : 1	Liquid nitrogen
Al alloy 7075	N/A	180	11.6	32 : 1	Stearic acid
Al-Mg alloy	8	180	6.4	32 : 1	Liquid nitrogen
Ni	15	180	6.4	30 : 1	Liquid nitrogen
Ni-Ti alloy	48, 1/2 h rest per h	400	20	13 : 4	Alcohol
Ni-20Cr alloy	20, 1/3 h rest per h	300	N/A	10 : 2	Toluene
Cu	12	200	6.4	30 : 1	Liquid nitrogen

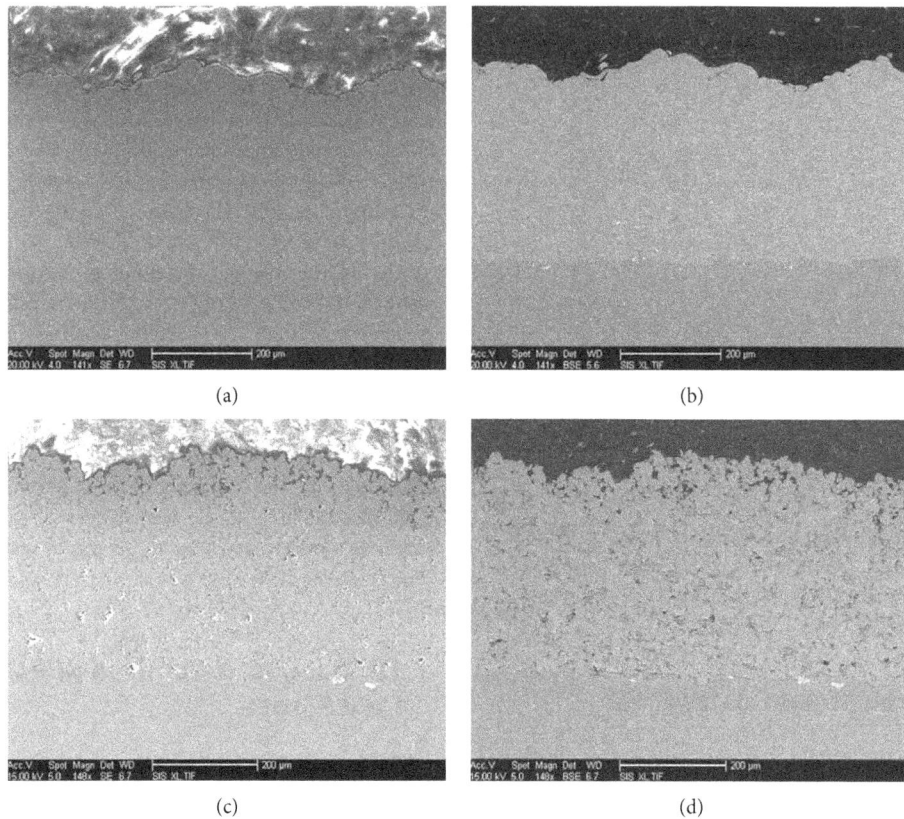

(a)

(b)

(c)

(d)

FIGURE 7: Comparison of the microstructure between cold-sprayed conventional and nanocrystalline Al coatings produced under same working parameters: (a) secondary electron imaging of conventional coating, (b) backscattered electron imaging of conventional coating, (c) secondary electron imaging of nanocrystalline coating, and (d) backscattered electron imaging of nanocrystalline coating [23].

milling as a robust technology was prevailingly used to prepare CNT-reinforced MMC powders. Figure 8 shows the cross-sectional view and EDX mapping of a ball-milled CNT-Cu MMC powder. As can be seen, CNT was successfully incorporated into the MMC powder and exhibited a homogenous distribution [41]. Following the preparation of the MMC powders, various CNT-reinforced MMC coatings were fabricated as listed in Table 2 [41–44, 65–68]. As an example, Figure 9 shows the TEM imaging of a cold-sprayed CNT-Cu MMC produced by low-pressure cold spray [65]. Clearly, CNT was successfully involved in the cold-sprayed Cu-based MMC coatings.

Although mechanical ball milling is promising for producing CNT-reinforced MMC powders, it also brings negative aspects. CNT reinforcements suffered from damage

during the milling process due to the plastic deformation of the binder phase [65]. In addition, the high-velocity impact happening during the coating deposition process also led to the fracture of CNT [44, 65]. Figure 10 shows the TEM imaging of damaged CNT in the cold-sprayed CNT-Cu coatings [44]. Two different damaging features can be noticed, which are impact-induced and shear-induced damages, respectively. The impact-induced damage was present in the form of a systematic fracture of the concentric tubes which progresses inward until the innermost tube has broken, while the shear-induced damage was featured uneven or asymmetric with respect to the tube axis [44]. Currently, prevention of damage of CNT during ball milling process is still a challenging work, which may be a research focus in the future work.

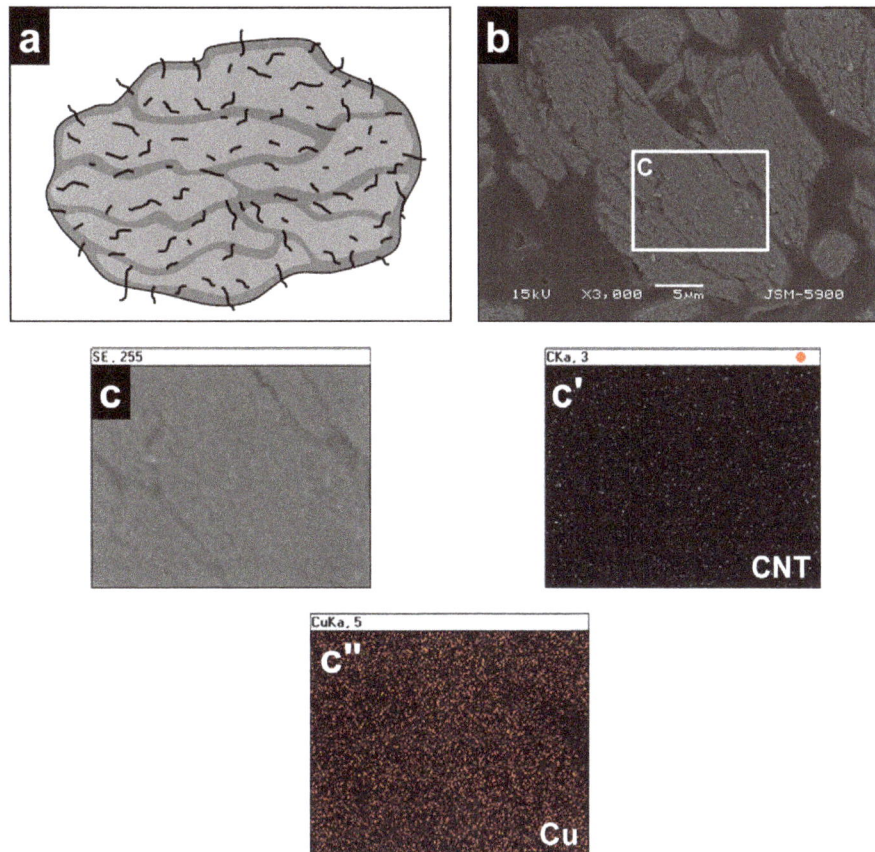

FIGURE 8: Cross-sectional view and EDX mapping of a ball-milled CNT-Cu MMC powder [41].

TABLE 2: CNT-reinforced MMC coatings produced by cold spray and their thermal properties [41–44, 65–68].

Composites	Contents of CNT	Thermal properties	References
CNT-Al	0.5–1.0% by weight	N/A	[44]
CNT-Cu	3% by volume	Higher thermal diffusivity	[65]
CNT-Cu	5–15% by volume	Better heat transfer performance	[41, 66]
CNT-Cu-SiC	4–4.5% by volume	Better heat transfer performance	[43, 67, 68]
CNT-Cu-AlN	4–4.5% by volume	Better heat transfer performance	[42, 67, 68]
CNT-Cu-BN	4% by volume	Better heat transfer performance	[67, 68]

In addition to mechanical ball milling, spray drying technology has also been used to disperse CNT within agglomerated metal powders. Spray drying is a process that produces powders from a liquid solution by rapid drying with hot gases. With this method, CNT was mainly embedded on the surface of agglomerated metal powders but hard to be homogenously incorporated inside the individual powder. Figure 11 shows the schematic of the coating fabrication procedure using spray drying MMC powders. As can be seen, a consequence of using spray drying powders is the inhomogeneous distribution of CNT in the cold-sprayed coatings as shown in Figure 12 [44].

As for the properties of cold-sprayed CNT-reinforced MMC coatings, it has been revealed that mechanical properties improved with additional CNT reinforcements. In the CNT-Al MMC coatings, local hardness was found to be higher at CNT-rich zone due to the higher stiffness of CNT [44]. Similar results were also concluded in the CNT-Cu MMC

coatings. CNT reinforcements were also found to lead to the improvement of thermal properties in terms of both heat transfer performance and thermal diffusivity [65, 66]. Table 2 lists the existing CNT-reinforced MMC coatings produced by cold spray and their improved thermal properties [41–44, 65–68]. The existing works clearly demonstrate that cold spray is a promising technology to produce CNT-reinforced MMC coatings.

5. 2D Material-Reinforced MMC Coatings via Cold Spray

5.1. Graphene-Reinforced MMC Coatings. Graphene, as the single layer of sp^2 bonded carbon atoms, has extraordinary mechanical, thermal, and electrical properties, attracting great attentions from both scientific and industrial communities. It is normally applied as reinforcements of MMCs to improve the matrix material properties. The

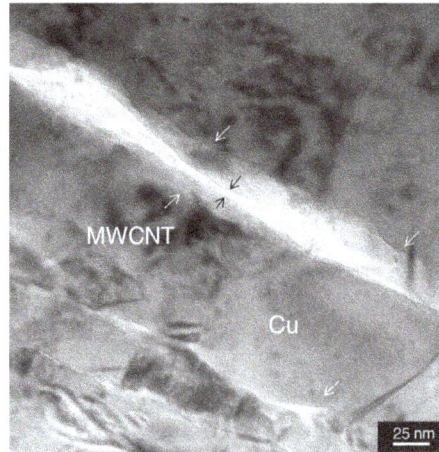

FIGURE 9: TEM imaging of a cold-sprayed CNT-Cu MMC coating produced by low-pressure cold spray [65].

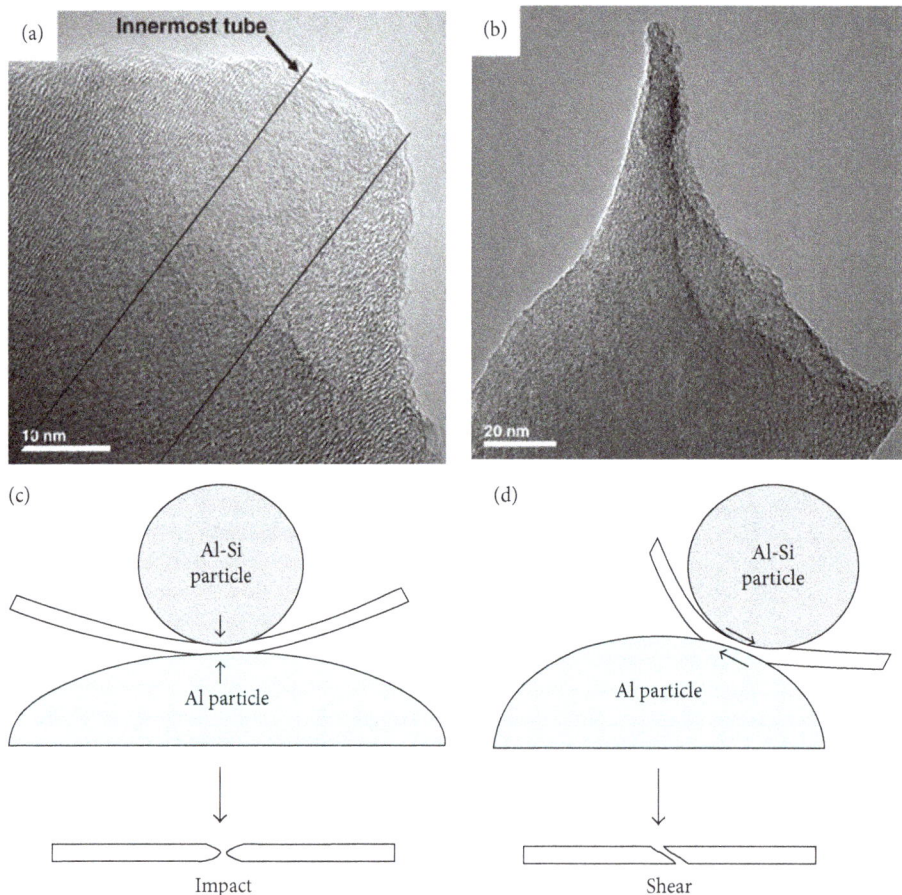

FIGURE 10: TEM imaging of damaged CNT in the cold-sprayed CNT-Cu coatings and their damaging mechanism: (a, c) impact-induced damage and (b, d) shear-induced damage [44].

graphene-reinforced MMCs have exhibited superior properties over the pure metals [69]. Currently, the most common ways for fabricating graphene-reinforced MMCs are a number of powder metallurgy techniques, for example, spark plasma sintering, laser sintering, and hot pressing [70–75]. The existing studies showed great capability of graphene-reinforced

MMCs to improve the material properties, for example, strength [70–72], Young's modulus [72, 76], hardness [72, 76], wear resistance [72–74], and electrical conductivity [75].

Graphene-reinforced MMCs were successfully produced via cold spray in very recent years [32, 33]. As a key step, preparation of MMC powders is of great importance to the

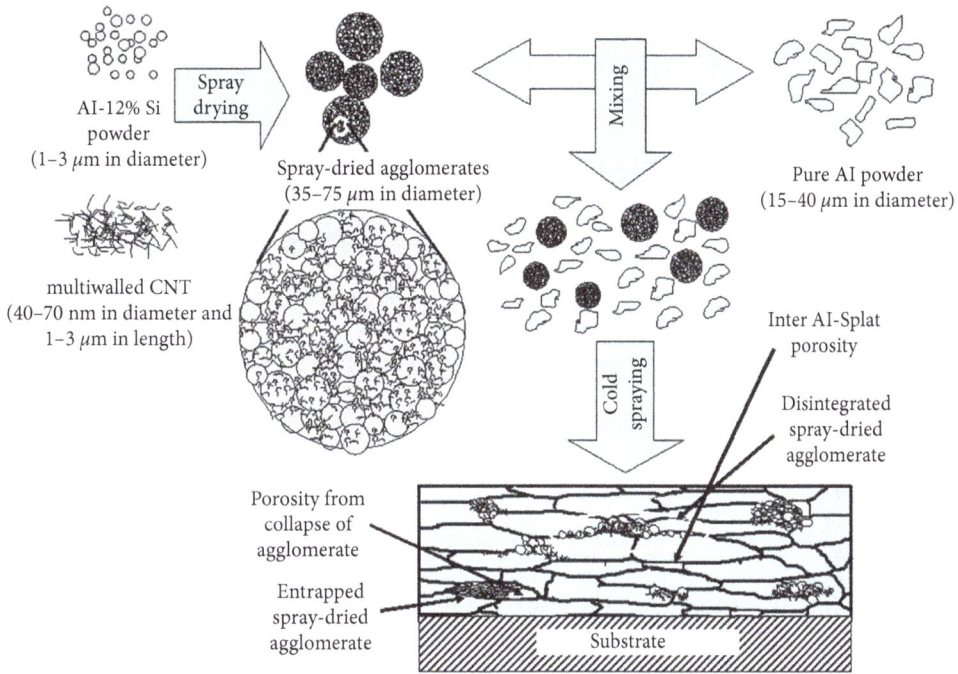

FIGURE 11: Schematic of the coating fabrication procedure using spray drying MMC powders [44].

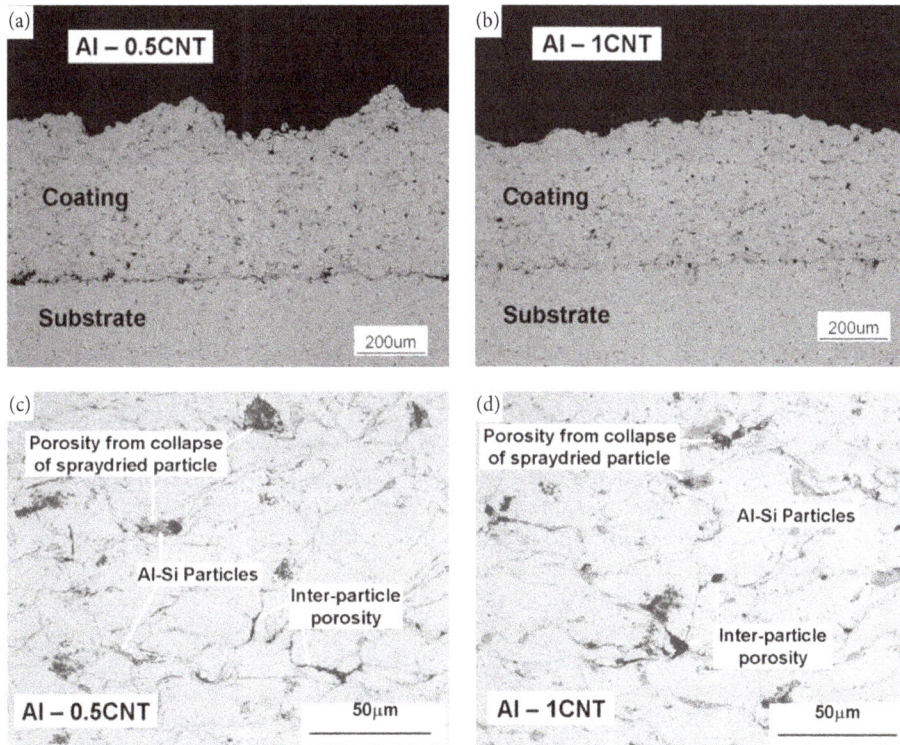

FIGURE 12: SEM imaging of the cross section of cold-sprayed CNT-Al MMC coatings with different CNT contents: (a) CNT content of 0.5%, (b) CNT content of 1%, and (c, d) high-magnification view [44].

coating deposition process and final coating quality. So far, two manufacturing methods have been used for MMC powder fabrication. In Yin et al.'s work, mechanical ball milling was applied to incorporate graphene nanosheets into spherical Cu particles [32]. Because the energy for ball milling is not high in that work, graphene nanosheets were mainly embedded on the Cu particle surface rather than homogenously distributed inside the Cu particle. Alternatively, in the work of Dardona et al., the graphene-reinforced MMC powders were synthesized through electroless plating

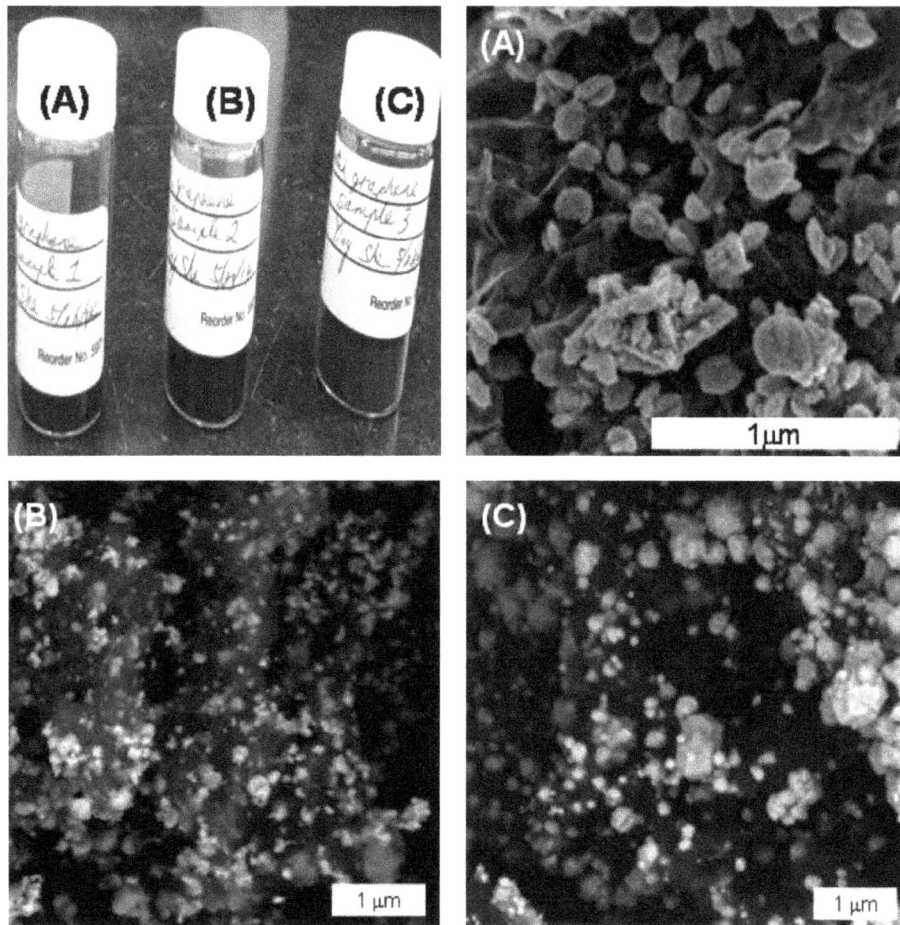

FIGURE 13: Morphology of the graphene-Cu MMC powders produced through electorless plating under different plating time. Digital photo of all samples, (A) plating for 1.5 h, (B) plating for 3 h, and (C) plating for 4.5 h [33].

of Cu film on the surface of graphene nanosheets [33]. Electroless plating is a method that deposits metals onto a solid piece by chemical approach. The piece to be plated is immersed in a reducing agent where metal ions can be changed to metal when catalyzed by certain materials to form a deposit. Figure 13 shows the morphology of the graphene-reinforced Cu MMC powders produced through electorless plating under different plating time. Powders produced in this way basically have a graphene core and an outside Cu film. Longer plating time would result in more Cu phase on the graphene surface as can be seen from Figure 13.

As for the coating microstructure, ball-milled MMC powders resulted in dense and thick coating with uniformly distributed graphene nanosheets as shown Figure 14. However, the electroless-plated MMC powders seemed to produce a low-quality coating with insufficient cohesion strength and inhomogeneous distribution of graphene nanosheets in the coating, as shown in Figure 15 [32]. This may be due to the low fraction of the Cu phase in the MMC powders, which significantly limits the effective metallic bonding between interparticles [33]. In terms of the coating properties, currently, only the wear resistance and electrical conductivity have been tested. In Yin et al.'s work, the

graphene-reinforced MMC coatings exhibited excellent wear-resistance performance, better than those produced by spark plasma sintering [32]. In the work of Dardona et al., the coating demonstrated worse electrical conductivity than bulk Cu, which was probably due to the low Cu thickness on the graphene powder surface, poor interparticle bonding, and inhomogeneous distribution of graphene nanosheets in the coating [33].

5.2. WS$_2$-Reinforced MMC Coatings. Monolayer tungsten disulfide (WS$_2$) has great potential in the optical sector due to its direct band gap and high photoluminescence intensity [34]. It also possesses excellent solid lubrication properties due to the 2D layered structure and easy interlayer sliding. In addition, the special structure of WS$_2$ allows it to be usable in high temperature, high pressure, high vacuum, high load, and with radiation and corrosive media environments [35, 36]. Moreover, because WS$_2$ has an excellent adsorption capacity on the metal surface, it can be used as reinforcements in MMCs to improve the lubrication performance [77]. The WS$_2$-Al MMCs made by sparkling plasma sintering have shown better wear-resistance performance than graphene-Al MMCs under ball-on-disk wear tests, which clearly demonstrates the superiority of WS$_2$ than other lubricants [36].

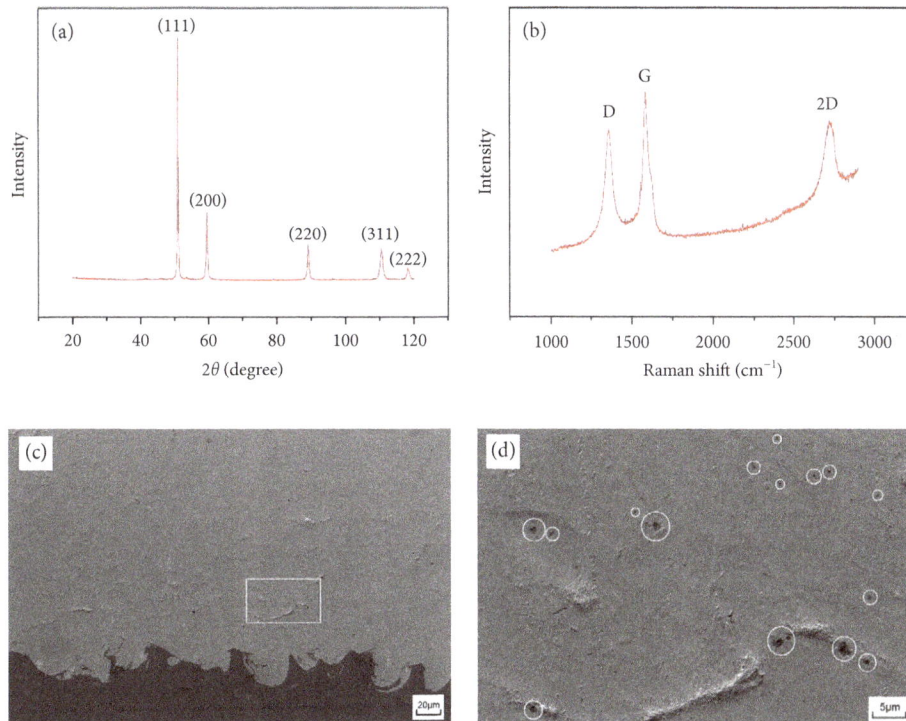

Figure 14: Characterization of graphene-Cu MMC coating produced using ball-milled powders: (a) XRD spectrum, (b) Raman spectrum, (c) coating cross-sectional view, and (d) a magnified view [32].

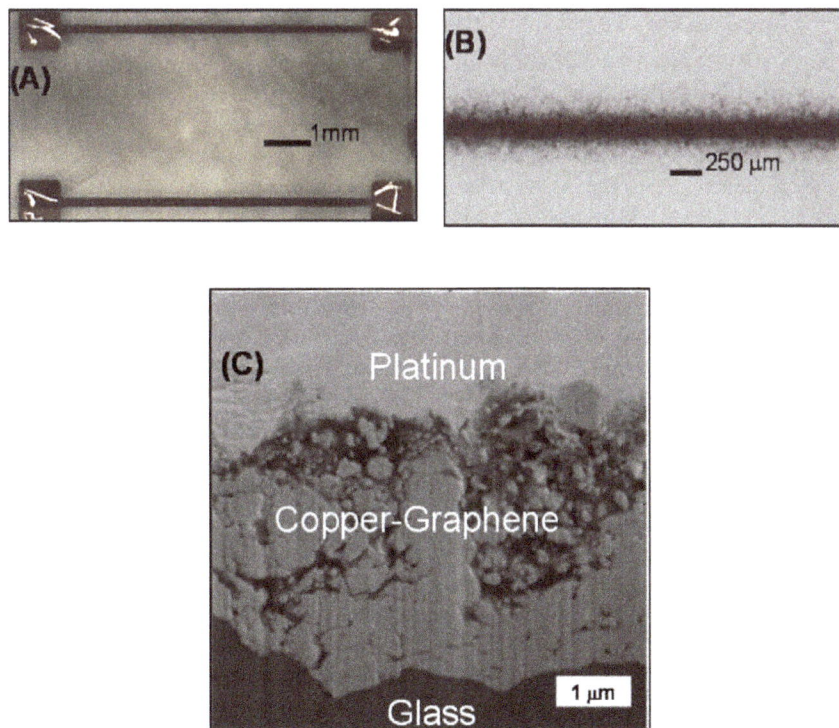

Figure 15: Characterization of graphene-Cu MMC coating produced using electroless-plated powders: (A) optical view of two tracks, (B) single-line track, and (C) coating cross-sectional view [33].

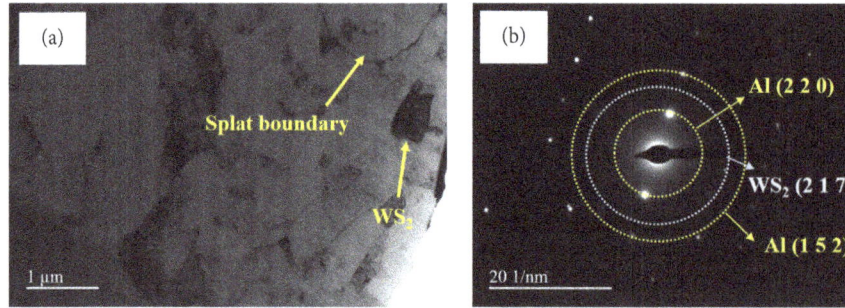

FIGURE 16: TEM imaging of WS$_2$-Al MMC coating: (a) bright field imaging with multiple splats, subgrains, and (b) SAED pattern [78].

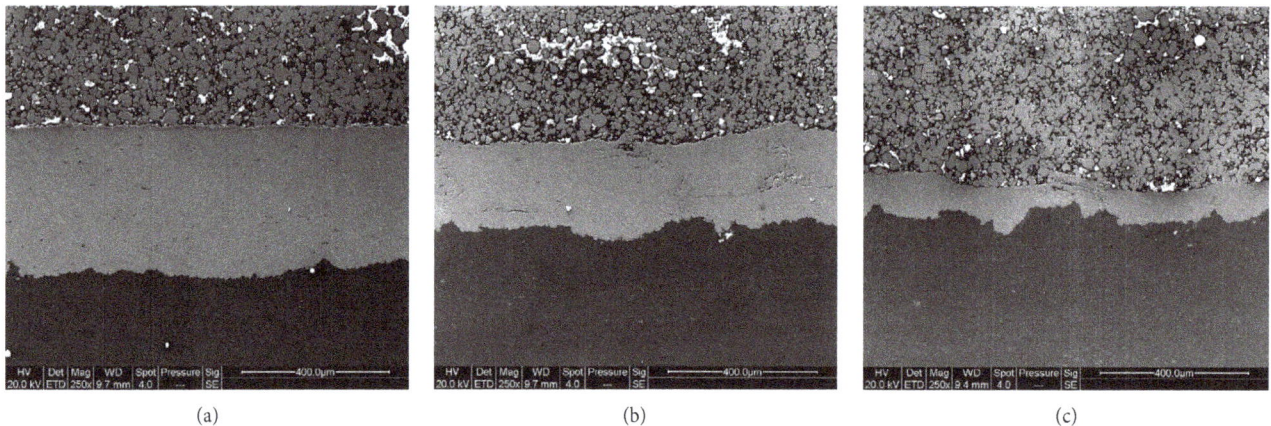

FIGURE 17: SEM imaging of the cross section of cold-sprayed hBN-Ni MMC coatings produced using (a) unaltered powders, (b) ball-milled powders for 1 h, and (c) ball-milled powders for 2 h [39].

WS$_2$-Al MMC coatings were produced via cold spray for wear resistance in a recent work [78]. Ball milling technology was used to produce the MMC powders. After low-energy ball milling, Al particles still remained spherical shape, and WS$_2$ was mostly attached on the surface of the Al particles. The surface morphology of the WS$_2$-Al MMC powders is quite similar with the graphene-Cu powders used in Yin et al.'s study [32]. The coatings were then produced using nitrogen under the pressure of 3.8 MPa and temperature of 400°C. Figure 16 shows the TEM imaging of the coating cross section. As can be seen, WS$_2$ was successfully deposited with Al onto the carbon steel substrate and uniformly distributed within the coating, which demonstrates the feasibility of cold spray to produce WS$_2$-reinforced MMC coatings. The wear test revealed that the WS$_2$-reinforced MMC coating had an outstanding wear-resistance performance due to the presence of 2D layered WS$_2$ which aids in shearing of WS$_2$ layers and uniform tribofilm formation comprised of WS$_2$ and WO$_3$.

5.3. hBN-Reinforced MMC Coatings. Hexagonal boron nitride (hBN) nanosheets are 2D crystalline form of hBN, which have a thickness of one to few atomic layers. It is similar in geometry to graphene but having completely different chemical, thermal, and electronic properties. Cold spray has been successfully used for producing hBN-reinforced MMC coatings, mainly hBN-Ni coatings [37–40].

Electroless plating was used to encapsulate hBN powders and to produce hBN-reinforced MMC powders. During the encapsulation process, the MMC particles tended to agglomerate and form large clusters. Therefore, as a comparison, following the electroless plating, ball milling was employed to de-agglomerate the clustered feedstock powders [38]. Both low- and high-energy ball milling methods were employed: low energy can eliminate voids inside the clusters and leads to higher density and uniform particle size; high energy resulted in breakup of agglomerations and destroyed the Ni encapsulant. Figure 17 shows the SEM imaging of the cross section of cold-sprayed hBN-Ni MMC coatings using various powders. As can be seen, the unaltered hBN-Ni MMC powders resulted in the thickest and densest coatings due to the significant plastic deformation as compared with ball-milled powders which experienced work hardening during the powder preparation process [39]. The tribological study demonstrates that the cold-sprayed hBN-Ni MMC coatings had a very promising wear-resistance performance. They can reduce the friction coefficient by almost 50% and significantly increase the wear resistance as compared with pure Ni [40].

6. 3D Nanoparticle-Reinforced MMC Coatings via Cold Spray

6.1. WC-Reinforced MMC Coatings. Tungsten carbide (WC) has very high hardness and stiffness, which has been widely

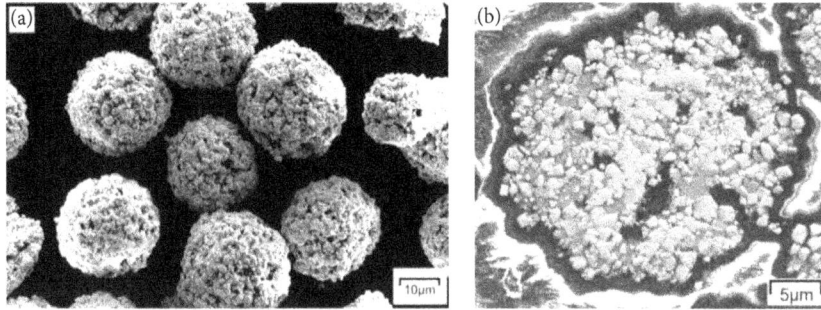

FIGURE 18: Morphology of WC-Co powder: (a) surface morphology and (b) cross-sectional view [84].

TABLE 3: Review of working parameters used for producing cold-sprayed WC-Co coatings [84].

Feedstock	Gas	Pressure (MPa)	Temperature (°C)	Preheating (°C)	Hardness (Hv)
	He	3.0	600	500	2053
	He	2.0	600	No	1812 ± 121
	He	2.0	650	No	N/A
WC-12Co	He	2.0	600	No	1600–2000
	He	2.0	650	No	1800
	N2	2.5–3.5	800	Yes	1419 ± 93
	N2	4.5	700	250	984
	N2	2.4	750 ± 30	Yes	1525 ± 143
WC-15Co	He	1.7	550	No	462 ± 92
	N2	4.5	800	250	1480
	He	3.5	600	200	918
WC-17Co	He	1.2–1.7	600	No	$1312 \pm 39, 1094 \pm 51$
	N2	2.5–3.5	800	Yes	1223 ± 59
	N2	2.4	750 ± 30	Yes	$1316 \pm 80, 1625 \pm 115$
WC-25Co	N2	4.0	800	No	$845 \pm 55, 981 \pm 58$
WC-10Co	N/A	2.2	650	No	893 ± 75

used in industrial machinery, cutting tools, and abrasives. It is normally present in the form of nanosized powder, sintered or agglomerated in soft Co to generate WC-Co MMCs. WC-Co protecting coating is one of the most important products of WC-Co MMCs, commonly used for preventing the underlying base materials from serious wear in aggressive environments [79]. The WC-Co coating fabrication mostly relies on high-temperature thermal spray processes to melt the Co matrix phase in the agglomerated WC-Co powder feedstock for achieving superior cohesion [79–83]. Figure 18 shows the typical morphology of WC-Co powders used for coating fabrications [84]. However, high deposition temperature frequently results in decarburization, phase transformation, and oxidation of hard WC reinforcements or soft Co matrix phase, significantly deteriorating the coating mechanical properties and wear-resistance performance [80–83, 85, 86]. Cold spray can effectively prevent the coating defects in relation to high-temperature processes and thereby has been found to be promising for fabricating WC-Co wear-resistance coatings.

Investigations have revealed that WC-Co particles were difficult to deposit with cold spray due to the lack of the sufficient binder phase to induce the particle plastic deformation. Thereby, the working parameters for fabricating WC-Co coatings must be extremely high so that WC-Co particles can obtain sufficient kinetic energy to promote the metallic bonding between Co matrix phases during the particle deposition. Table 3 lists the working parameters ever used for producing cold-sprayed WC-Co coatings [87]. As can be seen, cold-sprayed WC-Co coatings were mostly achieved by using compressed high-temperature helium aspropulsive gas or using nitrogen as propulsive gas combined with powder preheating treatment. Although the manufacturing cost is relatively high, cold-sprayed WC-Co coatings have shown several incomparable advantages. It has been proved that cold spray can produce fully dense WC-Co coatings as shown in Figure 19 [87]. The cold-sprayed coatings experienced no decarburization and phase transformation during the fabrication process due to the low working temperature [88–91]. This is a unique advantage that other thermal spray processes cannot achieve. In addition, it is also demonstrated that the Vickers hardness of cold-sprayed WC-Co coatings mostly fell into the range between 800 and 2000 Hv as listed in Table 3 [84]. The hardness is much higher than most metals, and comparable to or even higher than thermal-sprayed coatings. In terms of tribological performance, cold-sprayed WC-Co coatings showed lower wear

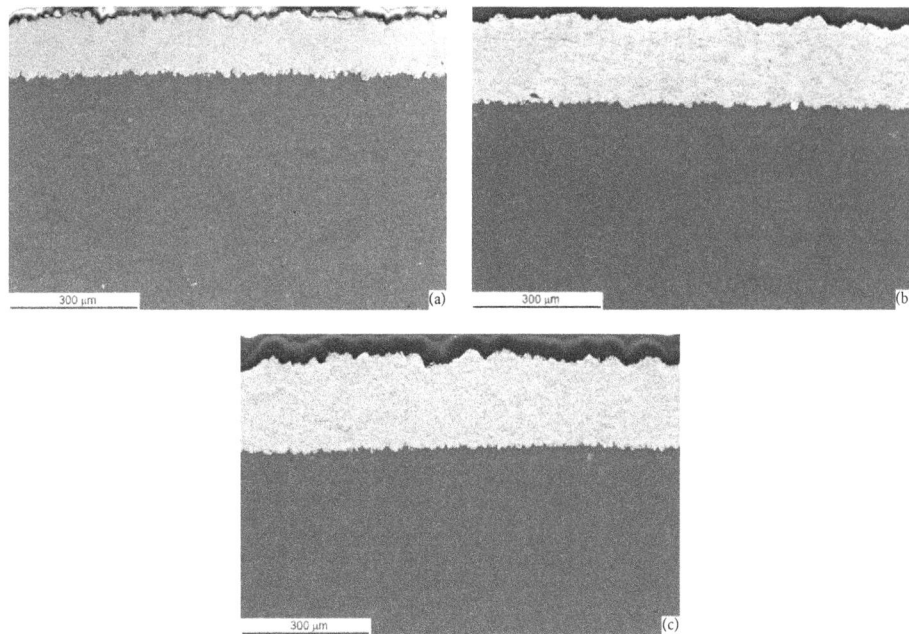

FIGURE 19: SEM imaging of the cross section of WC-Co coatings: (a)WC-12Co, (b)WC-17Co, and (c)WC-25Co coatings deposited onto the Al 7075-T6 substrate [87].

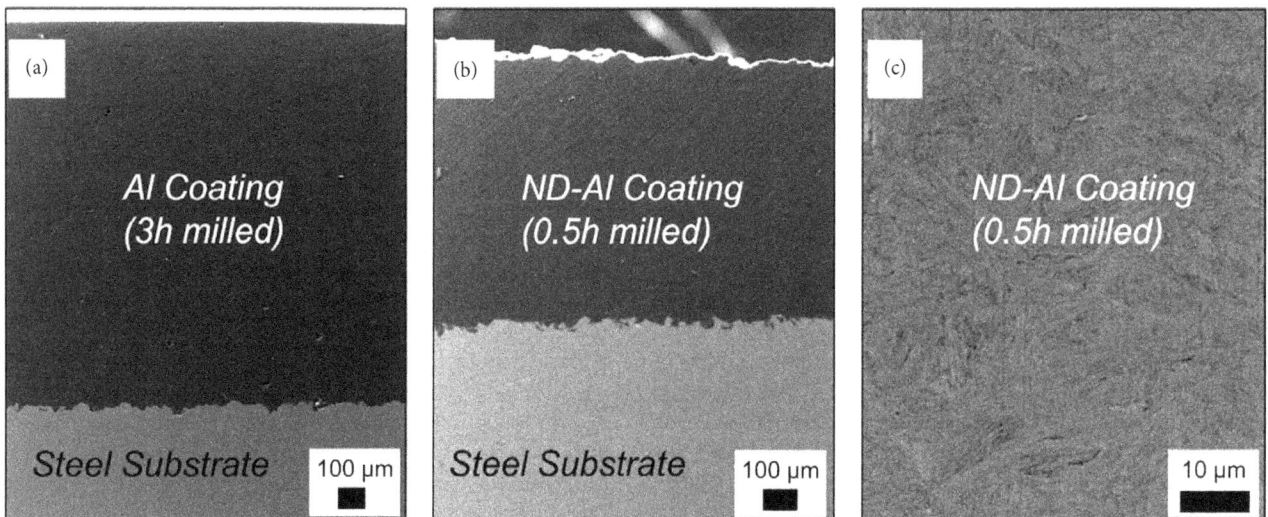

FIGURE 20: SEM imaging of the cross section of cold-sprayed nanodiamond-Al MMC coatings: (a) pure Al coating, (b) nanodiamond-Al MMC coating, and (c) magnified view.

rate than high-velocity oxy-fuel-sprayed coatings under both ball-on-disk sliding and dry abrasion tests, exhibiting superior wear-resistance performance [90, 92–94].

6.2. Diamond-Reinforced MMC Coatings. Diamond is known to possess extremely high hardness, allowing it to be used as an excellent wear-resistance material. However, for the same reason, it is difficult to be machined, which in turn limits its direct applications. Diamond-reinforced MMCs are novel materials in which the metallic phase acts as a binder, while the diamond phase helps to improve the

material properties. Currently, the common ways to fabricate bulk diamond-reinforced MMCs or MMC coatings are powder metallurgy [95–99], pressure infiltration techniques [100–104], and thermal spray techniques [105–109]. These methods mostly require extremely high processing temperatures to melt the metal binder, thereby significantly increasing the risk of the metal phase transformation and diamond graphitization [97]. Cold spray would greatly avoid the risk of high-temperature-induced diamond phase graphitization and simultaneously retain high diamond contents [110, 111].

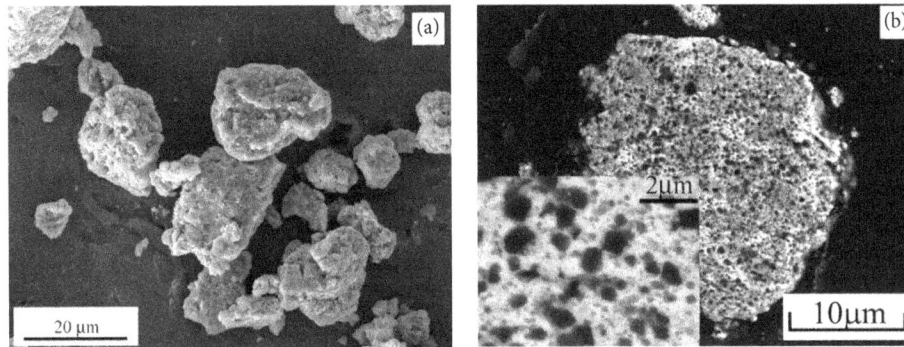

FIGURE 21: SEM imaging of the cBN-NiCrAl MMC powder after ball milling for 40 h: (a) morphology and (b) cross section. The bright regions and dark dots in (b) correspond to NiCrAl alloy and cBN particles.

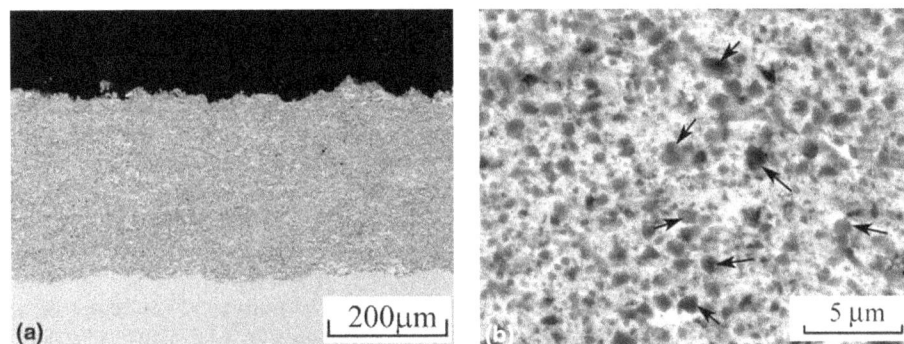

FIGURE 22: SEM imaging of cold sprayed cBN-NiCrAl MMC coatings with 40 vol.% cBN [49]. The bright regions and dark dots in (b) correspond to NiCrAl alloy and cBN particles.

Nanodiamond-reinforced MMC coatings have been successfully fabricated via cold spray [45, 46]. Analogous to most MMC powders, ball milling technology was used to produce the nanodiamond-reinforced MMC powders. In order to maximize the performance of the MMC powders, the effect of ball milling parameters and nanodiamond content on the powder properties were studied. The results revealed that powder properties including particle size distribution, hardness, and uniformity of reinforcements can be well controlled through modifying the nanodiamond content, milling time, and BPR ratio [45]. Figure 20 shows the SEM imaging of cold-sprayed nanodiamond-Al MMC coatings. The MMC coatings exhibited thick and dense features with homogenously dispersed nanodiamond reinforcements. In addition, coatings showed significant strengthening as compared with the pure Al coating. The reason for the mechanical performance strengthening was attributed to the dispersion strengthening, grain refinement, and strain hardening [46]. So far, property testing on the nanodiamond-reinforced MMC coatings was still very limited. As diamond also possesses high hardness and thermal and electrical conductivity, such coating may also possess high wear-resistance performance and thermal performance. Therefore, further investigations are encouraged in the future work.

6.3. cBN-Reinforced MMC Coatings. Cubic boron nitride (cBN), having similar crystal structure with diamond, is the second-known hardest material after diamond. It is synthesized from hBN under conditions similar to those used to produce synthetic diamond from graphite. It has been increasingly used as cutting and drilling tools in substitution for diamond-based tools owing to its superior thermal stability and chemical inertness. It is suitable especially for processing hard ferrous materials to which diamond is not applicable since diamond reacts with these materials at high temperature [112].

Cold spray has been used to fabricate cBN-NiCrAl MMC coatings [47–50]. As the most prevailing technology for MMC powder production, ball milling was also used to produce the cBN-reinforced MMC powders. Figure 21 shows the SEM imaging of the cBN-NiCrAl MMC powders after ball milling for 40 h. As can be seen, the MMC powders exhibited a near spherical shape; the cBN particles were uniformly distributed in the NiCrAl alloy matrix [47]. Figure 22 shows a SEM imaging of cold-sprayed cBN-NiCrAl MMC coatings with 40 vol.% of cBN particles [49]. Experimental results clearly showed that the as-sprayed coating had a rather dense microstructure with uniformly dispersed nano-cBN particles. In addition, no phase transformation and grain growth of the NiCrAl matrix occurred during the spraying process. The hardness of the 20 vol.% and 40 vol.% cBN-NiCrAl coatings were 1063 and 1175 Hv, respectively [47–49].

In addition, annealing treatment was found to significantly affect the microstructure of the cBN-reinforced NiCrAl MMC coatings. The nanostructure in the MMC coatings could be retained when the annealing temperature was below 825°C. However, a significant growth of dispersion reinforcements due to the occurrence of reaction between cBN particles and the NiCrAl matrix was observed at an annealing temperature higher than 825°C. This phenomenon led to a reduction of hardness as the annealing temperature increased. Furthermore, the tribological performance of the cold sprayed cBN-NiCrAl MMC coatings was also investigated. The as-sprayed coatings exhibited excellent wear-resistance performance. Coatings with 20 vol.% nano-cBN resulted in a wear resistance which is comparable to the HVOF-sprayed WC-12Co. Low-temperature heat treatment (750°C for 5 h) would further improve the wear-resistance performance due to the promoted interparticle bonding strength [48].

7. Summary and Perspectives

Cold spray is a solid-state coating deposition technology which can retain the original properties of feedstock in the final coating and prevent the adverse influence caused by high working temperature. As a low-temperature process, cold spray has been showing great potential in producing high-performance metal coatings with nanostructure. This paper provides an overview of cold-sprayed metal coatings with nanostructure. Basically, cold-sprayed coatings with nanostructure can be produced in the following three ways. Firstly, nanocrystallization in the form of grain refinement always occurs at the interparticle and coating-substrate interfacial regions during the deposition process due to the dynamic recrystallization, which can result in nanostructured grains within the cold-sprayed coating. This theory has been well understood so far. Secondly, the starting feedstock for cold spray is nanocrystalline powders; then the coating can retain the nanostructure of the starting powders. In this aspect, cold-sprayed nanocrystalline coatings are very promising as cold spray can completely retain the nanostructure of the starting powders in the coating. In the future, the relevant work should be continued, and special attention can be paid on the coating densification and coating property exploration. Thirdly, cold-sprayed nanomaterial-reinforced MMC coatings also exhibit nanostructure. In this field, although ball milling has been accepted as the most commonly used method for producing nanomaterial-reinforced MMC powders, the effect of milling parameters on the powder property is still now well clarified. Also, new technologies for producing more uniform MMC powders are needed. Therefore, preparation of nanomaterial-reinforced MMC powders will be a research focus in the future work. Moreover, investigation on the properties of cold-sprayed nanomaterial-reinforced MMC coatings will be another highlight as currently such property tests are still very limited.

Conflicts of Interest

The authors declare that they have no conflicts of interest.

References

[1] A. Papyrin, "Cold spray technology," *Advanced Materials and Processing*, vol. 159, pp. 49–51, 2001.

[2] P. C. King, S. H. Zahiri, and M. Jahedi, "Focused ion beam micro-dissection of cold-sprayed particles," *Acta Materialia*, vol. 56, no. 19, pp. 5617–5626, 2008.

[3] N. M. Melendez, V. V. Narulkar, G. A. Fisher, and A. G. McDonald, "Effect of reinforcing particles on the wear rate of low-pressure cold-sprayed WC-based MMC coatings," *Wear*, vol. 306, no. 1-2, pp. 185–195, 2013.

[4] S. Dosta, M. Couto, and J. M. Guilemany, "Cold spray deposition of a WC-25Co cermet onto Al7075-T6 and carbon steel substrates," *Acta Materialia*, vol. 61, no. 2, pp. 643–652, 2013.

[5] M. Grujicic, C. L. C. Zhao, W. S. W. DeRosset, and D. Helfritch, "Adiabatic shear instability based mechanism for particles/substrate bonding in the cold-gas dynamic-spray process," *Materials and Design*, vol. 25, no. 8, pp. 681–688, 2004.

[6] H. J. Choi, M. Lee, and J. Y. Lee, "Application of a cold spray technique to the fabrication of a copper canister for the geological disposal of CANDU spent fuels," *Nuclear Engineering and Design*, vol. 240, no. 10, pp. 2714–2720, 2010.

[7] V. K. Champagne, "The repair of magnesium rotorcraft components by cold spray," *Journal of Failure Analysis and Prevention*, vol. 8, no. 2, pp. 164–175, 2008.

[8] S. Yin, X. Wang, X. Suo et al., "Deposition behavior of thermally softened copper particles in cold spraying," *Acta Materialia*, vol. 61, no. 14, pp. 5105–5118, 2013.

[9] T. Klassen, H. Assadi, H. Kreye, and F. Gartner, "Cold spraying—a materials perspective," *Acta Materialia*, vol. 116, pp. 382–407, 2016.

[10] M. R. Rokni, S. R. Nutt, C. A. Widener, V. K. Champagne, and R. H. Hrabe, "Review of relationship between particle deformation, coating microstructure, and properties in high-pressure cold spray," *Journal of Thermal Spray Technology*, vol. 26, no. 6, pp. 1–48, 2017.

[11] S. Yin, M. Meyer, W. Li et al., "Gas flow, particle acceleration, and heat transfer in cold spray: a review in cold spray: a review," *Journal of Thermal Spray Technology*, vol. 25, no. 5, pp. 1–23, 2016.

[12] S. Yin, X. F. Wang, W. Y. Li, and H. E. Jie, "Effect of substrate hardness on the deformation behavior of subsequently incident particles in cold spraying," *Applied Surface Science*, vol. 257, no. 17, pp. 7560–7565, 2011.

[13] T. C. Jen, L. Li, W. Cui, Q. Chen, and X. Zhang, "Numerical investigations on cold gas dynamic spray process with nano- and microsize particles," *International Journal of Heat and Mass Transfer*, vol. 48, no. 21-22, pp. 4384–4396, 2005.

[14] J. Liu, H. Cui, X. Zhou, X. Wu, and J. Zhang, "Nanocrystalline copper coatings produced by cold spraying," *Metals and Materials International*, vol. 18, no. 1, pp. 121–128, 2012.

[15] K. Kim, M. Watanabe, J. Kawakita, and S. Kuroda, "Grain refinement in a single titanium powder particle impacted at high velocity," *Scripta Materialia*, vol. 59, no. 7, pp. 768–771, 2008.

[16] C. Lee and J. Kim, "Microstructure of kinetic spray coatings: a review," *Journal of Thermal Spray Technology*, vol. 24, no. 4, pp. 592–610, 2015.

[17] P. C. King, S. H. Zahiri, and M. Jahedi, "Microstructural refinement within a cold-sprayed copper particle," *Metallurgical and Materials Transactions A*, vol. 40, no. 9, pp. 2115–2123, 2009.

[18] M. R. Rokni, C. A. Widener, and V. R. Champagne, "Microstructural evolution of 6061 aluminum gas-atomized powder and high-pressure cold-sprayed deposition," *Journal of Thermal Spray Technology*, vol. 23, no. 3, pp. 514–524, 2014.

[19] X.-T. Luo, C.-X. Li, F.-L. Shang, G.-J. Yang, Y.-Y. Wang, and C.-J. Li, "High velocity impact induced microstructure evolution during deposition of cold spray coatings: a review," *Surface and Coatings Technology*, vol. 254, pp. 11–20, 2014.

[20] Y. Zou, W. Qin, E. Irissou, J. G. Legoux, S. Yue, and J. A. Szpunar, "Dynamic recrystallization in the particle/particle interfacial region of cold-sprayed nickel coating: electron backscatter diffraction characterization," *Scripta Materialia*, vol. 61, no. 9, pp. 899–902, 2009.

[21] Y. Zou, D. Goldbaum, J. A. Szpunar, and S. Yue, "Microstructure and nanohardness of cold-sprayed coatings: electron backscattered diffraction and nanoindentation studies," *Scripta Materialia*, vol. 62, no. 6, pp. 395–398, 2010.

[22] L. Ajdelsztajn, B. Jodoin, G. E. Kim, and J. M. Schoenung, "Cold spray deposition of nanocrystalline aluminum alloys," *Metallurgical and Materials Transactions A*, vol. 36, no. 3, pp. 657–666, 2005.

[23] L. Ajdelsztajn, A. Zúñiga, B. Jodoin, and E. J. Lavernia, "Cold-spray processing of a nanocrystalline Al-Cu-Mg-Fe-Ni alloy with Sc," *Journal of Thermal Spray Technology*, vol. 15, no. 2, pp. 184–190, 2006.

[24] L. Ajdelsztajn, B. Jodoin, and J. M. Schoenung, "Synthesis and mechanical properties of nanocrystalline Ni coatings produced by cold gas dynamic spraying," *Surface and Coatings Technology*, vol. 201, no. 3-4, pp. 1166–1172, 2006.

[25] P. Richer, B. Jodoin, L. Ajdelsztajn, and E. J. Lavernia, "Substrate roughness and thickness effects on cold spray nanocrystalline Al-Mg coatings," *Journal of Thermal Spray Technology*, vol. 15, no. 2, pp. 246–254, 2006.

[26] P. Sudharshan Phani, V. Vishnukanthan, and G. Sundararajan, "Effect of heat treatment on properties of cold sprayed nanocrystalline copper alumina coatings," *Acta Materialia*, vol. 55, no. 14, pp. 4741–4751, 2007.

[27] Y. Y. Zhang, X. K. Wu, H. Cui, and J. S. Zhang, "Cold-spray processing of a high density nanocrystalline aluminum alloy 2009 coating using a mixture of as-atomized and as-cryomilled powders," *Journal of Thermal Spray Technology*, vol. 20, no. 5, pp. 1125–1132, 2011.

[28] S. Tria, O. Elkedim, R. Hamzaoui et al., "Deposition and characterization of cold sprayed nanocrystalline NiTi," *Powder Technology*, vol. 210, no. 2, pp. 181–188, 2011.

[29] J. Liu, X. Zhou, X. Zheng, H. Cui, and J. Zhang, "Tribological behavior of cold-sprayed nanocrystalline and conventional copper coatings," *Applied Surface Science*, vol. 258, no. 19, pp. 7490–7496, 2012.

[30] R. Ghelichi, S. Bagherifard, D. Mac Donald et al., "Fatigue strength of Al alloy cold sprayed with nanocrystalline powders," *International Journal of Fatigue*, vol. 65, pp. 51–57, 2014.

[31] M. Kumar, H. Singh, N. Singh et al., "Development of nanocrystalline cold sprayed Ni-20Cr coatings for high temperature oxidation resistance," *Surface and Coatings Technology*, vol. 266, pp. 122–133, 2015.

[32] S. Yin, Z. Zhang, E. J. Ekoi et al., "Novel cold spray for fabricating graphene-reinforced metal matrix composites," *Materials Letters*, vol. 196, pp. 172–175, 2017.

[33] S. Dardona, J. Hoey, Y. She, and W. R. Schmidt, "Direct write of copper-graphene composite using micro-cold spray," *AIP Advances*, vol. 6, no. 8, p. 085013, 2016.

[34] K. M. McCreary, A. T. Hanbicki, G. G. Jernigan, J. C. Culbertson, and B. T. Jonker, "Synthesis of large-area WS2 monolayers with exceptional photoluminescence," *Scientific Reports*, vol. 6, no. 1, p. 19159, 2016.

[35] X. Zhang, H. Xu, J. Wang et al., "Synthesis of ultrathin WS2 nanosheets and their tribological properties as lubricant additives," *Nanoscale Research Letters*, vol. 11, no. 1, p. 442, 2016.

[36] S. Rengifo, *A Comparison between Graphene and WS2 as Solid Lubricant Additives to Aluminum for Automobile Applications*, M.S. thesis, Florida International University, Miami, FL, USA, 2015.

[37] I. Smid, A. E. Segall, P. Walia, G. Aggarwal, T. J. Eden, and J. K. Potter, "Cold-sprayed Ni-hBN self-lubricating coatings," *Tribology Transactions*, vol. 55, no. 5, pp. 599–605, 2012.

[38] M. Neshastehriz, I. Smid, and A. E. Segall, "In-situ agglomeration and de-agglomeration by milling of nano-engineered lubricant particulate composites for cold spray deposition," *Journal of Thermal Spray Technology*, vol. 23, no. 7, pp. 1191–1198, 2014.

[39] M. Neshastehriz, *Influence of Pre-Process Work Hardening of Nickel Encapsulated Hexagonal Boron Nitride Powders on Cold Spray Coatings*, The Pennsylvania State University, State College, PA, USA, 2014.

[40] L. M. Stark, *Engineered Self-Lubricating Coatings Utilizing Cold Spray Technology*, The Pennsylvania State University, State College, PA, USA, 2010.

[41] E. J. T. Pialago and C. W. Park, "Cold spray deposition characteristics of mechanically alloyed Cu-CNT composite powders," *Applied Surface Science*, vol. 308, pp. 63–74, 2014.

[42] E. J. T. Pialago, O. K. Kwon, M. S. Kim, and C. W. Park, "Ternary Cu-CNT-AlN composite coatings consolidated by cold spray deposition of mechanically alloyed powders," *Journal of Alloys and Compounds*, vol. 650, pp. 199–209, 2015.

[43] E. J. T. Pialago, O. K. Kwon, and C. W. Park, "Cold spray deposition of mechanically alloyed ternary Cu-CNT-SiC composite powders," *Ceramics International*, vol. 41, no. 5, pp. 6764–6775, 2015.

[44] S. R. Bakshi, V. Singh, K. Balani, D. G. McCartney, S. Seal, and A. Agarwal, "Carbon nanotube reinforced aluminum composite coating via cold spraying," *Surface and Coatings Technology*, vol. 202, no. 21, pp. 5162–5169, 2008.

[45] D. J. Woo, B. Sneed, F. Peerally et al., "Synthesis of nanodiamond-reinforced aluminum metal composite powders and coatings using high-energy ball milling and cold spray," *Carbon*, vol. 63, pp. 404–415, 2013.

[46] D. J. Woo, F. C. Heer, L. N. Brewer, J. P. Hooper, and S. Osswald, "Synthesis of nanodiamond-reinforced aluminum metal matrix composites using cold-spray deposition," *Carbon*, vol. 86, pp. 15–25, 2015.

[47] X. T. Luo, G. J. Yang, and C. J. Li, "Multiple strengthening mechanisms of cold-sprayed cBNp/NiCrAl composite coating," *Surface and Coatings Technology*, vol. 205, no. 20, pp. 4808–4813, 2011.

[48] X. T. Luo, E. J. Yang, F. L. Shang, G. J. Yang, C. X. Li, and C. J. Li, "Microstructure, mechanical properties, and two-body

abrasive wear behavior of cold-sprayed 2 vol.% cubic BN-NiCrAl nanocomposite coating," *Journal of Thermal Spray Technology*, vol. 23, no. 7, pp. 1181–1190, 2014.

[49] X. T. Luo and C. J. Li, "Thermal stability of microstructure and hardness of cold-sprayed cBN/NiCrAl nanocomposite coating," *Journal of Thermal Spray Technology*, vol. 21, no. 3-4, pp. 578–585, 2012.

[50] X. T. Luo, G. J. Yang, C. J. Li, and K. Kondoh, "High strain rate induced localized amorphization in cubic BN/NiCrAl nanocomposite through high velocity impact," *Scripta Materialia*, vol. 65, no. 7, pp. 581–584, 2011.

[51] H. Assadi, F. Gärtner, T. Stoltenhoff, and H. Kreye, "Bonding mechanism in cold gas spraying," *Acta Materialia*, vol. 51, no. 15, pp. 4379–4394, 2003.

[52] M. R. Rokni, C. A. Widener, and G. A. Crawford, "Microstructural evolution of 7075 Al gas atomized powder and high-pressure cold sprayed deposition," *Surface and Coatings Technology*, vol. 251, pp. 254–263, 2014.

[53] K. Kim, M. Watanabe, and S. Kuroda, "Thermal softening effect on the deposition efficiency and microstructure of warm sprayed metallic powder," *Scripta Materialia*, vol. 60, no. 8, pp. 710–713, 2009.

[54] K. Kang, H. Park, G. Bae, and C. Lee, "Microstructure and texture of Al coating during kinetic spraying and heat treatment," *Journal of Materials Science*, vol. 47, no. 9, pp. 4053–4061, 2012.

[55] Q. Wang, N. Birbilis, and M. X. Zhang, "Interfacial structure between particles in an aluminum deposit produced by cold spray," *Materials Letters*, vol. 65, no. 11, pp. 1576–1578, 2011.

[56] M. A. Meyers, A. Mishra, and D. J. Benson, "Mechanical properties of nanocrystalline materials," *Progress in Materials Science*, vol. 51, no. 4, pp. 427–556, 2006.

[57] A. I. Gusev and A. A. Rempel, *Nanocrystalline Materials*, Cambridge International Science Publishing, Great Abington, Cambridge, UK, 2004.

[58] X. Wang, Q. Li, J. Xie et al., "Fabrication of ultralong and electrically uniform single-walled carbon nanotubes on clean substrates," *Nano Letters*, vol. 9, no. 9, pp. 3137–3141, 2009.

[59] J. L. Song, W. G. Chen, L. L. Dong, J. J. Wang, and N. Deng, "An electroless plating and planetary ball milling process for mechanical properties enhancement of bulk CNTs/Cu composites," *Journal of Alloys and Compounds*, vol. 720, pp. 54–62, 2017.

[60] B. Cheng, R. Bao, J. Yi et al., "Interface optimization of CNT/Cu composite by forming TiC nanoprecipitation and low interface energy structure via spark plasma sintering," *Journal of Alloys and Compounds*, vol. 722, pp. 852–858, 2017.

[61] D. Kaewsai, A. Watcharapasorn, P. Singjai, S. Wirojanupatump, P. Niranatlumpong, and S. Jiansirisomboon, "Thermal sprayed stainless steel/carbon nanotube composite coatings," *Surface and Coatings Technology*, vol. 205, no. 7, pp. 2104–2112, 2010.

[62] S. Moonngam, P. Tunjina, D. Deesom, and C. Banjongprasert, "Fe-Cr/CNTs nanocomposite feedstock powders produced by chemical vapor deposition for thermal spray coatings," *Surface and Coatings Technology*, vol. 306, pp. 323–327, 2016.

[63] W. A. Curtin and B. W. Sheldon, "CNT-reinforced ceramics and metals," *Materials Today*, vol. 7, no. 11, pp. 44–49, 2004.

[64] A. Dorri Moghadam, E. Omrani, P. L. Menezes, and P. K. Rohatgi, "Mechanical and tribological properties of self-lubricating metal matrix nanocomposites reinforced by carbon nanotubes (CNTs) and graphene-a review," *Composites Part B: Engineering*, vol. 77, pp. 402–420, 2015.

[65] S. Cho, K. Takagi, H. Kwon et al., "Multi-walled carbon nanotube-reinforced copper nanocomposite coating fabricated by low-pressure cold spray process," *Surface and Coatings Technology*, vol. 206, no. 16, pp. 3488–3494, 2012.

[66] E. J. T. Pialago, O. K. Kwon, and C. W. Park, "Nucleate boiling heat transfer of R134a on cold sprayed CNT-Cu composite coatings," *Applied Thermal Engineering*, vol. 56, no. 1-2, pp. 112–119, 2013.

[67] E. J. T. Pialago, O. K. Kwon, J. S. Jin, and C. W. Park, "Nucleate pool boiling of R134a on cold sprayed Cu-CNT-SiC and Cu-CNT-AlN composite coatings," *Applied Thermal Engineering*, vol. 103, pp. 684–694, 2016.

[68] E. J. Pialago, X. R. Zheng, K. A. Ada, et al., "A study on the effects of ceramic content of CNT/metal composite surface coatings fabricated by cold spray for boiling heat transfer enhancement," in *Proceedings of the ASME 2012 3rd Micro/Nanoscale Heat & Mass Transfer International Conference*, Atlanta, GA, USA, 2012.

[69] D. Lin, C. Richard Liu, and G. J. Cheng, "Single-layer graphene oxide reinforced metal matrix composites by laser sintering: microstructure and mechanical property enhancement," *Acta Materialia*, vol. 80, pp. 183–193, 2014.

[70] L. Y. Chen, H. Konishi, A. Fehrenbacher et al., "Novel nanoprocessing route for bulk graphene nanoplatelets reinforced metal matrix nanocomposites," *Scripta Materialia*, vol. 67, no. 1, pp. 29–32, 2012.

[71] F. Chen, J. Ying, Y. Wang, S. Du, Z. Liu, and Q. Huang, "Effects of graphene content on the microstructure and properties of copper matrix composites," *Carbon*, vol. 96, pp. 836–842, 2016.

[72] W. Zhai, X. Shi, M. Wang et al., "Grain refinement: a mechanism for graphene nanoplatelets to reduce friction and wear of Ni3Al matrix self-lubricating composites," *Wear*, vol. 310, no. 1-2, pp. 33–40, 2014.

[73] Z. Xu, X. Shi, W. Zhai, J. Yao, S. Song, and Q. Zhang, "Preparation and tribological properties of TiAl matrix composites reinforced by multilayer graphene," *Carbon*, vol. 67, pp. 168–177, 2014.

[74] Y. Fan, L. Kang, W. Zhou, W. Jiang, L. Wang, and A. Kawasaki, "Control of doping by matrix in few-layer graphene/metal oxide composites with highly enhanced electrical conductivity," *Carbon*, vol. 81, pp. 83–90, 2015.

[75] X. Huang, X. Qi, F. Boey, and H. Zhang, "Graphene-based composites," *Chemical Society Reviews*, vol. 41, no. 2, pp. 666–686, 2012.

[76] J. Dutkiewicz, P. Ozga, W. Maziarz et al., "Microstructure and properties of bulk copper matrix composites strengthened with various kinds of graphene nanoplatelets," *Materials Science and Engineering: A*, vol. 628, pp. 124–134, 2015.

[77] S. Xu, X. Gao, M. Hu et al., "Dependence of atomic oxygen resistance and the tribological properties on microstructures of WS2 films," *Applied Surface Science*, vol. 298, pp. 36–43, 2014.

[78] A. Loganathan, S. Rengifo, A. F. Hernandez et al., "Effect of 2D WS2 addition on cold-sprayed aluminum coating," *Journal of Thermal Spray Technology*, vol. 26, no. 7, pp. 1585–1597, 2017.

[79] A. Rahbar-kelishami, A. Abdollah-zadeh, M. M. Hadavi, A. Banerji, A. Alpas, and A. P. Gerlich, "Effects of friction stir processing on wear properties of WC-12%Co sprayed on 52100 steel," *Materials and Design*, vol. 86, pp. 98–104, 2015.

[80] C. J. Li and G. J. Yang, "Relationships between feedstock structure, particle parameter, coating deposition, microstructure and properties for thermally sprayed conventional

and nanostructured WC-Co," *International Journal of Refractory Metals and Hard Materials*, vol. 39, pp. 2–17, 2013.

[81] S. Al-Mutairi, M. S. J. Hashmi, B. S. Yilbas, and J. Stokes, "Microstructural characterization of HVOF/plasma thermal spray of micro/nano WC-12%Co powders," *Surface and Coatings Technology*, vol. 264, pp. 175–186, 2015.

[82] H. L. D. V. Lovelock, "Powder/processing/structure relationships in WC-Co thermal spray coatings: a review of the published literature," *Journal of Thermal Spray Technology*, vol. 7, no. 3, pp. 357–373, 1998.

[83] J. He and J. M. Schoenung, "A review on nanostructured WC-Co coatings," *Surface and Coatings Technology*, vol. 157, no. 1, pp. 72–79, 2002.

[84] S. Yin, E. J. Ekoi, T. L. Lupton, D. P. Dowling, and R. Lupoi, "Cold spraying of WC-Co-Ni coatings using porous WC-17Co powders: formation mechanism, microstructure characterization and tribological performance," *Materials and Design*, vol. 126, pp. 305–313, 2017.

[85] K. Kumari, K. Anand, M. Bellacci, and M. Giannozzi, "Effect of microstructure on abrasive wear behavior of thermally sprayed WC-10Co-4Cr coatings," *Wear*, vol. 268, no. 11-12, pp. 1309–1319, 2010.

[86] A. A. Burkov and S. A. Pyachin, "Formation of WC-Co coating by a novel technique of electrospark granules deposition," *Materials and Design*, vol. 80, pp. 109–115, 2015.

[87] S. Dosta, G. Bolelli, A. Candeli, L. Lusvarghi, I. G. Cano, and J. M. Guilemany, "Plastic deformation phenomena during cold spray impact of WC-Co particles onto metal substrates," *Acta Materialia*, vol. 124, pp. 173–181, 2017.

[88] H.-J. Kim, C.-H. Lee, and S.-Y. Hwang, "Fabrication of WC–Co coatings by cold spray deposition," *Surface and Coatings Technology*, vol. 191, no. 2-3, pp. 335–340, 2005.

[89] H.-J. Kim, C.-H. Lee, and S.-Y. Hwang, "Superhard nano WC–12%Co coating by cold spray deposition," *Materials Science and Engineering: A*, vol. 391, no. 1-2, pp. 243–248, 2005.

[90] M. Couto, S. Dosta, and J. M. Guilemany, "Comparison of the mechanical and electrochemical properties of WC-17 and 12Co coatings onto Al7075-T6 obtained by high velocity oxy-fuel and cold gas spraying," *Surface and Coatings Technology*, vol. 268, pp. 180–189, 2014.

[91] A. S. M. Ang, C. C. Berndt, and P. Cheang, "Deposition effects of WC particle size on cold sprayed WC-Co coatings," *Surface and Coatings Technology*, vol. 205, no. 10, pp. 3260–3267, 2011.

[92] M. Couto, S. Dosta, J. Fernández, and J. M. Guilemany, "Comparison of the mechanical and electrochemical properties of WC-25Co coatings obtained by high velocity oxy-fuel and cold gas spraying," *Journal of Thermal Spray Technology*, vol. 23, no. 8, pp. 1251–1258, 2014.

[93] H.-T. Wang, X. Chen, X.-B. Bai, G.-C. Ji, Z.-X. Dong, and D.-L. Yi, "Microstructure and properties of cold sprayed multimodal WC–17Co deposits," *International Journal of Refractory Metals and Hard Materials*, vol. 45, pp. 196–203, 2014.

[94] G.-J. Yang, P.-H. Gao, C.-X. Li, and C.-J. Li, "Mechanical property and wear performance dependence on processing condition for cold-sprayed WC-(nanoWC-Co)," *Applied Surface Science*, vol. 332, pp. 80–88, 2015.

[95] Q. L. Che, J. J. Zhang, X. K. Chen et al., "Spark plasma sintering of titanium-coated diamond and copper–titanium powder to enhance thermal conductivity of diamond/copper composites," *Materials Science in Semiconductor Processing*, vol. 33, pp. 67–75, 2015.

[96] A. M. Abyzov, M. J. Kruszewski, Ł. Ciupiński, M. Mazurkiewicz, A. Michalski, and K. J. Kurzydłowski, "Diamond–tungsten based coating–copper composites with high thermal conductivity produced by pulse plasma sintering," *Materials & Design*, vol. 76, pp. 97–109, 2015.

[97] W. Z. Shao, V. V. Ivanov, L. Zhen, Y. S. Cui, and Y. Wang, "A study on graphitization of diamond in copper-diamond composite materials," *Materials Letters*, vol. 58, no. 1-2, pp. 146–149, 2004.

[98] K. Chu, Z. Liu, C. Jia et al., "Thermal conductivity of SPS consolidated Cu/diamond composites with Cr-coated diamond particles," *Journal of Alloys and Compounds*, vol. 490, no. 1-2, pp. 453–458, 2010.

[99] T. Schubert, Ł. Ciupiński, W. Zieliński, A. Michalski, T. Weißgärber, and B. Kieback, "Interfacial characterization of Cu/diamond composites prepared by powder metallurgy for heat sink applications," *Scripta Materialia*, vol. 58, no. 4, pp. 263–266, 2008.

[100] A. M. Abyzov, S. V. Kidalov, and F. M. Shakhov, "High thermal conductivity composite of diamond particles with tungsten coating in a copper matrix for heat sink application," *Applied Thermal Engineering*, vol. 48, pp. 72–80, 2012.

[101] H. Feng, J. K. Yu, and W. Tan, "Microstructure and thermal properties of diamond/aluminum composites with TiC coating on diamond particles," *Materials Chemistry and Physics*, vol. 124, no. 1, pp. 851–855, 2010.

[102] Q. Kang, X. He, S. Ren et al., "Preparation of copper-diamond composites with chromium carbide coatings on diamond particles for heat sink applications," *Applied Thermal Engineering*, vol. 60, no. 1-2, pp. 423–429, 2013.

[103] J. Li, H. Zhang, Y. Zhang, Z. Che, and X. Wang, "Microstructure and thermal conductivity of Cu/diamond composites with Ti-coated diamond particles produced by gas pressure infiltration," *Journal of Alloys and Compounds*, vol. 647, pp. 941–946, 2015.

[104] Y. Dong, R. Zhang, X. He, Z. Ye, and X. Qu, "Fabrication and infiltration kinetics analysis of Ti-coated diamond/copper composites with near-net-shape by pressureless infiltration," *Materials Science and Engineering: B*, vol. 177, no. 17, pp. 1524–1530, 2012.

[105] K. Venkateswarlu, A. K. Ray, M. K. Gunjan, D. P. Mondal, and L. C. Pathak, "Tribological wear behavior of diamond reinforced composite coating," *Materials Science and Engineering: A*, vol. 418, no. 1-2, pp. 357–363, 2006.

[106] A. F. Richardson, A. Neville, and J. I. B. Wilson, "Developing diamond MMCs to improve durability in aggressive abrasive conditions," *Wear*, vol. 255, no. 1–6, pp. 593–605, 2003.

[107] K. Venkateswarlu, V. Rajinikanth, T. Naveen, D. P. Sinha, Atiquzzaman, and A. K. Ray, "Abrasive wear behavior of thermally sprayed diamond reinforced composite coating deposited with both oxy-acetylene and HVOF techniques," *Wear*, vol. 266, no. 9-10, pp. 995–1002, 2009.

[108] L. Yang, B. Li, J. Yao, and Z. Li, "Effects of diamond size on the deposition characteristic and tribological behavior of diamond/Ni60 composite coating prepared by supersonic laser deposition," *Diamond and Related Materials*, vol. 58, pp. 139–148, 2015.

[109] J. Yao, L. Yang, B. Li, and Z. Li, "Beneficial effects of laser irradiation on the deposition process of diamond/Ni60 composite coating with cold spray," *Applied Surface Science*, vol. 330, pp. 300–308, 2015.

[110] S. Yin, Y. Xie, J. Cizek et al., "Advanced diamond-reinforced metal matrix composites via cold spray: properties and

deposition mechanism," *Composites Part B: Engineering*, vol. 113, pp. 44–54, 2017.

[111] Y. Xie, S. Yin, C. Chen, M. Planche, H. Liao, and R. Lupoi, "New insights into the coating/substrate interfacial bonding mechanism in cold spray," *Scripta Materialia*, vol. 125, pp. 1–4, 2016.

[112] W. J. Zhang, Y. M. Chong, I. Bello, and S. T. Lee, "Nucleation, growth and characterization of cubic boron nitride (cBN) films," *Journal of Physics D: Applied Physics*, vol. 40, no. 20, pp. 6159–6174, 2007.

15

Effects of Aluminum Powder on Ignition Performance of RDX, HMX, and CL-20 Explosives

Xiaoxiang Mao, Longfei Jiang⑩, Chenguang Zhu⑩, and Xiaoming Wang

Nanjing University of Science and Technology, Nanjing 210094, China

Correspondence should be addressed to Chenguang Zhu; zhuchen1967@gmail.com

Academic Editor: Fernando Lusquiños

As a kind of high explosives, aluminized explosive cannot release the energy maximumly, which is a key problem. Using DTA-TG equipment, the ignition performance of three kinds of aluminized explosives (RDX, HMX, and CL-20) with different mass percentages of aluminum powder (0%, 10 wt.%, 20 wt.%, and 30 wt.%) was investigated. The results showed that the energy release of the HMX/Al composite explosive with 10 wt.%, 20 wt.%, and 30 wt.% aluminum powder was only equivalent to 80%, 65%, and 36% of pure HMX, respectively. It was similar to RDX/Al and CL-20/Al composite explosives, except the CL-20/Al mixture with 10% aluminum powder. Rather than participating in the ignition and combustion, the aluminum powder does effect the complete reaction of RDX, HMX, and CL-20 in the initial stage of ignition or in the lower temperature area of the boundary.

1. Introduction

Aluminum powder can be incorporated into explosives to raise the reaction temperature, enhance the heat of detonation, increase bubble energies in underwater weapons, improve air blast, and create an incendiary effect [1]. Therefore, the aluminized explosive is usually called high explosive and has attracted much attention. Vadhe et al. [2] summarized the development trend of aluminized explosives and considered adequately their low sensitivity and good mechanical properties. Tao et al. [3] studied the early dynamic characteristics of Al powder which was heated after the detonation of aluminized explosives. Stromsoe et al. [4] proposed that both the bubble energy and the shock wave energy of aluminized explosives were higher than those of pure explosives in a certain range of Al content. Peng et al. [5] found that the size, the activity, and the shape of Al powder could significantly affect the energy level of aluminized explosives. Hwang et al. [6] reported the application characteristics of nickel-coated Al powder in the explosives, whose results showed that nickel was able to reduce the ignition temperature of Al powder and improved the impulse, the pressure, and the temperature field. However, the above researches highlight a problem that aluminized

explosives show nonideal behavior. The anticipated energy of aluminized explosives is very hard to be utilized fully [7–9]. The combustion of Al powder is incomplete in the actual explosion process, resulting in the incomplete release of energy. Thus, Gogulya et al. [10] tried to promote the reaction of Al powder using Teflon and Viton.

The purpose of the present work is to investigate the effect of Al powder on the ignition property based on thermogravimetric analysis and differential thermal analysis (TG-DTA). The samples are cyclotrimethylene trinitramine (RDX), cyclotetramethylenetetranitramine (HMX), and hexanitrohexaazaisowurtzitane (CL-20), with different mass percentages of Al powder. On this basis, the underlying reason is explored that aluminized explosives showed nonideal behavior in the process of practical application.

2. Experimental

2.1. Reagents and Instruments

2.1.1. Reagents. Al powder (350 mesh, particle size < 40 μm) was obtained from Tangshan Weihao Magnesium Powder Co., Ltd. RDX (80 mesh), HMX (80 mesh), and CL-20 (80 mesh) were obtained from Qingyang Chemical Co., Ltd.

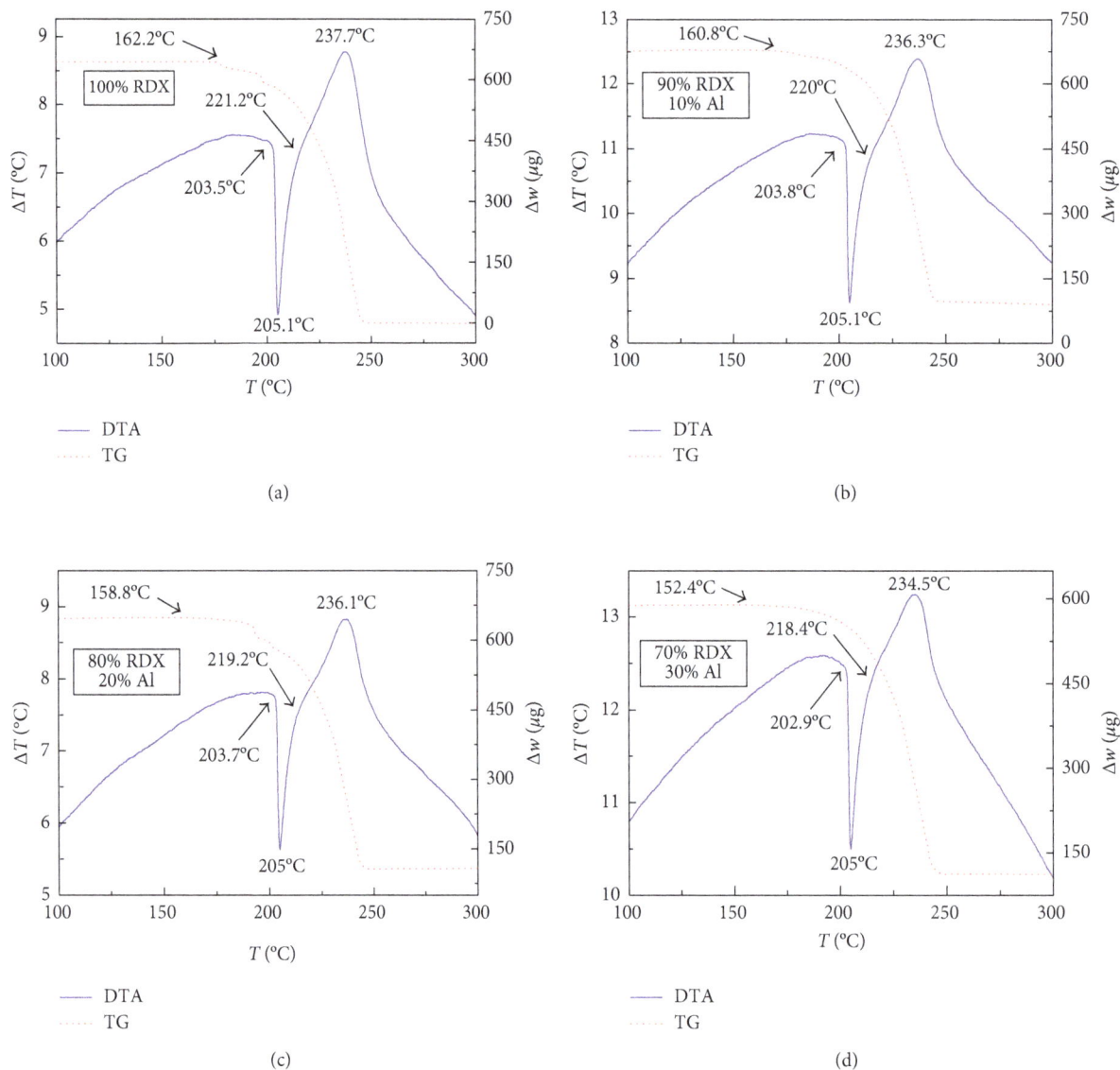

Figure 1: TG-DTA curves of RDX/Al. (a) Pure RDX; (b) RDX 90 wt.% and Al 10 wt.%; (c) RDX 80 wt.% and Al 20 wt.%; (d) RDX 70 wt.% and Al 30 wt.%.

2.1.2. Instrument. DTA-50-type differential thermal analyzer (NETZSCH-Gerätebau GmbH) and Hot Disk TPS 2500S thermal conductivity meter (Swedish Kaigenasi Company) were used.

2.2. Sample Preparation. Mix different mass percentages of Al powder (0%, 10 wt.%, 20 wt.%, and 30 wt.%), respectively, with RDX, HMX, and CL-20 uniformly.

2.3. TG-DTA Parameters. Differential thermal analysis was performed on STA449C high-temperature thermal analyzer. Experimental conditions are as follows: sample weight was 0.5 mg, carrier gas was air, gas velocity was 100 mL/min, Al_2O_3 crucibles, and gradient heating at a constant rate of 10°C/min ranging from normal temperature to 600°C. The α-Al_2O_3 was used as the reference sample to correct the reference

instrument, and the crucibles were exposed during the whole process. The equipment was corrected before each test.

3. Results and Discussion

3.1. Ignition Temperature. Figures 1–3 show the TG-DTA curves of the three groups of samples, corresponding to the formulations of RDX/Al, HMX/Al, and CL-20/Al, respectively.

According to the TG-DTA curves, the first strong exothermic peak of the DTA curves generally corresponds to a significant loss of weight in TG curve, which can be extrapolated by ignition temperature (Te). The Te of the three groups of samples is shown in Table 1 based on the results of Figures 1–3. As shown in Table 1, the changes of Te are not obvious after adding different mass percentages of Al powder, which means that Al powder does not change the Te of these three kinds of explosives.

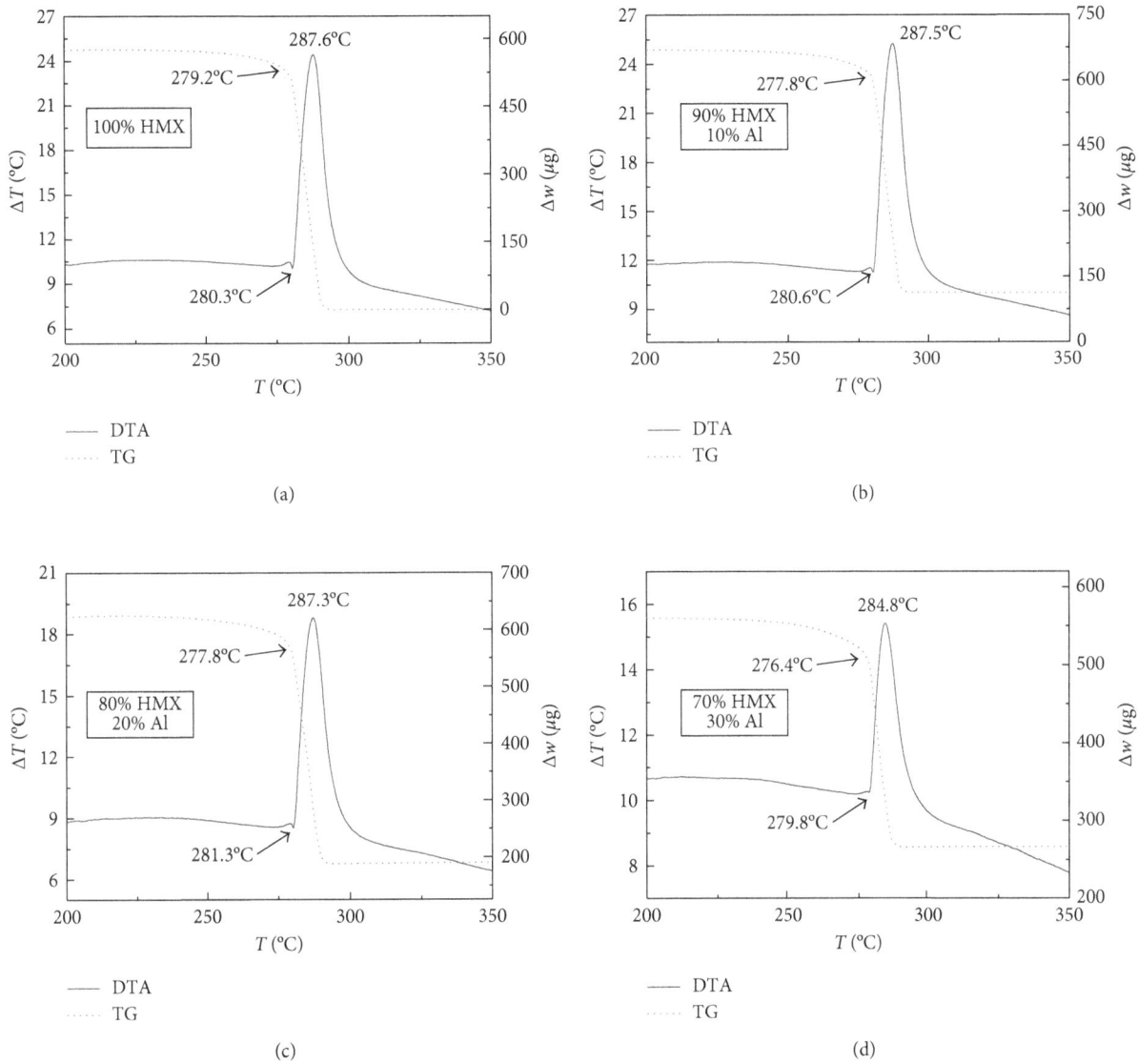

FIGURE 2: TG-DTA curves of HMX/Al. (a) Pure HMX; (b) HMX 90 wt.% and Al 10 wt.%; (c) HMX 80 wt.% and Al 20 wt.%; (d) HMX 70 wt.% and Al 30 wt.%.

3.2. Thermal Conductivity.

The thermal conductivity tests of samples were carried out after the samples were pressed into flakes. The thermal conductivity of pure Al and shell steel was measured simultaneously, and the testing values were 226 $W \cdot m^{-1} \cdot K^{-1}$ and 8.25 $W \cdot m^{-1} \cdot K^{-1}$.

The fitting results of testing values are shown in Figure 4, indicating the variation trend of the thermal conductivity with the mass percentage change of Al powder. We find that the thermal conductivities of different explosives are all improved gradually as the mass percentage of Al powder increases, originating from the higher thermal conductivities of Al.

3.3. Calculation Formula.

The weight of the samples was normalized to directly evaluate the effects of different mass percentages of Al powder on the energy release of explosives.

3.3.1. Mass Normalization Formula

$$M_s = \frac{1000}{n} \times \sum_{i=1}^{n} \frac{M_i}{M_0}, \qquad (1)$$

where M_s is the normalized mass, M_0 is the original mass, M_i is the dynamic mass measured by TG-DTA device, n is the number of data points, and 1000 is the reference value.

3.3.2. Theoretical Exothermic Enthalpy Percentage of Composite Explosives

$$\Delta H\% = 1 + \left(\frac{\Delta H_{Al}}{\Delta H_{explosive}} - 1 \right) \times p, \qquad (2)$$

where ΔH_{Al} and $\Delta H_{explosive}$ are the standard exothermic enthalpy of Al and pure explosives, respectively (Table 2) and p is the mass percentage of Al.

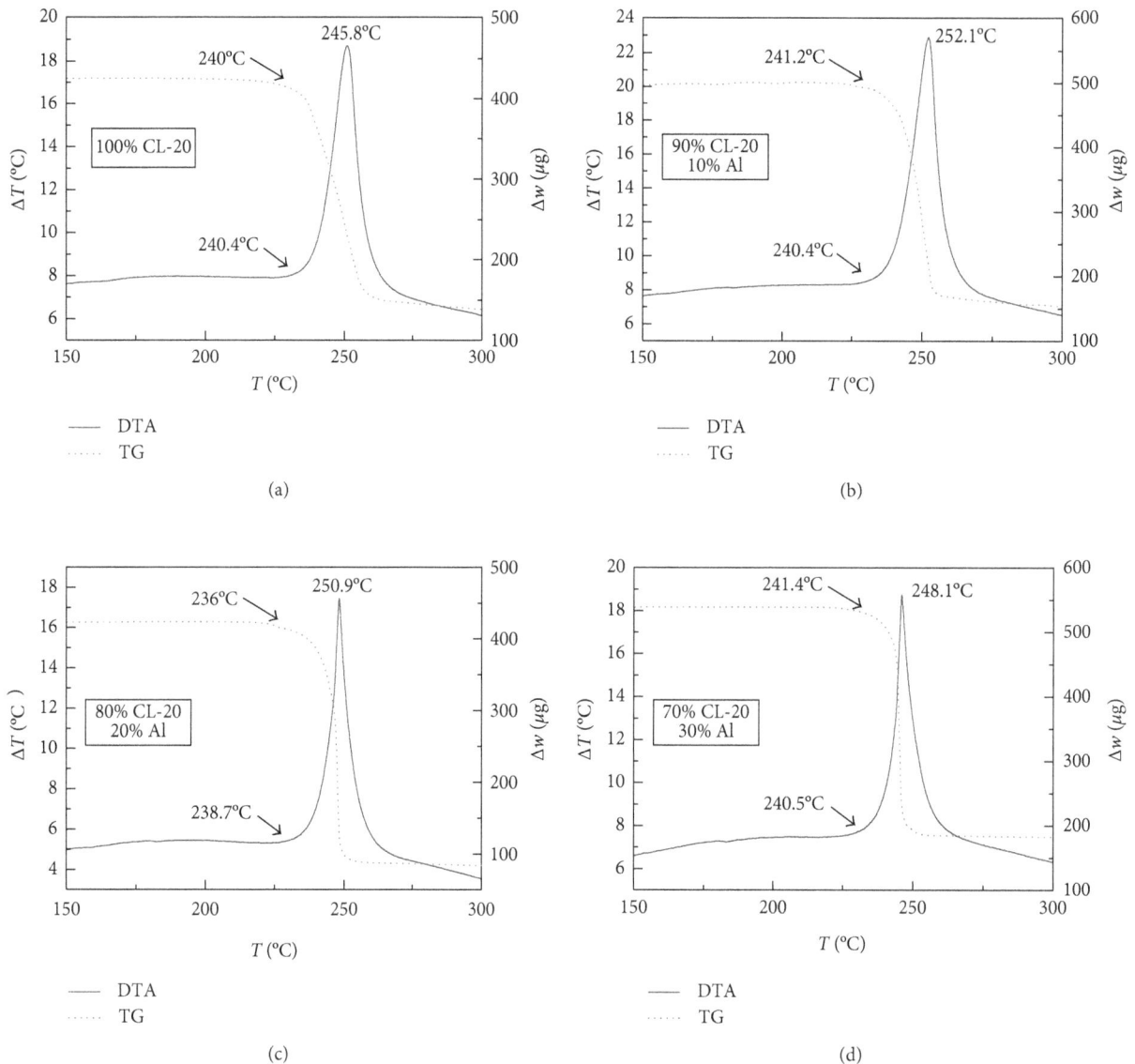

FIGURE 3: TG-DTA curves of CL-20/Al. (a) Pure CL-20; (b) CL-20 90 wt.% and Al 10 wt.%; (c) CL-20 80 wt.% and Al 20 wt.%; (d) CL-20 70 wt.% and Al 30 wt.%.

TABLE 1: Ignition temperature of the three groups of samples.

Sample	Te (°C)			
	Al-0%	Al-10 wt.%	Al-20 wt.%	Al-30 wt.%
RDX/Al	205.1	205.1	205.0	205.0
HMX/Al	280.3	280.6	281.3	279.8
CL-20/Al	240.4	240.4	238.7	241.5

3.4. Ignition Release Energy. After correcting the baseline of curves in Figures 1–3 and normalizing the mass of samples, the results are shown in Figures 5–7. Notably, Figure 5 presents the DTA curves of RDX mixed with different mass percentages of Al powder, in which the exothermic peaks reduce gradually with the increasing mass percentages of Al powder. The declining trend of exothermic peaks is more obvious in the DTA curve of HMX/Al shown in Figure 6.

However, CL-20 exhibits different characteristics compared with those of RDX and HMX shown in Figure 7, of which exothermic peaks change irregularly.

Table 3 lists the exothermic enthalpy data (ΔH) [11, 12] of the three groups of samples which can be calculated by integral processing of exothermic peaks in DTA curves.

For comparison, the relative exothermic enthalpy percentage of samples is calculated based on the exothermic

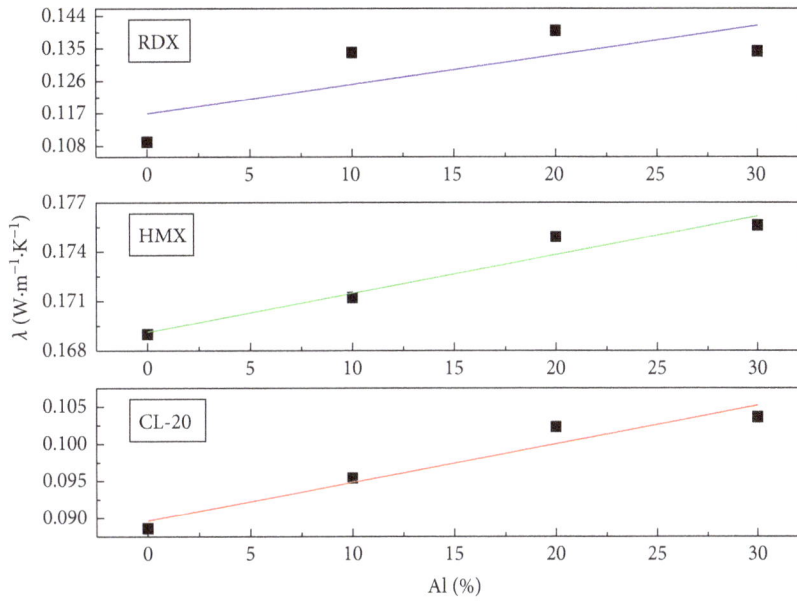

FIGURE 4: Thermal conductivity of the three groups of samples (testing temperature was 25°C).

TABLE 2: Standard exothermic enthalpy of pure substances.

Pure substance	ΔH (kJ·kg^{-1})
RDX	5400
HMX	5673
CL-20	7100
Al	30,222.22

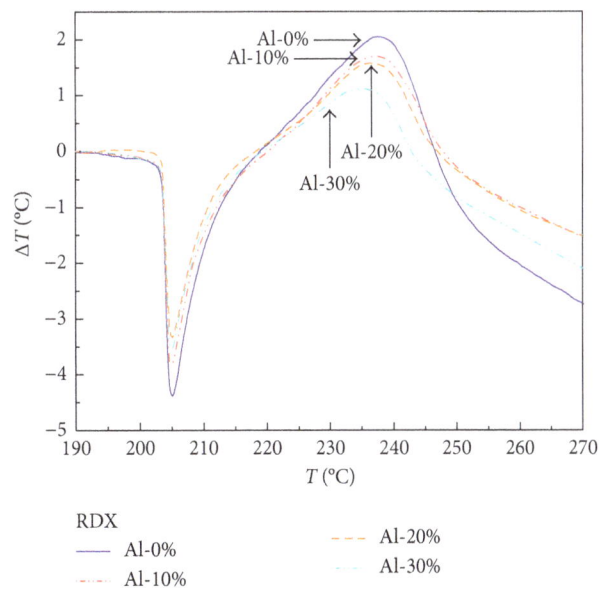

RDX

FIGURE 5: DTA curves of RDX with different mass percentages of Al powder.

enthalpy of pure explosives listed in Table 4. The data in parentheses are theoretical exothermic enthalpy percentages of composite explosives.

It can be seen from Table 4 that the released energy of the RDX/Al composite explosive with 10 wt.% Al powder (i.e., containing 90 wt.% RDX) is merely 74.4% of pure RDX,

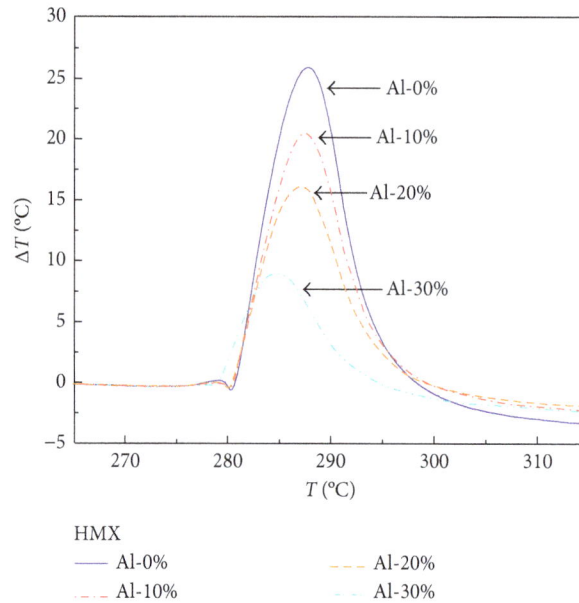

FIGURE 6: DTA curves of HMX with different mass percentages of Al powder.

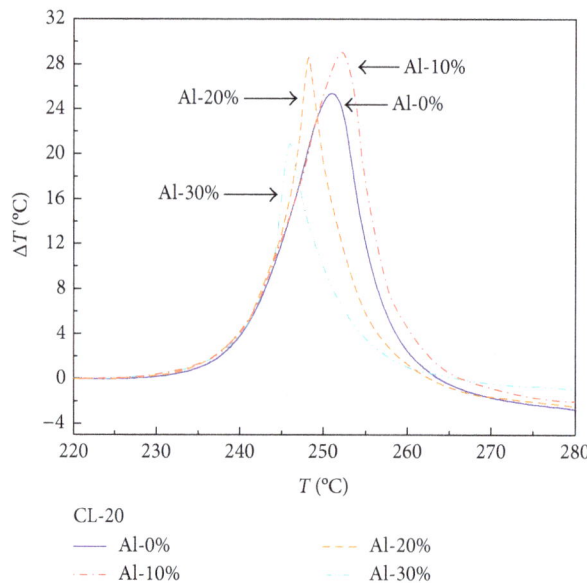

FIGURE 7: DTA curves of CL-20 with different mass percentages of Al powder.

which means the Al powder does not react and even block the complete reaction of RDX. The gap between the actual value and the theoretical value widens as the mass percentage of Al powder increases. The released energy of RDX/Al with 20 wt.% Al is merely 61.5% of pure RDX, and the value of RDX/Al with 30 wt.% Al is only 40.8%. The HMX/Al composite explosive exhibits the same rule as RDX/Al. The released energy of HMX/Al with 10 wt.% Al is merely 80.3% of pure HMX, not even reaching 90%, and the values of HMX/Al with 20 wt.% and 30 wt.% Al are 64.8% and 36.1%, respectively. These results illustrate that the Al powder in RDX/Al and HMX/Al composite explosives

TABLE 3: Exothermic enthalpy data of the three groups of samples.

Sample	ΔH (uV·s·mg^{-1})			
	Al-0%	Al-10 wt.%	Al-20 wt.%	Al-30 wt.%
RDX/Al	−332.2	−247.3	−204.3	−135.4
HMX/Al	−1483	−1191	−960.6	−535.0
CL-20/Al	−2120	−2230	−1684	−1169

cannot burn at the initial stage of ignition but impact the burning property of explosives around the Al particles.

Unlike RDX and HMX, the released energy of CL-20/Al composite explosive with 10 wt.% Al powder is 105.2% of

TABLE 4: Exothermic enthalpy percentage of the three groups of samples.

Sample	$\Delta H\%$			
	Al-0%	Al-10 wt.%	Al-20 wt.%	Al-30 wt.%
RDX/Al	100 (100)	74.4 (145.97)	61.5 (191.93)	40.8 (237.9)
HMX/Al	100 (100)	80.3 (143.27)	64.8 (186.55)	36.1 (229.82)
CL-20/Al	100 (100)	**105.2 (132.57)**	79.4 (165.13)	55.1 (197.7)

pure CL-20, exceeding 90% but lower than the theoretical value (132.57%), indicating that part of the Al powder joined the reaction and released energy. However, as the mass percentage of Al powder increases, the released energy is reduced obviously. The released energy of CL-20/Al with 20 wt.% and 30 wt.% Al is merely equal to 79.4% and 55.1% of pure CL-20, respectively, which suggests that part of CL-20 does not react. This may be attributed to the higher thermal conductivity of Al, which makes part of Al particles dissipate the heat by themselves and CL-20 around them simultaneously, resulting in unachievable reaction temperature condition and the end of the reaction.

The result of this experiment indicates that Al particles in RDX and HMX cannot react under low-temperature condition or the ignition temperature condition. Due to the high melting point (660.4°C), boiling point (2467°C), and low saturated vapor pressure, Al is difficult to volatilize [13], which makes Al nonreactive in gaseous form under the low-temperature condition. Although explosives release energy in the form of high temperature, high pressure, and high-speed detonation, the process could still be regarded as a continuous ignition process. Due to low temperature, the Al particles far from the explosion core do not react, and at the same time, it dissipates the heat around, which makes incomplete combustion of reactants.

Therefore, to make Al particles and explosives around them combust completely and fast in the microscale, they should be under the high-temperature condition or in the existence of an urge medium such as CL-20 and ammonium perchlorate [10, 12]. A further study can be carried out to investigate the mechanism of CL-20 promoting the combustion of Al particles in ignition process.

4. Conclusions

This study focuses on the ignition performance of RDX, HMX, and CL-20 with different mass percentages of Al powder. It is found that only 80%, 65%, and 36% of HMX reacted for the energy release from the HMX/Al composite explosive with 90%, 80%, and 70% HMX, respectively. Moreover, the RDX/Al composite explosive presents even lower energy release. In the initial stage of ignition, aluminum powder can obviously affect the reaction of RDX or HMX in the composite explosives. The experiment found that the energy release of the CL-20/Al composite explosive with 10 wt.% Al powder was 5% higher than that of pure CL-20, which meant part of aluminum powder participated in chemical reactions in the ignition stage. The conclusions are as follows:

(1) There are no obvious effects of Al powder ($\leq$30 wt.%) on the ignition temperature of RDX, HMX, and CL-20.

(2) Al powder blocks the energy release processes of RDX and HMX evidently and may not react at the initial stage of ignition.

(3) The CL-20/Al composite explosive with 10 wt.% Al powder can release more energy than that of pure CL-20, which may be caused by the energy release of partial Al powder promoted by CL-20.

The thermal conductivity of aluminized explosives increases accordingly with the increasing mass percentage of Al powder, which accelerates the heat dissipation. Therefore, part of the explosives in touch with the Al particle is not easy to approach spontaneous reaction temperature compared with the others, resulting in termination of reaction or time difference of energy release on the chemical reaction.

Conflicts of Interest

The authors declare that they have no conflicts of interest.

Acknowledgments

This work was supported by the National Science Foundation of China (approval number 51676100).

References

[1] R. A. Schaefer and S. M. Nicolich, "Development and evaluation of new high blast explosives," in *Proceeding of the 36th International Annual Conference of ICT*, Karlsrube, Germany, June-July 2005.

[2] P. P. Vadhe, R. B. Pawar, R. K. Sinha, S. N. Asthana, and A. Subhananda Rao, "Cast aluminized explosives (review)," *Combustion, Explosion, and Shock Waves*, vol. 44, no. 4, pp. 461–477, 2008.

[3] W. C. Tao, C. M. Tarver, J. W. Kury, and D. L. Ornellas, "Understanding composite explosive energetic: 4. Reactive flow modeling of aluminium reaction kinetics using normalized product equation of state," in *Proceedings of Tenth Symposium (International) on Detonation*, Annapolis, MD, USA, Naval Surface Weapons Center, 1993.

[4] E. Stromsoe and S. Eriksen, "Performance of high explosives in underwater applications. Part 2: aluminized explosives," *Propellants, Explosives, Pyrotechnics*, vol. 15, no. 2, pp. 52-53, 1990.

[5] J. Peng, W. Chen, H. Su et al., "The effect of aluminum powder upon aluminized explosive underwater detonation

performance," *Journal of Safety and Environment*, vol. 4, pp. 177–179, 2004.

[6] J. S. Hwang, C. K. Kim, and J. R. Cho, "Development of new thermobaric explosive composition using nickel coated aluminium powder," in *Proceedings of the 38th International Annual Conference of ICT*, Karlsruhe, Germany, June 2007.

[7] X. Wang, "Developmental trends in military composite explosive," *Chinese Journal of Explosives and Propellants*, vol. 34, no. 4, pp. 1–4, 2011.

[8] W. Arnold and E. Rottenkolber, "Thermobaric charges: modelling and testing," in *Proceedings of the 38th International Annual Conference of ICT*, Karlsruhe, Germany, June 2007.

[9] W. Arnold and E. Rottenkolber, "Combustion of an aluminized explosive in a detonation chamber," in *Proceedings of the 39th International Annual Conference of ICT*, Karlsruhe, Germany, June 2008.

[10] M. F. Gogulya, M. N. Makhov, M. A. Brazhnikov, and A. Y. Dolgoborodov, "Detonation-like process in Teflon/Al-based explosive mixture," in *Proceedings of the 40th International Annual Conference of ICT*, Karlsruhe, Germany, June 2009.

[11] P. B. Joshi, G. R. Marathe, N. S. S. Murti, V. K. Kaushik, and P. Ramakrishnan, "Reactive synthesis of titanium matrix composite powders," *Materials Letters*, vol. 56, no. 3, pp. 322–328, 2002.

[12] C. Zhu, H. Wang, and M. Li, "Ignition temperature of magnesium powder and pyrotechnic composition," *Journal of Energetic Materials*, vol. 32, no. 3, pp. 219–226, 2014.

[13] M. M. Avedesian and H. Baker, *Magnesium and Magnesium Alloys*, ASM International, Novelty, OH, USA, 1999.

16

Interactive Relationship between Silver Ions and Silver Nanoparticles with PVA Prepared by the Submerged Arc Discharge Method

Kuo-Hsiung Tseng ⓘ,[1] **Chih-Ju Chou,**[1] **To-Cheng Liu,**[1] **Der-Chi Tien** ⓘ,[1] **Tong-chi Wu,**[1] **and Leszek Stobinski**[2]

[1]*Department of Electrical Engineering, National Taipei University of Technology, Taipei 10608, Taiwan*
[2]*Materials Chemistry, Warsaw University of Technology, Warynskiego 1, 00-645 Warsaw, Poland*

Correspondence should be addressed to Kuo-Hsiung Tseng; khtseng@ee.ntut.edu.tw

Academic Editor: Claudio Pettinari

This study uses the submerged arc discharge method (SADM) and the concentrated energy of arc to melt silver metal in deionized water (DW) so as to prepare metal fluid with nanoparticles and submicron particles. The process is free from any chemical agent; it is rapid and simple, and rapid and mass production is available (0.5 L/min). Aside from the silver nanoparticle (Ag^0), silver ions (Ag^+) exist in the colloidal Ag prepared by the system. In the preparation of colloidal Ag, polyvinyl alcohol (PVA) is used as an additive so that the Ag^0/Ag^+ concentration, arcing rate, peak, and scanning electron microscopic (SEM) images in the cases with and without PVA can be analyzed. The findings show that the Ag^0/Ag^+ concentration increases with the addition level of PVA, while the nano-Ag and Ag^+ electrode arcing rate rises. The UV-Vis absorption peak increases Ag^0 absorbance and shifts as the dispersity increases with PVA addition. Lastly, with PVA addition, the proposed method can prepare smaller and more amounts of Ag^0 nanoparticles, distributed uniformly. PVA possesses many distinct features such as cladding, dispersion, and stability.

1. Introduction

The production of nanosized metallic silver particles with different morphologies and sizes using different methods has been reported in previous studies, such as the electric spark discharge system (ESDS) [1–3]. PVA (Mw 89,000–98,000, 99+% hydrolyzed, 341584, 9002-89-5, MDL: MFCD00081922) was the eligible polymer since it stands out for its viscoelastic behavior, hydrophilicity, chemical stability [4] and biocompatibility [5]. PVA contains a large amount of –OH functional groups, which can form chelate composed of metal ions [6]. It is a white powdered resin polymer, with features of viscoelasticity, hydrophilicity, chemical stability, and biocompatibility [7–9]. PVA is used in many medical devices approved by the FDA (Food and Drug Administration, USA) such as contact lenses, membranes, drug delivery systems, and orthopedic devices [10], It is extensively used in biomedical and pharmaceutical applications [11, 12], waste water treatment, and artificial articular cartilages [13]. Initially, we used a biocompatible polymer, PVA, as a reducing agent to convert silver salt ($AgNO_3$) to Ag-NP. Then, we obtained PVA/Ag-NP composite nanofibers via electrospinning [14]. A series of monodispersed Ag-NPs with 25, 35, 45, 60, and 70 nm sizes was reported by using PVP as the surfactant [15]. Also, Ag-NPs can be incorporated within biodegradable poly(lactic acid) [16] or deposited onto modified titanium surfaces [17] as an antibacterial scaffold for tissue engineering and medical applications. Poly(vinyl alcohol) (PVA) is a biodegradable polyester that has been investigated extensively as a biomedical material. Many researchers have utilized electrospinning to fabricate nanofibrous PVA scaffolds for use in wound healing [18–20]. PVA and its nanocomposites have found a wide range of industrial applications such as fiber and textile sizing, coating, adhesives, emulsifiers, and film packaging in food and optical holographic industries [21]. Nevertheless, there were only two articles about PVA/TiO_2 nanocomposite prepared by the ultrasonic irradiation

FIGURE 1: The color of the nanosilver particle suspension changed from yellow to dark brown.

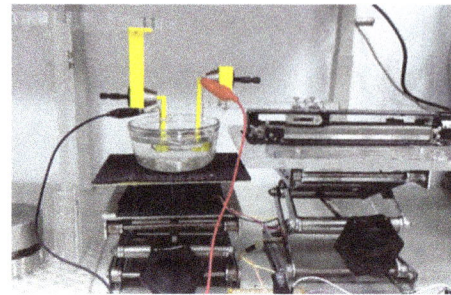

FIGURE 2: Micro-EDM system.

method. To the best of our knowledge [22, 23], high molecular materials, such as PVA [24], by stabilizing with polymer matrix, the silver nanoparticles, are surface modified. Hence, they can also act as capping agents. The homogeneous distribution of silver nanoparticles into the polymer matrix will also increase the surface area and make them fit for catalytic applications. There are several methods to fabricate silver nanoparticle polymer composites [25–27]. Due to PVA's special characteristics, the nano-Ag will distribute uniformly on antibacterial dressings such that the index of the antibacterial dressing becomes more smooth and consistent. The products can be made as a film or gel, or weaved into a fiber.

In recent researches of nanoparticles, it is shown that the color of the Ag nanoparticle, produced by different methods such as presented in literature [28], is getting darker with the increase in concentration, usually from light yellow (since the peak absorption of Ag nanoparticle is 395 nm) to dark brown. In this study, the color of the 10 ppm, 35 ppm, and 60 ppm nanoparticles is light yellow, yellow, and dark brown, respectively, as shown in Figure 1.

2. Research Method and Process

The microelectric discharge machine system comprises the overall mechanism, hardware circuit, power supply, chuck, and so on. The motion control adapter card of the computer is used as the global core. The process performance of parameter setting is observed by changing the arcing rate, thus obtaining the parameter setting for the optimum process efficiency and quality of the nano-Ag colloid. Figure 2 shows the micro-EDM system [29, 30].

2.1. Preparation of Nanosilver by Electrical Spark. This study used SADM to split silver material into nanosized particles by arc discharge. The process was free from chemical agents. Pure solutions, such as pure water, were used as the medium [1].

The silver (99.99% pure) wires with a diameter of 1 mm are used as anode and cathode and are submerged in DW or ethanol. The 200 ml DW is loaded, the positive and negative electrodes are manually fixed and aligned, and then the discharge parameters are set before discharge. The discharge is finished after a period of time, and the product is taken out. The system framework is shown in Figure 3 [3].

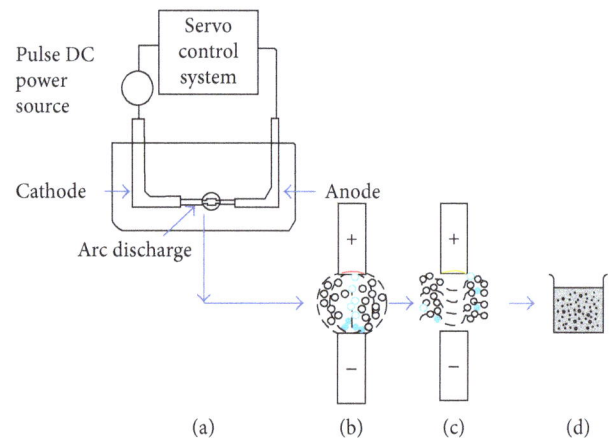

FIGURE 3: The submerged arc discharge method (SADM) system. (a) SADM system configuration. (b) Ionization and plasma formation. (c) Metal bursts out. (d) Metal particle suspended or precipitated in the dielectric liquid.

TABLE 1: EDX application for the analysis of the proportion of elements in silver.

Element	Weight (%)	Atomic (%)
C	44.63	57.26
Si	50.64	27.18
O	13.12	12.63
Ag	11.21	1.60
Na	1.08	0.73

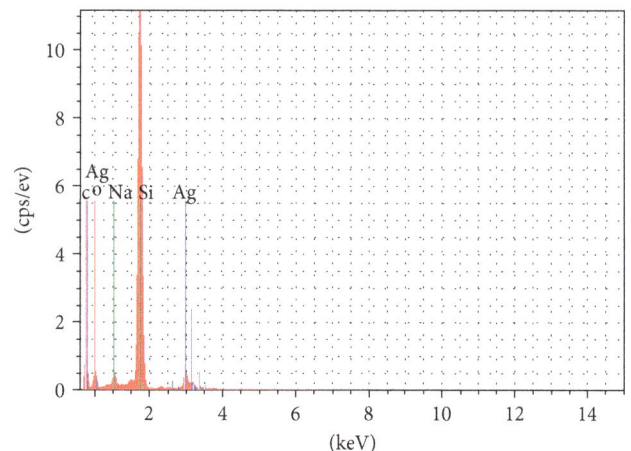

FIGURE 4: Silver nanoparticle EDX diagram.

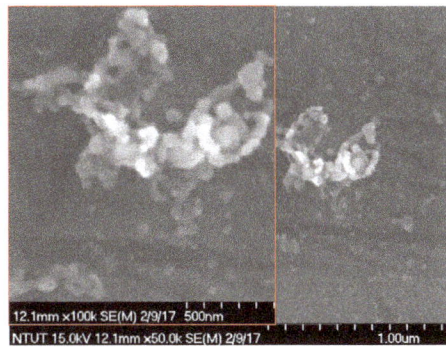

(a)

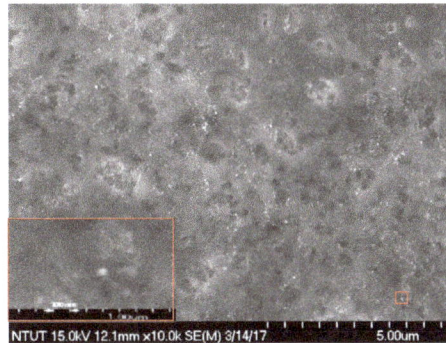

(b)

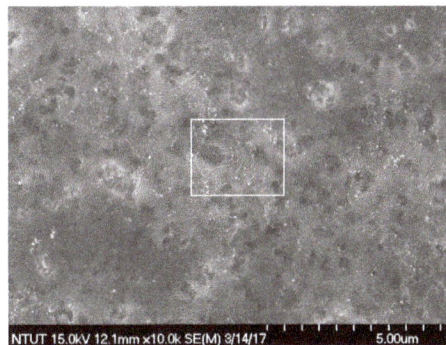

(c)

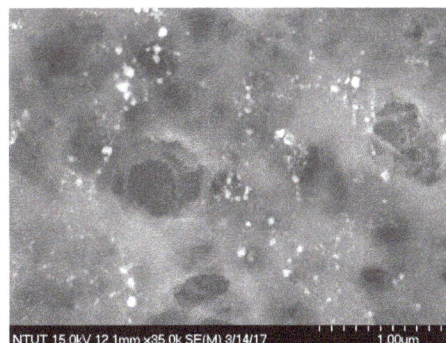

(d)

FIGURE 5: (a) SEM diagram of aggregated silver nanoparticles. (b) SEM diagram of silver + 0.5% PVA particles. (c) 3D structure of nanocolloid distribution. (d) Zoomed in picture of the 3D structure of the nanocolloid.

TABLE 2: The measured electrical conductivity versus UV-Vis Ag^+ absorbance at 190–300 nm band.

+PVA (ml)	Electrical conductivity (μs/cm)			UV-Vis silver ion peak UV (190–300 nm) band Ag^+ absorbance
	A (0 min)	B (6 min)	B − A	
2	1.5	2.4	0.9	0.23
4	12.6	21.6	9.0	2.25
6	19.5	33.66	14.2	2.57
8	34.5	52.7	18.2	2.88
10	34.5	59.5	24.9	3.15

TABLE 3: 2.3 μs/cm (DW) versus 20 μs/cm (with PVA) electrical conductivity basic parameters.

Time/item	0 min (μs/cm)	6 min (μs/cm)	12 min (μs/cm)
Electrical conductivity (water base)	1.5	4.5	—
Electrical conductivity (water + PVA)	17.3	44	60
UV absorbance	0 min	6 min	12 min
UV-Vis 190–300 nm (Ag^+ in PVA base)	1.87	1.2	2.032
UV-Vis 300–600 nm (Ag^0 in PVA base)	0.03	0.814	—
UV-Vis 190–300 nm (Ag^+ peak in water)	0	<190	—
UV-Vis 190–300 nm (Ag^+-PVA^- peak in water)	<190	194	—
UV-Vis 300–600 nm (Ag^0 peak in water)	0	—	394
UV-Vis 300–600 nm (Ag^0-PVA peak in water)	0	—	410
UV-Vis 190–300 nm (Ag^+ in water base)	0	0.227	—
UV-Vis 190–300 nm (Ag^+ in PVA base)	0	2.85	—
UV-Vis 300–600 (Ag^0 in water base)	0	0.247	—
UV-Vis 300–600 (Ag^0 in PVA base)	0	0.899	2.88

2.2. Silver Nanofluid Product Analysis. This study uses DW as a liquid medium to prepare Ag particles. The prepared sample is analyzed by spectrophotometry for the spectral characteristic of the product. The absorbance of nano-Ag colloid is analyzed by spectrophotometry based on the concentration index of nanoparticles. The conductivity of dielectric fluid is measured by a conductivity meter, the purity of dielectric fluid is guaranteed, and the conductivity of general DW is lower than 5.00 μs/cm.

Nanosilver fluid is fabricated. For energy dispersive X-ray spectroscopy (EDX) analysis, a silicon wafer was used as a carrier. Carbon, oxygen, and sodium were typically accompanied with the PVA, and therefore, Ag, C, O, and Na were present. Table 1 and Figure 4 display the EDX analysis results, and SEM analysis is shown in Figures 5(a)–5(d).

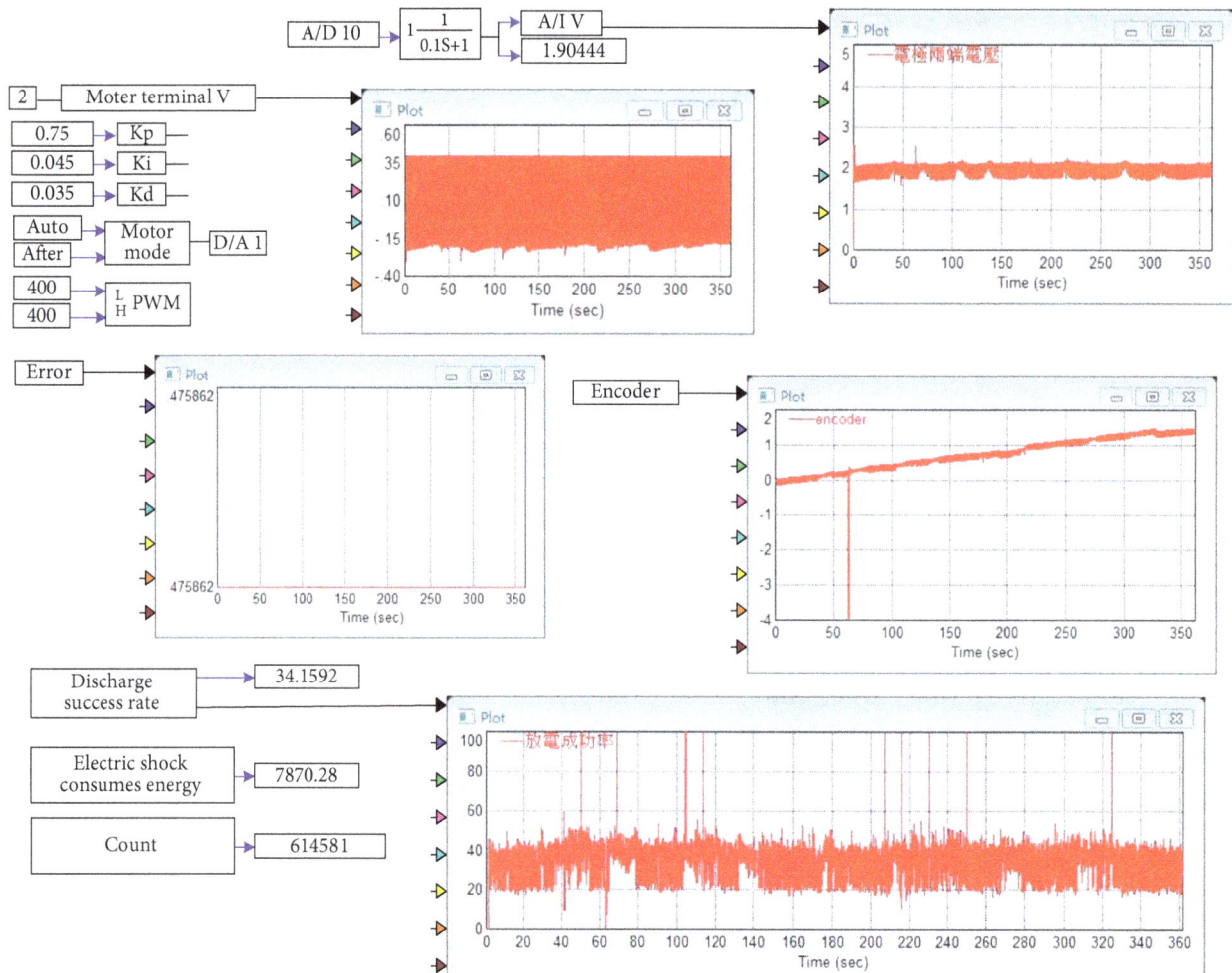

FIGURE 6: Diagram of Vissim prog. (arcing in DW for 6 min).

3. Experimental Results and Discussion

3.1. Experimental Procedure. The 2.3 μs/cm Ag⁺ solution (without PVA) and 20 μs/cm Ag⁺ solution (with 10 ml PVA) preparation processes are designed, and the nano-Ag is then prepared at intervals of 6 min. The correlation between Ag⁺ and nano-Ag of the nano-Ag colloid and the concentration in the preparation process are defined by electrical conductivity and spectrophotometry (UV-Vis).

3.1.1. Silver Ions. 200 ml DW is mixed with 2 ml, 4 ml, 6 ml, 8 ml, and 10 ml of the PVA, respectively. Ag is driven in 5 min, and the electrical conductivity is measured after the PVA is added to DW. When the PVA addition is 8 ml and 10 ml, the electrical conductivity does not distinctly increase, approaching the saturation concentration, as shown in Table 2.

3.1.2. Silver Ion Solution (with/without PVA). The 2.3 μs/cm Ag⁺ solution (DW) and 20 μs/cm Ag⁺ solution (with 10 ml PVA) preparation processes are designed. Then, the nano-Ag is prepared at intervals of 6 min. The correlation between Ag⁺ and nano-Ag of the nano-Ag colloid and the

concentration in the preparation process are defined by electrical conductivity and spectrophotometry (UV-Vis), as shown in Table 3.

3.2. Test Result Analysis

3.2.1. Vissim and UV-Vis. The laboratory report chart of PVA additive in Ag nanoparticle preparation (Vissim and UV-Vis) is shown in Figures 6–11.

3.2.2. Zeta Potential and Size Distribution. The laboratory report chart of PVA additive in Ag nanoparticle preparation (zeta potential and size distribution) is shown in Figures 12–15.

3.3. Results and Discussion. The experimental discussion about the interactive relationship between Ag⁺/nano-Ag and PVA is described below.

(1) With PVA, when 8 ml (52.7 μs/cm) or 10 ml (59.5 μs/cm) of DW (200 ml) is added, the electric conductivity has no significant change, meaning that it has reached saturation.

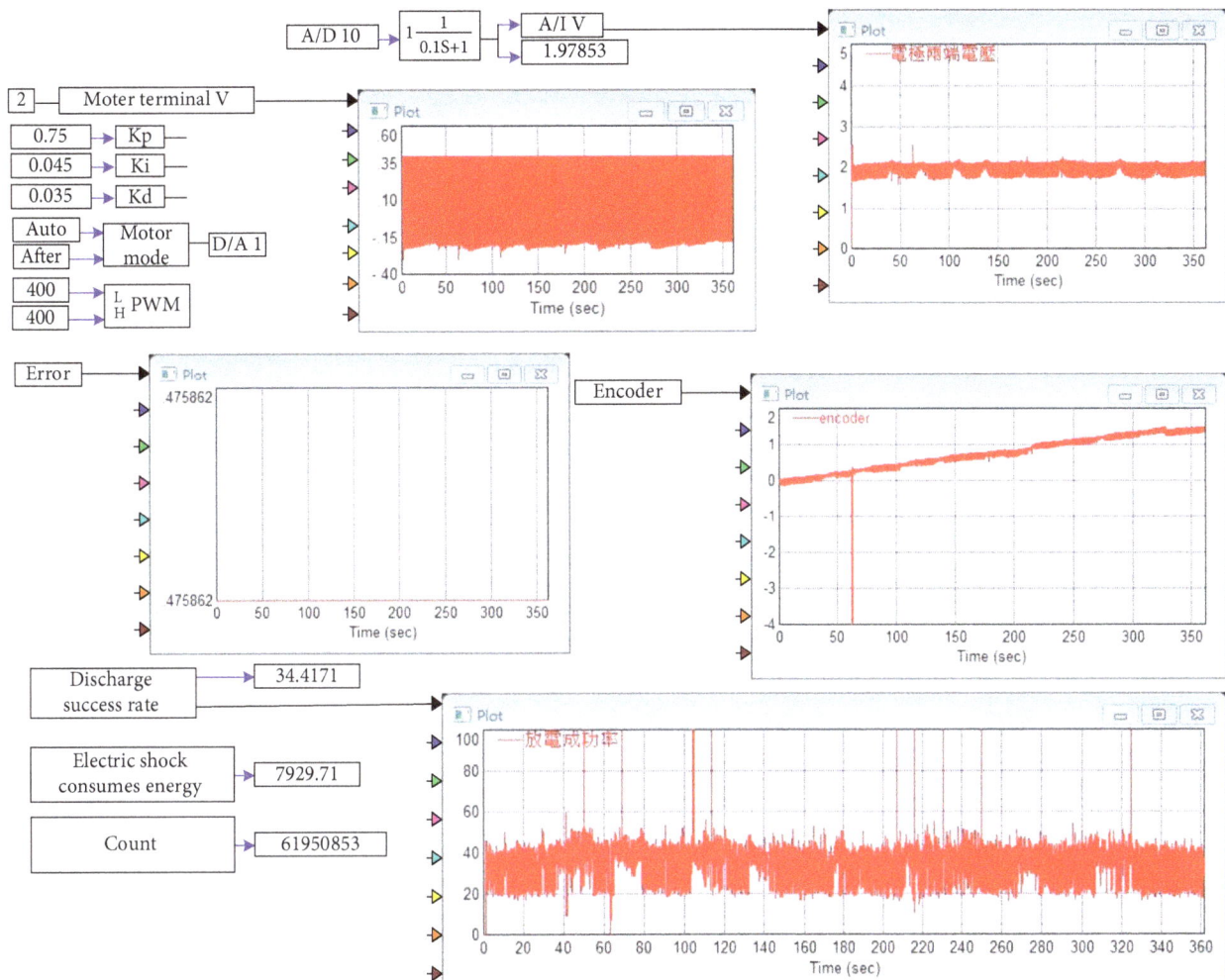

FIGURE 7: Diagram of Vissim prog. (arcing in DW + PVA for 12 min).

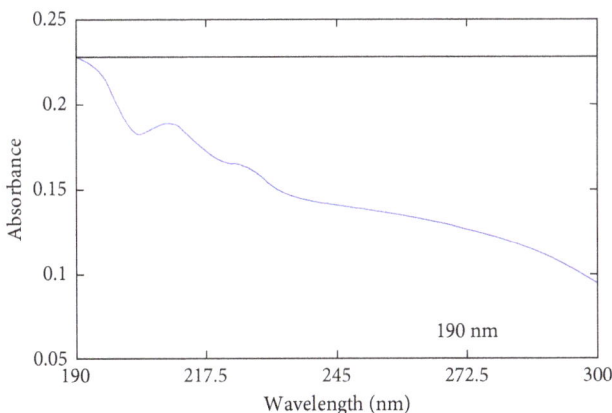

FIGURE 8: UV-Vis (arcing in DW for 6 min, peak at 190 nm).

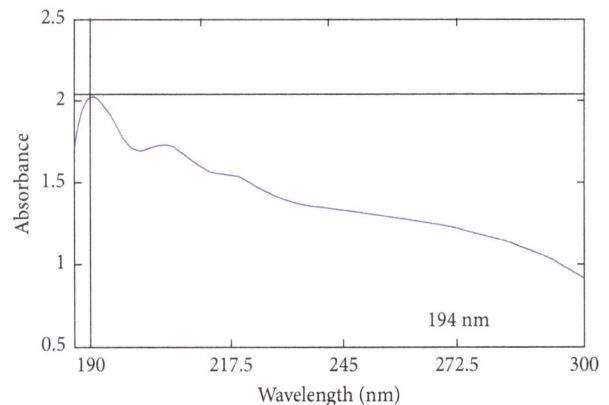

FIGURE 9: UV-Vis (arcing in DW + PVA for 12 min, peak at 194 nm).

(2) During the discharge process, since the PVA covers the Ag particle and increases the surface charge of the Ag particle, it is seen that, from Figures 6 and 7, the discharging efficiency rises with the increase in the PVA concentration. The diameter of the Ag particle is

30 nm, and it remains 30 nm after the PVA covers the Ag particle; the overall complex may be larger than 30 nm after cladding. This complex is relatively big, and the distance between particles is relatively long. The overall complex is connected after cladding, the complex distance increases, two complexes cannot be

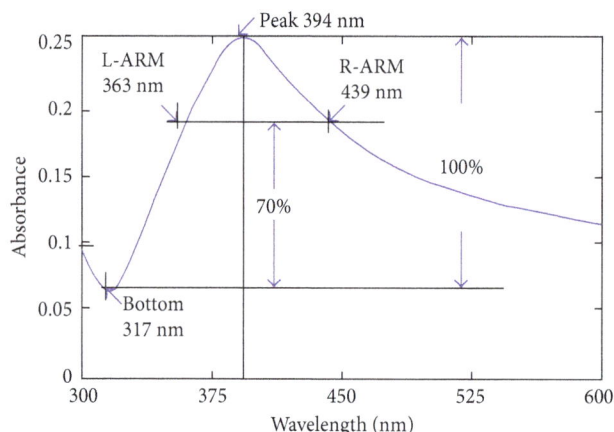

FIGURE 10: UV-Vis (arcing in DW for 6 min, band width = 76 nm, at absorbance 70% height).

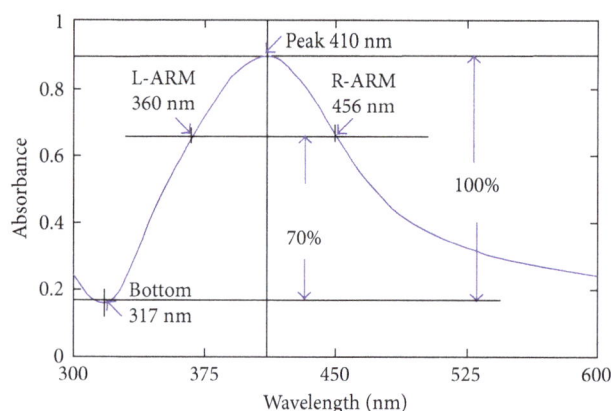

FIGURE 11: UV-Vis (arcing in DW + PVA for 12 min, band width = 96 nm, at absorbance 70% height).

Results			
	Mean (mV)	Area (%)	Width (mV)
Zeta potential (mV) : −26.4	Peak 1: −27.3	90.2	13.0
Zeta deviation (mV): 21.7	Peak 2: 5.70	8.1	4.96
Conductiviy (mS/cm): 0.0186	Peak 3: −82.4	1.0	6.48
Result quality: see result quality report			

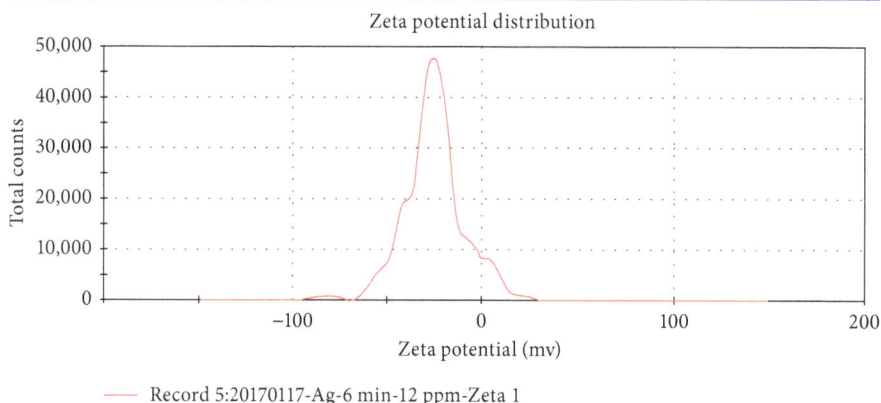

—— Record 5:20170117-Ag-6 min-12 ppm-Zeta 1

FIGURE 12: Zeta of aqueous colloid silver (arcing in DW for 6 min).

connected, the covered complex appearance is negatively charged, and each covered complex is negatively charged. The repulsive effect is generated during mutual collision, whereby the negative electric field on the covered complex appearance is higher than the original surface electric field, from 30 mv to 40–45 mv; and the negative electric field on the covered complex appearance is strong so that the covered nano-Ag particles are farther from each other in liquid or colloid. The following situation is tenable. In the same electric field, the farther the Ag particle is, the more unlikely the Ag particle is to cause a short-circuit bridge. As long as the short-circuit bridge does not occur in the discharge process, the controller makes a continuous electrode discharge, and thus, a lot of nano-Ag^0 and Ag^+ are generated so that the electrode consumption per unit time increases.

(3) According to Figure 8, the UV-Vis absorbance peak of DW discharge is driven in $Ag^+ < 190$ nm and the peak of absorption is a waveform < 190 nm (near 190 nm). According to Figure 9, the UV-Vis absorption peak of DW + PVA discharge is driven in $Ag^+ = 194$ nm. According to Figures 8 and 9, the UV absorbance peak of DW + PVA shifts right. Carefully inspecting Figures 8 and 9, it is obvious that Ag^+ may shift right under the addition of PVA introduced. This is the first breakthrough of the current study.

(4) According to Figure 10, in the case of DW-Ag^0 (6 min), Ag^0 surrounds the AgOH and Ag_2O, and the corresponding wavelength of UV-Vis absorbance peak is 394 nm. As shown in Figure 11, when the Ag^0 surrounds the Ag^+-PVA^- compound, the corresponding wavelength of absorption peak in the UV-Vis (300–600) spectrum is 410 nm, and the UV-Vis

Results

	Mean (mV)	Area (%)	Width (mV)
Zeta potential (mV): −38.6	Peak 1: −38.2	98.5	11.0
Zeta deviation (mV): 11.6	Peak 2: −74.3	1.5	4.10
Conductivity (mS/cm): 0.0514	Peak 3: 0.00	0.0	0.00
Result quality: good			

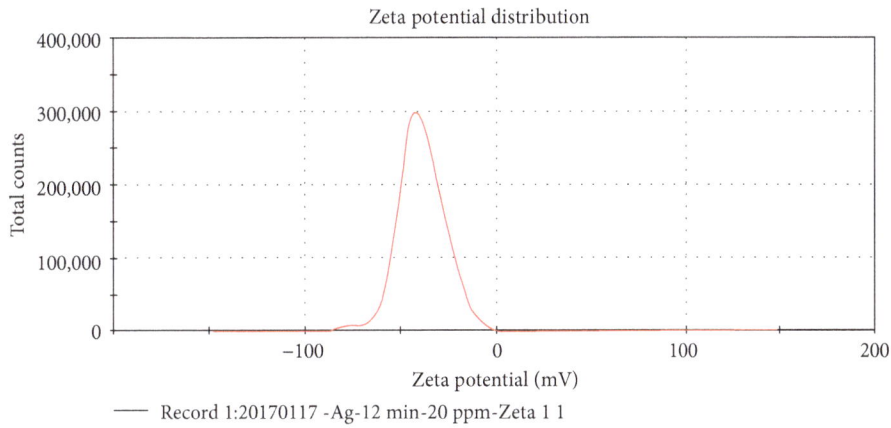

FIGURE 13: Zeta of aqueous colloid silver (arcing in DW + PVA for 12 min).

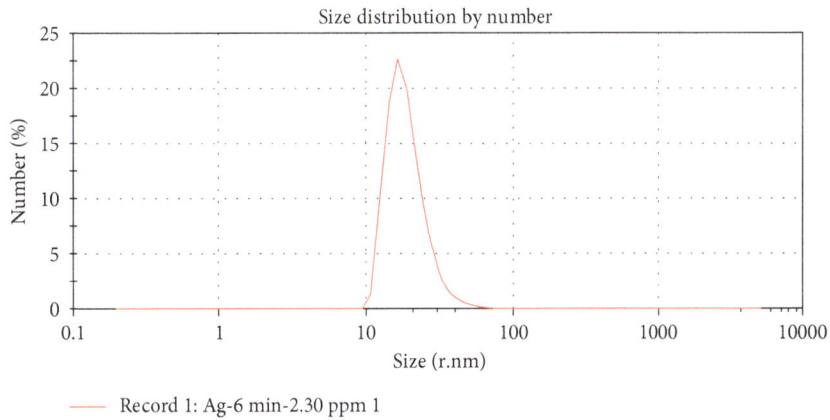

FIGURE 14: Particle size distribution of aqueous colloid silver (arcing in DW for 6 min).

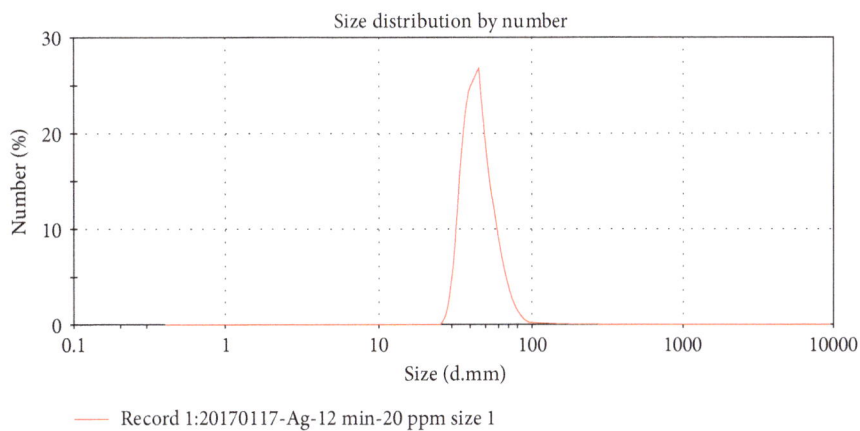

FIGURE 15: Particle size distribution of aqueous colloid silver (arcing in DW + PVA for 12 min).

(190–300) absorption peak increases to at least 1.2. This suggests that the Ag^+ concentration rises to 12 ppm, and 1 ppm in general approximates to 1 μm/cm. The total electrical conductivity is 24 μm/cm, meaning at least 12 μm/cm of the concentration results from Ag^+ and the rest of 12 μm/cm results from the PVA derivant. Figures 10 and 11 indicate that the DW-Ag^0 (6 min) wavelength is 76 nm at absorbance 70%, and the PVA (6 min) wavelength is 94 nm at absorbance 70%. Another new finding is that the particle size is measured by the Zetasizer. It is deduced that the Ag^0 complex is combined with PVA, which leads to increase in the equivalent diameter, thus causing the peak of Ag^0 shifting from 394 nm to 410 nm. Moreover, at the neck of the UV-Vis absorbance wavelength 70% in Figures 10 and 11, the wavelength of DW-Ag^0 (6 min) is 76 nm. The width of the neck of DW + 0.5%/w/w-Ag^0 (6 min) UV-Vis wavelength at 70% is 96 nm. This is because the (Ag^+-PVA^-) complex makes the overall Ag^0 UV-Vis spectrum neck width wavelength increase from 76 nm to 96 nm.

(5) According to Figures 14 and 15, the particle size and distribution are 50–100 nm in the DW solution without PVA. When (0.05%/w/w) PVA is added, the particle diameter is 25–75 nm.

4. Conclusion

The experimental results about the interactive relationship between Ag^+/Nano-Ag and PVA are described below.

(1) According to the increment ratio of electrical conductivity and absorption peak in the experiment, the PVA is correlated to the Ag^0/Ag^+ concentration.

(2) As the electrode discharges continuously, a lot of Ag^0 and Ag^+ are generated, and the electrode consumption per unit time increases, so that the Ag^0 and Ag^+ arcing rate rises.

(3) When Ag^+ is combined with PVA (0.05%/w/w), it forms a ($Ag^+ \cdot PVA^-$) compound. The molecular weight of this compound is much heavier than $(Ag^+)_2O^-$ and Ag^+OH^-, and the absorption peak of Ag^+ shifts right to 194 nm.

(4) The PVA not only helps the formation of Ag^+ but also increases the formation of Ag^0. After the PVA is added, the concentration of Ag^+ increases 5 times (1.2/0.227 = 5.28) and Ag^0 increases 3 times (0.899/0.247 = 3.63). Here, Ag^0 is covered by PVA, increasing its dispersity and forming the Ag^+-PVA^- compound. The neck width of Ag^0 on the UV-Vis absorbance wavelength at the height of 70% shifts from 76 nm to 96 nm, with its Ag^0 peak shifting from 394 nm to 410 nm.

(5) According to SEM image in Figures 5(a) and 5(b), the particle agglomeration is severe before the Ag^0 is mixed with PVA. The particles have irregular shape, and the particle size and distribution are nonuniform.

When 0.5% PVA is added in, the particle size becomes smaller and is distributed over the PVA surface uniformly. Meanwhile, the surface zeta potential of Ag^0 increases, and the catalytic activity rises, so that the Ag-PVA complex is easier to disperse.

Conflicts of Interest

The authors declare that there are no conflicts of interest regarding the publication of this paper.

References

[1] K.-H. Tseng, H.-L. Lee, D.-C. Tien, Y.-L. Tang, and Y.-S. Kao, "A study of antibioactivity of nanosilver colloid and silver ion solution," *Advances in Materials Science and Engineering*, vol. 2014, Article ID 371483, 6 pages, 2014.

[2] K.-H. Tseng, H.-L. Lee, C.-Y. Liao, K.-C. Chen, and H.-S. Lin, "Rapid and efficient synthesis of silver nanofluid using electrical discharge machining," *Journal of Nanomaterials*, vol. 2013, Article ID 174939, 6 pages, 2013.

[3] K.-H. Tseng, C.-J. Chou, T.-C. Liu, Y.-H. Haung, and M.-Y. Chung, "Preparation of Ag-Cu composite nanoparticles by the submerged arc discharge method in aqueous media," *Japan Institute of Metals and Materials Transactions*, vol. 57, no. 3, pp. 294–301, 2016.

[4] R. Tamaki and Y. Chujo, "Synthesis of poly(vinyl alcohol)/silica gel polymer hybrids by in-situ hydrolysis method," *Applied Organometallic Chemistry*, vol. 12, no. 10-11, pp. 755–762, 1998.

[5] S. R. Sudhamani, M. S. Prasad, and K. U. Sankar, "DSC and FTIR studies on gellan and polyvinyl alcohol (PVA) blend films," *Food Hydrocolloids*, vol. 17, no. 3, pp. 245–250, 2003.

[6] L. A. García-Cerda, M. U. Escareno-Castro, and M. Salazar-Zertuche, "Preparation and characterization of polyvinyl alcohol-cobalt ferrite nanocomposites," *Journal of Non-Crystalline Solids*, vol. 353, no. 8–10, pp. 808–810, 2007.

[7] M. I. Baker, S. P. Walsh, Z. Schwartz, and B. D. Boyan, "A review of polyvinyl alcohol and its uses in cartilage and orthopedic applications," *Journal of Biomedial Materials Research Part B: Applied Biomaterials*, vol. 100, no. 5, pp. 1451–1457, 2012.

[8] K. Ng, P. A. Torzilli, R. F. Warren, and S. A. Maher, "Characterization of a macroporous polyvinyl alcohol scaffold for the repair of focal articular cartilage defects," *Journal of Tissue Engineering and Regenerative Medicine*, vol. 8, no. 2, pp. 164–168, 2014.

[9] S. Jiang, S. Liu, and W. Feng, "PVA hydrogel properties for biomedical application," *Journal of the Mechanical Behaviour of Biomedical Materials*, vol. 4, no. 7, pp. 1228–1233, 2011.

[10] B. Bolto, T. Tran, M. Hoang, and Z. Xie, "Crosslinked poly (vinyl alcohol) membranes," *Progress in Polymer Science*, vol. 34, no. 9, pp. 969–981, 2009.

[11] C. M. Hassan and N. A. Peppas, "Structure and applications of poly(vinyl alcohol) hydrogels produced by conventional crosslinking or by freezing/thawing methods," *Advances in Polymer Science*, vol. 153, pp. 37–65, 2000.

[12] S. Mollazadeh, J. Javadpour, and A. Khavandi, "In situ synthesis and characterization of nano-size hydroxyapatite in poly(vinyl alcohol) matrix," *Ceramics International*, vol. 33, no. 8, pp. 1579–1583, 2007.

[13] L. I. Xinming and C. U. I. Yingde, "Study on synthesis and chloramphenicol release of poly(2-hydroxyethylmethacrylate-

co-acrylamide) hydrogels," *Chinese Journal of Chemical Engineering*, vol. 16, no. 4, pp. 640–645, 2008.

[14] A. Celebioglu, Z. Aytac, O. C. O. Umu, A. Dana, T. Tekinay, and T. Uyar, "One-step synthesis of size-tunable Ag nanoparticles incorporated in electrospun PVA/cyclodextrin nanofibers," *Carbohydrate Polymers*, vol. 99, pp. 808–816, 2014.

[15] G. K. Vertelov, Y. A. Krutyakov, O. V. Efremenkova, A. Y. Olenin, and G. V. Lisichkin, "A versatile synthesis of highly bactericidal Myramistin® stabilized silver nanoparticles," *Nanotechnology*, vol. 19, no. 35, p. 355707, 2008.

[16] Y. Zhang, H. Peng, W. Huang, Y. Zhou, and D. Yan, "Facile preparation and characterization of highly antimicrobial colloid Ag or Au nanoparticles," *Journal of Colloid and Interface Science*, vol. 325, no. 2, pp. 371–376, 2008.

[17] L. Li, J. Sun, X. Li et al., "Controllable synthesis of monodispersed silver nanoparticles as standards for quantitative assessment of their cytotoxicity," *Biomaterials*, vol. 33, no. 6, pp. 1714–1721, 2012.

[18] N. Charernsriwilaiwat, P. Opanasopit, T. Rojanarata, and T. Ngawhirunpat, "Lysozyme-loaded, electrospun chitosan-based nanofiber mats for wound healing," *International Journal of Pharmaceutics*, vol. 427, no. 2, pp. 379–384, 2012.

[19] C. Cencetti, D. Bellini, A. Pavesio et al., "Preparation and characterization of antimicrobial wound dressings based on silver, gellan, PVA and borax," *Carbohydrate Polymers*, vol. 90, no. 3, pp. 1362–1370, 2012.

[20] T. Nitanan, P. Akkaramongkolporn, T. Rojanarata, T. Ngawhirunpat, and P. Opanasopit, "Neomycin-loaded poly(styrene sulfonic acid-co-maleic acid) (PSSA-MA)/polyvinyl alcohol (PVA) ion exchange nanofibers for wound dressing materials," *International Journal of Pharmaceutics*, vol. 448, no. 1, pp. 71–78, 2013.

[21] V. Goodship and E. Ogur, *Polyvinyl Alcohol: Materials, Processing and Applications*, Smithers Rapra Technology, vol. 16, Shrewsbury, UK, 2005.

[22] Y. Lou, M. Liu, X. Miao, L. Zhang, and X. Wang, "Improvement of the mechanical properties of nano-TiO_2/poly (vinyl alcohol) composites by enhanced interaction between nanofiller and matrix," *Polymer Composites*, vol. 31, no. 7, pp. 1184–1193, 2010.

[23] M. Sairam, M. B. Patil, R. S. Veerapur, S. A. Patil, and T. Aminabhavi, "Novel dense poly(vinyl alcohol)–TiO_2 mixed matrix membranes for pervaporation separation of water-isopropanol mixtures at 30°C," *Journal of Membrane Science*, vol. 281, no. 1-2, pp. 95–102, 2006.

[24] G. A. Gaddy, A. S. Korchev, J. L. Mclain, B. L. Slaten, E. S. Steige rWalt, and G. Mills, "Light-induced formation of silver particles and clusters in cross linked PVA/PAA films," *Journal of Physical Chemistry B*, vol. 108, no. 39, pp. 14850–14857, 2004.

[25] P. K. Khanna, N. Singh, S. Charan, V. V. V. S. Subbarao, R. Gokhale, and U. P. Mulik, "Synthesis and characterization of Ag/PVA nanocomposite by chemical reduction method," *Materials Chemistry and Physics*, vol. 93, no. 1, pp. 117–122, 2005.

[26] S. Park, D. Seo, and J. Lee, "Preparation of Pb-free silver paste containing nanoparticles," *Colloids and Surfaces A: Physicochemical and Engineering Aspects*, vol. 313-314, pp. 197–201, 2008.

[27] K. Park, D. Seo, and J. Lee, "Conductivity of silver paste prepared from nanoparticles," *Colloids and Surfaces A: Physicochemical and Engineering Aspects*, vol. 313-314, pp. 351–354, 2008.

[28] Z. Zengsheng, L. Xianxue, G. Zhenyu, and L. Yujie, "Silver silver," *Scientific Development*, vol. 408, pp. 32–39, 2006.

[29] K.-H. Tseng, C.-Y. Liao, J.-C. Huang, D.-C. Tien, and T.-T. Tsung, "Characterization of gold nanoparticles in organic or inorganic medium (ethanol/water) fabricated by spark discharge method," *Materials Letters*, vol. 62, no. 19, pp. 3341–3344, 2008.

[30] K.-H. Tseng, Y.-S. Kao, and C.-Y. Chang, "Development and implementation of a micro-electric discharge machine: real-time monitoring system of fabrication of nanosilver colloid," *Journal of Cluster Science*, vol. 27, no. 2, pp. 763–773, 2016.

17

Solid Solubility in $Cu_5Gd_{1-x}Ca_x$ System: Structure, Stability, and Hydrogenation

Andraž Kocjan,[1] Luka Kelhar,[1,2] Anton Gradišek,[1] Blaž Likozar,[3] Kristina Žagar,[1,2] Jaafar Ghanbaja,[2,4] Spomenka Kobe,[1,2] and Jean-Marie Dubois[2,4]

[1]*Jožef Stefan Institute, Jamova cesta 39, SI-1000 Ljubljana, Slovenia*
[2]*International Associated Laboratory and PACS2 and CNRS Nancy and JSI, Ljubljana, Slovenia*
[3]*Department of Catalysis and Chemical Reaction Engineering, National Institute of Chemistry, Hajdrihova 19,*
 SI-1001 Ljubljana, Slovenia
[4]*Institut Jean Lamour (UMR 7198 CNRS-Université de Lorraine), Parc de Saurupt, CS50840, 54011 Nancy Cedex, France*

Correspondence should be addressed to Anton Gradišek; anton.gradisek@ijs.si

Academic Editor: Pavel Lejcek

We report on synthesis and characterization of a novel group of compounds based on copper, gadolinium, and calcium. Cu-Ca and Cu-Gd binaries were previously studied while Ca and Gd are known to be immiscible themselves. The effects of substituting Gd with Ca in $Cu_5Gd_{1-x}Ca_x$ compounds ($0 \leq x \leq 1$) were studied by investigating the phase stability and crystal structure of the resulting new compounds in five specimens with $x = 0$, 0.33, 0.50, 0.66, and 1, respectively. The samples produced by melt-spinning had hexagonal P6/mmm structure, irrespective of Ca amount (x), where lattice parameters varied with x linearly. This is an indication of good solid solubility under the preparation conditions. A slower cooling upon arc-melting caused the liquid phase separation into $Cu_{4.5}Gd$ and Cu-Ca compounds. Using TEM, rapidly solidified ribbons ($Cu_5Gd_{0.5}Ca_{0.5}$) were investigated and the formation of a homogeneous ternary phase with a nearly nominal stoichiometric composition and minor amounts of Cu-Ca secondary phase was observed. Using DSC and HT XRD, we found that these systems are stable at least up to 400°C. Upon a 16-hour hydrogenation at 1 bar and 300°C, all specimens absorbed about 0.5 wt.% of hydrogen. This caused changes in structure with the formation of pure Cu and $H_2Gd_{1-x}Ca_x$ solid solution.

1. Introduction

A number of empirical rules based on thermodynamic criteria have been developed to predict the formation of solid solutions in metallic systems. Hume-Rothery rules [1] were used for decades in traditional alloy design, as they postulate the conditions for substitutional solid solubility in binary systems, that is, the same crystal structure and valence state, similar electronegativity, and a difference in atom sizes between solute and solvent atoms that should be less than 15%. However, for alloys with more than two constituent elements, especially for the so-called high entropy alloys (HEA), which contain five or more elements in equimolar ratio, other rules and parameters must be applied, as reported by Takeuchi and Inoue [2], Zhang et al. [3], and Guo et al. [4, 5]. These authors derived several empirical criteria which can

be applied for phase selection and stability in multielement metallic systems: atomic size mismatch (δ), electronegativity difference ($\Delta\chi$), valence electron concentration (VEC or e/a), and enthalpy of mixing (ΔH_{mix}), etc.

Mizutani and his collaborators carried out an in-depth revision of the Hume-Rothery rules [6–8]. This work was based upon calculations of the electronic structure and densities of states, using a full-potential linearized augmented plane wave (FPLAPW) approximation. A very large series of binary compounds, with one having up to 1168 atoms per unit cell, was scrutinized. First, the authors found that contact between the Fermi sphere and specific facets of the Brillouin zone was responsible for the formation of a pseudogap whose depth and location in energy space in the vicinity of the Fermi energy were responsible for the stability of each specific compound. Second, hybridization between sp and d orbitals adds to that selection, an effect that could be exemplified in many compounds containing transition metals. Finally, the respective contribution of the many elemental constituents to the valence band could be precisely evaluated, with results that depart significantly from data published earlier in relation to the discovery of stable Al-Cu-Fe quasicrystals [9]. Such systems are found in ternary, or multinary, alloys of generic composition $A_xB_yC_z$ and are characterized by two important key facts: (i) the binary components A-B and A-C form well-defined compounds (negative formation enthalpy) and (ii) the B-C pair forms no compound nor solid solution, as B and C are immiscible due to the positive formation or mixing enthalpy. Such alloys are coined "push-pull" alloys [10]. A well-known example is the already mentioned, Al-Cu-Fe system [11], in which several binary Al-Cu and Al-Fe compounds are known, with some being rather complex. Based on the Cu-Fe phase diagram, these two constituents are known to be immiscible. A quasicrystal is found in the Al-rich corner of the phase diagram, in the vicinity of the simple ω-Al_7Cu_2Fe crystal having 48 atoms in its unit cell. From a macroscopic standpoint, such ternary alloys act as complexity amplifiers, roughly analogic to push-pull amplifiers in electronics. The same situation is observed in other well-known quasicrystal forming alloys, such as Al-Cu-Li, Al-Mg-Zn, Al-Co-Cu, and a periodic complex metallic alloy Al-Cu-Ta with more than 27,000 atoms in a unit cell.

Here, we focused on the ternary Cu-Gd-Ca system, which shows great potential for the formation of complex compounds via the push-pull mechanism. Namely, there is more than a 99.6% miscibility gap for the elements Ca and Gd which means that they form neither a stable compound nor a solid solution. Even more so, up to date, no phase diagram is available for the Gd-Ca system. Immiscibility in the liquid phase has been observed by Kato and Copeland [12] and Gschneidner and Calderwood [13]. At the same time, Cu-Gd [14–16] and Cu-Ca [17, 18] form stable compounds, which was already reported back in the 1980s.

Up to date, we discovered no quasicrystal or complex structure in the Cu-Gd-Ca system. Instead, intermixing of Cu_5Gd and Cu_5Ca compounds in the solid state was observed, yet only through rapid quenching. Thus, in this paper, we report on the synthesis, structure, and stability of the $Cu_5Gd_{1-x}Ca_x$ ($x = 0, 0.33, 0.5, 0.66, 1$) system. We also provide results about loading the samples with hydrogen, which changes the crystal structure.

2. Experimental

Our experimental study investigated five samples with the general formula $Cu_5Gd_{1-x}Ca_x$ ($x = 0, 0.33, 0.5, 0.66,$ and 1), which were prepared by arc-melting and melt-spinning in high-purity (99.9999%) argon gas. First, copper and gadolinium pieces were arc-melted to form a eutectic alloy into which small pieces of pure Ca were added, carefully mastering the dissolution of this element into the liquid bath since calcium is extremely volatile. The arc-melted precursors were then spun on a rotating copper wheel at 40 m/s tangential speed using boron-nitride-coated graphite

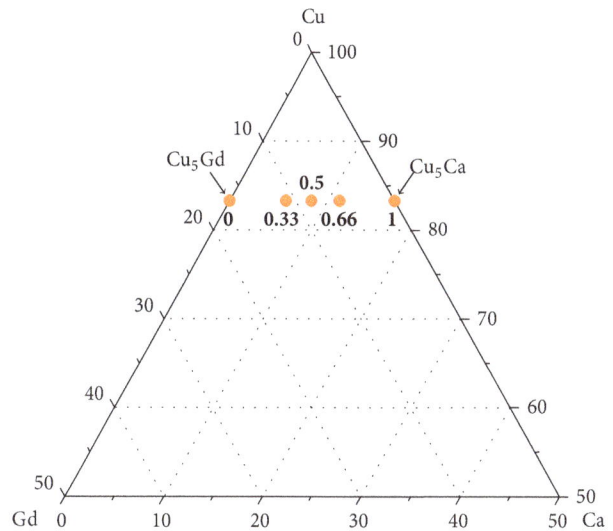

FIGURE 1: The Cu-rich half of the Cu-Gd-Ca ternary concentration chart with the studied $Cu_5Gd_{1-x}Ca_x$ compositions marked by dots, near to which x values are given.

crucibles to prevent the formation of any carbide phases in the ribbons. In order to minimize Ca evaporation, we increased the pressure in the melt-spinning chamber from the standard 700 mbar value to a pressure of 1500 mbar. In spite of this precaution, 2-3 wt.% of excess Ca had to be added to maintain the stoichiometry. The X-ray diffraction (XRD) patterns were recorded using a PANalytical diffractometer with Cu-Kα radiation in the 2Θ interval 2–60°. In situ XRD experiments were performed at set temperatures of 400, 580, and 700°C, using a heating rate of 20°C min^{-1}. Anton Paar high temperature oven chamber 1200 N was used for high temperature measurements. Differential scanning calorimetry (DSC) was performed using a Netzsch STA 449 C/6/G Jupiter® apparatus and a sample mass of the order of few tens of milligrams. The temperature scheme was as follows: 20°C min^{-1} up to 800°C, natural cooling down to 200°C, and then heating up again at 20°C min^{-1} to 800°C in order to investigate the reversibility of the observed phenomena. TEM investigations were performed on a Jeol JEM ARM 200 CF microscope. The sample preparation included mechanical grinding and dimpling using isopropanol to avoid oxidation and subsequent Ar ion-milling. Hydrogenation was performed under 1 bar of 99.999% hydrogen gas at 300°C for 16 hours. Gas Analytical System QMS 403 C Aëolos (STA 449 C/6/G Jupiter, QMS 403 C, Netzsch, Germany) thermal analyzer with an attached mass spectrometer was used to perform mass-spectrometry of desorbed hydrogen from hydrogenated samples by 20°C min^{-1} heating up to 800°C.

3. Results and Discussion

The compositions of the prepared samples are plotted on a partial Cu-Gd-Ca ternary concentration chart in Figure 1, that is, $Cu_5Gd_{1-x}Ca_x$ ($x = 0, 0.33, 0.5, 0.66,$ and 1).

(a)

(b)

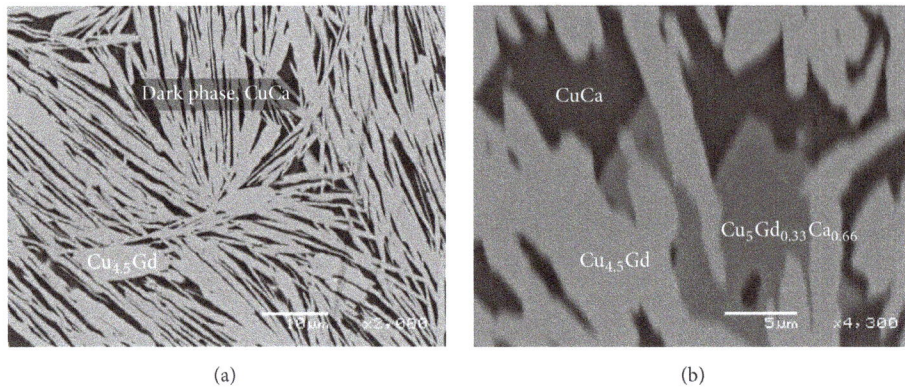

FIGURE 2: Backscattered SEM images of polished surface of (a) arc-melted $Cu_5Gd_{0.5}Ca_{0.5}$ sample, and (b) the same sample after 7 days of homogenization at $700°C$ in an argon-filled sealed silica tube and subsequently ice-water quenched.

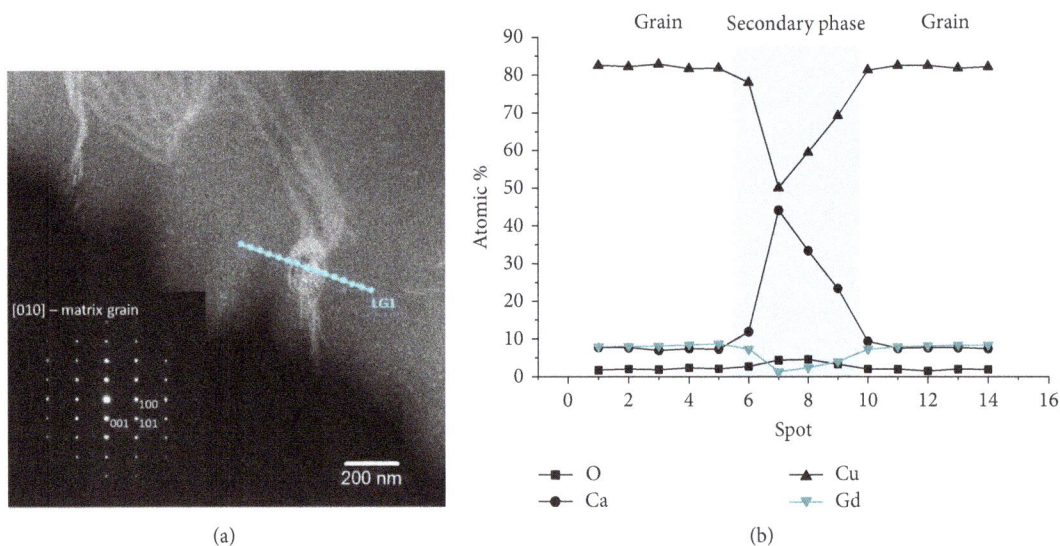

(a)

(b)

FIGURE 3: (a) Cross section image in STEM mode of a $Cu_5Gd_{0.5}Ca_{0.5}$ melt-spun ribbon showing that the matrix grains have hexagonal symmetry (SAED). Also, a negligible amount of secondary Cu-Ca phase was found; (b) EDS line scan (14 points) confirmed stoichiometric, that is, nominal, composition of the matrix phase, low oxygen level, and segregation of Cu-Ca phase.

3.1. Microstructure.

Figure 2(a) shows a microstructure of the arc-melted $Cu_5Gd_{0.5}Ca_{0.5}$ sample. There is clear evidence in favor of a strong phase separation at slow cooling rate, since the two immiscible Gd and Ca constituents do not coexist in a single phase after a large amount of Cu has been added. The bright phase is $Cu_{4.5}Gd$, whereas the dark phase is the equiatomic Cu-Ca compound.

In an attempt to produce the very first ternary phase of this system, we homogenized the arc-melted button in a sealed silica tube containing a Ti-getter (which reacts with residual oxygen in the Ar gas) under highly pure argon gas at $700°C$ for 7 days, and subsequently ice-water quenched the ampoule. A backscattered SEM image with the associated EDS analyses is shown in Figure 2(b). Beside the bright $Cu_{4.5}Gd$ phase, we observed ~1 at.% of Gd dissolved in a dark Cu-Ca phase and, more importantly, a new ternary intermetallic compound with an intermediate Z-contrast with composition $Cu_5Gd_{0.33}Ca_{0.66}$ was found. This result was used in another study to produce a single grain of that phase

and study its intricate magnetic structure [19]. Thus, in order to obtain a single-phase $Cu_5Gd_{0.33}Ca_{0.66}$ sample, we used a rapid quenching technique, that is, melt-spinning, for its preparation. This method proved successful since backscattered SEM imaging and EDS analysis revealed a single-phase material with nominal stoichiometric composition. The same results were obtained for the other four samples dealt with in this article.

In order to check whether the melt-spun ribbons were completely homogeneous on the nanoscale, we performed a TEM characterization of the as-spun sample with $x = 0.5$. Our data are shown in Figure 3. Minor amount of Cu-Ca secondary phase was found to coexist in the matrix phase that had a stoichiometric $Cu_5Gd_{0.5}Ca_{0.5}$ composition. Using SAED we confirmed the hexagonal structure of the matrix phase, which is in agreement with the XRD results. The population of the secondary phase was below the detection limit of XRD.

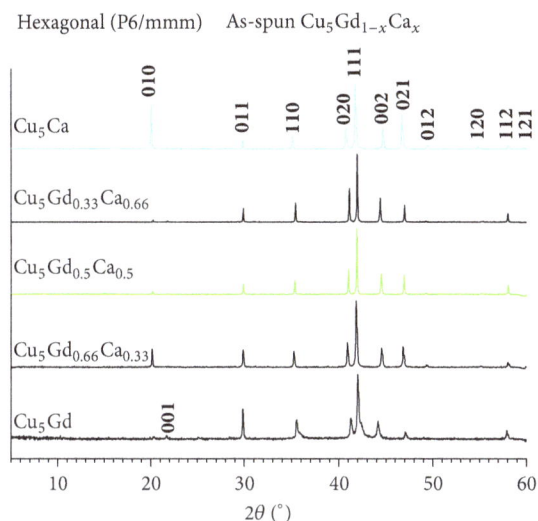

FIGURE 4: XRD patterns of $Cu_5Gd_{1-x}Ca_x$ ($x = 0$, 0.33, 0.5, 0.66, and 1) ribbons, melt-spun at a tangential speed of the copper wheel of 40 m/s.

The amount of oxygen detected in the matrix phase was 2 at.%, while in the Cu-Ca secondary phase it was 5 at.%, which both can be attributed to surface oxidation during the TEM sample preparation.

3.2. X-Ray Diffraction.

XRD patterns of the as-spun ribbons in Figure 4 indicate good mixing of the copper, gadolinium, and calcium constituents in the $Cu_5Gd_{1-x}Ca_x$ system as pointed out by the preservation of the hexagonal structure as Gd is replaced by increasing amounts of Ca ($x > 0$). For Cu_5Gd, two crystal modifications have been found in the phase diagram [14–16], namely, the high temperature (HT) hexagonal β-Cu_5Gd phase and the low temperature (LT) cubic α-Cu_5Gd phase, whereas Cu_5Ca exists only in hexagonal polymorph [18]. Thus, it is clear that the large cooling rate associated with melt-spinning is crucial for the preparation of a single-phase material, that is, solid solutions of Cu_5Gd and Cu_5Ca compounds with hexagonal structure. Slight variations of peak positions in the XRD patterns, as well as differences in the relative intensities, were found along the sample series, particularly for the 010 peak. Peak shifts can be attributed to atom size effects which modify the lattice parameters, whereas the intensity changes reflect the respective influence of Ca replacing Gd onto the structure factor. Experimental data revealed that the unit cell volume of Cu_5Ca is 1.65% larger than that of Cu_5Gd. The apparently broader peaks in Cu_5Gd as opposed to other compositions may be related to smaller grain size and/or to disorder due to mechanical strains in the material.

When describing the dependence of lattice parameters and compositional factor x, we used Vegard's law [20], which holds that a linear relation exists, at constant temperature, between the crystal lattice parameters of a solid solution and the concentrations of the constituent elements (or binary compounds). Experimental dependence of hexagonal P6/mmm lattice parameters a and c on the compositional factor x was found to be fairly linear. The lattice parameter a was decreasing with the addition of Ca while c was increasing (Figure 5(a)). This is a good example of Vegard's law, which demonstrates that replacement of Gd by Ca in the hexagonal compounds is random and substitutional. The unit cell volume increases linearly with x (Figure 5(b)). Figure 5(c) shows the P6/mmm unit cell of $Cu_5Gd_{0.5}Ca_{0.5}$, with equal occupancy of Gd and Ca atoms on the Ca1 sites in the Cu_5Ca prototype cell.

3.3. Phase Stability, DSC, and HT XRD.

In order to assess the stability of these rapidly solidified alloys, we performed DSC and HT XRD measurements as previously described. Here, we show the results for $x = 0$, 0.5, and 1 samples only, as they are representative for the whole system. Figure 6 shows the DSC data and Figure 7 shows the HT XRD. All the recorded DSC peaks were due to irreversible transformations since none of them was observed again upon cooling and, later, upon a second heating run. A weak low-T exothermic signal at 367°C ($x = 0$) or 350°C ($x = 1$) was found. Due to the exothermal nature of those peaks and the fact that the samples were rapidly solidified, they could be attributed to the relaxation of structural stresses.

Additionally, two endothermic DSC peaks were determined at 560 and 657°C for pure Cu_5Ca. According to the Cu-Ca phase diagram reported by Chakrabarti and Laughlin [18], we could attribute the 560°C peak to the peritectic reaction $Cu_5Ca + L \leftrightarrow \beta$-Cu-Ca, which occurs at 567°C. The peak at 657°C was most likely due to the melting of aluminum, which formed upon reduction of the alumina pot by segregated calcium metal. The exothermic Ca oxidation peak must therefore be hidden behind this peak, which results in a lower apparent melting enthalpy of aluminum.

In order to check if any structural changes occurred at elevated temperatures, we performed an in situ XRD analysis at the temperatures at which each DSC peak was ending, that is, 400, 580, and 700°C (Figures 7(a)–7(c)).

The high temperature XRD data shown in Figure 7 point out that the two diffraction peaks of the Fm-3m cubic structure appear above 400°C. Its lattice parameter linearly increases with the temperature due to thermal expansion. Extrapolation to room temperature gave a value for the cubic lattice parameter $a = 3.615$ Å, which corresponds to the lattice parameter of pure copper. Since the relative intensity of Cu peaks increases with temperature, we concluded that we deal with diffusion-controlled copper segregation where the DSC signal of this process is smeared over a wide temperature range and thus cannot be observed. Also, in the case of the Cu_5Ca sample, the hexagonal phase peaks were still present at 700°C, suggesting that Cu segregation is inhibited due to higher stability of hexagonal Cu_5Ca. In the Cu_5Gd XRD, no LT phase peaks were observed upon heating. Note that the broad XRD peaks observed at low angles are not related to the sample but are caused by the kapton foil that the X-ray beam crosses on the way to the sample chamber.

Based on the combination of the DSC and the XRD data, we concluded that the ternary alloys are stable at least until 400°C since the XRD showed no phase separation to Cu-Gd and Cu-Ca binaries. However, at higher temperatures,

(a)

(b)

P6/mmm

(c)

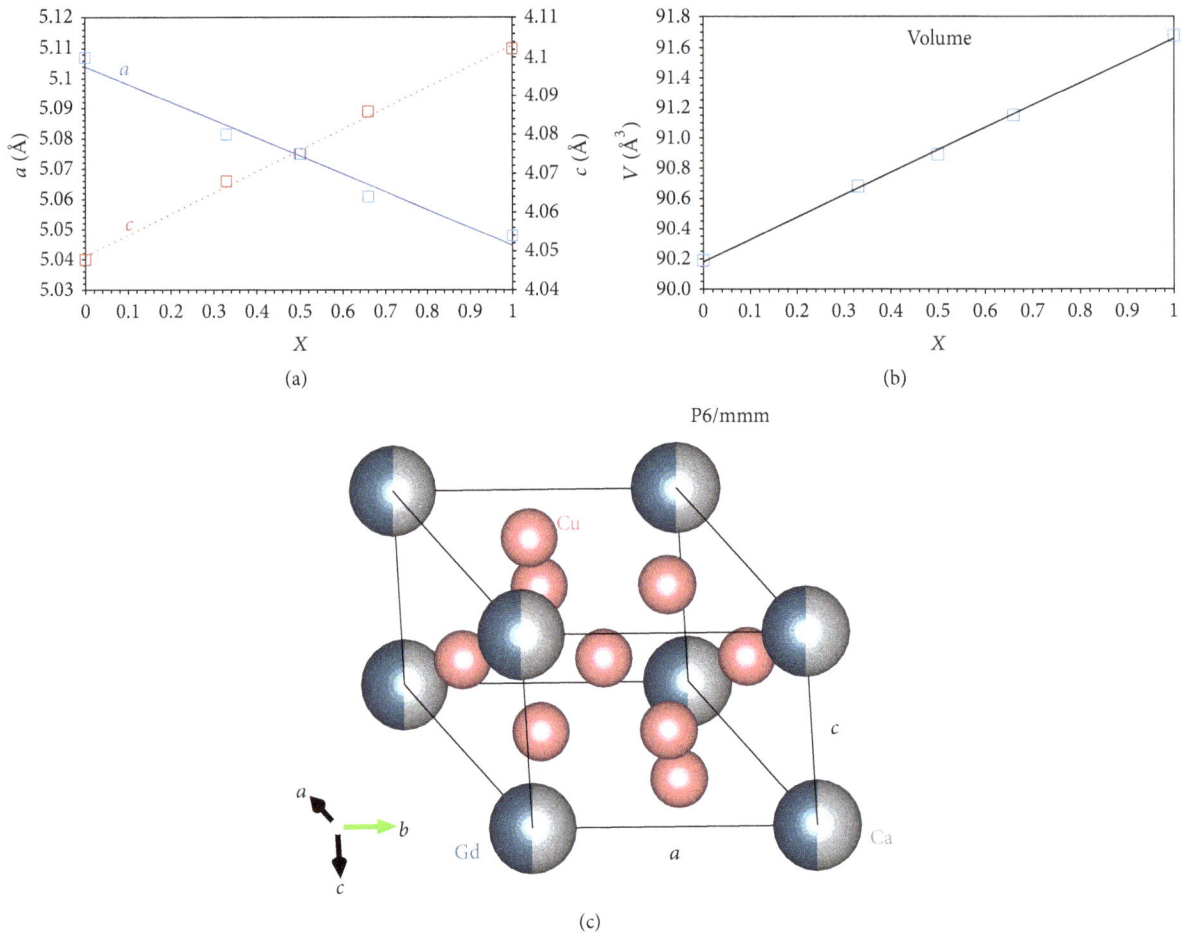

FIGURE 5: (a) P6/mmm lattice parameters a and c as functions of Ca concentration x in $Cu_5Gd_{1-x}Ca_x$ compounds and (b) corresponding unit cell volume. The lines serve to guide the eyes. Error bars along x- and y-axes are within the size of the data points; (c) P6/mmm unit cell of $Cu_5Gd_{0.5}Ca_{0.5}$.

severe copper segregation to the surface of the ribbons occurred, particularly in the $x = 0$ sample, and consequently oxidation of residual Gd as decomposition product, upon which cubic Gd_2O_3 (Ia-3, $a = 1.0790$ nm) has formed. Additional indication for Gd-oxide formation is that the corresponding diffraction peaks positions are independent of the temperature, which is related to low thermal expansion coefficient of ceramic materials. On the contrary, peaks of metallic Cu shift to higher d-values with temperature. The existence of other secondary phases like Cu_6Gd (orthorhombic, Pnma, $a = 0.8315$ nm, $b = 0.5983$ nm, $c = 0.9949$ nm) and (nanostructured) Cu-oxides cannot be excluded, but the exact phase identification is complicated due to poor definition of the background axis very common in this type of experiments in which the primary, rather than the diffracted, beam has been collimated by a crystal monochromator [21].

In the case of the $x = 1$ sample, no peritectically formed β-Cu-Ca phase could be found because it melts at 580°C (and forms peritectically at 567°C). Instead, XRD peaks of CaO were found. This phase is cubic Fm-3m, $a = 0.481$ nm, and most likely formed from segregated Ca from the liquid phase and oxygen as impurity in the Ar carrier gas flowing in the HT XRD chamber. Since the $x = 0.5$ sample contains a lower

amount of Gd_2O_3 and no CaO, it can be concluded that (i) the addition of Ca to Cu_5Gd stabilizes the hexagonal phase and (ii) the addition of Gd to the Cu_5Ca system prevents liquid phase formation during heating up to 700°C and thus Ca oxidation. This way, ternary alloys turn out to be more stable than both binaries.

3.4. Hydrogenation. After exposing the as-spun hexagonal samples to pure hydrogen atmosphere at 300°C for 16 hours, approximately 0.5 wt.% of desorbed hydrogen was released from all samples upon subsequent dehydrogenation. XRD analysis of $Cu_5Gd_{1-x}Ca_x$ hydrides (Figure 8(a)) revealed a multiphase formation rather than hexagonal crystal lattice expansion upon hydrogenation. In case of binary alloys, we determined pure Cu and GdH_2 ($x = 0$) or CaH_2 ($x = 1$) peaks. Formation of Gd hydride (Gd + H_2 → GdH_2) has already been observed by Hruška et al. [22], whereas Ca reaction with hydrogen (Ca + H_2 → CaH_2) was investigated by Sturdy and Mulford [23], and Rittmeyer and Wietelmann [24]. Copper hydride has been previously found to form from elemental copper only under severe pressure around 20 GPa [25], so it is reasonable to assume that protons do not react with copper atoms in our alloys at given hydrogenation

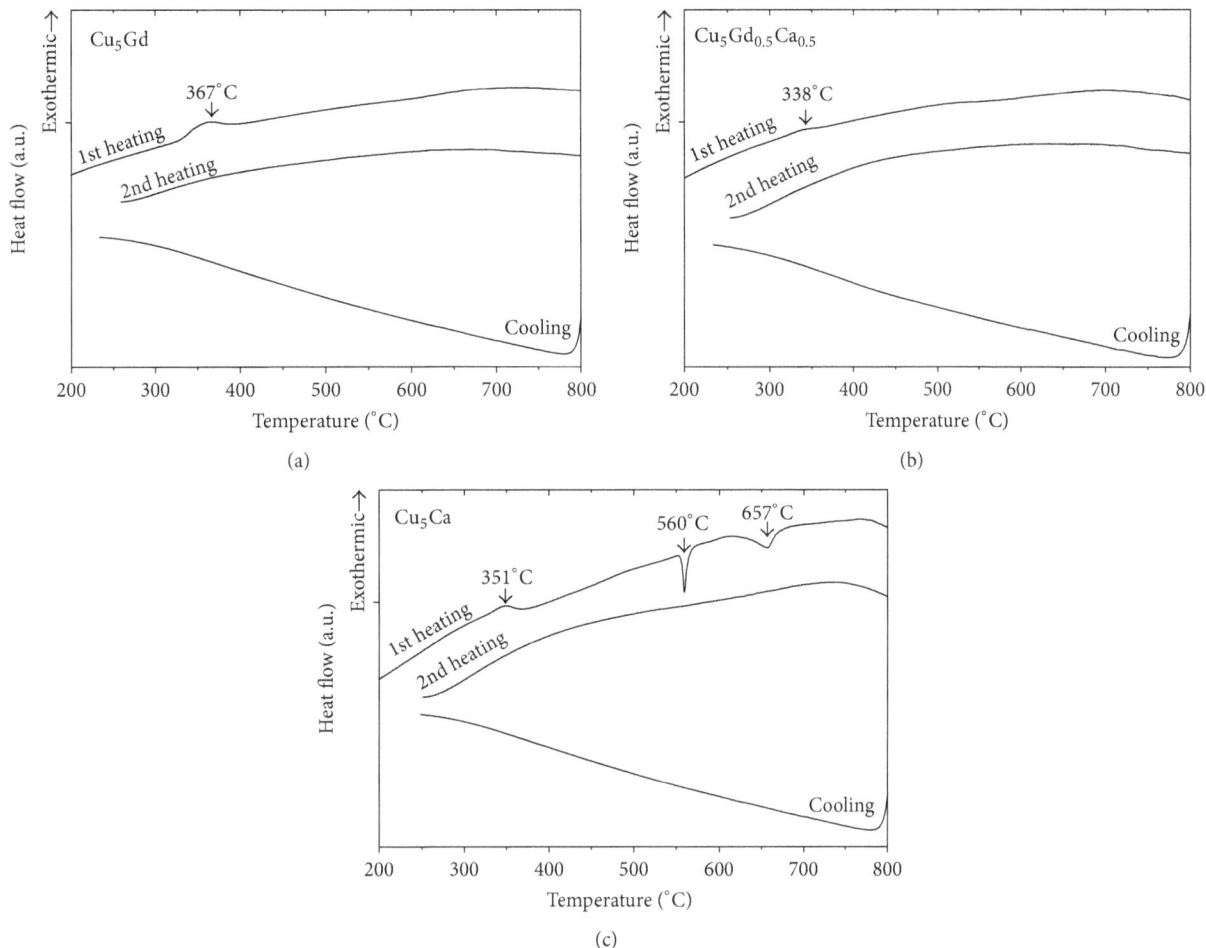

FIGURE 6: DSC analysis of $Cu_5Gd_{1-x}Ca_x$ ($x = 0$, 0.5, and 1) melt-spun ribbons during two-stage heating experiments.

conditions. In fact, this might be the driving force behind the severe phase separation we observed.

In all three ternary systems, we noticed the appearance of a novel phase which appears to be a $H_2Gd_{1-x}Ca_x$-based solid solution, coexisting with pure Fm-3m Cu. As seen in Figure 8(a), all XRD peaks of GdH_2 were shifting to lower angles as the amount of Ca was increasing. Thus, a cubic solid solution of GdH_2 and CaH_2 compounds must have formed; otherwise, their separated peaks would appear with peak intensity depending on x. In addition, Figure 8(b) shows that the H_2 mass-spectra of $x = 0.33$, 0.5, and 0.66 samples are not linear combinations of those of GdH_2 and CaH_2 (which would be the case if the system consisted of grains of individual hydrides). In order to check if the hydride solid solution contains any Cu, high-resolution EDS mapping or STEM analysis should be performed. However, in case of hydrogenated Cu_5Gd, we found pure Cu and GdH_2. If the latter contained some copper, it would definitely be seen as a shift of XRD peaks. The same conclusion holds for Cu_5Ca and corresponding CaH_2.

Nuclear magnetic resonance (NMR) is a technique per excellence to study proton mobility in a material. Therefore, to investigate how the Ca concentration in the $Cu_5Gd_{1-x}Ca_x$

system (or rather $H_2Gd_{1-x}Ca_x$) affects the interaction of protons with the host metallic lattice and interpret Figure 8(b) through proton diffusivity and hopping activation energies, we measured the proton spin-lattice relaxation by means of ^{1}H NMR at proton resonance frequency of 500 MHz, corresponding to a magnetic field of 11.75 Tesla. We used a superconducting Bruker magnet and an ^{1}H probe, with samples being placed in 5 mm glass tubes. Proton hopping between the interstitial sites causes fluctuations in the interproton spin interactions and activation energy for these jumps can be obtained from the temperature dependence of the proton spin-lattice relaxation time T_1 [26]. However, it turns out that proton relaxation time T_1 is roughly temperature-independent in the temperature range from room temperature to 120°C (the highest temperature achievable in the setup). In this temperature range, rapid proton hopping is expected. However, the temperature-independent spin-lattice relaxation indicates that the dynamic contribution to proton relaxation is masked by another contribution, perhaps due to proton coupling to paramagnetic centers in the sample, such as defects in the crystal structure or individual Gd atoms with their paramagnetic moment unshaded by the free

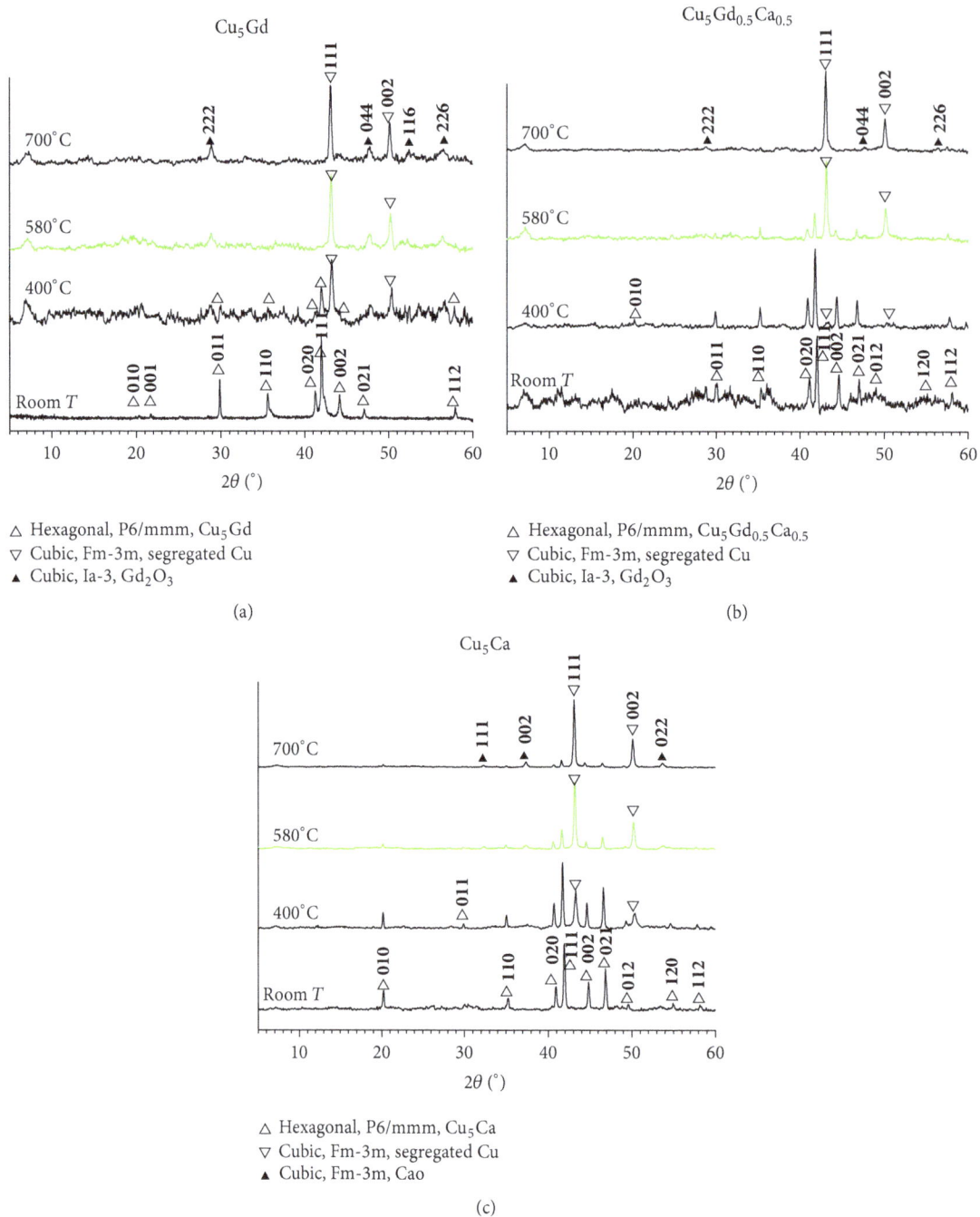

FIGURE 7: (a–c) High temperature XRD patterns of $Cu_5Gd_{1-x}Ca_x$ melt-spun ribbons recorded at room temperature, 400, 580, and 700°C; (a) $x = 0$, (b) $x = 0.5$, and (c) $x = 1$.

electron gas. Therefore, no information about the activation energies for jumps could be obtained by ^{1}H NMR.

4. Conclusion

We prepared a series of melt-spun ribbons by substitution of Gd with Ca in a Cu_5Gd master alloy. After rapid quenching by melt-spinning, all samples exhibited hexagonal P6/mmm structure regardless of the Gd/Ca ratio, which implies good

substitutional solid solubility of Ca and Gd in the system with 83.3 at.% of Cu. This result is in accordance with results published by Carnasciali et al. [15] and Subramanian and Laughlin [16] who reported that the high temperature polymorph of the Cu_5Gd compound has hexagonal P6/mmm Cu_5Ca-like structure ($a = 0.504$ nm, $c = 0.411$ nm). No ternary Cu-Gd-Ca compound synthesis was reported so far, except the $Cu_5Gd_{0.54}Ca_{0.42}$ single crystal, which was grown by parallel collaborating group [19] upon findings reported here and published prior to the present publication.

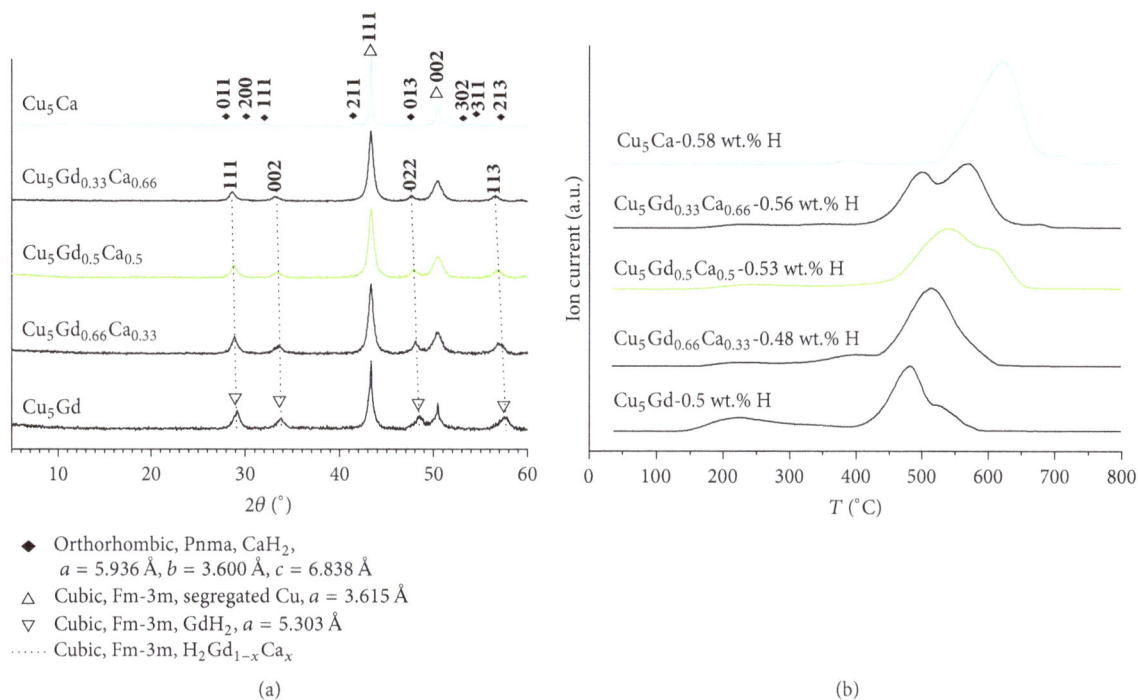

FIGURE 8: (a) XRD patterns of hydrogenated $Cu_5Gd_{1-x}Ca_x$ ($x = 0$, 0.33, 0.5, 0.66, and 1) melt-spun ribbons and (b) corresponding mass-spectra of thermally desorbed hydrogen gas.

Transmission electron microscopy of the melt-spun ribbon with $x = 0.5$ revealed homogeneity on the nanoscale of $Cu_5Gd_{0.5}Ca_{0.5}$ matrix phase, although minor amounts of Cu-Ca precipitates were discovered, which did not influence significantly the stoichiometry of the matrix phase. Copper segregation was observed in all investigated samples upon heating by in situ XRD. Regardless of Cu segregation, the ternary alloys seem more stable or corrosion resistant, as the amounts of Gd_2O_3 and CaO secondary phases were suppressed. The later was found only in $x = 1$ sample.

Copper segregation was also observed upon loading the samples with hydrogen. Cubic Fm-3m $H_2Gd_{1-x}Ca_x$-based solid solution was identified from the XRD measurements and the mass-spectra of thermally desorbed hydrogen gas. EDS measurements uncovered the presence of oxide species; however, their size was beneath the resolution of XRD. A proton mobility study in the hydride phase was attempted by means of NMR but the activation energy for proton hopping was not possible to obtain since the proton spin-lattice relaxation is completely dominated by paramagnetic centers in the system.

Conflicts of Interest

The authors declare no conflicts of interest.

Acknowledgments

This research was part of the bilateral French-Slovenian PROTEUS project CalGad-X. The authors are grateful to Dr. Marjeta Maček for assistance in DSC and mass-spectrometry experiments and to Ms. Urška Kavčič for performing catalytic tests under controlled atmosphere. The authors acknowledge the financial support from the Slovenian Research Agency (research core funding no. P2-0152) and CNRS PICS "CalGad-X".

References

[1] B. S. Murty, J.-W. Yeh, and S. Ranganathan, *High-Entropy Alloys*, Butterworth-Heinemann (Elsevier), London, UK, 1st edition, 2014.

[2] A. Takeuchi and A. Inoue, "Quantitative evaluation of critical cooling rate for metallic glasses," *Materials Science and Engineering: A*, vol. 304-306, pp. 446–451, 2001.

[3] Y. Zhang, Y. J. Zhou, J. P. Lin, G. L. Chen, and P. K. Liaw, "Solid-solution phase formation rules for multi-component alloys," *Advanced Engineering Materials*, vol. 10, no. 6, pp. 534–538, 2008.

[4] S. Guo, C. Ng, J. Lu, and C. T. Liu, "Effect of valence electron concentration on stability of fcc or bcc phase in high entropy alloys," *Journal of Applied Physics*, vol. 109, no. 10, Article ID 103505, 2011.

[5] S. Guo, Q. Hu, C. Ng, and C. T. Liu, "More than entropy in high-entropy alloys: forming solid solutions or amorphous phase," *Intermetallics*, vol. 41, pp. 96–103, 2013.

[6] U. Mizutani, M. Inukai, H. Sato, and E. S. Zijlstra, "Hume-Rothery stabilization mechanism and e/a determination for RT- and MI-type 1/1-1/1-1/1 approximants studied by FLAPW-Fourier analyses," *Chemical Society Reviews*, vol. 41, no. 20, pp. 6799–6820, 2012.

[7] U. Mizutani, M. Inukai, H. Sato, and E. S. Zijlstra, "Hume-Rothery stabilization mechanism and e/a determination in MI-type Al–Mn, Al–Re, Al–Re–Si, Al–Cu–Fe–Si and Al–Cu–Ru–Si 1/1-1/1-1/1 approximants—a proposal for a new Hume-Rothery electron concentration rule," *Philosophical Magazine*, vol. 92, no. 13, pp. 1691–1715, 2012.

[8] U. Mizutani, *Hume-Rothery Rules for Structurally Complex Alloy Phases*, CRC Press, Boca Raton, Fla, USA, 2010.

[9] A.-P. Tsai, "Discovery of stable icosahedral quasicrystals: progress in understanding structure and properties," *Chemical Society Reviews*, vol. 42, no. 12, pp. 5352–5365, 2013.

[10] J.-M. Dubois, E. Belin-Ferré, and A. P. Tsai, "Quasicrystals and complex metallic alloys," in *Kirk-Othmer Encyclopedia of Chemical Technology*, pp. 1–19, John Wiley & Sons, 2016.

[11] F. Faudot, A. Quivy, Y. Calvayrac, D. Gratias, and M. Harmelin, "About the Al-Cu-Fe icosahedral phase formation," *Materials Science and Engineering: A*, vol. 133, pp. 383–387, 1991.

[12] H. Kato and M. I. Copeland, "USBM-U-952," Metallurgical Progress Report 15, National Technical Information Service, Springfield, Va, USA, 1962.

[13] K. A. Gschneidner and F. W. Calderwood, "The Ca-Gd (Calcium-Gadolinium) system," *Bulletin of Alloy Phase Diagrams*, vol. 8, no. 6, article 515, 1987.

[14] H. Okamoto, *Desk Handbook: Phase Diagrams for Binary Alloys*, ASM International, Geauga County, Ohio, USA, 2nd edition, 2010.

[15] M. M. Carnasciali, S. Cirafici, and E. Franceschi, "On the Gd-Cu system," *Journal of The Less-Common Metals*, vol. 92, no. 1, pp. 143–147, 1983.

[16] P. R. Subramanian and D. E. Laughlin, "The Cu-Hf (copper-hafnium) system," *Bulletin of Alloy Phase Diagrams*, vol. 9, no. 1, pp. 51–56, 1988.

[17] F. Merlo and M. L. Fornasini, "The structures of α-CaCu, β-CaCu, SrAg and BaAg: four different stacking variants based on noble-metal-centred trigonal prisms," *Acta Crystallographica B*, vol. 37, no. 3, pp. 500–503, 1981.

[18] D. J. Chakrabarti and D. E. Laughlin, "The Ca-Cu (Calcium-Copper) system," *Bulletin of Alloy Phase Diagrams*, vol. 5, no. 6, pp. 570–576, 1984.

[19] M. Krnel, S. Vrtnik, P. Koželj et al., "Random-anisotropy ferromagnetic state in the $Cu_5 Gd_{0.54} Ca_{0.42}$ intermetallic compound," *Physical Review B*, vol. 93, no. 9, Article ID 094202, 2016.

[20] A. R. Denton and N. W. Ashcroft, "Vegard's law," *Physical Review A*, vol. 43, no. 6, pp. 3161–3164, 1991.

[21] V. K. Pecharsky and P. Y. Zavalij, *Fundamentals of Powder Diffraction and Structural Characterization of Materials*, Springer, Berlin, Germany, 2nd edition, 2009.

[22] P. Hruška, J. Čížek, P. Dobroň et al., "Investigation of nanocrystalline Gd films loaded with hydrogen," *Journal of Alloys and Compounds*, vol. 645, no. 1, pp. S308–S311, 2015.

[23] G. E. Sturdy and R. N. R. Mulford, "The gadolinium-hydrogen system," *Journal of the American Chemical Society*, vol. 78, no. 6, pp. 1083–1087, 1956.

[24] P. Rittmeyer and U. Wietelmann, "Hydrides," in *Ullmann's Encyclopedia of Industrial Chemistry*, Wiley-VCH, Weinheim, Germany, 2002.

[25] C. Donnerer, T. Scheler, and E. Gregoryanz, "High-pressure synthesis of noble metal hydrides," *Journal of Chemical Physics*, vol. 138, no. 13, Article ID 134507, 2013.

[26] A. Gradišek and T. Apih, "Hydrogen dynamics in partially quasicrystalline $Zr_{69.5}Cu_{12}Ni_{11}Al_{7.5}$: a fast field-cycling relaxometric study," *Journal of Physical Chemistry C*, vol. 119, no. 19, pp. 10677–10681, 2015.

18

The Effect of Fe Doping on the Magnetic and Magnetocaloric Properties of $Mn_{5-x}Fe_xGe_3$

Jeffrey Brock, Nathanael Bell-Pactat, Hong Cai, Timothy Dennison, Tucker Fox, Brandon Free, Rami Mahyub, Austin Nar, Michael Saaranen, Tiago Schaeffer, and Mahmud Khan

Department of Physics, Miami University, Oxford, OH 45056, USA

Correspondence should be addressed to Mahmud Khan; khanm2@miamioh.edu

Academic Editor: Jamal Berakdar

The magnetic and magnetocaloric properties of a series of minutely doped $Mn_{5-x}Fe_xGe_3$ compounds that exhibit the $D8_8$-type hexagonal crystal structure at room temperature have been investigated. For all Fe concentrations, the alloys are ferromagnetic and undergo a second-order ferromagnetic-to-paramagnetic transition near room temperature. Although the small Fe doping had little effect on the ferromagnetic transition temperatures of the system, changes in the saturation magnetization and magnetic anisotropy were observed. For $x \leq 0.15$, all compounds exhibit nearly the same magnetic entropy change of ~7 J/kg K, for a field change of 50 kOe. However, the magnitude of the refrigerant capacities increased with Fe doping, with values up to 108.5 J/kg and 312 J/kg being observed for field changes of 20 kOe and 50 kOe, respectively. As second-order phase transition materials, the $Mn_{5-x}Fe_xGe_3$ compounds are not subject to the various drawbacks associated with first-order phase transition materials yet exhibit favorable magnetocaloric effects.

1. Introduction

Motivated by the energy-inefficiency, environmental impacts, and poor durability of current gas-compression based refrigeration technologies, there has been a great deal of interest in the development of alternative technologies that are more energy efficient and environmentally friendly. One such solution that has spurred considerable interest is the magnetocaloric effect (MCE) [1, 2]. In brief, the MCE occurs when a magnetic material is adiabatically magnetized in a changing magnetic field. As the individual magnetic moments align themselves with the applied field, there is a decrease in the number of energy microstates available to the system, commensurate to an isothermal magnetic entropy change (ΔS_M). This change in a thermodynamic state variable can lead to an adiabatic temperature change (ΔT_{Ad}). The ΔT_{Ad} generated via the MCE can enable the transport of a significant amount of heat at greater efficiency than current technologies, attracting interest for solid-state cooling applications [3]. First observed in Ni by Weiss in 1917, research interest in MCE burgeoned

after Pecharsky and Gschneidner Jr. published their seminal report regarding so-called giant MCE in $Gd_5(Si_2Ge_2)$, attainable using magnetic field changes within the reach of permanent magnets [4].

Generally, large MCE is associated with materials that undergo a coupled first-order magnetostructural phase transition (FOMPT), as the simultaneous change in both the magnetic and structural phase is conducive to larger ΔS_M values and potentially larger ΔT_{Ad} values [5–7]. Strong MCE has been observed in many Mn-based FOMPT materials, including MnFe(P, As) [3, 8–10], Mn(As, Sb) [5], and MnFe(P, As, Ge) [11] compounds. Unfortunately, the heightened MCE of these materials is offset by the downfalls that all FOMPT materials are susceptible to, including thermal and magnetic hysteresis losses and irreproducibility due to mechanical fatigue [12]. Moving towards the realization of robust magnetic refrigeration technologies, it is clear that these FOMPT pitfalls should be addressed in FOMPT materials, or that FOMPT-comparable MCE should be sourced from materials exhibiting second-order phase transitions.

The Mn_5Ge_3 compound has attracted significant research interest because it exhibits a second-order ferromagnetic phase transition with a Curie temperature (T_C) near room temperature (296 K) in addition to a moderately large ΔS_M (9.2 J/kg K for a field change of 50 kOe) [13, 14]; both of these quantities are characteristics that should be sought from candidate MCE materials. Given that the magnetic properties of transition metal-based systems are heavily sensitive to the interatomic distances that governs the magnetic exchange mechanisms [15, 16], attempts have been made to incorporate other transition metals within basic systems in order to modify the lattice parameters and exchange interactions and thus optimize the magnetic and MCE properties of the Mn_5Ge_3 parent compound [15]. One such system is the $Mn_{5-x}Fe_xGe_3$ series previously investigated by Zhang et al. [17]. In their report, they employed substitutions with x ranging from 0 to 1, in steps of 0.25, and found that larger refrigerant capacities were observed in samples with smaller x (less Fe). This system is worth reexamination using smaller substitutions, given that the previously reported replacements displaced T_C from 300 K in the $x = 0$ compound to 330 K in the $x = 1$ compound, as well as the aforementioned findings regarding the refrigerant capacity.

Keeping this in mind, we have performed an experimental study of the $Mn_{5-x}Fe_xGe_3$ ($0 \leq x \leq 0.15$) system, exploring the effect of smaller Fe substitutions on the structural, magnetic, and magnetocaloric properties of the compounds. In limiting investigation to small substitutions of Fe, our goals were to observe whether a changing nature of the magnetic interactions was apparent across the series and to assess the MCE potential of the compounds by means of ΔS_M and refrigerant capacity calculations.

2. Experimental Details

Polycrystalline buttons of $Mn_{5-x}Fe_xGe_3$ ($0 \leq x \leq 0.15$) weighing approximately 2 g each were fabricated in Ar atmosphere using the vacuum arc melting technique. The virgin metals used in fabrication were of at least 3 N purity, and an additional 2 wt.% of Mn was incorporated to account for Mn loss during the melting process. The samples were flipped and remelted several times in order to ensure homogeneity. Next, the as-casted buttons were wrapped in Ta foil, sealed in partially evacuated (Ar partial pressure) vycor tubes, and annealed at 750°C for four days in an electric furnace to promote further homogenization. After annealing, the samples were immediately quenched in cold water.

In order to explore the structure and phase purity of the alloys, powder X-ray diffraction (XRD) measurements were performed using a Scintag PAD X, a device which employs a monochromatic Cu-$K\alpha_1$ radiation source and a theta-theta geometry. The powder patterns were indexed using PowderCell [18]. Compositional properties were examined using a Zeiss Supra 35 FEG SEM, equipped with energy-dispersive X-ray spectroscopic (EDS) capabilities. Magnetization as a function of temperature and applied magnetic field ($M(T)$ and $M(H)$, resp.) were measured using a Physical Property Measurement System (PPMS), manufactured by Quantum Design, Inc. $M(T)$ and $M(H)$ measurements were

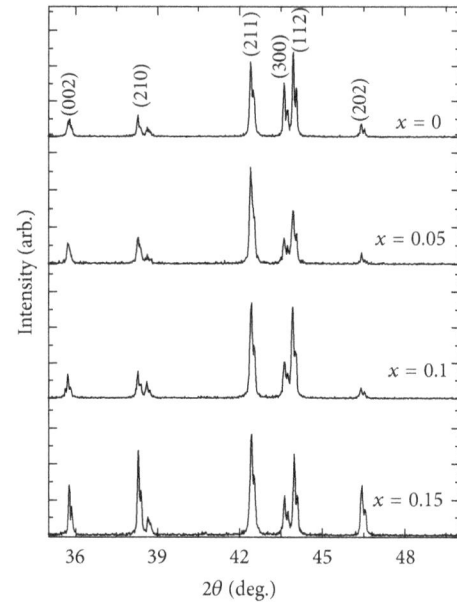

FIGURE 1: Room temperature XRD patterns for the $Mn_{5-x}Fe_xGe_3$ system.

performed in temperatures ranging from 5 to 400 K and in applied magnetic fields up to 50 kOe. $M(T)$ data was collected under both the zero-field-cooled and field-cooled protocols (ZFC and FC, resp.), and each sample's T_C was determined by inspecting the derivative of the $M(T)$ curves for the minima. From $M(H)$ measurements collected isothermally at temperatures near the samples' T_C, ΔS_M was determined using the well-known Maxwell relation [19]:

$$\Delta S_M = - \int_0^H \left(\frac{\partial M}{\partial T} \right)_H dH. \tag{1}$$

Furthermore, as ΔS_M may not necessarily be the best metric by which to judge the appropriateness of a material for MCE applications, we have calculated the refrigerant capacity (RC) of our compounds using the RC_A expression of Gschneidner and Pecharsky, given as follows [20]:

$$RC_A = \int_{T_{Cold}}^{T_{Hot}} \Delta S_M (T, \Delta H) \, dT, \tag{2}$$

where T_{Hot} and T_{Cold} are defined as the temperature bounds forming the full-width at half-maximum of each sample's ΔS_M curve, for a given field change (ΔH).

3. Results and Discussion

The room temperature XRD patterns of the four fabricated $Mn_{5-x}Fe_xGe_3$ compounds are shown in Figure 1. In agreement with previous literature on the $Mn_{5-x}Fe_xGe_3$ system, the compounds were found to crystallize in the hexagonal $D8_8$-type crystalline structure (CSG: P63/mcm). The lattice parameters of the $x = 0$ sample and all samples' c/a ratios match the literature as well. Rietveld refinement of the powder patterns corroborates past neutron diffraction studies

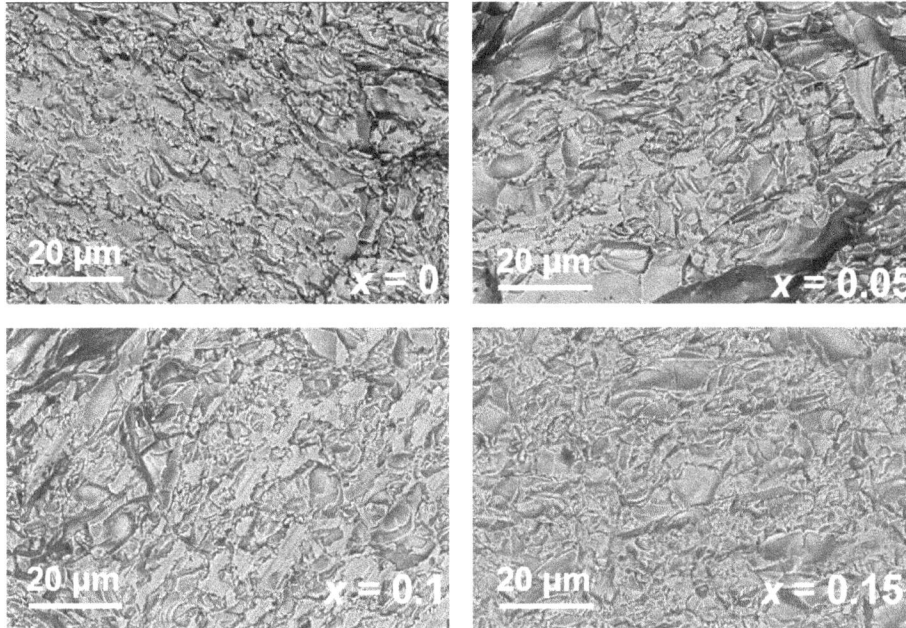

FIGURE 2: Backscatter SEM micrographs of the $Mn_{5-x}Fe_xGe_3$ compounds, collected using a 30 keV primary beam.

of the Mn_5Ge_3 compound, suggesting that Mn occupies the $4d$ and $6g$ crystallographic sites while Ge is localized to the $6f$ sites [24]. The lattice parameters decrease minutely with x, from $a = b = 7.2058$ Å and $c = 5.0398$ Å in the $x = 0$ sample to $a = b = 7.2023$ Å and $c = 5.0375$ Å in the $x = 0.15$ sample. The static c/a ratio across the system indicates a uniform cell contraction, suggesting that the substitutional Fe assumes the $4d$ site in $Mn_{5-x}Fe_xGe_3$. SEM backscatter micrographs of the compounds are shown in Figure 2. The lack of contrast in the micrographs demonstrates that a single compositional phase is formed in the $Mn_{5-x}Fe_xGe_3$ compounds, suggesting that a nearly ideal solid-state solution is formed. EDS analysis (not shown) showed that the fabricated alloys suitably matched their respective targeted compositions.

The $M(T)$ data for the $Mn_{5-x}Fe_xGe_3$ compounds is shown in Figure 3. The $M(T)$ curves were collected in a 1 kOe magnetic field. For each sample, the only anomalous behavior observed is the sudden drop in magnetization corresponding to the ferromagnetic-to-paramagnetic phase transition at T_C. The lack of thermal hysteresis at the phase transition between the ZFC and FC curves indicates that it is a second-order transition. Interestingly, T_C is nearly constant across the series (~296 K). It was previously argued that Fe and Mn randomly occupy the $4d$ sites in $Mn_{5-x}Fe_xGe_3$ compounds [17]. It is possible that the maximum concentration of Fe ($x \leq 0.15$, where 0.15 implies only 3 at. %) in our compounds is not large enough to cause any change in the exchange interactions, as far as T_C is concerned.

To further explore the magnetic properties of the $Mn_{5-x}Fe_xGe_3$ compounds, $M(H)$ curves were collected at $T = 10$ K and the results are shown in Figure 4. As shown in the figure, all samples exhibit typical ferromagnetic behavior and attain magnetic saturation (M_S) well below 50 kOe. As shown in the inset of Figure 4, M_S initially increases with

x until $x = 0.05$ and decreases with further increase of x. The initial increase in M_S may be attributed to the occupancy considerations discussed above as well as the comparatively larger magnetic moment of the Fe atoms versus the $4d$ Mn atoms (2.6 μ_B to 1.85 μ_B, resp.) [25]. The decreased M_S in the higher-Fe samples may be explained by an appeal to the XRD findings, which demonstrated a uniform unit cell contraction with Fe content. As the lattice parameters decrease, it is possible that the ferrimagnetic ordering of the Fe_5Ge_3 phase may become more dominant in determining the bulk magnetic properties of the $Mn_{5-x}Fe_xGe_3$ compounds [17]. Additionally, it is noted that the $M(H)$ curves for samples with $x \leq 0.1$ exhibit relatively "harder" magnetic behavior, whereas the samples with $x > 0.1$ exhibit "softer" magnetic behavior [26]. This indicates that a greater degree of magnetic anisotropy exists in the higher-Fe samples and further confirms a change in magnetic properties across the relatively small compositional span of the $Mn_{5-x}Fe_xGe_3$ ($0 \leq x \leq 0.15$) system.

In order to assess the magnetic entropy changes (ΔS_M) in the $Mn_{5-x}Fe_xGe_3$ compounds, isothermal $M(H)$ data was collected at multiple temperatures in the vicinity of the samples' T_C. As an illustration, isothermal $M(H)$ data for the $x = 0.05$ compound is shown in Figure 5. It is clear that the sample with $x = 0.05$ exhibits typical ferromagnetic behavior (i.e., paramagnetic for isotherms above T_C, ferromagnetic below T_C [27]). From each sample's isothermal $M(H)$ curves, the respective $\Delta S_M(T)$ curve was determined by applying (1) to each of the five field changes, and the results are shown in Figure 6. It is evident that $\Delta S_M(T)$ initially increases with temperature until each sample's respective T_C, at which point it begins to decay. The largest peak ΔS_M values are observed in the sample with $x = 0$—7.0 J/kg K and 3.65 J/kg K for field changes of 50 kOe and 20 kOe, respectively.

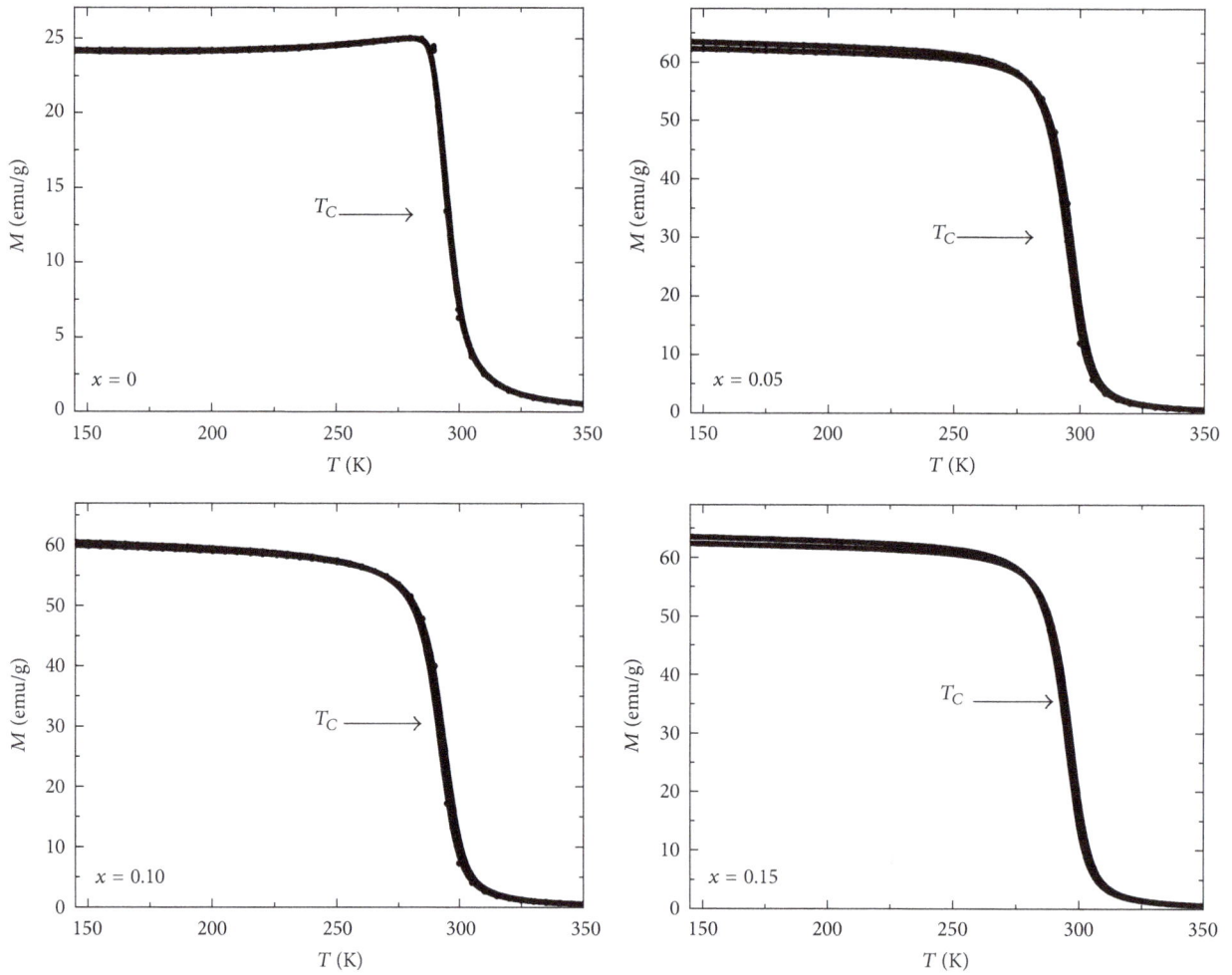

FIGURE 3: Temperature dependence of magnetization, $M(T)$, for the $Mn_{5-x}Fe_xGe_3$ system. The data was collected in an applied magnetic field of 1 kOe.

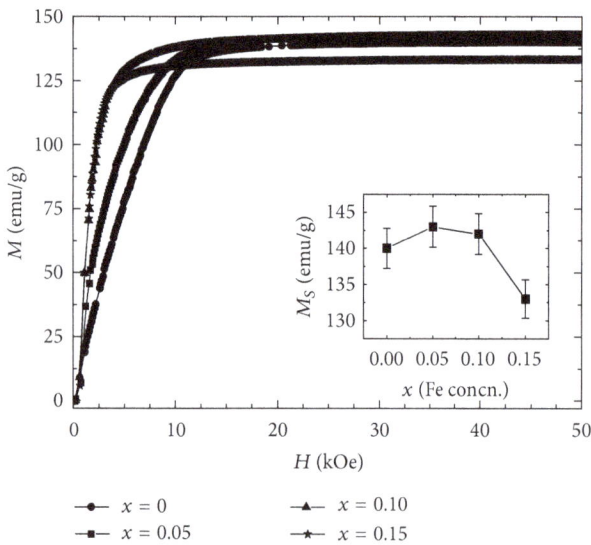

FIGURE 4: Magnetization as a function of applied magnetic field, $M(H)$, for the $Mn_{5-x}Fe_xGe_3$ system, measured at a temperature of 10 K. The inset shows the saturation magnetization, M_S, as a function of Fe concentration.

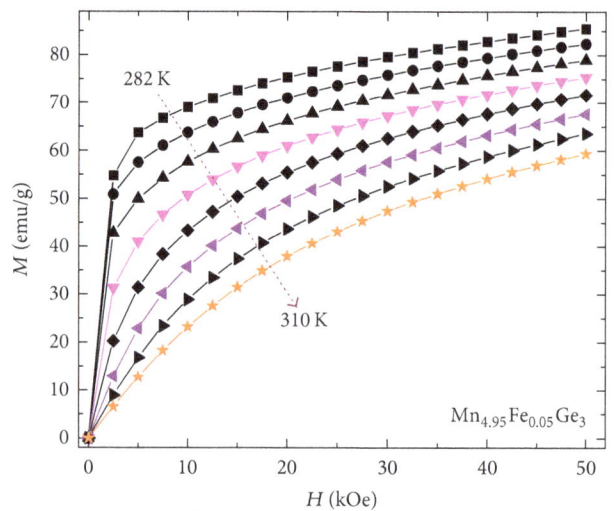

FIGURE 5: Isothermal magnetization as a function of applied magnetic field of the $Mn_{5-x}Fe_xGe_3$ ($x = 0.05$) sample measured at isotherms in the vicinity of the sample's Curie temperature.

As mentioned earlier, the peak ΔS_M values are not the only or most appropriate quantity by which to judge the

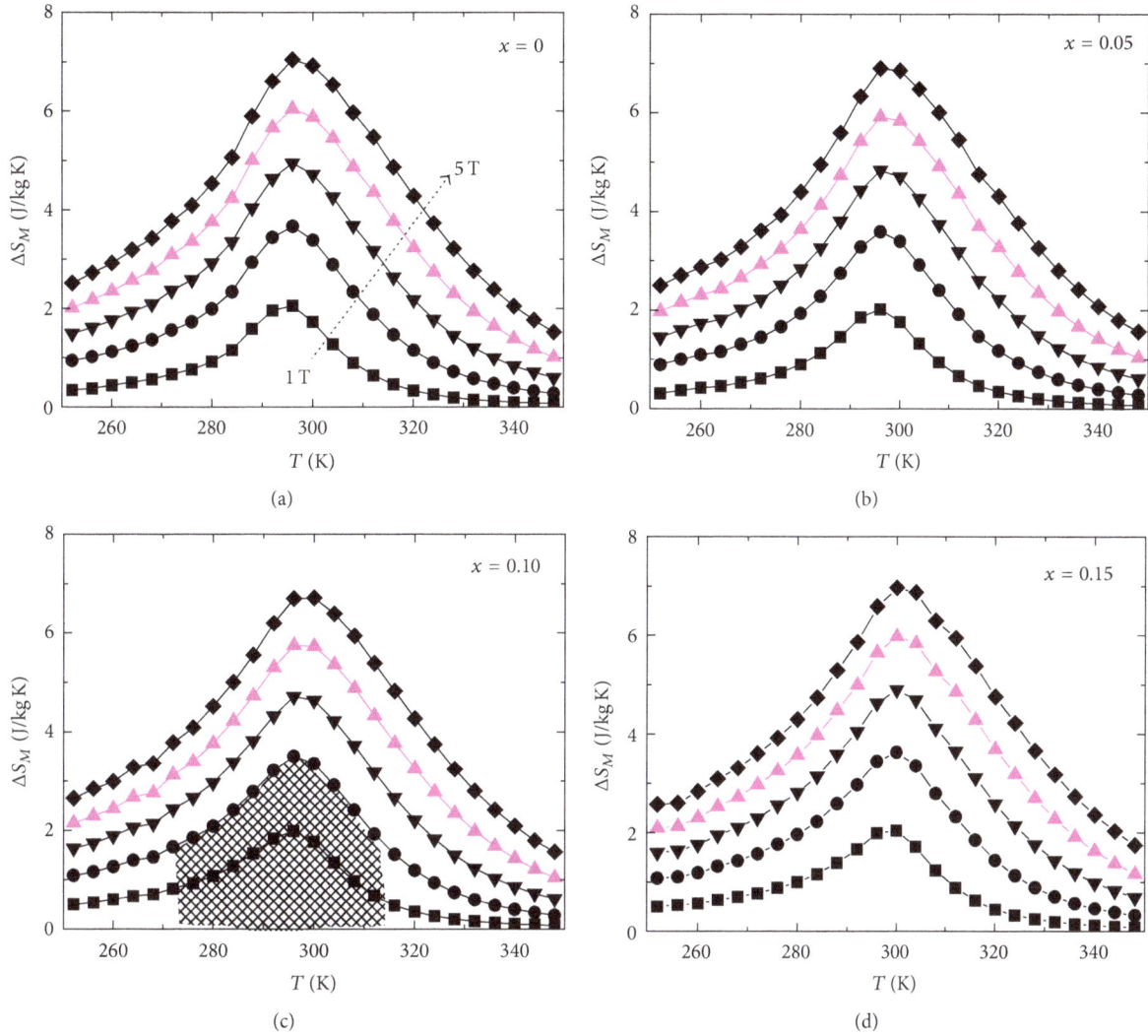

FIGURE 6: The magnetic entropy change as a function of temperature of the $Mn_{5-x}Fe_x Ge_3$ compounds, for magnetic field changes of 10 kOe (1 T) to 50 kOe (5 T).

MCE potential of a material. Using the well-established protocol enumerated in (2), we calculated each sample's RC_A for the explored field changes, and the results are shown in Table 1. A visual representation of this integration for a field change of 20 kOe is provided in Figure 6(c) for the $x = 0.1$ sample. From Table 1, it is apparent that the $x = 0.15$ sample exhibits the largest RC_A values—108.5 J/kg and 312 J/kg for field changes of 20 kOe and 50 kOe, respectively. The heightened MCE properties of the $x = 0.15$ sample may be explained by the broadness of its ΔS_M full-width at half-maximum. While the RC values reported in [17] were calculated using the Wood/Potter technique, recent literature overwhelmingly utilizes RC_A as the MCE assessment parameter; thus, we have chosen to use this metric as it is more suitable for comparison. The RC_A values calculated for the $Mn_{5-x}Fe_x Ge_3$ ($0 \leq x \leq 0.15$) system favorably compare to a variety of recently reported second-order phase transition compounds, including $Mn_5 Ge_{2.7} Ga_{0.3}$ [21], $Mn_5 Ge_{2.8} Sb_{0.2}$ [13], $Gd(Co_{0.35}Mn_{0.65})_2$ [22], and $Ni_2 Mn_{0.95} Cr_{0.05} In$ [23],

as demonstrated in Table 1. The large RC values of the $Mn_{5-x}Fe_x Ge_3$ system suggest that it is worthy of further MCE explorations, including ΔT_{Ad} measurements.

4. Conclusion

In brief, we have performed an experimental study of the structural, magnetic, and magnetocaloric properties of the $Mn_{5-x}Fe_x Ge_3$ ($0 \leq x \leq 0.15$) system. All samples were found to crystallize in the $D8_8$-type hexagonal structure at room temperature and in a single compositional phase. Increasing Fe content has little effect on the Curie temperatures but has a noticeable effect on the saturation magnetization and magnetic anisotropy of the samples, indicating a change in the magnetic properties. For all Fe concentrations, large refrigerant capacities are observed in the vicinity of room temperature; maximal refrigerant capacities are observed in the sample with the highest Fe co™ntent, which has the magnetic anisotropy. The experimental findings

TABLE 1: The magnetocaloric properties of the $Mn_{5-x}Fe_xGe_3$ compounds compared to several other second-order phase transition compounds.

Compound	Peak ΔS_M (J/kg K)		RC_A (J/kg)		Reference
	20 kOe	50 kOe	20 kOe	50 kOe	
Mn_5Ge_3	3.65	7.0	97	298	This work
$Mn_{4.95}Fe_{0.05}Ge_3$	3.58	6.9	95	294	This work
$Mn_{4.9}Fe_{0.1}Ge_3$	3.50	6.96	104.5	305	This work
$Mn_{4.85}Fe_{0.15}Ge_3$	3.63	6.9	108.50	312	This work
$Mn_5Ge_{2.7}Ga_{0.3}$	3.2	6.1	84	256.2	[21]
$Mn_5Ge_{2.8}Sb_{0.2}$	—	6.2	—	320	[13]
$Gd(Co_{0.35}Mn_{0.65})_2$	—	3.4	—	289	[22]
$Ni_2Mn_{0.95}Cr_{0.05}In$	1.69	3.37	59.79	167.5	[23]

suggest that additional magnetocaloric investigations of the $Mn_{5-x}Fe_xGe_3$ system may be warranted.

Conflicts of Interest

The authors declare that there is no conflicts of interest regarding the publication of this paper.

Acknowledgments

The XRD measurements were performed by Professor Shane Stadler's group at the Louisiana State University. The SEM micrographs and related EDS data were collected at the Miami University Center for Advanced Microscopy (CAMI), Oxford, OH 45056, USA.

References

[1] S. Ashley, "Fridge of the future," *Mechanical Engineering*, vol. 116, p. 76, 1994.

[2] C. Zimm, A. Jastrab, A. Sternberg et al., "Description and performance of a near-room temperature magnetic refrigerator," in *Advances in Cryogenic Engineering*, vol. 43 of *Advances in Cryogenic Engineering*, pp. 1759–1766, Springer, Berlin, Germany, 1998.

[3] E. Brück, M. Ilyn, A. Tishin, and O. Tegus, "Magnetocaloric effects in MnFeP1−xAsx-based compounds," *Journal of Magnetism and Magnetic Materials*, vol. 290-291, pp. 8–13, 2005.

[4] V. K. Pecharsky and K. A. Gschneidner Jr., "Giant magnetocaloric effect in $Gd_5(Si_2Ge_2)$," *Physical Review Letters*, vol. 78, no. 23, pp. 4494–4497, 1997.

[5] H. Wada and Y. Tanabe, "Giant magnetocaloric effect of $MnAs_{1-x}Sb_x$," *Applied Physics Letters*, vol. 79, no. 20, 2001.

[6] O. Tegus, E. Brück, L. Zhang, Dagula, K. H. J. Buschow, and F. R. De Boer, "Magnetic-phase transitions and magnetocaloric effects," *Physica B: Condensed Matter*, vol. 319, no. 1-4, pp. 174–192, 2002.

[7] F. Hu, B. Shen, J. Sun, Z. Cheng, G. Rao, and X. Zhang, "Influence of negative lattice expansion and metamagnetic transition on magnetic entropy change in the compound $LaFe_{11.4}Si_{1.6}$," *Applied Physics Letters*, vol. 78, no. 23, pp. 3675–3677, 2001.

[8] O. Tegus, E. Brück, K. H. J. Buschow, and F. R. De Boer, "Transition-metal-based magnetic refrigerants for room-temperature applications," *Nature*, vol. 415, no. 6868, pp. 150–152, 2002.

[9] O. Tegus, G. X. Lin, W. Dagula et al., "A model description of the first-order phase transition in $MnFeP_{1-x}As_x$," *Journal of Magnetism and Magnetic Materials*, vol. 290-291, pp. 658–660, 2005.

[10] E. Brück, O. Tegus, L. Zhang, X. W. Li, F. R. De Boer, and K. H. J. Buschow, "Magnetic refrigeration near room temperature with Fe_2P-based compounds," *Journal of Alloys and Compounds*, vol. 383, no. 1-2, pp. 32–36, 2004.

[11] O. Tegus, B. Fuquan, W. Dagula et al., "Magnetic-entropy change in $Mn_{1.1}Fe_{0.9}P_{0.7}As_{0.3--x}Ge_x$," *Journal of Alloys and Compounds*, vol. 396, no. 1-2, pp. 6–9, 2005.

[12] E. Brück, "Developments in magnetocaloric refrigeration," *Journal of Physics D: Applied Physics*, vol. 38, no. 23, pp. R381–R391, 2005.

[13] Songlin, Dagula, O. Tegus, E. Brück, F. R. De Boer, and K. H. J. Buschow, "Magnetic and magnetocaloric properties of $Mn_5Ge_{3-x}Sb_x$," *Journal of Alloys and Compounds*, vol. 337, no. 1-2, pp. 269–271, 2002.

[14] T. Toliński and K. Synoradzki, "Specific heat and magnetocaloric effect of the Mn_5Ge_3 ferromagnet," *Intermetallics*, vol. 47, pp. 1–5, 2014.

[15] F. Q. Zhao, W. Dagula, O. Tegus, and K. H. J. Buschow, "Magnetic-entropy change in $Mn_5Ge_{3-x}Si_x$ alloys," *Journal of Alloys and Compounds*, vol. 416, no. 1-2, pp. 43–45, 2006.

[16] X. B. Liu and Z. Altounian, "Exchange interaction in GdT_2 (T=Fe,Co,Ni) from first-principles," *Journal of Applied Physics*, vol. 107, no. 9, 2010.

[17] Q. Zhang, J. Du, Y. B. Li et al., "Magnetic properties and enhanced magnetic refrigeration in $(Mn_{1-x}Fe_x)_5Ge_3(Mn_{1-x}Fe_x)_5Ge_3$ compounds," *Journal of Applied Physics*, vol. 101, no. 12, Article ID 123911, 2007.

[18] http://www.bam.de/de/service/publikationen/powder_cell.htm.

[19] T. Hashimoto, T. Numasawa, M. Shino, and T. Okada, "Magnetic refrigeration in the temperature range from 10 K to room temperature: the ferromagnetic refrigerants," *Cryogenics*, vol. 21, no. 11, pp. 647–653, 1981.

[20] K. A. Gschneidner and V. K. Pecharsky, "Magnetocaloric materials," *Annual Review of Materials Science*, vol. 30, no. 1, pp. 387–429, 2000.

[21] L. Xi-Bin, Z. Shao-Ying, and S. Bao-Gen, "Magnetic properties and magnetocaloric effects of $Mn_5Ge_{3-x}Ga_x$," *Chinese Physics*, vol. 13, no. 3, pp. 397–400, 2004.

[22] J. Y. Zhang, J. Luo, J. B. Li et al., "Magnetocaloric effect of $Gd(Co_{1-x}Mn_x)_2$ compounds," *Solid State Communications*, vol. 143, no. 11-12, pp. 541–544, 2007.

[23] J. Brock and M. Khan, "Large refrigeration capacities near room temperature in $Ni_2Mn_{1-x}Cr_xIn$," *Journal of Magnetism and Magnetic Materials*, vol. 425, pp. 1–5, 2017.

[24] J. B. Forsyth and P. J. Brown, "The spatial distribution of magnetisation density in Mn_5Ge_3," *Journal of Physics: Condensed Matter*, vol. 2, no. 11, pp. 2713–2720, 1990.

[25] N. Yamada, S. Shibasaki, K. Asai, Y. Morii, and S. Funahashi, "Magnetic properties of $(Mn_{1-x}TE_x)_5$ Ge_3: (TE = Cr and Fe)," *Physica B: Physics of Condensed Matter*, vol. 213-214, pp. 357–359, 1995.

[26] M. Khan, I. Dubenko, S. Stadler, and N. Ali, "Magnetic and structural phase transitions in Heusler type alloys Ni 2MnGa1-xinx," *Journal of Physics Condensed Matter*, vol. 16, no. 29, pp. 5259–5266, 2004.

[27] B. Arayedh, S. Kallel, N. Kallel, and O. Peña, "Influence of non-magnetic and magnetic ions on the MagnetoCaloric properties of $La_{0.7}Sr_{0.3}Mn_{0.9}M_{0.1}O_3$ doped in the Mn sites by M=Cr, Sn, Ti," *Journal of Magnetism and Magnetic Materials*, vol. 361, pp. 68–73, 2014.

19

Analysis of Metal Flow Behavior and Residual Stress Formation of Complex Functional Profiles under High-Speed Cold Roll-Beating

Fengkui Cui,[1,2] **Yongxiang Su** ⓘ**,**[1,2] **Kege Xie,**[1,2] **Wang Xiaoqiang,**[1,2] **Xiaolin Ruan,**[1,2] **and Fei Liu** ⓘ[1,2]

[1]*School of Mechatronics Engineering, Henan University of Science and Technology, Luoyang 471003, China*
[2]*Collaborative Innovation Center of Machinery Equipment Advanced Manufacturing of Henan Province, Luoyang 471003, China*

Correspondence should be addressed to Yongxiang Su; syxsuper@163.com

Academic Editor: Antonino Squillace

To obtain a good surface layer performance of the complex functional profile during the high-speed cold roll-beating forming process, this paper analyzed the metal plastic flow and residual stress-formed mechanism by using a theoretical model of the metal flow and residual stress generation. By using simulation software, the cold roll-beating forming process of a spline shaft was simulated and analyzed. The metal flow and residual stress formation law in the motion were researched. In a practical experiment, the changes in the grains in the spline tooth profile section and the residual stress distribution on the tooth profile were studied. A microcorrespondence relationship was established between the metal plastic flow and the residual stress generation. The conclusions indicate that the rate at which the metal flow decreases changes gradually at different metal layers. The residual stress value is directly related to the plastic flow difference. As the roll-beating speed increases, the uneven degree of plastic deformation at the workpiece surface increases, and the residual stress in the tooth profile is generally greater. At the same roll-beating speed, the rate change trend of the metal flow decreases gradually from the surface to the inner layer and from the dedendum to the addendum. The residual stress distribution on the surface of the tooth profile decreases from the dedendum to the addendum. These findings provide a basis and guidance for the controlled use of residual stress, obtaining better surface layer quality in the high-speed cold roll-beating process of the complex functional profile.

1. Introduction

The high-speed cold roll-beating precision formation of a complex functional profile is a sophisticated, efficient, green, and advanced manufacturing technology [1, 2]. It uses the plasticity of metal materials to plastically deform the workpiece and maintain its shape after deformation. The metal plastic forming process enables the formation of a continuous fibrous tissue inside the workpiece to strengthen the material and improve the mechanical properties of the workpiece. However, metal plastic forming technology will inevitably produce residual stress in the workpiece interior [3, 4]. On one hand, the residual stresses can reduce the expansion of surface microcracks and improve the workpiece fatigue strength; on the other hand, the stresses decrease the stability of the workpiece size [5–7]. In addition, recent studies on cold roll-beating formation have shown that changes in the surface roughness and surface hardening of the workpiece are closely related to the generation of residual stress [8–10]. Therefore, research on the formation, control, and utilization of residual stress in the high-speed cold roll-beating forming process has high academic significance and high value for engineering applications.

In previous research, Smirnov proposed the feasibility of using splines formed by cold roll-beating based on the experimental data. The inhomogeneity and complexity of the deformation extent and stress distribution in the forming process were verified in an experimental data analysis [11]. Kurz established a simulation model of plastic deformation in the cold roll-beating process that used finite element theory to calculate the value of forming force as well as stress and strain on the workpiece in the cold roll-beating process. A comparison of the simulation results and the experimental data showed that the simulation results were reliable [12]. Cui studied the dynamic response of the cold roll-beating process and the stress wave. The metal flow law and deformation mechanism of the involute spline were revealed from the macroperspective by analyzing the involute spline cold roll-beating process [13]. Zhang et al. used the slab method to analyze the deformation force and established an analytical equation by a simulation analysis [14].

In research on forming accuracy, Grob and Krapfenbauer studied the installation angle problem of a roll-beating wheel. The roll-beating wheel was fixed on the rolling head at the appropriate installation angle by adding a thrusting ring to improve the accuracy of the workpiece [15]. Cui modified the profile of the roller based on the relationship of the relative motion between the roll-beating wheel and the workpiece. A higher surface accuracy of the involute spline can be obtained by cold roll-beating than that can be obtained by milling [13].

In research on the forming surface quality, Weck et al. noted that the fatigue strength increased by 25%–35% due to the hardening layer on the surface of the cold roll-beating forming products by comparing the products of cold roll-beating and cutting [16]. Cui et al. studied the fiber tissue and the tooth surface hardness distribution of the spline in the process of cold roll-beating through a simulation analysis and experimental observation. The plastic deformation and the phenomenon of surface hardening during the process of cold roll-beating were explained from the microscopic perspective. The effect and properties of metal plastic deformation on the structure were analyzed. The effect of thermomechanical coupling on the metal surface hardening and the effect of forming parameters on the surface roughness of the workpiece in the cold roll-beating process were obtained [8, 9, 17, 18].

In summary, recent studies on the surface quality have concentrated mainly on examining the workpiece roughness, work hardening, and other factors. Little research has been done on the problem of residual stress being caused by uneven plastic deformation in the process of cold roll-beating to date. Therefore, the research on residual stress has important significance for determining the process parameters and further improving the surface quality of cold roll-beating.

In this paper, the metal plastic flow law of spline shafts in the process of cold roll-beating was studied based on the law of minimum resistance and volume invariance. The effect of uneven extent of the plastic deformation with different cold roll-beating speeds on the overall workpiece and the same rotation speed at different parts of the spline tooth profile was analyzed. The forming process of cold roll-beating was simulated, and the displacement of nodes was tracked by finite element software to verify the theoretical part of this

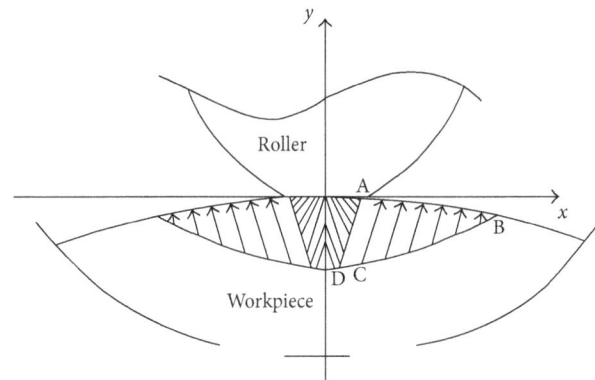

FIGURE 1: Contact model of cold rolling.

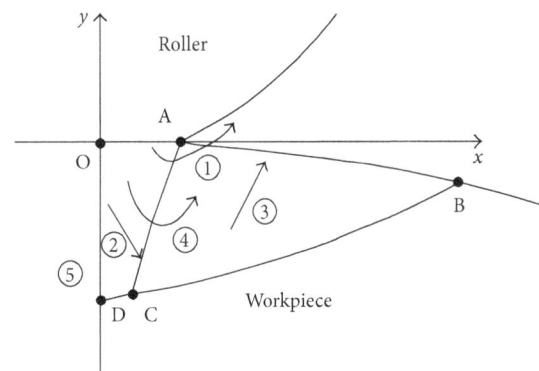

FIGURE 2: Analytic result of metal flowing trend.

paper regarding the metal plastic flow law. By extracting the residual stress from the surface of the spline tooth profile, this paper clarified the effect of the metal plastic flow process on the residual stress formation in the profile of spline teeth. At the end, the forming mechanism of the residual stress under high-speed cold roll-beating was revealed, and the experiment verified the theoretical and finite element simulation parts of this paper.

2. Metal Flowing and Residual Stress Forming Mechanism of the Complex Functional Profile under High-Speed Cold Roll-Beating

2.1. Metal Flowing of the Complex Functional Profile under Cold Roll-Beating. By analyzing the process of cold roll-beating, Cui et al. [19] established the contact model of a roller and a workpiece, as shown in Figure 1. The trend of metal flow at the involute spline tooth profile is determined by the law of least resistance, as shown in Figure 2.

It can be seen from the metal flow trends shown in Figure 2 that with the invasion of the roller, the metals in the ACDO region gradually flow into the ABC region and the region outside the AB border. The new inflowing metals force the metal in the ABC region to flow to the AB boundary. In the flowing process, because the metals near the top position of the roller (especially near the C point on the CD online) flow under the direct compression force of the roller, the flow rate is the largest. As a result, the most

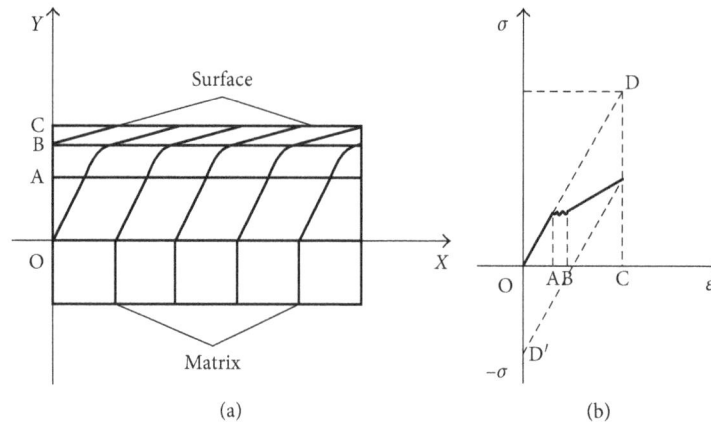

FIGURE 3: Schematic form of residual stress. (a) Metal flowing. (b) Generation model of residual stress.

uneven plastic flow occurred in this region. The CA line, due to the metals near point C, suffered much greater rolling stress than did the area near point A, and the resistance of the metal flow in the ABC area is greater than that to the flow along the line CA. Therefore, when deformation occurs, parts of the metals at C point in the ACDO area flow into the ABC region, and other parts flow along the CA line direction. However, close to point A, the flow pattern weakens gradually. Therefore, the degree of unevenness on the metal plastic flow gradually decreases along the CA line direction, as does the residual stress. However, in the ABC region, since the metal is subjected to indirect forces of the roller from flowing, the force and metal plastic flow are relatively uniform.

2.2. Residual Stress Forming of the Complex Functional Profile under High-Speed Cold Roll-Beating. It is known from the above analysis that due to the nature of the metal material and the effect of the changing force during the cold roll-beating forming process, the metallic material flow of the workpiece cannot be uniform, leading to the interaction of the plastic deformation portion and the elastic deformation portion inside the workpiece structure. Finally, an elastic stress field is formed. Under the dual action of the elastic stress field recovery and plastic deformation hindrance, the internal residual stress is produced. The formation diagram is shown in Figure 3. The metal flowing at different depth layers during the forming process is described in Figure 3(a), and the generation model of residual stress is described in Figure 3(b).

In Figure 3(a), the *OX* axis represents the flow direction of the metal surface layer, and the *OY* axis represents the thickness of the workpiece metal layer. The OA segment is the elastic deformation area. The AB segment shows the elastic-plastic deformation zone. The BC section is the plastic deformation zone. In Figure 3(b), the $O\varepsilon$ and $O\sigma$ axes indicate the workpiece material strain and stress, respectively. OA, AB, and BC represent the elastic deformation stage, yield stage, and effective plastic strain-strengthening stage, respectively. Without considering the plastic deformation of the material, point D is the maximum stress that the material can achieve before reaching the material's

strain limit. Point D′ is the residual stress produced by the ideal elastic recovery action after the external force is removed.

The schematic of the residual stress formed is shown in Figure 3(a). When the metal plastic flow occurs on the surface layer of the workpiece, the deformation resistance of the matrix layer (the sections below the *OX* axis) is greater, but hardly any deformation appears. However, the elastic deformation gradually increases in zone OA, which is far from the matrix layer and near the surface layer. Then, some metals yield and enter the transition area AB of elastic-plastic deformation. After the external force is removed, the elastic deformation has a tendency to recover but is still affected by plastic deformation in the surface area; the matrix area and the existing state of elastic deformation will be maintained. The stored deformation can be expressed as residual stress.

The value of the residual stress formed is shown in Figure 3(b). Workpiece materials under the action of the external force first produce an elastic deformation (OA). With an increase in external force, workpiece materials enter the yield stage (AB). As the increases continue, materials enter the effective strain-strengthening phase (BC). If the plastic deformation has not occurred and the strain reaches point C, the material stress reaches point D, and the stress and strain return to the origin after removing the external force. In fact, because of the plastic deformation of the materials and the deformation field recovering along the dotted line, in parallel with the elastic line, strain still exists in the material when the stress has decreased to zero. Hence, to restore the strain to zero, the stress values should change continually along the dotted line until reaching the intersection with the negative side of the longitudinal axis, as shown at point D′ in Figure 3(b). The stress value at point D′ is the size of the residual stress value.

3. Simulation Analysis of Metal Flow and Residual Stress Forming of a Complex Functional Profile under Cold Roll-Beating

Taking the involute spline for high-speed cold roll-beating forming as an example and using finite element simulation

TABLE 1: J-C material model parameters.

A (MPa)	B (MPa)	C	m	n	E (MPa)	μ
213	53	0.051	0.76	0.345	2.13e5	0.282

TABLE 2: Other physical parameters.

E (MPa)	μ	T_r (K)	T_m (K)	K (W/(m·K))	SPH (J/(kg·K))	α
2.13e5	0.286	1623	293	48.00	440	1.19e−5

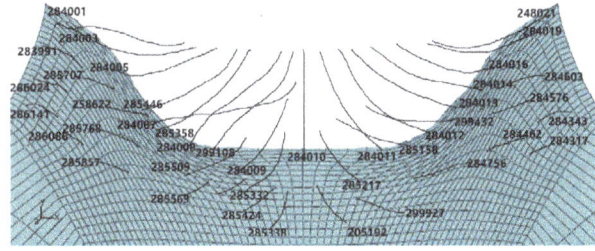

FIGURE 4: Metal flow trajectory in the simulation.

software, the metal flow conditions and residual stress on the tooth profile during the forming process were simulated. A schematic of the cold roll-beating forming spline was established by Cui [13, 19].

3.1. Model Building

3.1.1. Finite Element Model.
According to the cold roll-beating forming movement process given in [13] and [19], the roll-beating model is simplified by analyzing half of the symmetric model. To improve the efficiency of the analysis without affecting the results, the workpiece's self-rotation was ignored, and a single tooth profile surface formed under roll-beating was analyzed. The axial dimension of the semi-finished shaft is set as 60 mm, and the semi-finished shaft diameter is 35.15 mm, as obtained by the estimating equation. According to the standard inverse involute equation, the tooth profile of the roller is established; its radius is 19 mm, thickness is 8 mm, modulus is 2.5, and rounded value is 0.5 mm. The gyration radius is 55 mm, and the strike depth is 1.95 mm.

Since the roller partially contacts the workpiece during the roll-beating forming process, the workpiece mesh block, and the partial contact mesh refinement, the element type selected is the Solid 164 entity structure unit.

3.1.2. Material Model.
Through dynamic compression tests on 1020 steel with the cold roll-beating experimental platform, the stress-strain data were obtained at different temperatures and strain rates. The J-C material model parameters were established by the data, listed in Table 1, and other physical parameters, listed in Table 2.

Relative to the workpiece material (1020 steel), the deformation of the roller with a material of Cr12MoV does not need to be considered in the high-speed cold roll-beating forming process. In this situation, the roller is defined as rigid, according to the movement of the roller. The rotational freedom along the Z-axis is retained only with the definition of the rigid material model.

3.2. Key Simulation Parameters.
The symmetry constraint is applied on the symmetry plane of the workpiece, and displacement constraints are applied on the rear and perpendicular to the workpiece feed direction. A surface-to-surface eroding contact is used between the roller and the spline

shaft blank. The dynamic and static friction coefficients are set as 0.12 and 0.2, respectively. The hourglass control and volume viscosity are set as the system defaults. The feed rate of the workpiece is 0.5 mm/s, and different rotational speeds of the roller are set as 188.5 rad/s (1800 r/min) and 235.6 rad/s (2250 r/min) for comparison.

3.3. Simulation Results and Analysis

3.3.1. Metal Flowing Analysis.
The metal flow patterns obtained by simulation under different roll-beating speeds are very similar, so the metal flow trajectory under the 188.5 rad/s speed is depicted in Figure 4.

As seen in Figure 4, the contact area near the peripheral area of the node (e.g., nodes 284019, 284016, and 284014) flows to the blank area in the upper right. The flow path can be attributed to the number 1 trace line in Figure 2. In the area near the center line, the main flow direction of the node (e.g., nodes 285217 and 285192) is obliquely downward, with an angle toward the center line. The flow trajectory is nearly linear and can be attributed to the number 2 trace line in Figure 2. The nodes on both sides (e.g., nodes 284603, 284576, 284462, 284343, and 284317) that are far from the deformation zone flow to the region outside the shaft blank in a certain direction and along an approximately linear trajectory that can be attributed to the number 3 trace line in Figure 2. The node (e.g., nodes 299432 and 285158) is located in the middle of the right half of the gullet flow in the forming region and has difficulty flowing to the region outside the shaft blank. The flow trajectory is a certain curve and can be attributed to the number 4 track line in Figure 2. The nodes represented by node 284010 on the center line flow down along the center line of the cogging that are summed in Figure 2 on the number 5 track line.

Comparing the simulation results of the metal overall flow trend (Figure 4) and the theoretical analysis of the metal flow trend (Figure 2), the conclusions are consistent.

3.3.2. Residual Stress Analysis.
The main failure modes of spline teeth are fatigue fracture and tooth surface pitting; fatigue fractures occur at the root of the tooth, and the dedendum is exactly the stress concentration point. In the case of continuous stress shocks, fatigue cracks are the most likely to occur, resulting in broken dedendum. Residual compressive stress will inhibit the generation and propagation of cracks, thereby increasing the fatigue strength at

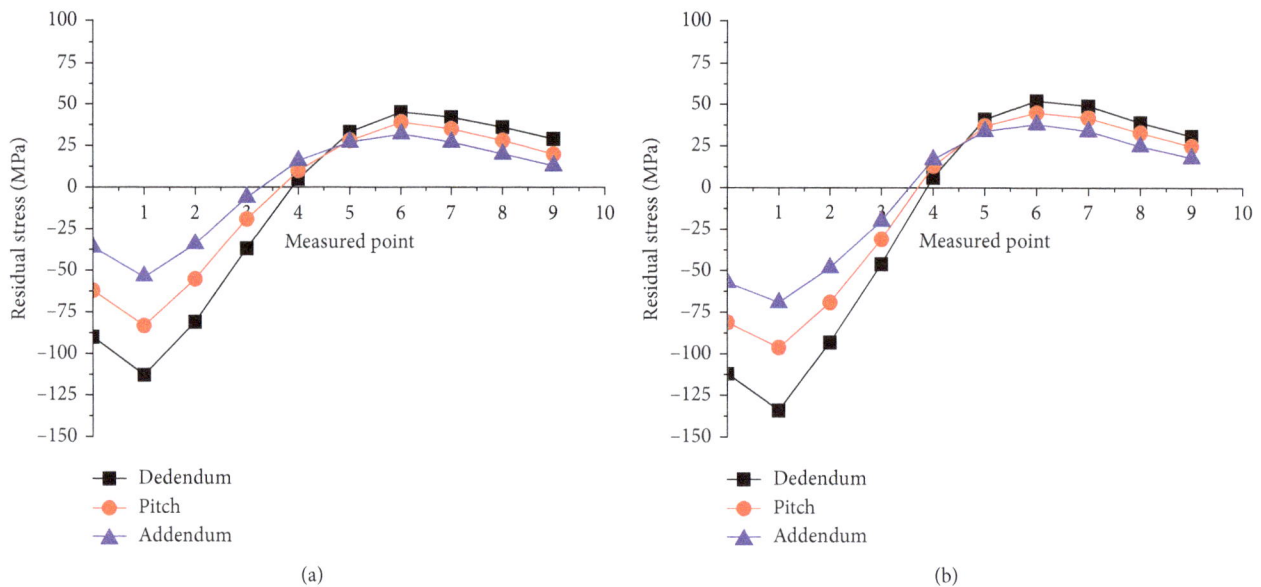

FIGURE 5: Residual stress distribution curve under different rotation speeds: (a) 188.5 rad/s and (b) 235.6 rad/s.

the tooth roots. Therefore, the use of splined dedendum is selected as a research point. The tooth surface pitting occurs mostly at the pitch of the splines, and the main contact portion of the two spline teeth in the meshing process is the pitch; in the meshing process, spline teeth have both relative rolling and relative sliding, and the relative sliding friction force is opposite in direction of the two sides of the pitch circle node; therefore, a pulsating load is generated, and the effect of these two forces results in pulsating cyclically varying shear stress at the depth of the gear index circle. When the shear stress exceeds the shear fatigue limit of the gear material, the surface will generate fatigue cracks, and the crack propagation will eventually cause the tooth surface to flake off and form a small crater on the tooth surface; therefore, it is selected as the research point to select the pitch of spline. During the cold roll forming process, the splined addendum is generated by the metal flow, so the splined addendum is selected as the research point, and the fatigue strength at the tooth tip is mainly checked by measuring the residual stress.

The internal residual stress distribution curves for dedendum, pitch, and addendum of the spline tooth profile under different speeds are shown in Figure 5.

As seen in Figure 5, the residual stress distribution curve trends of tooth profiles of three positions under the two different speeds are roughly the same. However, as the speed increases, there is an increasing trend of residual stress on the whole tooth profile. The residual stress distribution curve from different locations on tooth profiles at the same speed shows that the surface of the tooth profile is compressive residual stress. With a measurement depth to the tooth surface (measured point increasing), the internal tooth profile gradually shows the tensile residual stress, and the maximum value and gradient of the compressive stress are larger than those of the tensile residual stress. Residual stress

reaches the maximum at the dedendum and the minimum at the addendum.

Combined with the theoretical section of this paper, as the roll-beating speed increases, the strain rate of the workpiece increases, leading to an increase in the material flow stress in the deformation. Thereby, the deformation depth of the workpiece layer is affected; however, because the deformation resistance of metal at depth is larger than that at the surface, deformation does not occur easily at depth, and the plastic flow on the surface of workpiece is more uneven. Therefore, as the rotating speed increases, the residual stress on the whole tooth profile also has an increasing trend. In the root of the spline tooth profile part, because of the direct contact with the outer edge of the roller, the strain, strain rate, and deformation resistance are larger than those at the pitch and addendum position during deformation. As a result, the degree of unevenness of the metal flow increases, and the final residual stress formed here is the largest. In the addendum position, the force of the workpiece material caused by the roller is reduced, and the metal's flow decreases. Residual stress internal to the metal is gradually released at the same time; therefore, the residual stress is less than that at the pitch position.

4. Experimental Study of Metal Flow and Residual Stress Formation

4.1. Experimental Scheme. The 1020 steel blanks come from the same batch. The samples are machined by a Swiss ZRMe9 roll-beating machine from Grob Inc. The modulus of the involute spline is 2.5. There are 14 teeth. The forming method is pulling out down-beating and continuous indexing. The samples are divided into two groups. The processing parameters of the first group are as follows: the spindle speed is 1809 r/min and the feed rate is

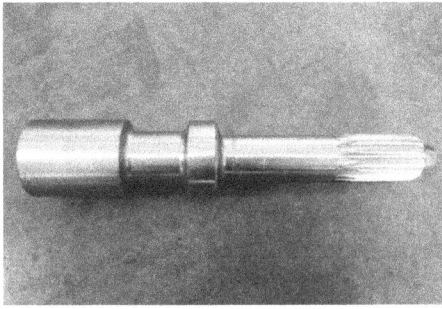

FIGURE 6: Processed involute spline of cold roll-beating.

28 mm/min. The parameters of the second group are as follows: the spindle speed is 2258 r/min and the feed rate is 28 mm/min. The involute spline after processing is shown in Figure 6.

A tooth of the spline was cut by wire electric discharge machining (wire EDM). The surface of the tooth section was polished, and it was cut from the individual tooth along the tooth direction. After cleaning the surface with alcohol swabs and rinsing the surface with alcohol containing 4% concentrations of nitric acid for 50 seconds (until the surface of the sample was light brown), the samples were finally obtained.

Using a JSM-5610LV scanning electron microscope (SEM), the microstructures and morphologies at the position of the tooth top, the dividing circle, and the dedendum of the tooth section in two groups were observed. The electron acceleration voltage was 20 kV.

In the WEDM-LS procedure, using a 0.1 mm bronze wire, a tooth is cut off from the spline at a speed of 2 mm/min, and the specimen is then cut along the symmetrical surface shown in the shaded part of Figure 7(a) at a feed rate of 0.5 mm/min. The dimensions of each specimen are as follows: $l = 10$ mm, $w = 4.35$ mm, and $h_0 = 2.68$ mm. The point coordinates of the cutting plane of Figure 7(a) were measured using a Serein-CMM FUNCTION 1000 three-coordinate measuring machine (to reduce the error, both surfaces produced by cutting must be measured, for a total of four surfaces). Measurements were performed at 0.01×0.01 mm intervals using the reciprocating measurement method (a single measurement track along the direction parallel to the cutting line is required to cover the two surfaces). The annealed samples are considered not to contain residual stress, so after the cutting, any form of deformation of the unannealed specimen relative to the corresponding position of the annealed specimen can be attributed to the release of residual stress. After measurement, the measured data that correspond to the two planes are subtracted to obtain the measured point change amount (vector deformation), which is the amount of deformation caused by the release of the residual stress of the specimen. A curved surface for the small change (annealed specimen relative to the unannealed one) corresponding to each measured point is fitted by using three order spline fitting algorithm. Then, the surface was inverted and taken as a boundary condition to be applied in the finite element

model, with the same size and shape as the deformed specimen, using Abaqus software. The material model used in the simulation analysis was a static stress-strain simulation that was established using quasistatic compression tests on 20 steel samples on the experimental platform. Then, a finite element static solution was performed in the Abaqus software, and the deformed model was restored to the shape before cutting. To avoid rigid body displacement in the model analysis process, an additional constraint that did not affect the free deformation of the contour was imposed on the corner node at the other end of the model, consistent with the literature [20–23]. Finally, the stress on the cut surface of the model obtained after the solution was obtained was equivalent to the residual stress at the same position when the sample was not cut. The extraction direction and position are shown in Figure 7(b). The residual stress at the spline circle is extracted by the position and direction of b, and the extraction node spacing is approximately 0.05 mm; some distortion exists, so it is not possible to select nodes with strict regular spacing. In this case, $h_a = 2.20$ mm, $h_d = 1.39$ mm, and $h_f = 0.5$ mm.

4.2. Metal Plastic Flow Analysis of the Experimental Results.
The microstructures of the spline tooth section in the two experimental groups are shown in Figures 8(a) and 8(b), respectively. In the images, from top to bottom, the position of addendum, pitch, and dedendum are at the distances of 0.1 mm, 0.5 mm, and 0.9 mm, from left to right.

Compared with the two groups' microstructure morphology results of splines obtained from the experiments, the metal grains change significantly at different speeds (different strain rates). At a higher strain rate, a larger amount of grain deformation occurs inside the metal, and a higher degree of unevenness is produced. According to the direct invasion influence of the roller on the spline surface, metals flow rapidly along the roller surface, and a dense fibrous structure is formed. Since the internal metals of the spline teeth suffer an indirect force caused by the surface, the stress is relatively uniform, and the resistance of the movement is larger in this part than it is on the surface. Therefore, along the surface to the internal area, the elongated degree of the grain is reduced, and the metal plastic flows weaken gradually.

The microstructure topographies at three different positions of two spline sectionals show that, at the same rolling speed, the elongated degree of the grain in different positions of the spline cross section is also different. The nonuniform degree of grain is highest in the tooth position, and grains are stretched significantly. The elongation at the addendum is not as high, and the grain deformation is similar. The degree of grain deformation at the pitch position is between those at the dedendum and addendum. Comparing the grain deformation in different locations, the degree of nonuniform metal plastic flow is the highest in the root position, followed by the pitch, and the flow is the lowest at the addendum.

In summary, a larger strain rate leads to a larger nonuniform degree of metal plastic flow. At the same strain rate,

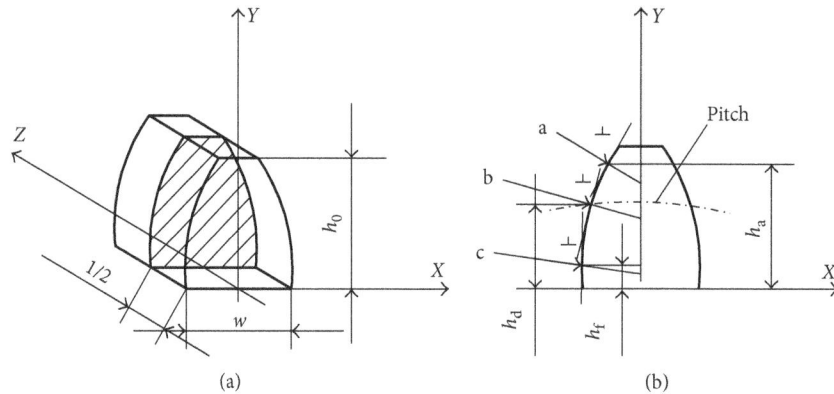

FIGURE 7: Measured scheme of residual stress of the contour method. (a) Sample shape and cutting position needed for measurement. (b) Measured position and direction.

the degree of incomplete plastic deformation is the highest in the root position, followed by the graduated circle, and is the lowest in the addendum position. This is entirely consistent with the previous analysis.

4.3. Residual Stress Analysis of the Experimental Results. The residual stresses in the root, pitch, and addendum, as measured by the contour method at 1809 r/min and 2258 r/min spline speeds, are shown in Figures 9(a) and 9(b), respectively. Only the residual stress on the section from the left tooth surface to the *YZ* plane symmetry is depicted because the single spline tooth is symmetrical with respect to the *YZ* plane. The distances of three positions on the left flank to center symmetry are 2.14 mm, 2.00 mm, and 1.82 mm. The stress measurement point is located on the sample l/2 symmetry plane with a 0.25 mm interval measurement to the *YZ* plane.

Compared with the residual stress size and distribution shown in Figures 9 and 5, the experimental and simulation results have a high degree of consistency in the overall curve trend. These results show that compressive residual stress is formed on the surface of the tooth profile and that tensile residual stress is formed deep below the surface. The residual stress is larger at the dedendum than are those at the pitch and the addendum, and it is smallest at the addendum. Compared with the results obtained at the two forming speeds, the residual stress tends to increase as the spindle speed increases. These results all coincide with the theory and simulation.

The sizes of the residual stress obtained by experiments and simulations show that the residual stress from the numerical simulation is slightly larger in the whole system. High-speed cold roll-beating is a formation process with large deformation, a high strain rate, a high instantaneous temperature, and high stress values. The process may lead to transformation in the area of the workpiece and cause the residual stress to change. Furthermore, after forming, residual stress on spline shaft tooth profile surface will gradually release and decrease over time. However, in the cold roll-beating forming simulation, to ensure that the

model is not too complex, the analysis failed to take the transformation and residual stress releasing gradually over time during the forming process into account (as the time increases, the releasing roughly obeys a logarithmic curve form). Thus, the final residual stress values that are from the experiments are slightly lower than those of the numerical simulation.

5. Discussion

Compared with the simulation (Figure 5) and experimental (Figure 9) residual stress results, at high spindle speeds, the value and depth of residual stress are both larger because the nature of the residual stress is generated by the nonuniform plastic deformation of the workpiece material. At a high spindle speed, an increase in the workpiece material strain rate causes a greater material flow stress, greater effect depth in the workpiece, and higher degree of inhomogeneous deformation of the workpiece material. As a result, the depth and value of residual stress increase finally. At a high spindle speed, the increase in the inhomogeneous deformation of the workpiece material could be determined by the experiment (Figure 8). During the forming process, the grains on the tooth profile are stretched, and the lattice is distorted. The elongated degree of the grains on the tooth profile of the workpiece and the depth of tensile deformation grains are both greater at high spindle speeds.

The residual stress distribution is different at different positions of the tooth profile during the high-speed cold roll-beating forming process. As seen from the simulation (Figure 4) and the experimental metal flow (Figure 8), the degree of metal plastic deformation caused by the metal flow is different in different parts of the tooth profile. The degree of metal plastic deformation is clearly the largest in fillet parts of the dedendum. In this position, the severe deformation element extrudes deeper into the element in the simulation. The grain is pulled into a dense fiber whose thickness is larger than elsewhere under a scanning electron microscope. The metal plastic deformation degree decreased at the position of the pitch circle and is smallest at the addendum.

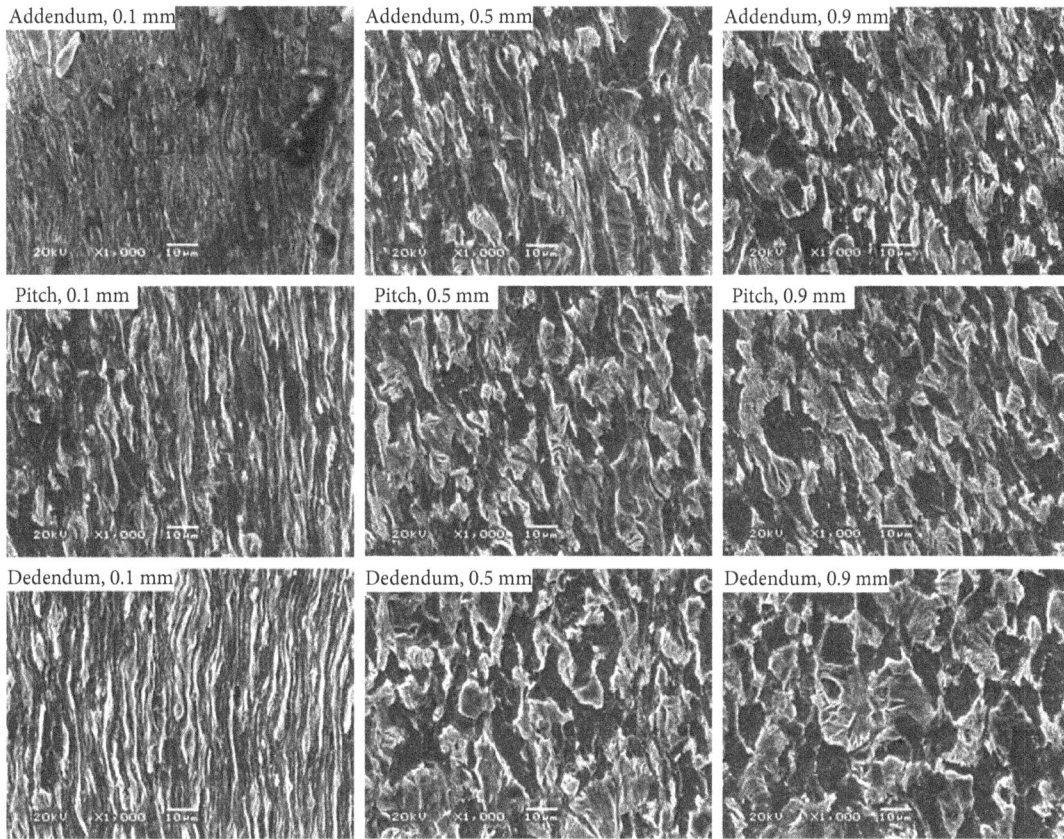

(a)

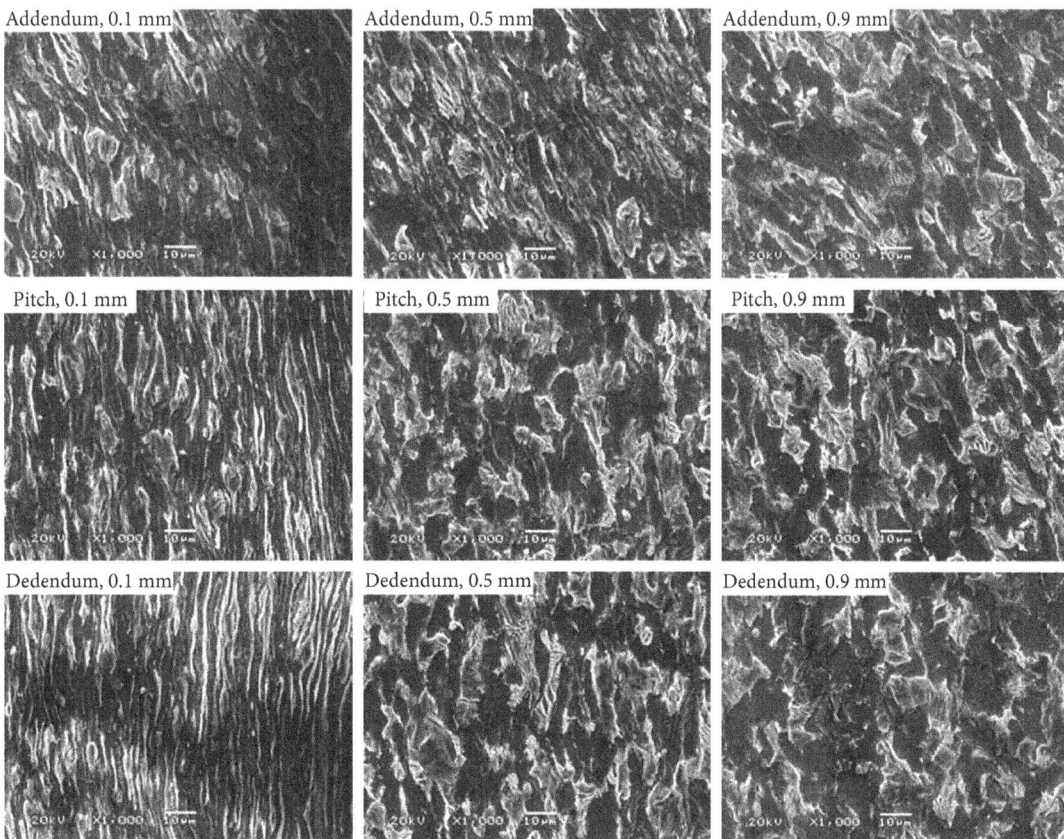

(b)

FIGURE 8: Microstructure morphology under different spindle rotation speeds: (a) 1809 r/min and (b) 2258 r/min.

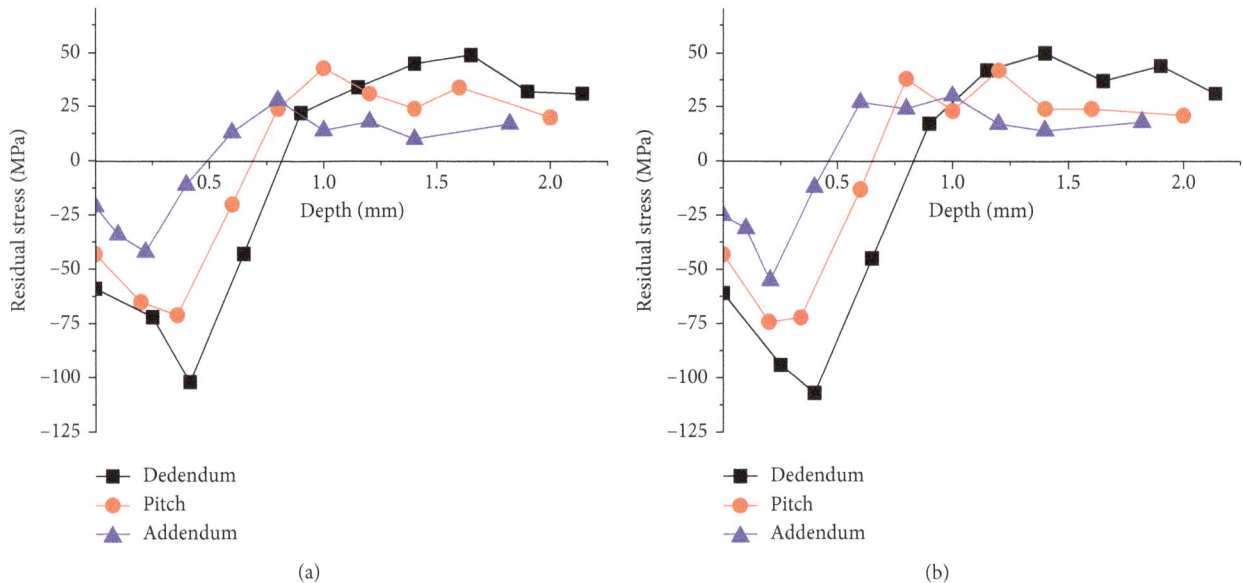

FIGURE 9: Distribution and size of the residual stress at three locations under two spindle speeds: (a) 1809 r/min and (b) 2258 r/min.

In summary, the residual stress of a cold roll-beating spline shaft occurs mainly because of the nonuniform plastic deformation caused by the plastic flow of the local metal in the workpiece that is extruded by the roller. In the residual stress formation process, the workpiece surface metal has the largest extrusion force and flow rate, which both decrease inwardly with a larger decreasing gradient. A large degree of uneven metal plastic deformation is produced, and residual compressive stress is formed on the workpiece's surface. Residual tensile stress is formed in the workpiece's interior because of the mutual balancing effect between the workpiece interior and the surface. Affected by the shape of the roller, the metal flow of the tooth profile is different. The degrees of metal plastic flow and uneven plastic deformation are the highest at the dedendum, followed by the pitch circle, and are the smallest at the tooth top. The largest residual stress is ultimately formed at the dedendum circle, followed by the pitch circle, and the smallest is found at the addendum.

6. Conclusion

(1) In the cold roll-beating process, the metal plastic flow rate is different between the surface and inner areas of the location. The metal flow rate of the surface is largest, and it hardly occurs at the base layer. The decreasing rate of the metal flow is gradually reduced.

(2) The flow deviation is caused by the changes in the metal flow rate. It leads to plastic deformation nonuniformity and forms the metal residual stress. The residual stress value is directly related to an uneven plastic flow.

(3) At a high roll-beating speed, the degree of plastic deformation at the workpiece surface is more uneven, and the residual stress in the tooth profile is larger overall.

(4) At the same roll-beating speed, the rate change trend of the metal flow gradually decreases along the direction of the surface to the inner layer from the dedendum to the addendum and is the highest at the fillet parts of the dedendum. The residual stress distribution on the surface of the tooth profile decreases from the dedendum to the addendum.

Conflicts of Interest

The authors declare that they have no conflicts of interest.

Acknowledgments

This project was sponsored by the National Natural Science Foundation of China (Grant nos. 51475146, 51475366, and 51075124).

References

[1] P. A. F. Martins, N. Bay, M. Skjoedt, and M. B. Silva, "Theory of single point incremental forming," *CIRP Annals-Manufacturing Technology*, vol. 5, no. 57, pp. 247–252, 2008.

[2] M. B. Silva, M. Skjoedt, A. G. Atkins, N. Bay, and P. A. F. Martins, "Single-point incremental forming and formability-failure diagrams," *Journal of Strain Analysis for Engineering Design*, vol. 1, no. 43, pp. 15–35, 2008.

[3] E. M. Viatkina, W. A. M. Brekelmans, and M. G. D. Geers, "Modelling of the internal stress in dislocation cell structures," *European Journal Mechanics–A/Solids*, vol. 5, no. 26, pp. 982–998, 2007.

[4] M. S. Abdul Aziz, T. Furumoto, K. Kuriyama et al., "Residual stress and deformation of consolidated structure obtained by layered manufacturing process," *Journal of Advanced Mechanical Design System and Manufacturing*, vol. 2, no. 7, pp. 244–256, 2013.

[5] L. Del Llano-Vizcaya, C. Rubio-Gonzalez, G. Mesmacque, and A. Banderas-Hemandez, "Stress relief effect on fatigue and relaxation of compression springs," *Materials and Design*, vol. 4, no. 28, pp. 1130–1134, 2006.

[6] T. Furumoto, T. Ueda, M. S. A. Aziz, A. Hosokawa, and R. Tanaka, "Study on reduction of residual stress induced during rapid tooling process: influence of heating conditions on residual stress," *Key Engineering Materials*, vol. 9, no. 447-448, pp. 785–789, 2010.

[7] C. J. Lammi and D. A. Lados, "Effects of residual stresses on fatigue crack growth behavior of structural materials: analytical corrections," *International Journal of Fatigue*, vol. 2, no. 33, pp. 858–867, 2011.

[8] F. K. Cui, W. Zhao, Y. F. Xu, and J. X. Yang, "Research on fibrous structure and surface work hardening test of cold rolling involute spline," *Machine Tool and Hydraulics*, vol. 3, no. 37, pp. 36–38, 2010.

[9] F. K. Cui, X. D. Dong, X. Q. Wang et al., "The analysis of macroscopic property changes and microstructural evolution of 1020 steel under cold beating," *Materials Research Innovations*, vol. 4, no. 19, pp. 56–61, 2015.

[10] H. Mizutani and M. Wakabayashi, "Influence of cutting edge shape on residual stresses of cut surface (effects of nose radius and cutting edge roundness)," *Journal of Advanced Mechanical Design System and Manufacturing*, vol. 4, no. 14, pp. 1201–1209, 2008.

[11] G. A. Smirnov, *Study of Finite Plastic Deformation in Cold Form Rolling of Splines on Shafts*, Macmillan, London, 1971.

[12] N. Kurz, "Theoretical and experimental investigations of the 'grob' cold shape-rolling process," in *Proceedings of the Twenty-Fifth International Machine Tool Design and Research Conference*, pp. 551–559, Birmingham, UK, April 1985.

[13] F. K. Cui, *Study of High-Speed Precise Forming with Cold Roll-Beating Technique*, Xi'an University of Technology, Xi'an, China, 2007.

[14] L. Zhang, M. S. Yang, Y. Li, and Q. L. Yuan, "Analytic method and its modification for deformation force of high-speed cold roll-beating forming," *Journal of Plasticity Engineering*, vol. 3, no. 18, pp. 1–7, 2012.

[15] E. Grob and H. Krapfenbauer, "Roller head for cold rolling of splined shafts or gears," US Patent 3818735A, 1973.

[16] M. Weck, W. Koenig, G. Bartsch, and K. Steffens, "Manufacture and load-bearing capacity of cold-rolled gears," in *Proceedings of 3rd International Conference on Rotary Metalworking Processes*, pp. 395–406, Kyoto, Japan, September 1984.

[17] F. K. Cui, K. G. Xie, Y. F. Xie, X. Q. Wang, W. J. Zhu, and Y. Li, "Analysis of coupled thermal–mechanical mechanism based on work hardening phenomenon in high-speed cold roll-beating," *Materials Research Innovations*, vol. 5, no. 19, pp. 1212–1218, 2015.

[18] F. K. Cui, Y. F. Xu, and W. Zhao, "Research on metal microstructure deformation of splines manufactured by cold rolling, milling and cutting processes," *Forging and Stamping Technology*, vol. 7, no. 32, pp. 70–74, 2008.

[19] F. K. Cui, Y. F. Xie, X. D. Dong, and L. M. Hou, "Simulation analysis of metal flow in high-speed cold roll-beating," *Journal of Henan Polytechnic University*, vol. 9, no. 33, pp. 467–471, 2014.

[20] M. B. Prime, "Cross-sectional mapping of residual stresses by measuring the surface contour after a cut," *Journal of Engineering Materials and Technology*, vol. 4, no. 123, pp. 162–168, 2001.

[21] P. Pagliaro, M. B. Prime, J. S. Robinson et al., "Measuring inaccessible residual stresses using multiple methods and superposition," *Experimental Mechanics*, vol. 9, no. 51, pp. 1123–1134, 2011.

[22] P. Pagliaro, M. B. Prime, H. Swenson, and B. Zuccarello, "Measuring multiple residual-stress components using the contour method and multiple cuts," *Experimental Mechanics*, vol. 50, no. 2, pp. 187–194, 2010.

[23] P. Pagliaro, M. B. Prime, M. L. Lovato, and B. Zuccarello, "Known residual stress specimens using opposed indentation," *Journal of Engineering Materials and Technology*, vol. 131, no. 3, pp. 0310021–03100210, 2009.

20

Frequency Dependence of C-V Characteristics of MOS Capacitors Containing Nanosized High-κ Ta$_2$O$_5$ Dielectrics

Nenad Novkovski[1,2] and Elena Atanassova[3]

[1]Institute of Physics, Faculty of Natural Sciences and Mathematics, University "Ss. Cyril and Methodius",
 Arhimedova 3, 1000 Skopje, Macedonia
[2]Research Center for Environment and Materials, Macedonian Academy of Sciences and Arts,
 Krste Misirkov 2, 1000 Skopje, Macedonia
[3]Institute of Solid State Physics, Bulgarian Academy of Sciences, 72 Tzarigradsko Chaussee Blvd., 1784 Sofia, Bulgaria

Correspondence should be addressed to Nenad Novkovski; nenad@iunona.pmf.ukim.edu.mk

Academic Editor: Antonio Riveiro

Capacitance of metal–insulator–Si structures containing high permittivity dielectric exhibits complicated behaviour when voltage and frequency dependencies are studied. From our study on metal (Al, Au, W)–Ta$_2$O$_5$/SiO$_2$–Si structures, we identify serial C-R measurement mode to be more convenient for use than the parallel one usually used in characterization of similar structures. Strong frequency dependence that is not due to real variations in the dielectric permittivity of the layers is observed. Very high capacitance at low frequencies is due to the leakage in Ta$_2$O$_5$ layer. We found that the above observation is mainly due to different leakage current mechanisms in the two different layers composing the stack. The effect is highly dependent on the applied voltage, since the leakage currents are strongly nonlinear functions of the electric field in the layers. Additionally, at low frequencies, transition currents influence the measured value of the capacitance. From the capacitance measurements several parameters are extracted, such as capacitance in accumulation, effective dielectric constant, and oxide charges. Extracting parameters of the studied structures by standard methods in the case of high-κ/interfacial layer stacks can lead to substantial errors. Some cases demonstrating these deficiencies of the methods are presented and solutions for obtaining better results are proposed.

1. Introduction

High permittivity dielectrics (high-κ) are nowadays extensively studied as a replacement of silicon dioxide in various microelectronics devices [1–4], like gate dielectrics [5], memory devices [6], and so forth.

Between the high-κ dielectrics tantalum pentoxide (Ta$_2$O$_5$) [7] has been identified as very good solution for dynamic random access memories (DRAM) [8]. In this work we specifically study the case of Ta$_2$O$_5$.

An unavoidable interfacial layer few nanometers thick (typically between 1 nm and 4 nm), grown between the Si substrate and Ta$_2$O$_5$ due to the thermodynamic instability of the Ta$_2$O$_5$/Si interface, appears [9, 10]. The situation is similar to most of the other high-κ dielectrics: TiO$_2$, HfO$_2$ [11], LaScO$_3$ [12], GdScO$_3$ [13], Er$_2$O$_3$ [14], and so forth.

Interfacial layer influences the properties of the dielectric film that has to be studied as a stacked layer (high-κ/interfacial layer). We previously developed a comprehensive model for I-V characteristics of Ta$_2$O$_5$/SiO$_2$ structures [15, 16]. In [17] we showed that the frequency dependence of the effective series capacitance of metal–Ta$_2$O$_5$/SiO$_2$–Si structures can be successfully described by a five-element model.

The main aim of this paper is to study the frequency dependence of the effective series capacitance of nanosized dielectrics (10 nm or thinner) and to test the applicability of the model previously developed and applied on thicker films (50 nm) [17]. C-V characteristics for various frequencies were measured both in serial and in parallel measurement mode. The observed differences between them are discussed. Previously [18], we reported preliminary results on the above discussed issues. In [19] we elaborated the issue of determination

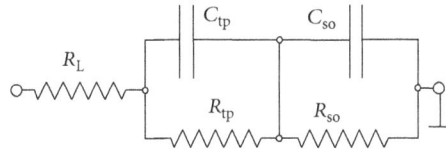

FIGURE 1: Equivalent circuit for metal-Ta_2O_5/SiO_2–Si structures [17].

of interface state densities in metal-dielectric-Si structures containing high-κ dielectrics. The issue of hysteresis-like flat band voltage instabilities in Al/Ta_2O_5–SiO_2/Si structures has been studied in detail in [19]. It has been shown that under defined conditions repeatable patterns of C-V characteristics are obtained. Thus obtained C-V characteristics can be effectively used in determination of equivalent oxide thickness and fast interface state densities using the C-V curves obtained when sweeping the voltage from negative to positive bias. In the present work we study in detail the issue of analysis of capacitance measurement data obtained in serial and parallel mode for the case of metal (Al, Au, W)–Ta_2O_5/SiO_2–Si structures.

2. Fabrication of the Samples

The samples studied here were fabricated on p-type (1 0 0) 15 Ω cm Si substrates. After chemical cleaning, a Ta film was deposited on Si by sputtering of a Ta target in Ar atmosphere. Subsequently, the Ta film was oxidized in dry O_2 at 600°C. More details on the sample preparation can be found in [8]. The above oxidation temperature was chosen so as to be low enough to minimize the substrate oxidation in order to prevent the formation of tantalum silicides. The thickness of thus obtained Ta_2O_5 films and the refractive index were measured ellipsometrically ($\lambda = 632.8$ nm). Layers with a thickness of about 10 nm were used in this study. The refractive index was found to be 2.1. The test structures were metal-insulator-silicon capacitors with three different metal gates: Al, W, and Au. The gate areas (S) were $1.96 \cdot 10^{-3}$ cm^2 for Au and $2.5 \cdot 10^{-3}$ cm^2 for Al and W. W layers were sputtered in Ar to a thickness of 300 nm under the following conditions: power density of 3.1 W cm^{-2} and gas pressure 3 Pa. Al and Au electrodes were obtained by thermal evaporation using a conventional technique.

In the first part of the study, we measured the capacitance in C_s–R_s (serial) mode in the frequency range from 100 Hz to 1 MHz, with the use of a HP 4284 A LCR-meter. The measurements were done with substrates in accumulation with an ac signal level of 24 mV at a gate bias ranging from −2 V to −3 V. For the second part of the study, high-frequency C-V measurements, both in serial and parallel (C_p–R_p) mode, were performed at frequencies ranging from 3 kHz to 1 MHz, in the voltage range from −3 V through +1 V, starting from the left (the most negative gate voltage) and ending at the right (maximum positive voltage). Repeated measurements under identical conditions gave practically the same results, showing that no substantial wearout took place during the measurements.

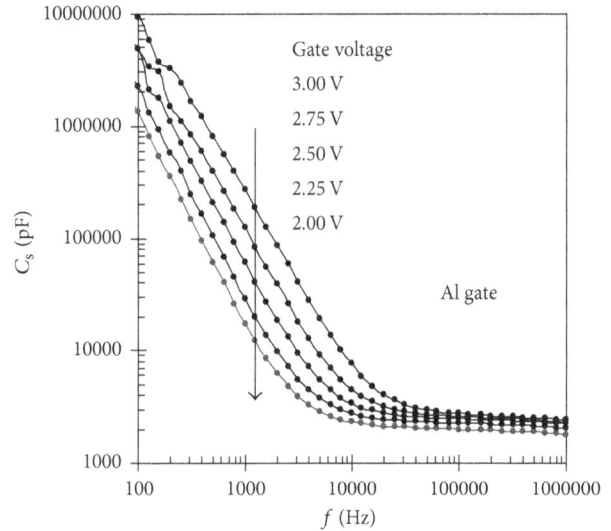

FIGURE 2: C_s-f characteristics in the case of an Al gate.

3. Frequency Dependence of the Effective Series Capacitance

First, we studied the dependence of the effective serial resistance of metal–Ta_2O_5/SiO_2–Si structures on the frequency, at a given gate bias voltage. Model developed in [17] was used to explain the obtained results. More details on the model and its application can be found in the same work. The equivalent circuit composed of five elements [17] is shown in Figure 1. C_{so} and the R_{so} are capacitance and parallel resistance of the interfacial silicon oxide layer, while C_{tp} and R_{tp} are capacitance and parallel resistance of the bulk tantalum oxide layer. Capacitances in the model are constant values, while the resistances depend on the bias voltage. R_L is the serial (load) resistance of the structure.

Effective serial capacitance of the considered five-element circuit is expressed by [17]

$$C_s(\omega)$$
$$= \left(\frac{1/C_{tp}}{1 + 1/\left(\omega C_{tp} R_{tp}\right)^2} + \frac{1/C_{so}}{1 + 1/\left(\omega C_{so} R_{so}\right)^2} \right)^{-1}, \quad (1)$$

where $\omega = 2\pi f$ is the angular frequency of the measurement signal (f is the signal frequency).

In Figure 2 the experimental results for the case of an Al gate structure are shown. Substantial increase of three

TABLE 1: Parameters used in the theoretical calculations shown in Figure 4.

Gate	R_L (Ω)	R_{tp} (kΩ)	R_{so} (kΩ)	C_{tp} (nF)	C_{so} (nF)
Al	160	6	7.0	9.8	3.1
W	160	6	7.5	11.0	2.9
Au	200	60	70	7.9	2.3

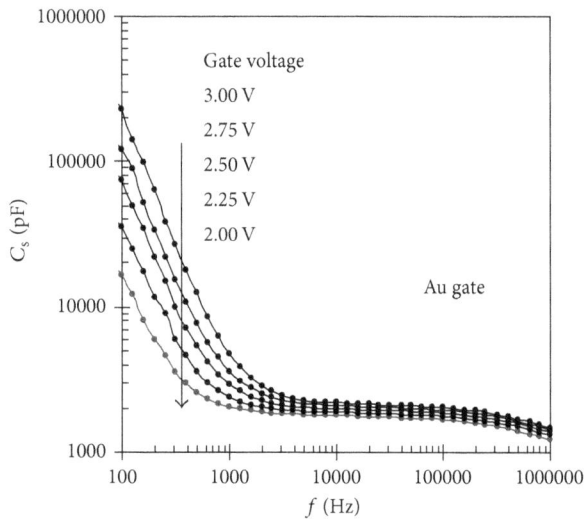

FIGURE 3: C_s-f characteristics in the case of an Au gate.

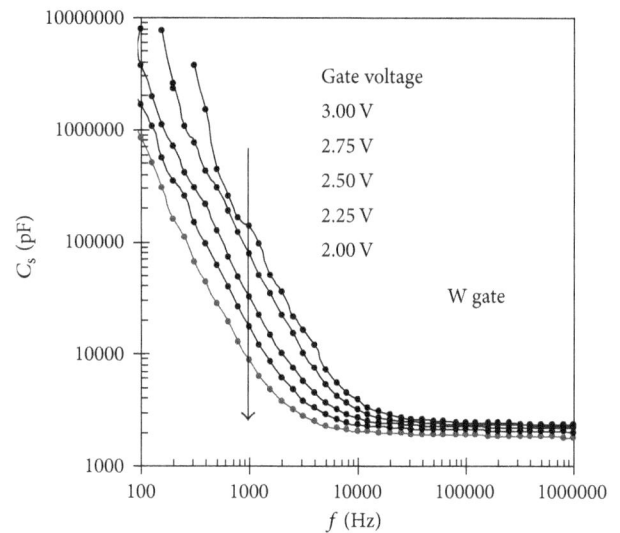

FIGURE 4: C_s-f characteristics in the case of a W gate.

orders of magnitude at 100 Hz is observed compared to high frequencies (50 kHz to 1 MHz range).

In Figure 3 the experimental results for the case of an Au gate structure are shown. Increase of two orders of magnitude at 100 Hz is observed compared to high frequencies. The increase is more important at higher gate voltages, attaining higher values at low frequencies and influencing the capacitance at higher frequencies.

In Figure 4 the experimental results for the case of a W gate structure are shown. Increase of three orders of magnitude at 100 Hz is observed compared to high frequencies (50 kHz to 1 MHz). In all three cases considered here, the increase of the capacitance is more important at higher gate voltages, attaining higher values at low frequencies and influencing the capacitance at higher frequencies.

In order to test the validity of the model, in Figure 5, we present the comparison of the experimental with theoretical results obtained using the five-element model. Only results for gate voltage −2.5 V are shown; other results being quite similar to the results presented here. Parameters used in the calculations are given in Table 1.

It is seen that in the case of Au gate leakage current is almost two orders of magnitude lower than these for other two metals. Since the differences between the capacitances of both the Ta_2O_5 and the SiO_2 layers are not substantial, it can be concluded that the main origin of the low leakage current for the Au gate is not related to differences in the thicknesses of the layers, but to lower density of defects, manifested in substantially higher resistances in the case of Au gate both of

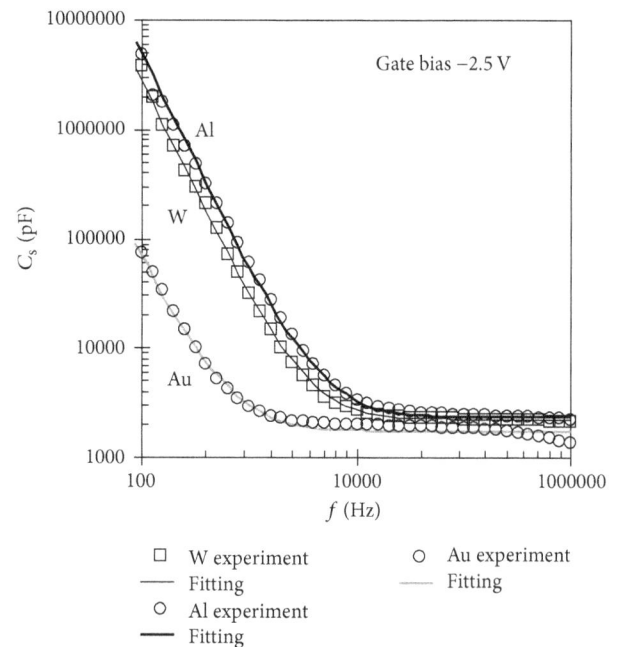

FIGURE 5: Comparison of experimental and theoretical results for effective series capacitances.

the Ta_2O_5 and the SiO_2 layers compared to the case of W and Al gates.

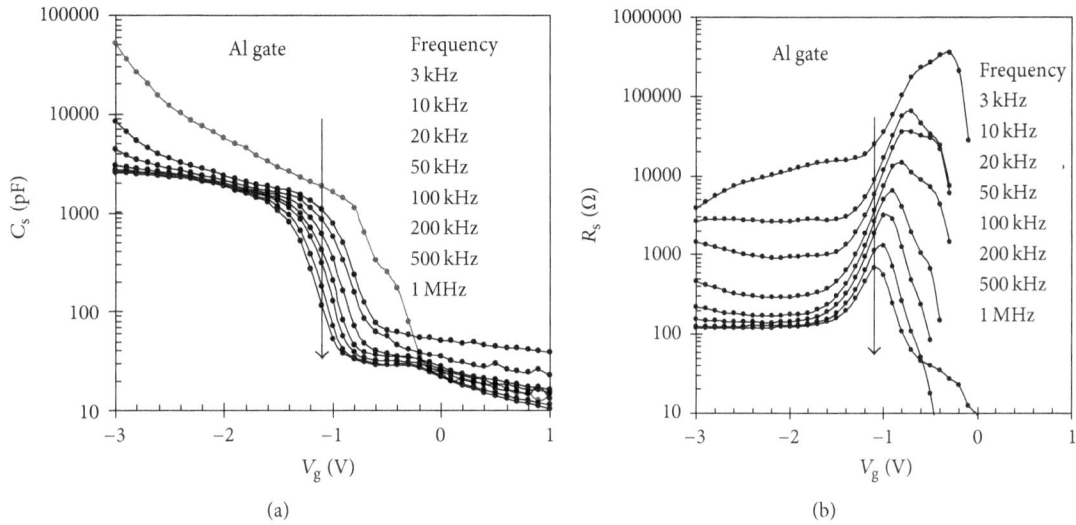

FIGURE 6: C_s-V and R_s-V characteristics for an Al gate.

4. Frequency Dependence of C-V Characteristics

The analysis done in previous section is applicable for C-f characteristics. For C-V characteristics it cannot be applied as is, since the parameters of model depend on the applied voltage. Therefore further development of characterization method is to be done. In this section we study the main aspects of the influence of leakage currents on measured C-V characteristics. The measurements of C-V characteristics are mainly done in the parallel mode. Above choice has been made assuming that the structure in accumulation can be described by a two-element model: metal-insulator-silicon capacitor (C_p) with a parallel resistance (R_p). Such an approach is justified for silicon dioxide insulating layers with low leakage. In the case of leaky and high-κ dielectrics, measurements in parallel mode have many disadvantages. A three-element model, including an additional serial resistance, was used to describe frequency dependence of the measured capacitance [20, 21]. In the case of high leakage thin dielectrics, it was observed that the C-V characteristics measured at higher frequencies are unrealistic; the measured capacitance decreases with the frequency and the gate voltage. In order to correct this deficiency, it was proposed to correct the C-V characteristics, by using the results of the measurements done at two different frequencies and the expressions obtained with the three-element model [22]. Using that method, more realistic results are obtained. The method was further improved by adding serial impedance [23]. Nevertheless, the method is based on the assumption that the difference between the C-V characteristics is due only to the leakage currents. Indeed, as we showed in [24], the flat band voltages are different for different frequencies (50 kHz, 100 kHz, and 1 MHz). The variations of the flat band voltage with frequency can be in great part accounted for by the effect of serial resistance (R_L). Before proceeding to

the method of correction of measured curves for the effect of serial resistance, appropriate choice of the measurement mode for high-frequency C-V characteristics is to be found. Therefore, as a crucial step in the procedure of MOS characterization, the best single frequency measurement method has to be adopted for obtaining correct high-frequency C-V characteristics. Further we shall show and discuss the results for C-V characteristics obtained at various frequencies both in serial and in parallel mode.

In Figure 6 the experimental results for the serial capacitance and serial resistance in the case of an Al gate structure are shown. It is seen that for high frequencies (50 kHz to 1 MHz), capacitance in accumulation at negative voltages higher than −2 V practically does not depend on the frequency. Therefore, the effect of the leakage does not influence the measured capacitances in the considered frequency range. A systematic shift to the left of the C-V curves with frequency is observed. It can not be explained as an artefact due to the leakage but appears to be connected with some real effects of oxide charge on the capacitance and the effect of serial resistance (R_L). In order to explain these effects, further investigations are required; they remain out of the scope of this study.

In inversion, at gate positively biased ($V_g > 0$), substantial decrease of the capacitance with increasing gate voltage is observed (Figure 6(a)). It can be explained by deep depletion resulting from exhausting of minority carriers (electrons) due to their tunnelling injection occurring at positive gate voltages [15]. In I-V characteristics this exhausting of minority carriers results in saturation of leakage currents [16].

A peak in the effective serial resistance is observed for all frequencies from 3 kHz to 1 MHz. Gradual shift to the left is observed as for the capacitances. Peak R_s values can be used for calculation of conductance and subsequently determination of interface state densities by the method using single G/ω–V curve measured at a given frequency ω [19].

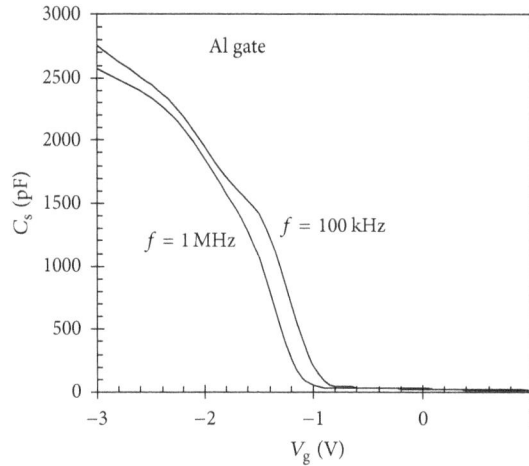

FIGURE 7: *C-V* characteristics at 100 kHz and 1 MHz, obtained in serial measurement mode for an Al gate.

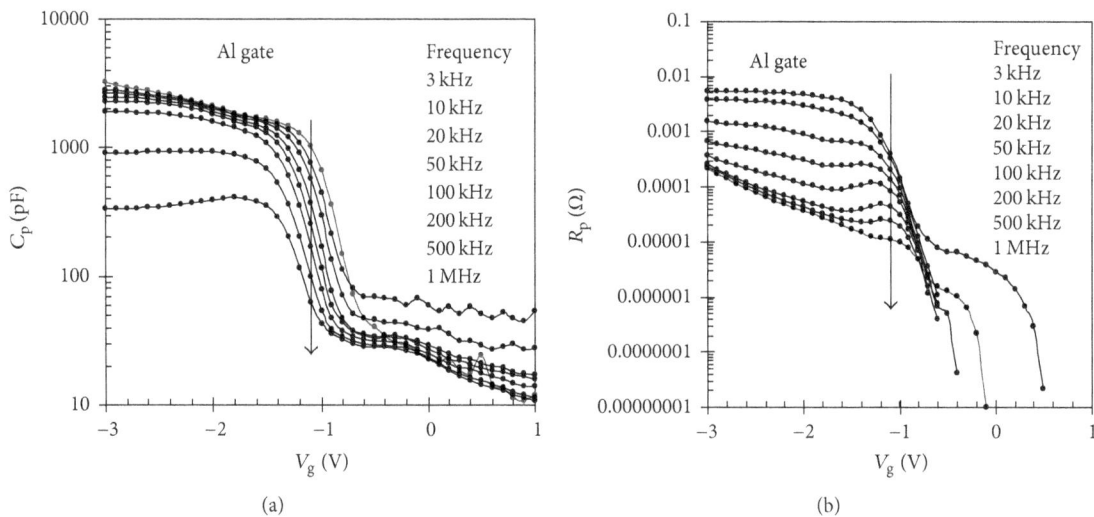

(a)   (b)

FIGURE 8: C_p-V and R_p-V characteristics for an Al gate.

In addition, the entire curves can be used for reliable determination of densities and energy positions of interface traps.

In Figure 7 the *C-V* characteristics obtained in serial measurement mode for two frequencies (100 kHz and 1 MHz, typically used for high-frequency *C-V* measurements) in a linear scale are shown. It is seen that the differences are smaller than 10%.

In Figure 8 the experimental results for the parallel capacitance and parallel resistance in the case of an Al gate structure are shown. It is seen that for high frequencies capacitance decreases for two orders of magnitude when the frequency increases from 50 kHz to 1 MHz. Therefore, the effect of the leakage influences critically the measured capacitances in the considered frequency range. A systematic shift to the left of the *C-V* curves with frequency is observed as in the case of serial measurement mode. Peaks in *R-V* curves are observed, being much less pronounced than for the serial mode.

Contrary to the case of a serial resistance, the parallel resistance decreases an order of magnitude when the frequency increases from 100 kHz to 1 MHz (Figure 8(a)). In Figure 9 the *C-V* characteristics for serial and parallel mode are compared for 100 kHz (Figure 9(a)) and 1 Mhz. Even if the difference between the capacitance values in the case of frequency 100 kHz is not high, the shape substantially differ in accumulation region. At 1 Mhz (Figure 9(b)) there are enormous differences between the curves.

Particular attention is to be paid to the shape of the curves C_p-V at highest frequencies. At 1 MHz (Figure 8(a)) it decreases from −1.5 V to −3 V instead of increasing. Therefore, the part in accumulation can not be used for further analysis, such as determination of interface states densities. At 100 kHz the curve looks much better, going to saturation at −3 V. However, this is a misleading fact. Namely, at −3 V one can not expect reaching saturation, since voltages of about −5 V are required for this. At such high voltages important wearout occurs, and hence it is not possible to

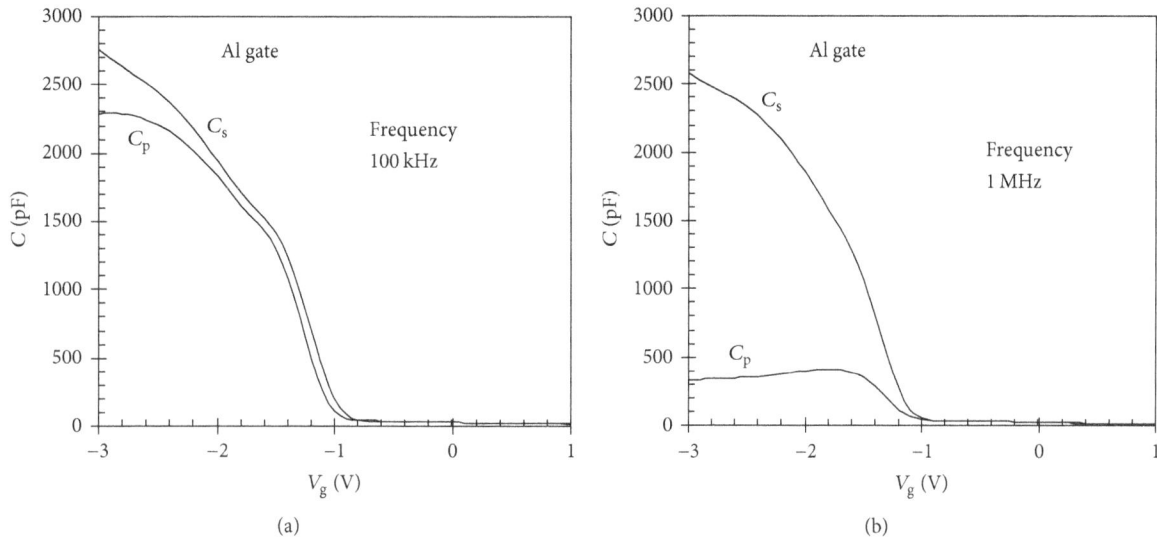

FIGURE 9: Comparison of C_s-V with C_p-V at two frequencies: 100 kHz (a) and 1 MHz (b) in the case of an Al gate.

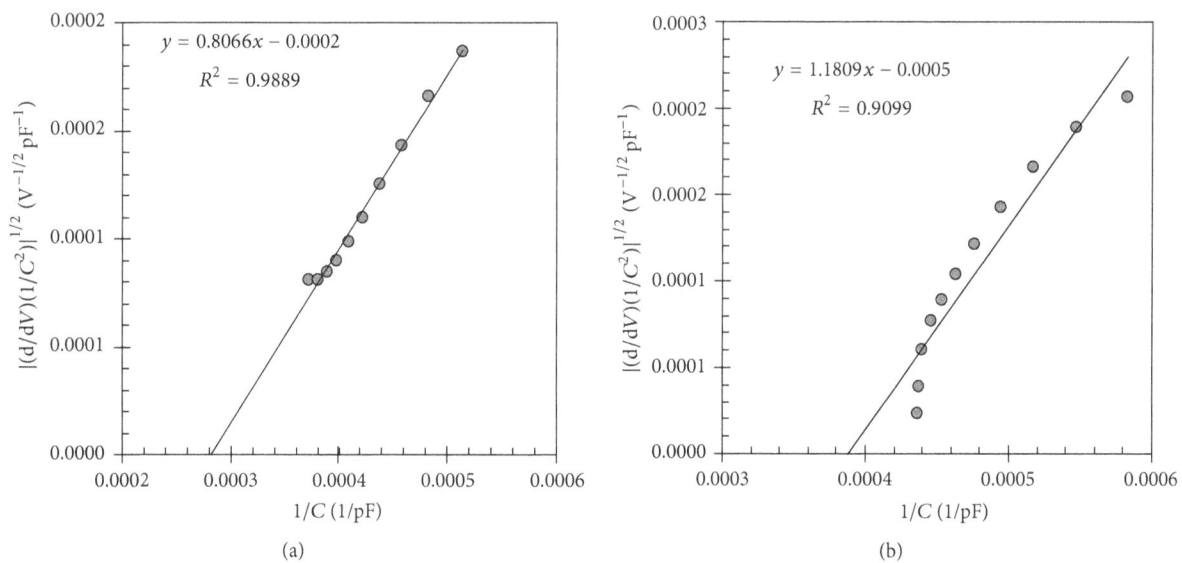

FIGURE 10: $d(1/C^2)/dV$ versus $1/C$ plot for 100 kHz for C_s (a) and C_p (b) in the case of an Al gate.

measure directly saturated capacitance. The observed flat part on the left is due to the compensation of the real increase of the capacitance with the decrease of the effective parallel capacitance due to the increase of the leakage (decrease of the parallel capacitances).

Above statement can be further supported by the analysis of the part of C-V curves in accumulation using the characterization method for extracting data on high-κ dielectrics proposed by Kar et al. [25]. If plotting $d(1/C^2)/dV$ versus $1/C$, straight line is to be obtained. As is seen from Figure 10, this is valid for C_s (a) but not for C_p (b). Therefore, measurements in serial mode can be used for further extraction of parameters, but not in parallel mode.

In Figure 11 the experimental results for serial capacitance and serial resistance in the case of an Au gate structure are shown. Similar results as in the case of Al gate are obtained.

In Figure 12 the experimental results for the parallel capacitance and parallel resistance in the case of an Au gate structure are shown. It is seen that for high frequencies capacitance decreases for an order of magnitude when the frequency increases from 50 kHz to 1 MHz. Therefore, the effect of the leakage influences less the measured capacitances in the considered frequency range than in the case of Al gates. A systematic shift to the left of the C-V curves with frequency is observed as in the case of serial measurement mode. Peaks in R-V curves are observed, being much more pronounced

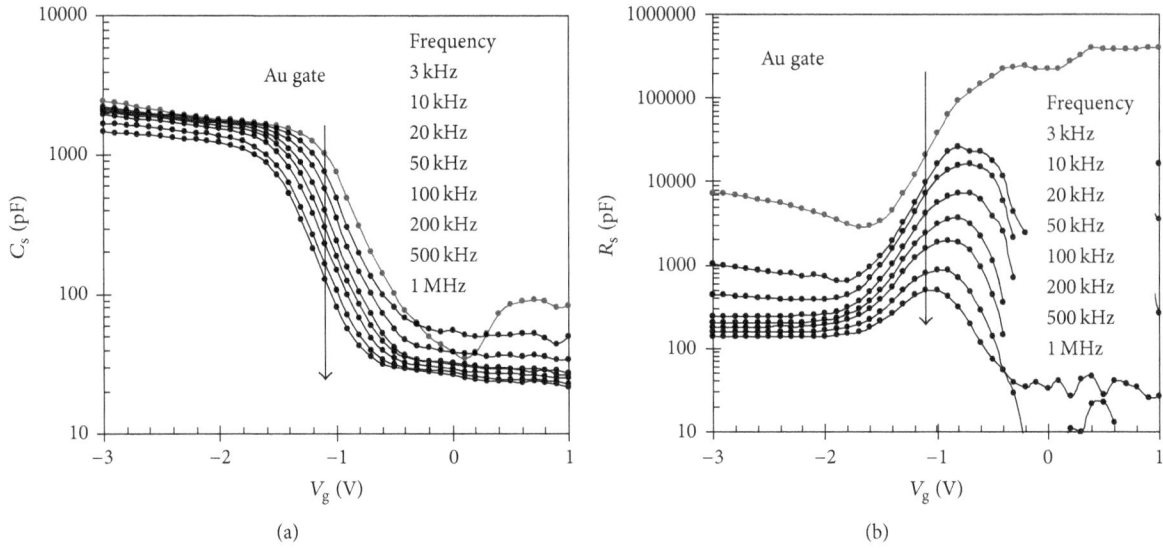

FIGURE 11: C_s-V and R_s-V characteristics for an Au gate.

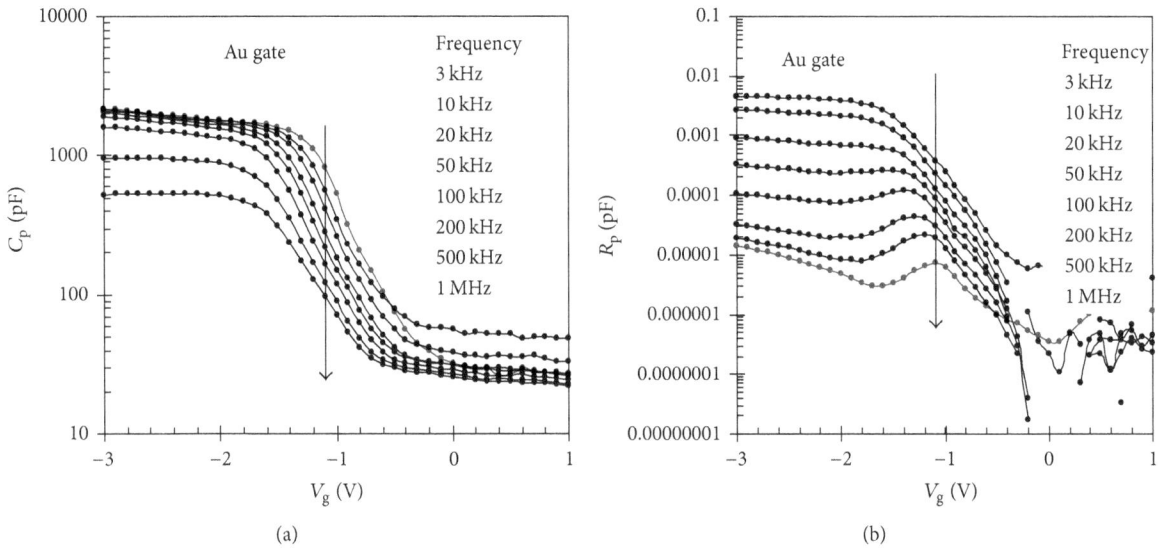

FIGURE 12: C_p-V and R_p-V characteristics for an Au gate.

than in the case of an Al gate. The differences between the results for Al and Au gates can be explained by lower leakage in the case of Au gates. This is in accordance with the values of the parallel resistances extracted from C_s-f measurements (Table 1).

In Figure 13 C_s-V curves are compared to C_p-V curves for f = 500 kHz. The frequency f = 500 kHz was so chosen because when making simple visual inspection, C_p-V curve looks like a curve for MOS capacitors containing low leakage dielectric, exhibiting clear saturation of the capacitance towards the strong accumulation. Even if the C_p-V curve seems to be rather good, in the d$(1/C^2)$/dV versus $1/C$ plot only for C_s a good straight line is obtained (Figure 14(a)) and

not for C_p (Figure 14(b)). Nevertheless, the deviation from a straight line is smaller than in the case of an Al gate, due to lower leakage in the case of Au.

In Figure 15 the experimental results for serial capacitance and serial resistance in the case of a W gate structure are shown. Similar results as in the case of Al gate are obtained.

In Figure 16 the experimental results for the parallel capacitance and parallel resistance in the case of a W gate structure are shown. Results similar to those obtained for Al and Au gates are obtained for a W gate.

Above is also valid for the comparison of the C_s-V with C_p-V curves (Figure 17) and the d$(1/C^2)$/dV versus $1/C$ plot (Figure 18). Therefore, the peculiarities of the C-V

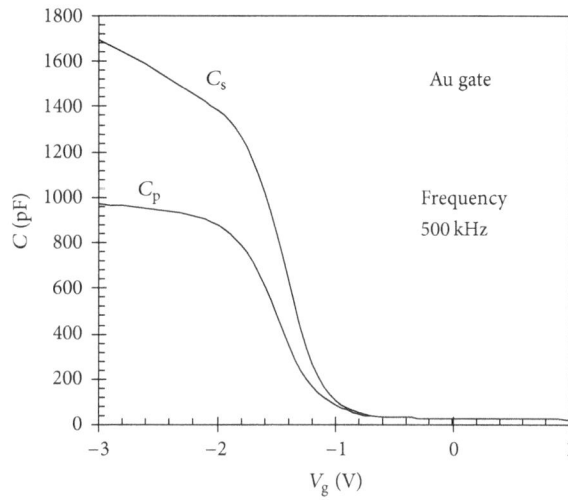

FIGURE 13: Comparison of C_s-V with C_p-V at 500 kHz in the case of an Au gate.

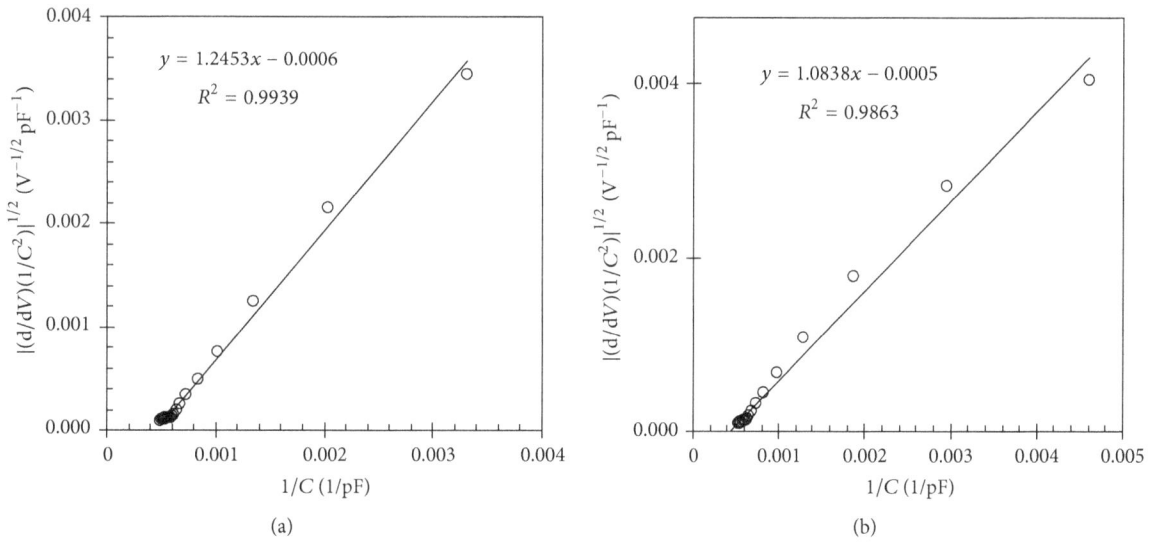

(a)

(b)

FIGURE 14: $d(1/C^2)/dV$ versus $1/C$ plot for 100 kHz for C_s (a) and C_p (b) in the case of an Au gate.

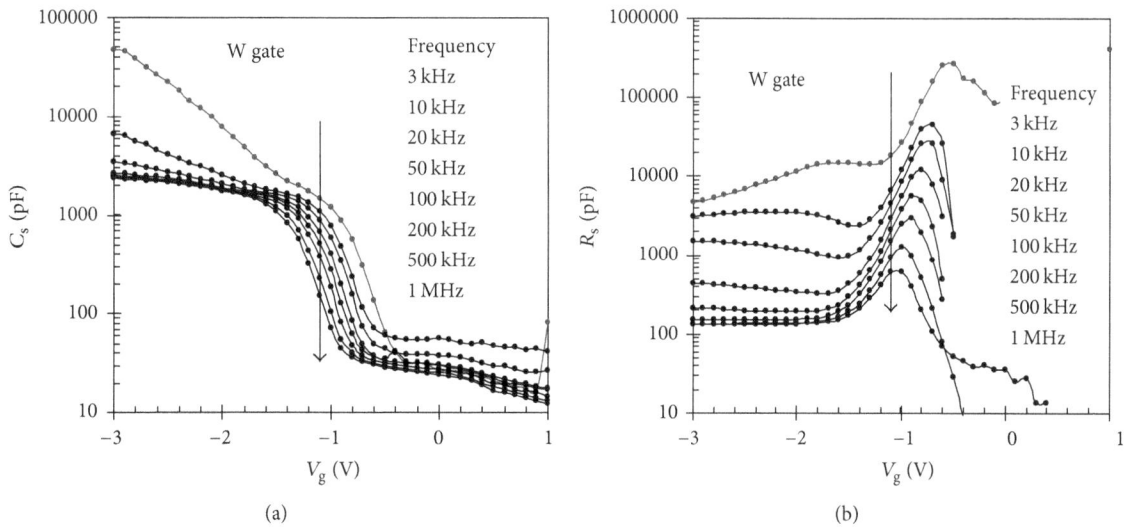

(a)

(b)

FIGURE 15: C_s-V and R_s-V characteristics for a W gate.

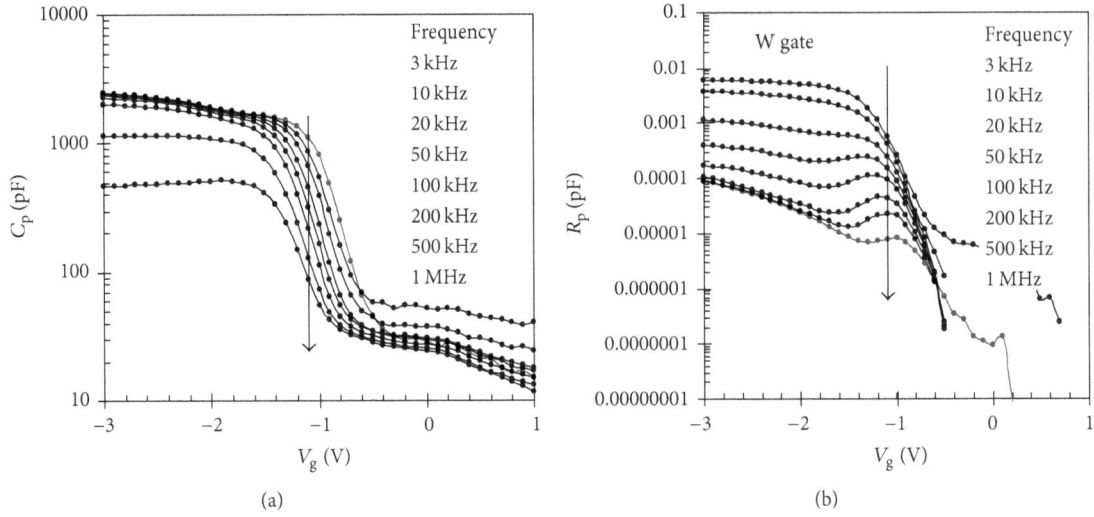

FIGURE 16: C_p-V and R_p-V characteristics for a W gate.

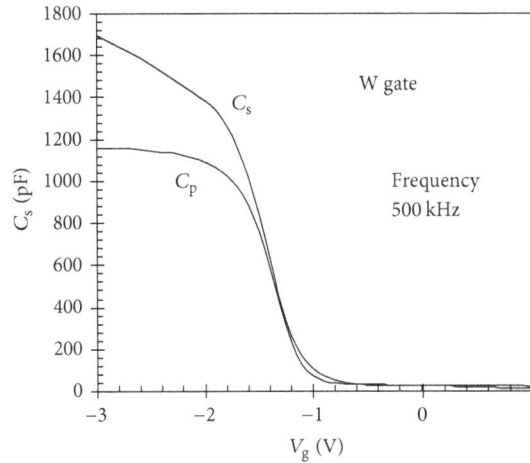

FIGURE 17: Comparison of C_s-V with C_p-V at 500 kHz in the case of a W gate.

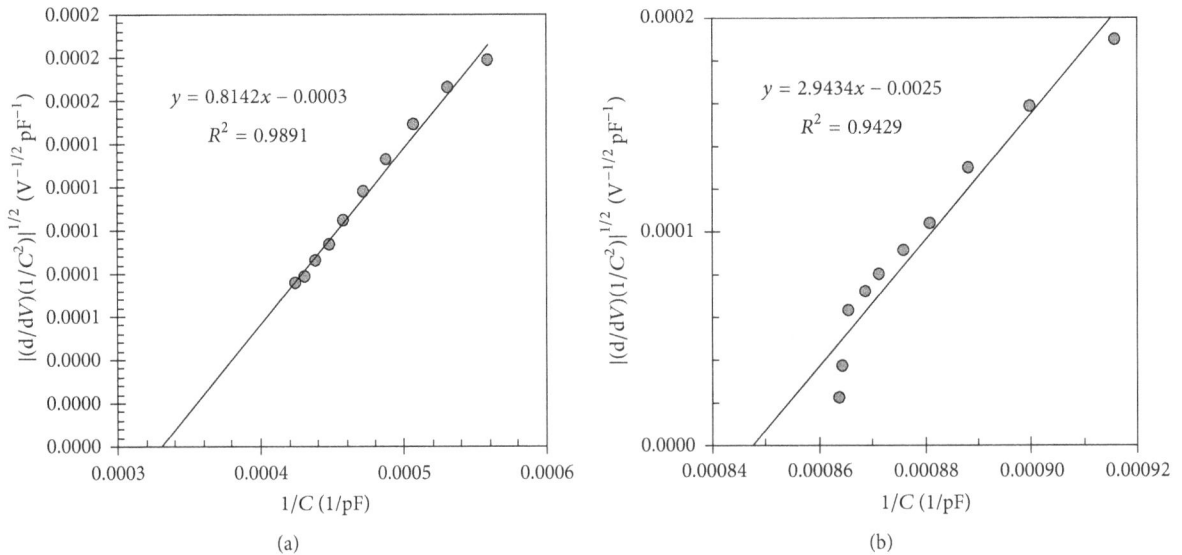

FIGURE 18: $d(1/C^2)/dV$ versus $1/C$ plot for 500 kHz for C_s (a) and C_p (b) in the case of a W gate.

TABLE 2: Equivalent oxide thickness (d_{eq}), flat band voltage (V_{fb}) obtained from C-V curves measured at 100 kHz, ideal flat band voltage ($V_{fb,id}$), and determined oxide charge (Q_{ox}).

Gate	d_{eq} (nm)	V_{fb} (V)	$V_{fb,id}$ (V)	Q_{ox} (10^{12} cm^{-2})
Al	2.27	−0.91	−0.59	2.8
W	2.71	−0.86	−0.29	4.6
Au	3.44	−0.78	0.18	6.0

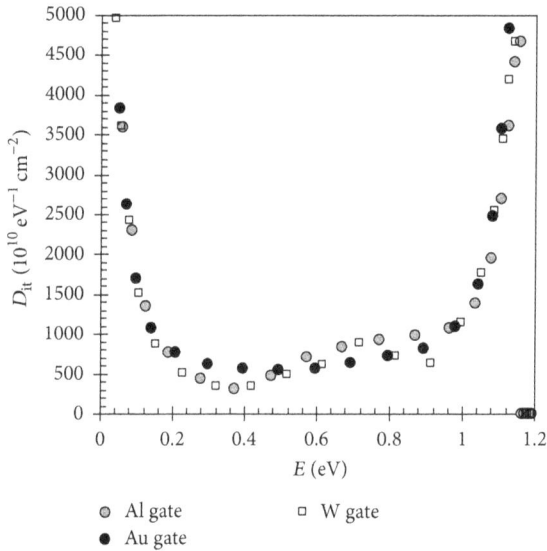

FIGURE 19: Interface state density versus energy over the entire Si bandgap.

measurements discussed above seem to be valid for different gate metals. Somehow different result is obtained in the case of lowest frequency in the considered range (3 kHz) for Al gate. This is a case where the gate is reactive with Ta$_2$O$_5$ [26] and the density of defects contributing to the leakage attains high values. Therefore, except for materials with high defect density, results obtained here are applicable in general. In addition, while limiting to frequencies in the range from 10 kHz 1 to MHz, even if using reactive gate metals, predictable results are obtained and the proposed method of characterization in this work is expected to provide highly reliable results.

For illustration, values of equivalent oxide thickness (d_{eq}), flat band voltage (V_{fb}), ideal flat band voltage ($V_{fb,id}$), and oxide charge (Q_{ox}), obtained from C-V curves measured at 100 kHz, using standard methods, are shown in Table 2. It is seen that oxide charges in all cases are comparable. Substantial difference in the equivalent oxide thickness is observed. The lowest value of 2.27 nm is obtained for Al gate and is attributed to the known reactivity of Al with Ta$_2$O$_5$ [26].

C-V characteristics obtained in serial mode can be used for effective determination of interface state densities (D_{it}). In Figure 19 the results obtained by standard single curve high-frequency method (Terman method) from characteristics measured at 100 kHz are shown. No marked differences for different gates are obtained, indicating that the SiO$_2$–Si interface properties strongly are not affected by the gate.

5. Conclusions

Capacitance of the metal–high-κ–semiconductor structures demonstrates complicated behaviour with voltage and frequency. Based on our studies on metal (Al, Au, W)–Ta$_2$O$_5$/SiO$_2$–Si structures described as well as on the experience from various other metal–high-κ–semiconductor structures, we found that serial C-R measurement mode is more convenient for characterization of such structures than the dominantly used parallel one. Strong frequency dependence that is not due to real variations in the dielectric permittivity of the layers is observed. Very high apparent capacitance at low frequencies is described to be due to the leakage in Ta$_2$O$_5$ layer. We found that the above observation is mainly due to different leakage current mechanisms in the two different layers in the dielectric stack. The effect is highly dependent on the applied voltage, since the leakage currents rapidly grow with the applied electric fields in the layers. Additionally, at low frequencies, the measured value of the capacitance is influenced by transition currents of various origins.

From the capacitance measurements several parameters of the structures are extracted: capacitance in accumulation, effective dielectric constant, oxide charges, and interface state densities. Extracting parameters of the studied structures by standard methods without proper modification in the case of high-κ/interfacial layer stacks can lead to substantial errors that can make in many cases the results unusable or misleading. Here, described effects, as the one with the effects found by other authors for nanosized and ultra-thin dielectric layers, are to be used in the analysis. The applicability of all standard methods is to be reconsidered in view of these new effects. Some of the most important ceases are presented in this work.

Conflicts of Interest

The authors declare that there are no conflicts of interest regarding the publication of this paper.

References

[1] J. Robertson, "High dielectric constant gate oxides for metal oxide Si transistors," *Reports on Progress in Physics*, vol. 69, no. 2, pp. 327–396, 2006.

[2] W. Hu, B. Frost, and R. L. Peterson, "Thermally stable yttrium-scandium oxide high-k dielectrics deposited by a solution process," *Journal of Physics D: Applied Physics*, vol. 49, no. 11, Article ID 115109, 2016.

[3] K. H. Goh, A. S. M. A. Haseeb, and Y. H. Wong, "Effect of oxidation temperature on physical and electrical properties of Sm$_2$O$_3$ thin-film gate oxide on si substrate," *Journal of Electronic Materials*, vol. 45, no. 10, pp. 5302–5312, 2016.

[4] J. Q. Song, L. X. Qian, and P. T. Lai, "Effects of Ta incorporation in Y$_2$O$_3$ gate dielectric of InGaZnO thin-film transistor," *Applied Physics Letters*, vol. 109, no. 16, article 163504, 2016.

[5] T. Ando, U. Kwon, S. Krishnan, M. M. Frank, and V. Narayanan, High-κ oxides on Si: MOSFET gate dielectrics," in *Thin Films on Silicon: Electronic And Photonic Applications*, pp. 323–367, 2016.

[6] J. A. Kittl, K. Opsomer, M. Popovici et al., "High-k dielectrics for future generation memory devices," *Microelectronic Engineering*, vol. 86, no. 7–9, pp. 1789–1795, 2009.

[7] D. Q. Yu, W. S. Lau, H. Wong, X. Feng, S. Dong, and K. L. Pey, "The variation of the leakage current characteristics of W/Ta_2O_5/W MIM capacitors with the thickness of the bottom W electrode," *Microelectronics Reliability*, 2016.

[8] E. Atanassova and T. Dimitrova, "Thin Ta_2O_5 layers on Si as an alternative to SiO_2 for high density DRAM applications," in *Handbook of Surfaces and Interfaces of Materials*, H. S. Naiwa, Ed., vol. 4, pp. 439–479, Academic, San Diego, Calif, USA, 2001.

[9] G. B. Alers, D. J. Werder, Y. Chabal et al., "Intermixing at the tantalum oxide/silicon interface in gate dielectric structures," *Applied Physics Letters*, vol. 73, no. 11, pp. 1517–1519, 1998.

[10] A. P. Baraban, V. A. Dmitriev, V. E. Drozd, V. A. Prokofiev, S. N. Samarin, and E. O. Filatova, "Interface properties of Si-SiO_2-Ta_2O_5 structure by cathodoluminescence spectroscopy," *Journal of Applied Physics*, vol. 119, no. 5, Article ID 055307, 2016.

[11] J. H. Choi, Y. Mao, and J. P. Chang, "Development of hafnium based high-k materials—a review," *Materials Science and Engineering R: Reports*, vol. 72, no. 6, pp. 97–136, 2011.

[12] F. Liu and G. Duscher, "Thermal annealing effect on the interface structure of high-κ $LaScO_3$ on silicon," *Applied Physics Letters*, vol. 91, no. 15, Article ID 152906, 2007.

[13] P. Myllymäki, M. Roeckerath, M. Putkonen et al., "Characterization and electrical properties of high-κ $GdScO_3$ thin films grown by atomic layer deposition," *Applied Physics A: Materials Science and Processing*, vol. 88, no. 4, pp. 633–637, 2007.

[14] M. Losurdo, M. M. Giangregorio, G. Bruno et al., "Er2 O_3 as a high-κ dielectric candidate," *Applied Physics Letters*, vol. 91, no. 9, Article ID 091914, 2007.

[15] N. Novkovski and E. Atanassova, "Injection of holes from the silicon substrate in Ta_2O_5 films grown on silicon," *Applied Physics Letters*, vol. 85, no. 15, pp. 3142–3144, 2004.

[16] N. Novkovski and E. Atanassova, "A comprehensive model for the i -V characteristics of metal-Ta_2O_5/SiO_2 -Si structures," *Applied Physics A: Materials Science and Processing*, vol. 83, no. 3, pp. 435–445, 2006.

[17] N. Novkovski and E. Atanassova, "Frequency dependence of the effective series capacitance of metal-Ta_2O_5/SiO_2-Si structures," *Semiconductor Science and Technology*, vol. 22, no. 5, article 013, pp. 533–536, 2007.

[18] N. Novkovski and E. Atanassova, "Peculiarities of capacitance measurements of nanosized high-κ dielectrics: case of Ta_2O_5," *Journal of Optoelectronics and Advanced Materials - Symposia*, vol. 1, no. 3, pp. 398–403, 2009.

[19] N. Novkovski, "Determination of interface states in metal(Ag, TiN,W)-Hf:Ta_2O_5/SiO_xN_y-Si structures by different compact methods," *Materials Science in Semiconductor Processing*, vol. 39, pp. 308–317, 2015.

[20] S. Ezhilvalavan, M. S. Tsai, and T. Y. Tseng, "Dielectric relaxation and defect analysis of Ta_2O_5 thin films," *Journal of Physics D: Applied Physics*, vol. 33, no. 10, pp. 1137–1142, 2000.

[21] Y. Fukuda, Y. Otani, H. Toyota, and T. One, "Electrical characterization techniques of dielectric thin films using metal-insulator-metal structures," *Japanese Journal of Applied Physics, Part 1: Regular Papers and Short Notes and Review Papers*, vol. 46, no. 10 B, pp. 6984–6986, 2007.

[22] K. J. Yang and C. Hu, "MOS capacitance measurements for high-leakage thin dielectrics," *IEEE Transactions on Electron Devices*, vol. 46, no. 7, pp. 1500-1501, 1999.

[23] H.-T. Lue, C.-Y. Liu, and T.-Y. Tseng, "An improved two-frequency method of capacitance measurement for $SrTiO_3$ as high-k gate dielectric," *IEEE Electron Device Letters*, vol. 23, no. 9, pp. 553–555, 2002.

[24] E. Atanassova, A. Paskaleva, and N. Novkovski, "Effects of the metal gate on the stress-induced traps in Ta_2O_5/SiO_2 stacks," *Microelectronics Reliability*, vol. 48, no. 4, pp. 514–525, 2008.

[25] S. Kar, S. Rawat, S. Rakheja, and D. Reddy, "Characterization of accumulation layer capacitance for extracting data on high-κ gate dielectrics," *IEEE Transactions on Electron Devices*, vol. 52, no. 6, pp. 1187–1193, 2005.

[26] R. M. Fleming, D. V. Lang, C. D. W. Jones et al., "Defect dominated charge transport in amorphous Ta_2O_5 thin films," *Journal of Applied Physics*, vol. 88, no. 2, pp. 850–862, 2000.

21

Effect of Indium Additions on the Formation of Interfacial Intermetallic Phases and the Wettability at Sn-Zn-In/Cu Interfaces

Janusz Pstruś, Tomasz Gancarz, and Przemyslaw Fima

Institute of Metallurgy and Materials Science, Polish Academy of Sciences, Ul. Reymonta 25, 30-059 Krakow, Poland

Correspondence should be addressed to Janusz Pstruś; j.pstrus@imim.pl

Academic Editor: Michael Aizenshtein

The wettability of copper substrates by Sn-Zn eutectic solder alloy doped with 0, 0.5, 1, and 1.5 at.% of indium was studied using the sessile drop method, with flux, in air, at 250°C and reflow time of 3, 8, 15, 30, and 60 min. Wetting tests were performed at 230, 250, 280, 320, and 370°C for an alloy containing 1.5 at.% of indium, in order to determine activation energy of diffusion. Solidified solder/substrate couples were studied using scanning electron microscopy (SEM), the intermetallic phases from Cu-Zn system which formed at the solder/substrate interface were identified, and their growth kinetics was investigated. The ε-CuZn$_4$ was formed first, as a product of the reaction between liquid solder and the Cu substrate, whereas γ-Cu$_5$Zn$_8$ was formed as a product of the reaction between ε-CuZn$_4$ and the Cu substrate. With increasing wetting time, the thickness of ε-CuZn$_4$ increases, while the thickness of ε-CuZn$_4$ does not change over time for indium-doped solders and gradually disappears over time for Sn-Zn eutectic solder.

1. Introduction

The continuous development of microelectronics in respect to miniaturisation and improved efficiency motivates the search for new materials and technologies capable of allowing both cost reductions and increased density of connections and efficiency, along with improved reliability. The European Parliament and the Council of Europe directives issued in [1–3] 2003 and in 2008 prohibited the use of solders containing metals such as lead and cadmium, which are detrimental to human health. Therefore, in recent years intense research has been conducted worldwide, the aim of which is to devise new alternatives to hazardous solders such as those containing lead.

Interest in alloys based on Sn-Zn eutectic is the result of the search for the optimum replacement of Pb-Sn solders. Sn-Zn eutectic's greatest advantages are its low melting temperature (198°C), which is close to the melting temperature of Pb-Sn eutectic, and the low cost per mass unit. In addition, Sn-Zn alloys exhibit relatively high mechanical strength, corrosion resistance, and resistance to thermal fatigue [4–8]. On the other hand, these alloys wet soldered surfaces poorly and are prone to oxidation unless soldering is carried out under a protective atmosphere or with a sufficiently aggressive flux [9–11]. Well-chosen alloying additions may improve joint properties, so in designing joints it is essential to understand the effect of alloying additions on processes occurring at the interface of solder and substrate. The addition of indium may lower the melting temperature of solders but also improve the wettability and durability of solder joints [12]. We earlier investigated the effect of indium addition on the properties of soldered joints: Cu/SnAgCu + In [13], Ni/Sn-Zn + In [14], and to some extent Cu/Sn-Zn + In [15, 16]. In the case of Cu/SnAgCu + In [13], it was found that the Cu$_6$Sn$_5$ was formed at the solder/Cu interface for solders with low In concentration, whereas for high In concentrations it was the Cu$_{41}$In$_{11}$ phase that developed. Also, higher In content leads to the formation of In-based phases (InSn$_4$), which in turn hinder the diffusion of Cu and resulted in the decreased thickness of the interfacial Cu-rich layer [13]. The addition of

In to Sn-Zn eutectic leads to a lower wetting angle on Ni and Cu substrates [14]. Earlier [15], we found that the spreading area increases with In content over a relatively short wetting time, but for longer wetting times this effect was absent.

The authors of [16] studied the effect of indium on the surface tension, electrical conductivity, and viscosity of liquid In-Sn, and the correlation between these properties and the structure of the liquid. They found anomalies in concentration dependence of viscosity near indium concentrations corresponding to intermetallic phases existing in the In-Sn system.

Taking the above into account, the aim of this work is to investigate the effect of indium addition to Sn-Zn eutectic on spreading on copper substrates, as well as on the phenomena occurring at solder/substrate interface during soldering.

2. Experimental

For wetting tests, copper substrates (99.9 wt.%, Alfa-Aesar) were used and their surfaces prepared in accordance with EN ISO 9455-10:2000. Copper plates ($40 \times 40 \times 0.25$ mm) were etched in 99.5% copper sulphide solution for 8 min, subsequently ground with sand paper (1,000), and degreased with methyl alcohol and acetone directly before wetting tests.

Alloys were prepared by melting carefully weighed amounts of pure (99.999 wt.%) Sn, Zn, and In in alumina crucibles under high purity (99.9999 vol.%) Ar atmosphere. The wetting tests were carried out with the use of the apparatus earlier described in [17], at 250°C, for different annealing times (3, 8, 15, 30, and 60 minutes), without a protective atmosphere and with the use of the Alu-33® flux produced by Amasan. According to ISO 9454-1, this is a 2.1.2-type flux, that is, organic, water-soluble, and activated with halides. In order to determine the activation energy of the growth of the Cu_5Zn_8 phase for an alloy containing 1.5 at.% of indium, wetting test were performed at 230, 250, 280, 320, and 370°C. For each set of the experimental conditions (sample composition, temperature, and time) 5–7 independent wetting tests were performed. Solidified samples were washed with tap water to remove flux residue and photographed from two perpendicular directions. Wetting angle data are the average of individual measurements for all samples. The wetting angles were determined from photographs using the application ImageJ. The spreading areas of solders were determined from photographs of the top view of drops. Selected solidified solder/substrate couples were cut perpendicular to the plane of the interface, mounted in conductive resin, polished for microstructural characterisation, and sputtered with a thin carbon layer. Microstructural and energy-dispersive spectroscopy (EDS) analysis was performed with the FEI E-SEM XL30 system, at 20 kV, with the use of the standardless Analysis EDAX System based on Genesis 2000 software. The intermetallic phases of $CuZn_4$ and Cu_5Zn_8 were identified based on analysis of results EDS and knowledge of the binary Cu-Zn phase diagram. The thickness of intermetallic (IMC) layers was determined from micrographs with the use of ImageJ software.

3. Results and Discussion

Figure 1(b) illustrates the wetting angle of Sn-Zn-In alloys on copper substrate at 250°C, for varying indium content and time. It is clear that the wetting angles of In-doped alloys are lower than those of the alloy without In addition. This difference is more pronounced than in our earlier work [14], where we also found that the wetting angle decreased with increasing temperature. However, this earlier investigation was carried out with different flux. For the alloys in the present study, the wetting angle seems to be practically independent of wetting time. Similarly, we earlier [15] observed, with the use of the same flux as in this study, that the spreading area does not increase over time. Figure 1(a), based on the example of an alloy containing 1.5% of In, illustrates that the spreading area increases with increasing temperature

$$\sigma_{SV} = \sigma_{SL} + \sigma_{LV} \cos \phi. \qquad (1)$$

According to the Young-Dupre equation (1), as illustrated in Figure 2, wettability can be improved (wetting angle ϕ lowered) by increasing the surface energy of the solid substrate σ_{SV}, by lowering the surface tension of the liquid alloy σ_{LV}, or by lowering solid-liquid interfacial tension σ_{SL}. The increase of surface energy of solid σ_{SV} occurs as the oxide layer is removed by flux action. The surface tension of the alloy, on the other hand, can be lowered by doping with surface active components or metals with substantially lower surface tension. From Figure 1 we can conclude that, with increasing In content, the wettability improves, despite the fact that authors of [16, 18] did not observe surface tension decreasing with indium concentration. The authors of [13] claim that increasing concentration of In in liquid solder (Sn-Ag-Cu alloy in their case) improves wettability on copper. This is because In lowers solid-liquid interfacial tension, even if it does not lower the surface tension of the solder.

Figure 1(b) presents the spreading area of $(SnZn)_{eut} + 1.5\%$ at. In solder on Cu after 8 min of wetting at 230, 250, 280, 320, and 370°C. It is clear that the spreading area increases with temperature, which could be a result of the surface tension of Sn-Zn-In alloy decreasing with temperature [18] and may also be due to changes in interfacial tension.

After wetting tests, the microstructure of the solder/Cu interface of selected samples was investigated. According to Lee et al. [19], Gibbs free energy of the formation of the γ-Cu_5Zn_8 phase is -212.10 kJ/mol (at 150°C), while Gibbs free energy of the formation of Cu_6Sn_5 is -26.26 kJ/mol, which explains why the intermetallic phase from the Cu-Zn system is formed at the liquid Sn-Zn/solid Cu interface. At 250°C, in the Sn-Zn-Cu system the liquid phase (composition close to eutectic Sn-Zn) coexists in equilibrium with two intermetallic phases: γ-Cu_5Zn_8 and ε-$CuZn_4$ [20, 21]. In a study of Sn-Zn-Cu/Cu interfaces over short wetting time [22], we found that, at the earliest stage of the soldering process, liquid solder reacts with the Cu substrate and ε-$CuZn_4$ is formed. After 60 s, as a result of the diffusion of elements, γ-Cu_5Zn_8 starts to appear. In the present study, both ε-$CuZn_4$ and γ-Cu_5Zn_8 were found at the solder/substrate interface.

Figures 3 and 4, respectively, illustrate the interfacial microstructure of Sn-Zn/Cu and Sn-Zn-1.5% In/Cu couples

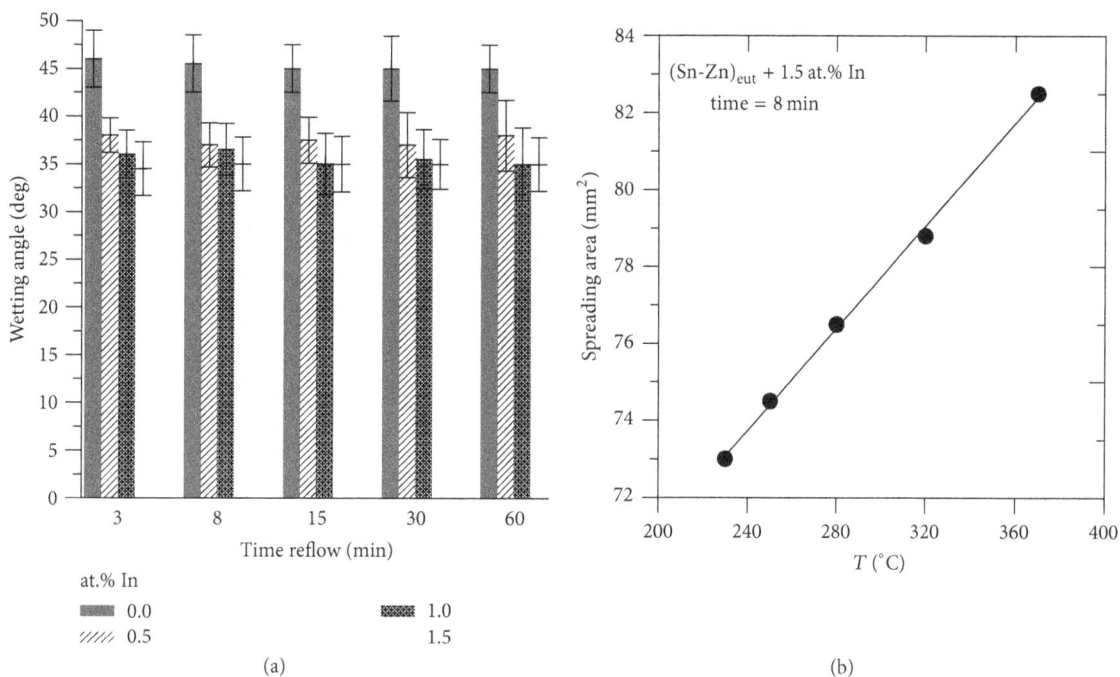

FIGURE 1: Wetting angle (a) and spreading area (b) of 0.5 g liquid Sn-Zn-xIn ($x = 0, 0.5, 1, 1.5$) on copper substrate: (a) 250°C, 3–60 min, (b) 230–370°C, 8 min.

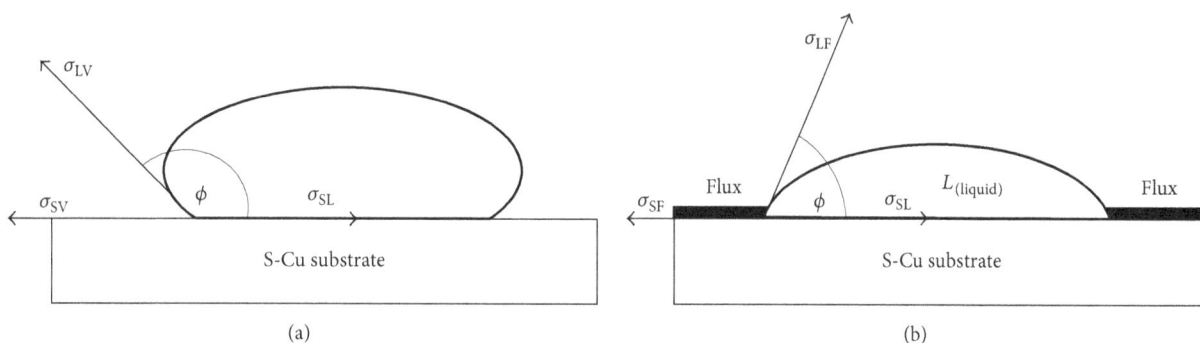

FIGURE 2: Sketch of Young-Dupre equation.

after (a) 3, (b) 15, and (c) 30 min at 250°C. The growth of intermetallic layers over time at the solder/substrate interface of the abovementioned couples is illustrated in Figure 5. For both examples, there are two intermetallic phases at the interface in the early stage (up to 15 min). These are ε-CuZn$_4$ on the side of the solder and γ-Cu$_5$Zn$_8$ adjacent to the Cu substrate. At this stage, the thickness of the γ-Cu$_5$Zn$_8$ phase increases over time, while the thickness of ε-CuZn$_4$ remains practically the same. For longer wetting times, the thickness of the γ-Cu$_5$Zn$_8$ phase continues to increase, while ε-CuZn$_4$ in the case of Sn-Zn/Cu couples begins to disappear and grows very little in the case of Sn-Zn-1.5% In/Cu couples. Moreover, just 0.5% of In is sufficient to keep the ε-CuZn$_4$ stable even after 60 min of wetting (Figures 6 and 7), as we observed earlier [15]. The bright spots between the ε-CuZn$_4$ and γ-Cu$_5$Zn$_8$ in Figure 6 are, according to EDS analysis,

composed mostly of Sn with In content somewhat higher than the In content in the Sn-Zn-In solder.

According to the Gibbs phase rule, a maximum of three phases can coexist in equilibrium under the conditions of constant pressure and temperature ($T, p = $ const) in a three-component system. Although the solder/substrate interfaces are not under equilibrium conditions, it can be assumed that, as wetting time increases, Sn-Zn/Cu couples are getting closer to equilibrium. This might explain why, after 60 min, three phases coexist in the Sn-Zn/Cu couples. These phases are liquid (solder), Cu$_5$Zn$_8$, and Cu (substrate). The addition of the fourth component, In in the present case, allows the fourth phase to coexist (CuZn$_4$ in the present case).

In the course of the experiment, Sn, Zn, and In diffuse from the liquid solder towards the solid Cu, while Cu diffuses in the opposite direction. As a result of diffusion and

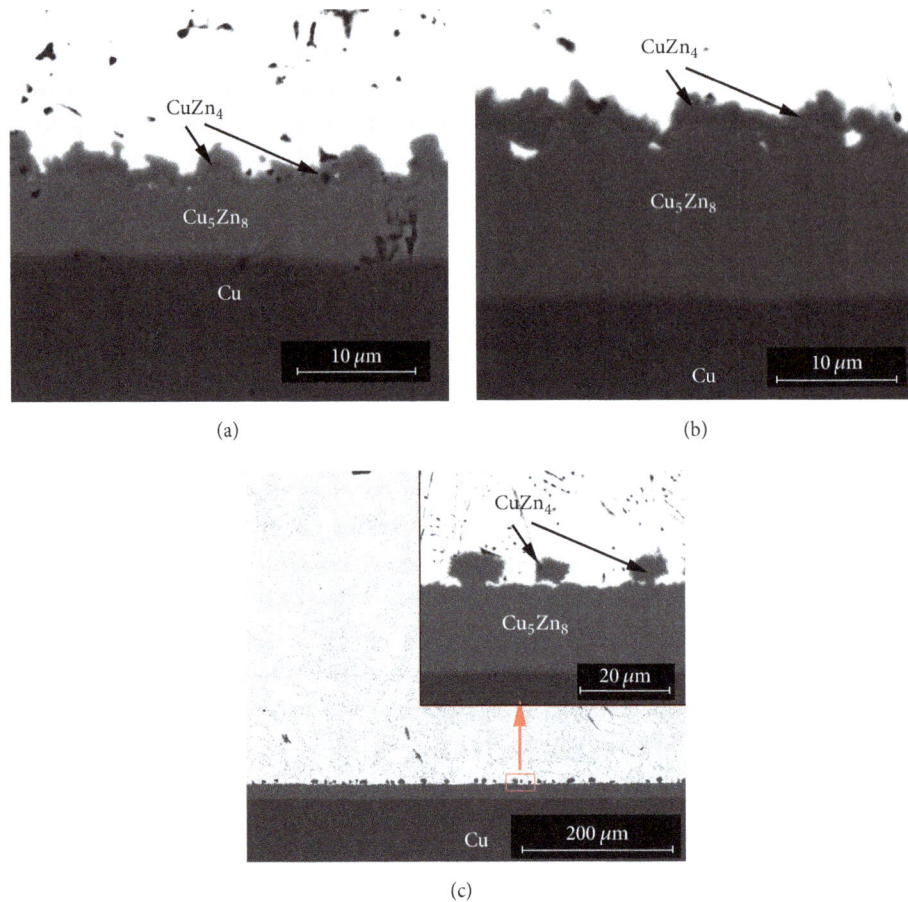

(a)

(b)

(c)

FIGURE 3: Interfacial microstructure of Sn-Zn/Cu couples, $T = 250°C$, reflow time: (a) 3 min, (b) 15 min, and (c) 30 min.

reactions between the liquid solder and solid substrate, one or more intermetallic layers are formed at the solder/substrate interface. As explained earlier [22], at the earliest stage of wetting ε-CuZn$_4$ appears first, and only after 30 s of wetting can the γ-Cu$_5$Zn$_8$ be distinguished. Diffusion processes continue and lead to the increased thickness of intermetallic layer(s). Moreover, the γ-Cu$_5$Zn$_8$ is thermodynamically more stable than ε-CuZn$_4$, so it can be speculated that, for Sn-Zn/Cu couples, after 15 min the ε-CuZn$_4$ transforms to γ-Cu$_5$Zn$_8$. This can be expressed as the reaction 5 CuZn$_4$ $\rightarrow$ Cu$_5$Zn$_8$ + 12 Zn. The freed Zn atoms go to the liquid solder, locally increasing Zn concentration near the interface. In the case of (Sn-Zn)$_{eut}$ + 1.5% In/Cu couples, In atoms diffusing through the ε-CuZn$_4$ layer towards substrate may stop at the ε-CuZn$_4$/γ-Cu$_5$Zn$_8$ interface. Earlier, we discussed the growth mechanism of intermetallic layers at the Sn-Zn-Cu/Cu interface [22] as a result of Cu and Zn diffusion in the opposite direction. It is possible that the indium atoms form a kind of diffusion barrier for Cu diffusing towards the solder, but not a barrier for Zn diffusing towards the substrate. As a result, the ε-CuZn$_4$ does not disappear after a prolonged time for Sn-Zn-In/Cu couples, but neither does its thickness increase substantially.

Evolution of intermetallic layer thickness with time is described with [23–25]

$$\delta = ke^{-Q/RT}t^n, \qquad (2)$$

where δ is the thickness of the intermetallic layer [μm], k is the growth constant, Q is the activation energy of growth [kJ/mol], R is the universal gas constant [8.314 J/mol], T is soldering temperature [K], t is soldering time [min.], and n is the time exponent.

Figure 8 illustrates the growth kinetics of ε-CuZn$_4$ and γ-Cu$_5$Zn$_8$ for (Sn-Zn)$_{eut}$ and (Sn-Zn)$_{eut}$ + 1.5 at% In in comparison to literature data [26, 27]. It is clear that the growth constant of the γ-Cu$_5$Zn$_8$ phase ($k = 3.18$) in the case of Sn-Zn/Cu is the same as in the case of Sn-Zn-1.5% In/Cu and is more than ten times higher than the growth constant of the ε-CuZn$_4$ phase. The present k of the γ-Cu$_5$Zn$_8$ phase is slightly higher than that reported in [26]. On the other hand, it can be assumed that growth constant k increases with temperature, since for Sn-Zn/Cu at 230°C the reported k is 1.72.

In order to determine the activation energy of the growth of the γ-Cu$_5$Zn$_8$ phase, samples were held at 230, 250, 280, and 320°C for 8 min. Figure 9 illustrates the relation between

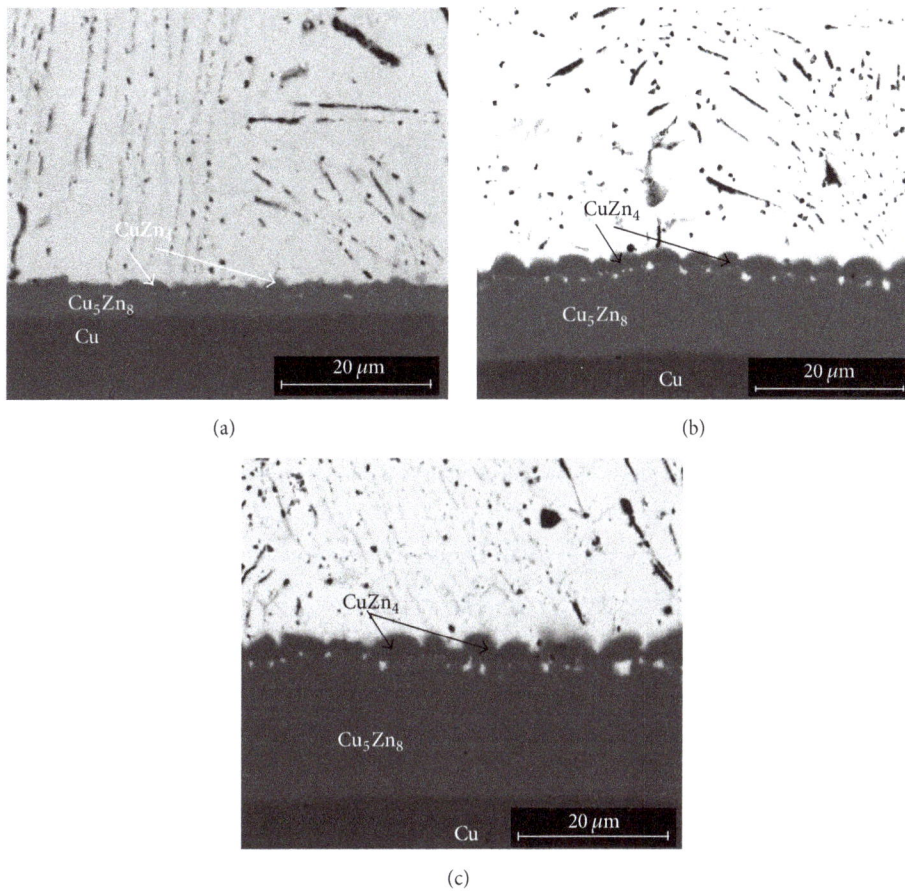

FIGURE 4: Interfacial microstructure of Sn-Zn-1.5% In/Cu couples, $T = 250°C$, reflow time: (a) 3 min, (b) 15 min, and (c) 30 min.

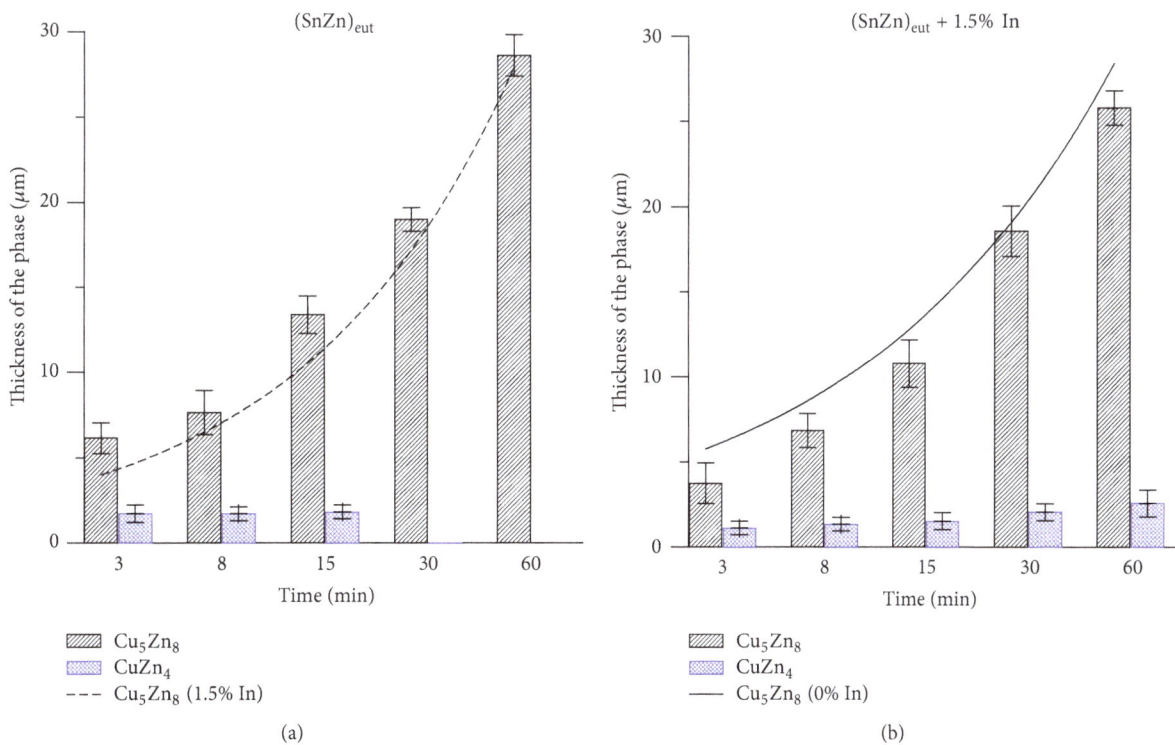

FIGURE 5: Thickness of intermetallic layers (Cu_5Zn_8, $CuZn_4$) at $(Sn-Zn)_{eut}$/Cu and $(Sn-Zn)_{eut} + 1.5\%$ In/Cu interface at 250°C.

(a)   (b)

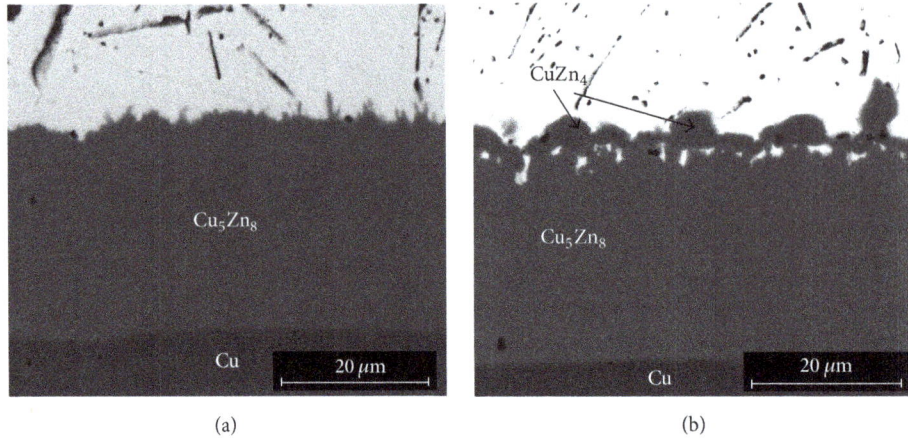

FIGURE 6: Interfacial microstructures of (a) Sn-Zn/Cu couples and (b) Sn-Zn-0.5% In/Cu couples, $T = 250°C$, reflow time 60 min.

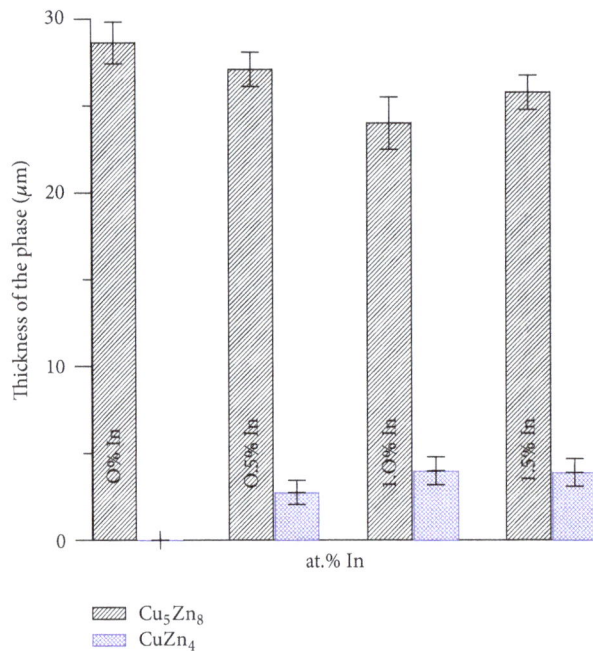

FIGURE 7: Thickness of the $CuZn_4$ and Cu_5Zn_8 intermetallic layers at the $(Sn-Zn)_{eut} + xIn/Cu$ ($x = 0, 0.5, 1, 1.5\%$) interfaces after 60 min at 250°C.

the growth rate and reciprocal soldering temperature, according to (3). The activation energy of growth of the phase is, according to our results, equal to 29.5 kJ/mol, which is close to 26 kJ/mol reported in [26].

$$\ln\delta = \ln k + n\ln t - \frac{Q}{RT}. \tag{3}$$

4. Conclusion

Wettability studies of Sn-Zn eutectic doped with In on Cu substrate were performed in the presence of flux over a temperature range of 230–370°C for wetting times in the of range 3–60 min. It was found that additions of indium lower the wetting angle, but the wetting time does not affect the wetting angle.

The microstructure of solidified solder/Cu interfaces was investigated with scanning electron microscopy. Two intermetallic phases are observed at the interface: ε-$CuZn_4$ on the side of the solder and γ-Cu_5Zn_8 adjacent to the Cu substrate. At the early stage (up to 15 min), the thickness of the γ-Cu_5Zn_8 phase increases much faster. For longer wetting times, greater than 15 min, the thickness of the γ-Cu_5Zn_8 phase continues to increase, while ε-$CuZn_4$ in the case of Sn-Zn/Cu couples begins to disappear and grows very little in the case of Sn-Zn-1.5% In/Cu couples. Therefore, the addition of indium to Sn-Zn solder stabilises ε-$CuZn_4$ at the interface of solder and substrate, even after 60 min of wetting.

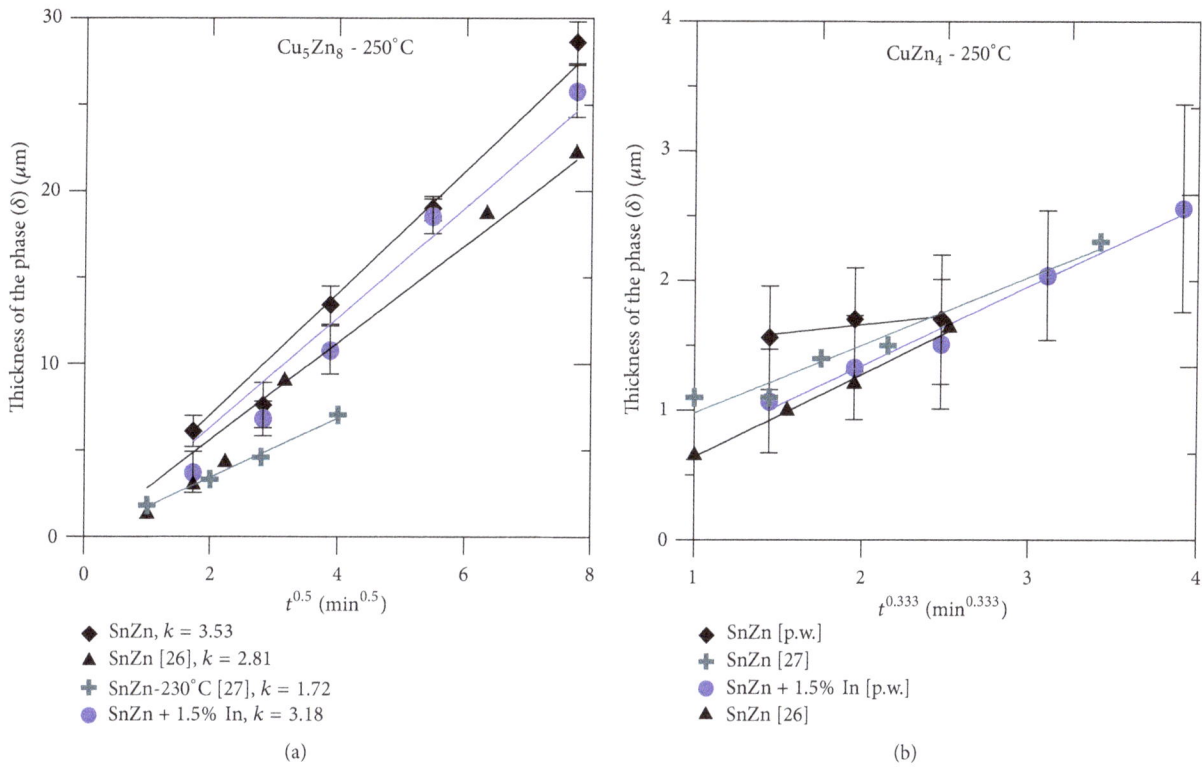

FIGURE 8: Growth kinetics of ε-CuZn$_4$ (a) and γ-Cu$_5$Zn$_8$ (b) at 250°C, for (Sn-Zn)$_{eut}$/Cu and (Sn-Zn)$_{eut}$ + 1.5% In/Cu interfaces. Present work (p.w.) and [26] at 250°C, [27] at 230°C.

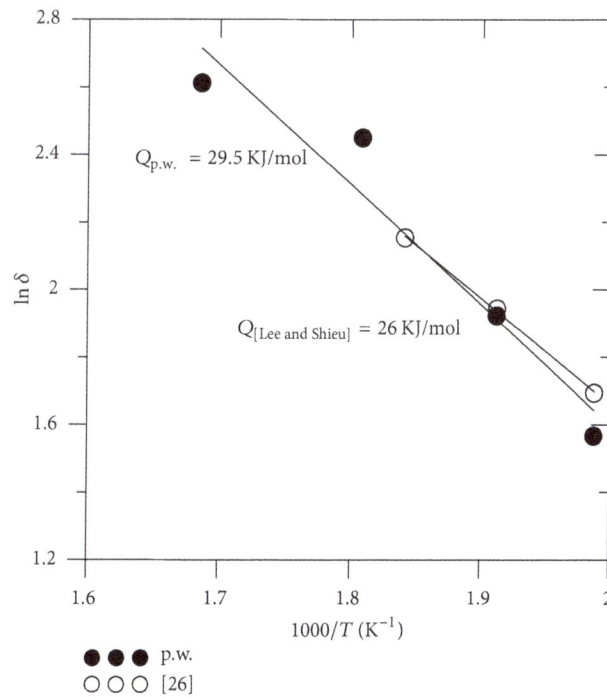

FIGURE 9: Arrhenius plot for the activation energy of Cu$_5$Zn$_8$ phase layer at the interface between molten (Sn-Zn)$_{eut}$ [26] or molten (Sn-Zn)$_{eut}$ + In [present work -p.w.] and Cu.

Conflicts of Interest

The authors declare that there are no conflicts of interest regarding the publication of this article.

Acknowledgments

This work was financed within the framework of the Project POIG.01.01.02-00-015/09 (advanced materials and their production technologies, ZAMAT), cofunded by the European Regional Development Fund (ERDF).

References

[1] Directive 2002/96/WE of European Parliament and Council of 27/01/2003,.

[2] Directive 2003/108/WE of European Parliament and Council of 08/12/2003,.

[3] Directive 2008/35/WE of European Parliament and Council of 11/03/2008,.

[4] L. Zhang and K. N. Tu, "Structure and properties of lead-free solders bearing micro and nano particles," *Materials Science and Engineering: R: Reports*, vol. 82, pp. 1–32, 2014.

[5] S. Liu, S.-B. Xue, P. Xue, and D.-X. Luo, "Present status of Sn–Zn lead-free solders bearing alloying elements," *Journal of Materials Science: Materials in Electronics*, vol. 26, no. 7, pp. 4389–4411, 2015.

[6] X. Wei, H. Huang, L. Zhou, M. Zhang, and X. Liu, "On the advantages of using a hypoeutectic Sn-Zn as lead-free solder material," *Materials Letters*, vol. 61, no. 3, pp. 655–658, 2007.

[7] P. Xue, S.-B. Xue, Y.-F. Shen, and H. Zhu, "Interfacial microstructures and mechanical properties of Sn-9Zn-0.5Ga-xNd on Cu substrate with aging treatment," *Materials and Design*, vol. 60, pp. 1–6, 2014.

[8] L. Zhang, S. B. Xue, L. L. Gao et al., "Development of Sn-Zn lead-free solders bearing alloying elements," *Journal of Materials Science: Materials in Electronics*, vol. 21, no. 1, pp. 1–15, 2010.

[9] J. Glazer, "Metallurgy of low temperature pb-free solders for electronic assembly," *International Materials Reviews*, vol. 40, no. 2, pp. 65–93, 1995.

[10] B. P. Richards and K. Nimmo, *Lead-free Soldering–Update*, Department of Trade and Industry, Lead-free Soldering–Update, UK, 2000.

[11] A. A. El-Daly and A. E. Hammad, "Elastic properties and thermal behavior of Sn-Zn based lead-free solder alloys," *Journal of Alloys and Compounds*, vol. 505, no. 2, pp. 793–800, 2010.

[12] S. Ganesan and M. Pecht, "Lead-free Electronics," *Lead-free Electronics*, pp. 1–766, 2006.

[13] P. Šebo, Z. Moser, P. Švec et al., "Effect of indium on the microstructure of the interface between Sn3.13Ag0.74CuIn solder and Cu substrate," *Journal of Alloys and Compounds*, vol. 480, no. 2, pp. 409–415, 2009.

[14] P. Fima, T. Gancarz, J. Pstruś, and A. Sypień, "Wetting of Sn-Zn-xIn (x = 0.5, 1.0, 1.5 wt%) alloys on cu and Ni substrates," *Journal of Materials Engineering and Performance*, vol. 21, no. 5, pp. 595–598, 2012.

[15] T. Gancarz, P. Fima, and J. Pstruś, "Thermal expansion, electrical resistivity, and spreading area of Sn-Zn-In alloys," *Journal of Materials Engineering and Performance*, vol. 23, no. 5, pp. 1524–1529, 2014.

[16] Z. Moser, W. Gąsior, J. Pstruś, I. Kaban, and W. Hoyer, "Thermophysical properties of liquid In-Sn alloys," *International Journal of Thermophysics*, vol. 30, no. 6, pp. 1811–1822, 2009.

[17] J. Pstruś, P. Fima, and T. Gancarz, "Wetting of Cu and Al by Sn-Zn and Zn-Al eutectic alloys," *Journal of Materials Engineering and Performance*, vol. 21, no. 5, pp. 606–613, 2012.

[18] J. Pstruś, "Surface tension and density of liquid In-Sn-Zn alloys," *Applied Surface Science*, vol. 265, pp. 50–59, 2013.

[19] H. M. O. Lee, S. W. Yoon, and B.-J. Lee, "Thermodynamic prediction of interface phases at Cu/solder joints," *Journal of Electronic Materials*, vol. 27, no. 11, pp. 1161–1166, 1998.

[20] M. Date, K. N. Tu, T. Shoji, M. Fujiyoshi, and K. Sato, "Interfacial reactions and impact reliability of Sn-Zn solder joints on Cu or electroless Au/Ni(P) bond-pads," *Journal of Materials Research*, vol. 19, no. 10, pp. 2887–2896, 2004.

[21] Y.-C. Huang, S.-W. Chen, C.-Y. Chou, and W. Gierlotka, "Liquidus projection and thermodynamic modeling of Sn-Zn-Cu ternary system," *Journal of Alloys and Compounds*, vol. 477, no. 1-2, pp. 283–290, 2009.

[22] P. Fima, J. Pstruś, and T. Gancarz, "Wetting and interfacial chemistry of SnZnCu alloys with Cu and Al substrates," *Journal of Materials Engineering and Performance*, vol. 23, no. 5, pp. 1530–1535, 2014.

[23] R. J. Klein Wassink, *Soldering in Electronics*, Electrochemical Publications Ltd., Ayr, Scotland, 1989.

[24] P. T. Vianco, J. A. Rejent, and P. F. Hlava, "Solid-state intermetallic compound layer growth between copper and 95.5Sn-3.9Ag-0.6Cu solder," *Journal of Electronic Materials*, vol. 33, no. 9, pp. 991–1004, 2004.

[25] A. R. Fix, G. A. López, I. Brauer, W. Nüchter, and E. J. Mittemeijer, "Microstructural development of Sn-Ag-Cu solder joints," *Journal of Electronic Materials*, vol. 34, no. 2, pp. 137–142, 2005.

[26] C. S. Lee and F. S. Shieu, "Growth of intermetallic compounds in the Sn-9Zn/Cu joint," *Journal of Electronic Materials*, vol. 35, no. 8, pp. 1660–1664, 2006.

[27] Y. C. Chan, M. Y. Chiu, and T. H. Chuang, "Intermetallic compounds formed during the soldering reactions of eutectic Sn-9Zn with Cu and Ni substrates," *Zeitschrift für Metallkunde*, vol. 93, no. 2, pp. 95–98, 2002.

Permissions

All chapters in this book were first published in AMSE, by Hindawi Publishing Corporation; hereby published with permission under the Creative Commons Attribution License or equivalent. Every chapter published in this book has been scrutinized by our experts. Their significance has been extensively debated. The topics covered herein carry significant findings which will fuel the growth of the discipline. They may even be implemented as practical applications or may be referred to as a beginning point for another development.

The contributors of this book come from diverse backgrounds, making this book a truly international effort. This book will bring forth new frontiers with its revolutionizing research information and detailed analysis of the nascent developments around the world.

We would like to thank all the contributing authors for lending their expertise to make the book truly unique. They have played a crucial role in the development of this book. Without their invaluable contributions this book wouldn't have been possible. They have made vital efforts to compile up to date information on the varied aspects of this subject to make this book a valuable addition to the collection of many professionals and students.

This book was conceptualized with the vision of imparting up-to-date information and advanced data in this field. To ensure the same, a matchless editorial board was set up. Every individual on the board went through rigorous rounds of assessment to prove their worth. After which they invested a large part of their time researching and compiling the most relevant data for our readers.

The editorial board has been involved in producing this book since its inception. They have spent rigorous hours researching and exploring the diverse topics which have resulted in the successful publishing of this book. They have passed on their knowledge of decades through this book. To expedite this challenging task, the publisher supported the team at every step. A small team of assistant editors was also appointed to further simplify the editing procedure and attain best results for the readers.

Apart from the editorial board, the designing team has also invested a significant amount of their time in understanding the subject and creating the most relevant covers. They scrutinized every image to scout for the most suitable representation of the subject and create an appropriate cover for the book.

The publishing team has been an ardent support to the editorial, designing and production team. Their endless efforts to recruit the best for this project, has resulted in the accomplishment of this book. They are a veteran in the field of academics and their pool of knowledge is as vast as their experience in printing. Their expertise and guidance has proved useful at every step. Their uncompromising quality standards have made this book an exceptional effort. Their encouragement from time to time has been an inspiration for everyone.

The publisher and the editorial board hope that this book will prove to be a valuable piece of knowledge for researchers, students, practitioners and scholars across the globe.

List of Contributors

Wang Chen and Yanhui Liu
State Key Laboratory for Seismic Reduction/Control & Structural Safety (Cultivation), Guangzhou University, Guangzhou 510006, China

Ruijie Li
State Key Laboratory for Seismic Reduction/Control & Structural Safety (Cultivation), Guangzhou University, Guangzhou 510006, China
School of Materials Science and Engineering, Guizhou Minzu University, Guiyang 550025, China

Laura Angélica Ardila Rodriguez and Dilermando Nagle Travessa
Federal University of São Paulo (UNIFESP), Institute of Science and Technology, Laboratory of Advanced Metals and Processing, São José dos Campos, SP, Brazil

Haruki Kinemuchi and Bungo Ochiai
Department of Chemistry and Chemical Engineering, Faculty of Engineering, Yamagata University, 4-3-16 Jonan, Yonezawa, Yamagata 992-8510, Japan

Periyasamy Manikandan and Gin Boay Chai
Aerospace Engineering Cluster, School of Mechanical and Aerospace Engineering, Nanyang Technological University, Singapore

Worawat Wattanathana and Apirat Laobuthee
Department of Materials Engineering, Faculty of Engineering, Kasetsart University, Chatuchak, Bangkok 10900, Thailand

Suttipong Wannapaiboon
Synchrotron Light Research Institute, 111 University Avenue, Suranaree, Muang, Nakhon Ratchasima 30000, Thailand
Chair of Inorganic and Metal-Organic Chemistry, Technical University of Munich, Lichtenbergstr. 4, 85748 Garching, Germany

Chatchai Veranitisagul
Department of Materials and Metallurgical Engineering, Faculty of Engineering, Rajamangala University of TechnologyThanyaburi, Klong 6, Thanyaburi, Pathumthani 12110, Thailand

Navadol Laosiripojana
The Joint Graduate School of Energy and Environment, CHE Center for Energy Technology and Environment, King Mongkut's University of TechnologyThonburi, Bangkok 10140, Thailand

Nattamon Koonsaeng
Department of Chemistry, Faculty of Science, Kasetsart University, Chatuchak, Bangkok 10900, Thailand

A. R. Sufizadeh and S. A. A. Akbari Mousavi
School of Metallurgy and Materials Engineering, College of Engineering, University of Tehran, Tehran, Iran

Yinuo Wang
China Academy of Building Research, Beijing 100013, China

Ying-Ji Chuang and Ching-Yuan Lin
Department of Architecture, National Taiwan University of Science and Technology, Taipei, Taiwan

Hao Zhang
China Academy of Building Research, Beijing 100013, China
University of Science and Technology, Beijing, China

E. Rendell
Mechanical Engineering Department, Faculty of Engineering and Applied Science, Memorial University of Newfoundland, St. John's, NL, Canada A1B 3X5

A. Hsiao
Mechanical Engineering Department, Faculty of Engineering and Applied Science, Memorial University of Newfoundland, St. John's, NL, Canada A1B 3X5
University of Prince Edward Island, Charlottetown, PEI, Canada C1A 4P3

J. Shirokoff
Process Engineering Department, Faculty of Engineering and Applied Science, Memorial University of Newfoundland, St. John's, NL, Canada A1B 3X5

Weilian Qu, Ernian Zhao and Qiang Zhou
Hubei Key Laboratory of Roadway Bridge & Structure Engineering, Wuhan University of Technology, Wuhan, Hubei 430070, China

Guojun Liu, Xiang Ma, Yanhui Jia and Tengfei Wang
College of Mechanical Science and Engineering, Jilin University, Changchun 130025, China

Xiaodong Sun
College of Communication Engineering, Jilin University, Changchun 130025, China

Sunil Babu Eadi, Sungjin Kim and Soon Wook Jeong
School of Advanced Materials & Engineering, Kumoh National Institute of Technology, 61 Daehak-Ro, Gumi 39177, Republic of Korea

Heung Woo Jeon
School of Electronic Engineering, Kumoh National Institute of Technology, 61 Daehak-Ro, Gumi 39177, Republic of Korea

Stefan Flege, Ruriko Hatada andreas Hanauer and Wolfgang Ensinger
Department of Materials Science, Technische Universität Darmstadt, Alarich-Weiss-Str. 2, 64287 Darmstadt, Germany

Takao Morimura
Nagasaki University, Graduate School of Engineering, 1-14 Bunkyo, Nagasaki 852-8521, Japan

Koumei Baba
Nagasaki University, Graduate School of Engineering, 1-14 Bunkyo, Nagasaki 852-8521, Japan
Industrial Technology Center of Nagasaki, Omura, Nagasaki 856-0026, Japan

P. A. Prates, A. F. G. Pereira, N. A. Sakharova, M. C. Oliveira and J. V. Fernandes
CEMUC, Department of Mechanical Engineering, University of Coimbra, Rua Luís Reis Santos, Pinhal de Marrocos, 3030-788 Coimbra, Portugal

Shuo Yin and Rocco Lupoi
Department of Mechanical and Manufacturing Engineering, Trinity College Dublin, e University of Dublin, Parsons Building, Dublin 2, Ireland

Chaoyue Chen
ICB UMR 6303, CNRS, Univ. Bourgogne Franche-Comté, UTBM, 90010 Belfort, France

Xinkun Suo
Key Laboratory of Marine Materials and Related Technologies, Ningbo Institute of Materials Technology and Engineering, Chinese Academy of Sciences, Ningbo 315201, China

Xiaoxiang Mao, Longfei Jiang, Chenguang Zhu and Xiaoming Wang
Nanjing University of Science and Technology, Nanjing 210094, China

Kuo-Hsiung Tseng, Chih-Ju Chou, To-Cheng Liu, Der-Chi Tien and Tong-chi Wu
Department of Electrical Engineering, National Taipei University of Technology, Taipei 10608, Taiwan

Leszek Stobinski
Materials Chemistry, Warsaw University of Technology, Warynskiego 1, 00-645 Warsaw, Poland

Andraž Kocjan and Anton Gradišek
Jŏzef Stefan Institute, Jamova cesta 39, SI-1000 Ljubljana, Slovenia

Luka Kelhar, Kristina Žagar and Spomenka Kobe
Jŏzef Stefan Institute, Jamova cesta 39, SI-1000 Ljubljana, Slovenia
International Associated Laboratory and PACS2 and CNRS Nancy and JSI, Ljubljana, Slovenia

Blaž Likozar
Department of Catalysis and Chemical Reaction Engineering, National Institute of Chemistry, Hajdrihova 19, SI-1001 Ljubljana, Slovenia

Jaafar Ghanbaja and Jean-Marie Dubois
International Associated Laboratory and PACS2 and CNRS Nancy and JSI, Ljubljana, Slovenia
Institut Jean Lamour (UMR 7198 CNRS-Universit´e de Lorraine), Parc de Saurupt, CS50840, 54011 Nancy Cedex, France

Jeffrey Brock, Nathanael Bell-Pactat, Hong Cai, Timothy Dennison, Tucker Fox, Brandon Free, Rami Mahyub, Austin Nar, Michael Saaranen, Tiago Schaeffer and Mahmud Khan
Department of Physics, Miami University, Oxford, OH 45056, USA

Fengkui Cui, Yongxiang Su, Kege Xie, Wang Xiaoqiang, Xiaolin Ruan and Fei Liu
School of Mechatronics Engineering, Henan University of Science and Technology, Luoyang 471003, China
Collaborative Innovation Center of Machinery Equipment Advanced Manufacturing of Henan Province, Luoyang 471003, China

Nenad Novkovski
Institute of Physics, Faculty of Natural Sciences and Mathematics, University "Ss. Cyril and Methodius", Arhimedova 3, 1000 Skopje, Macedonia
Research Center for Environment and Materials, Macedonian Academy of Sciences and Arts, Krste Misirkov 2, 1000 Skopje, Macedonia

Elena Atanassova
Institute of Solid State Physics, Bulgarian Academy of Sciences, 72 Tzarigradsko Chaussee Blvd., 1784 Sofia, Bulgaria

Janusz Pstruś, Tomasz Gancarz and Przemyslaw Fima
Institute of Metallurgy and Materials Science, Polish Academy of Sciences, Ul. Reymonta 25, 30-059 Krakow, Poland

Index